The Handbook of Environmental Chemistry

Editor-in-Chief: O. Hutzinger

Volume 5 Water Pollution
Part Q

The Handbook of Environmental Chemistry

Recently Published and Forthcoming Volumes

Environmental Specimen Banking
Volume Editors: S. A. Wise and P. P. R. Becker
Vol. 3/S, 2008

Polymers: Chances and Risks
Volume Editors: P. Eyerer, M. Weller
and C. Hübner
Vol. 3/V, 2008

The Black Sea Environment
Volume Editors: A. Kostianoy and A. Kosarev
Vol. 5/Q, 2008

Emerging Contaminants from Industrial and Municipal Waste
Volume Editors: D. Barceló and M. Petrovic
Vol. 5/S, 2008

Fuel Oxygenates
Volume Editor: D. Barceló
Vol. 5/R, 2007

The Rhine
Volume Editor: T. P. Knepper
Vol. 5/L, 2006

Persistent Organic Pollutants in the Great Lakes
Volume Editor: R. A. Hites
Vol. 5/N, 2006

Antifouling Paint Biocides
Volume Editor: I. Konstantinou
Vol. 5/O, 2006

Estuaries
Volume Editor: P. J. Wangersky
Vol. 5/H, 2006

The Caspian Sea Environment
Volume Editors: A. Kostianoy and A. Kosarev
Vol. 5/P, 2005

Marine Organic Matter: Biomarkers, Isotopes and DNA
Volume Editor: J. K. Volkman
Vol. 2/N, 2005

Environmental Photochemistry Part II
Volume Editors: P. Boule, D. Bahnemann
and P. Robertson
Vol. 2/M, 2005

Air Quality in Airplane Cabins and Similar Enclosed Spaces
Volume Editor: M. B. Hocking
Vol. 4/H, 2005

Environmental Effects of Marine Finfish Aquaculture
Volume Editor: B. T. Hargrave
Vol. 5/M, 2005

The Mediterranean Sea
Volume Editor: A. Saliot
Vol. 5/K, 2005

Environmental Impact Assessment of Recycled Wastes on Surface and Ground Waters
Engineering Modeling and Sustainability
Volume Editor: T. A. Kassim
Vol. 5/F (3 Vols.), 2005

Oxidants and Antioxidant Defense Systems
Volume Editor: T. Grune
Vol. 2/O, 2005

The Black Sea Environment

Volume Editors: Andrey G. Kostianoy · Aleksey N. Kosarev

With contributions by

V. S. Arkhipkin · V. K. Chasovnikov · D. Ya. Fashchuk · Z. Z. Finenko
A. I. Ginzburg · E. I. Ignatov · A. N. Kosarev · A. G. Kostianoy
V. N. Mikhailov · M. V. Mikhailova · J. W. Murray · L. N. Neretin
N. P. Nezlin · S. V. Pakhomova · O. I. Podymov · A. I. Ryabinin
N. A. Sheremet · T. A. Shiganova · P. A. Stunzhas · G. V. Surkova
A. A. Svitoch · V. S. Tuzhilkin · I. I. Volkov · E. V. Yakushev
A. G. Zatsepin · S. S. Zhiltsov · I. S. Zonn

Springer

Environmental chemistry is a rather young and interdisciplinary field of science. Its aim is a complete description of the environment and of transformations occurring on a local or global scale. Environmental chemistry also gives an account of the impact of man's activities on the natural environment by describing observed changes.
The Handbook of Environmental Chemistry provides the compilation of today's knowledge. Contributions are written by leading experts with practical experience in their fields. The Handbook will grow with the increase in our scientific understanding and should provide a valuable source not only for scientists, but also for environmental managers and decision-makers.
The Handbook of Environmental Chemistry is published in a series of five volumes:

Volume 1: The Natural Environment and the Biogeochemical Cycles
Volume 2: Reactions and Processes
Volume 3: Anthropogenic Compounds
Volume 4: Air Pollution
Volume 5: Water Pollution

The series Volume 1 The Natural Environment and the Biogeochemical Cycles describes the natural environment and gives an account of the global cycles for elements and classes of natural compounds.
The series Volume 2 Reactions and Processes is an account of physical transport, and chemical and biological transformations of chemicals in the environment.
The series Volume 3 Anthropogenic Compounds describes synthetic compounds, and compound classes as well as elements and naturally occurring chemical entities which are mobilized by man's activities.
The series Volume 4 Air Pollution and Volume 5 Water Pollution deal with the description of civilization's effects on the atmosphere and hydrosphere.
Within the individual series articles do not appear in a predetermined sequence. Instead, we invite contributors as our knowledge matures enough to warrant a handbook article.
Suggestions for new topics from the scientific community to members of the Advisory Board or to the Publisher are very welcome.

ISBN 978-3-642-09364-7 e-ISBN 978-3-540-74292-0

DOI 10.1007/978-3-540-74292-0

Springer Series in Advanced Manufacturing ISSN 1433-6863

Softcover reprint of the hardcover 1st edition 2008

Cover design: WMXDesign GmbH, Heidelberg

Printed on acid-free paper 02/3180 YL – 5 4 3 2 1 0

springer.com

Editor-in-Chief

Volume Editors

Advisory Board

Prof. Dr. A. H. Neilson
Swedish Environmental Research Institute
P.O. Box 21060
10031 Stockholm, Sweden
ahsdair@ivl.se

Prof. Dr. J. Paasivirta
Department of Chemistry
University of Jyväskylä
Survontie 9
P.O. Box 35
40351 Jyväskylä, Finland

Prof. Dr. Dr. H. Parlar
Institut für Lebensmitteltechnologie
und Analytische Chemie
Technische Universität München
85350 Freising-Weihenstephan, Germany

Prof. Dr. S. H. Safe
Department of Veterinary
Physiology and Pharmacology
College of Veterinary Medicine
Texas A & M University
College Station, TX 77843-4466, USA
ssafe@cvm.tamu.edu

Prof. P. J. Wangersky
University of Victoria
Centre for Earth and Ocean Research
P.O. Box 1700
Victoria, BC, V8W 3P6, Canada
wangers@telus. net

The Handbook of Environmental Chemistry Also Available Electronically

For all customers who have a standing order to The Handbook of Environmental Chemistry, we offer the electronic version via SpringerLink free of charge. Please contact your librarian who can receive a password or free access to the full articles by registering at:

springerlink.com

If you do not have a subscription, you can still view the tables of contents of the volumes and the abstract of each article by going to the SpringerLink Homepage, clicking on "Browse by Online Libraries", then "Chemical Sciences", and finally choose The Handbook of Environmental Chemistry.

You will find information about the

- Editorial Board
- Aims and Scope
- Instructions for Authors
- Sample Contribution

at springer.com using the search function.

Preface

Environmental Chemistry is a relatively young science. Interest in this subject, however, is growing very rapidly and, although no agreement has been reached as yet about the exact content and limits of this interdisciplinary discipline, there appears to be increasing interest in seeing environmental topics which are based on chemistry embodied in this subject. One of the first objectives of Environmental Chemistry must be the study of the environment and of natural chemical processes which occur in the environment. A major purpose of this series on Environmental Chemistry, therefore, is to present a reasonably uniform view of various aspects of the chemistry of the environment and chemical reactions occurring in the environment.

The industrial activities of man have given a new dimension to Environmental Chemistry. We have now synthesized and described over five million chemical compounds and chemical industry produces about hundred and fifty million tons of synthetic chemicals annually. We ship billions of tons of oil per year and through mining operations and other geophysical modifications, large quantities of inorganic and organic materials are released from their natural deposits. Cities and metropolitan areas of up to 15 million inhabitants produce large quantities of waste in relatively small and confined areas. Much of the chemical products and waste products of modern society are released into the environment either during production, storage, transport, use or ultimate disposal. These released materials participate in natural cycles and reactions and frequently lead to interference and disturbance of natural systems.

Environmental Chemistry is concerned with reactions in the environment. It is about distribution and equilibria between environmental compartments. It is about reactions, pathways, thermodynamics and kinetics. An important purpose of this Handbook, is to aid understanding of the basic distribution and chemical reaction processes which occur in the environment.

Laws regulating toxic substances in various countries are designed to assess and control risk of chemicals to man and his environment. Science can contribute in two areas to this assessment; firstly in the area of toxicology and secondly in the area of chemical exposure. The available concentration ("environmental exposure concentration") depends on the fate of chemical compounds in the environment and thus their distribution and reaction behaviour in the environment. One very important contribution of Environmental Chemistry to the above mentioned toxic substances laws is to develop laboratory test methods, or mathematical correlations and models that predict the environmental fate of new chemical compounds. The third purpose of this Handbook is to help in the basic understanding and development of such test methods and models.

The last explicit purpose of the Handbook is to present, in concise form, the most important properties relating to environmental chemistry and hazard assessment for the most important series of chemical compounds.

At the moment three volumes of the Handbook are planned. Volume 1 deals with the natural environment and the biogeochemical cycles therein, including some background information such as energetics and ecology. Volume 2 is concerned with reactions and processes in the environment and deals with physical factors such as transport and adsorption, and chemical, photochemical and biochemical reactions in the environment, as well as some aspects of pharmacokinetics and metabolism within organisms. Volume 3 deals with anthropogenic compounds, their chemical backgrounds, production methods and information about their use, their environmental behaviour, analytical methodology and some important aspects of their toxic effects. The material for volume 1, 2 and 3 was each more than could easily be fitted into a single volume, and for this reason, as well as for the purpose of rapid publication of available manuscripts, all three volumes were divided in the parts A and B. Part A of all three volumes is now being published and the second part of each of these volumes should appear about six months thereafter. Publisher and editor hope to keep materials of the volumes one to three up to date and to extend coverage in the subject areas by publishing further parts in the future. Plans also exist for volumes dealing with different subject matter such as analysis, chemical technology and toxicology, and readers are encouraged to offer suggestions and advice as to future editions of "The Handbook of Environmental Chemistry".

Most chapters in the Handbook are written to a fairly advanced level and should be of interest to the graduate student and practising scientist. I also hope that the subject matter treated will be of interest to people outside chemistry and to scientists in industry as well as government and regulatory bodies. It would be very satisfying for me to see the books used as a basis for developing graduate courses in Environmental Chemistry.

Due to the breadth of the subject matter, it was not easy to edit this Handbook. Specialists had to be found in quite different areas of science who were willing to contribute a chapter within the prescribed schedule. It is with great satisfaction that I thank all 52 authors from 8 countries for their understanding and for devoting their time to this effort. Special thanks are due to Dr. F. Boschke of Springer for his advice and discussions throughout all stages of preparation of the Handbook. Mrs. A. Heinrich of Springer has significantly contributed to the technical development of the book through her conscientious and efficient work. Finally I like to thank my family, students and colleagues for being so patient with me during several critical phases of preparation for the Handbook, and to some colleagues and the secretaries for technical help.

I consider it a privilege to see my chosen subject grow. My interest in Environmental Chemistry dates back to my early college days in Vienna. I received significant impulses during my postdoctoral period at the University of California and my interest slowly developed during my time with the National Research Council of Canada, before I could devote my full time of Environmental Chemistry, here in Amsterdam. I hope this Handbook may help deepen the interest of other scientists in this subject.

Amsterdam, May 1980 *O. Hutzinger*

Twenty-one years have now passed since the appearance of the first volumes of the Handbook. Although the basic concept has remained the same changes and adjustments were necessary.

Some years ago publishers and editors agreed to expand the Handbook by two new open-end volume series: Air Pollution and Water Pollution. These broad topics could not be fitted easily into the headings of the first three volumes. All five volume series are integrated through the choice of topics and by a system of cross referencing.

The outline of the Handbook is thus as follows:

1. The Natural Environment and the Biochemical Cycles,
2. Reaction and Processes,
3. Anthropogenic Compounds,
4. Air Pollution,
5. Water Pollution.

Rapid developments in Environmental Chemistry and the increasing breadth of the subject matter covered made it necessary to establish volume-editors. Each subject is now supervised by specialists in their respective fields.

A recent development is the accessibility of all new volumes of the Handbook from 1990 onwards, available via the Springer Homepage springeronline.com or springerlink.com.

During the last 5 to 10 years there was a growing tendency to include subject matters of societal relevance into a broad view of Environmental Chemistry. Topics include LCA (Life Cycle Analysis), Environmental Management, Sustainable Development and others. Whilst these topics are of great importance for the development and acceptance of Environmental Chemistry Publishers and Editors have decided to keep the Handbook essentially a source of information on "hard sciences".

With books in press and in preparation we have now well over 40 volumes available. Authors, volume-editors and editor-in-chief are rewarded by the broad acceptance of the "Handbook" in the scientific community.

Bayreuth, July 2001 *Otto Hutzinger*

Contents

Hdb Env Chem Vol. 5, Part Q (2008): 1–10
DOI 10.1007/698_5_092

Published online: 28 September 2007

Introduction

Aleksey N. Kosarev[1] · Andrey G. Kostianoy[2] (✉)

[1]Geographic Department, Lomonosov Moscow State University, Vorobievy Gory, 119992 Moscow, Russia

[2]P.P. Shirshov Institute of Oceanology, Russian Academy of Sciences, 36 Nakhimovsky Pr., 117997 Moscow, Russia
kostianoy@online.ru

Abstract The intensification of complex studies in the Black Sea at the end of the twentieth century was determined by three principal factors: the influence of the changes in the regional climate during the last decade of the century on the entire Black Sea ecosystem, the strongest impact of biological invaders on pelagic and bottom biocenoses of the basin, and the catastrophic reduction in its fishery potential and the large-scale construction and plans of construction of oil and gas complex structures in the sea area. An integrated and sustainable development of the Black Sea region will require an interdisciplinary approach, which the present book reflects.

Keywords Black Sea · Sea of Azov · Environment · Physico-geographical conditions · Anthropogenic pressure · Pollution

The Black Sea (together with the Sea of Azov) is located deeply inside the continent and represents the most isolated part of the World Ocean. In the southwest, it is connected with the Sea of Marmara via the Bosphorus Strait; the boundary between the seas runs along the line Cape Rumeli–Cape Anadolu. The Kerch Strait connects the Black Sea and the Sea of Azov; the boundary between them is defined along the line Cape Takyl–Cape Panagia (Fig. 1). The length of the Bosphorus Strait is 30 km at a width of 0.7–3.6 km; the depth of the navigation channel is 20–102 m. The depth of the swell on the Black Sea side reaches 50 m, while on the side of the Sea of Marmara it is 40 m. The length of the Kerch Strait is about 41 km at a width from 4.5 to 15 km; its maximal depth reaches 15 m.

The northernmost and the southernmost points of the sea lie at 46°33′N and 40°56′N, respectively. The westernmost and the easternmost points of the sea lie at 27°27′E and 41°42′E, respectively. The maximal length of the sea along the latitude 42°29′N is 1148 km, while its minimal width along the meridian from Cape Sarych (Crimea) to the coast of Turkey is only 258 km.

The principal morphometric characteristics of the Black Sea slightly differ in different publications. In this monograph, the authors accept the following values: 423 000 km^2 for the sea area, 555 000 km^3 for its volume, and 1315 and 2258 m for the mean and maximal depths, respectively.

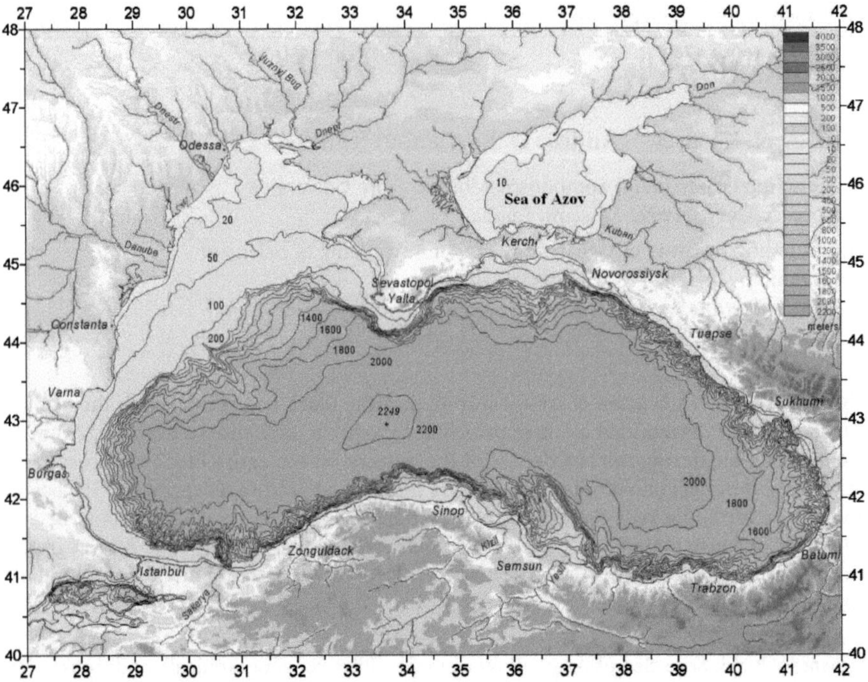

Fig. 1 Bottom topography (isobaths, m) of the Black Sea

In the northwestern part of the sea, major bays are located such as Odessa, Karkinit, and Kalamit bays. In addition, Samsun and Sinop bays are recognized on the southern coast and Burgas Bay in the west. Minor Zmeinyi and Berezan' islands are located in the northwestern part of the sea, and Kefken Island is located northeast of the Bosphorus Strait.

The bulk of the riverine runoff (up to 80%) is delivered to the northwestern part of the sea by the major rivers such as the Danube (200 km^3/year), the Dnieper (50 km^3/year), and the Dniester (10 km^3/year). On the Caucasian coast of the Black Sea, the sea accepts the waters of the Inguri, Rioni, Chorokh, and numerous minor rivers. Over the rest of the coast, the runoff is insignificant.

In the bottom topography of the sea, one can clearly distinguish three principal structures: the shelf, the continental slope, and the deep-water basin. The shelf occupies up to 25% of the total area of the seafloor and, on average, is restricted to sea depths of 100–200 m. It reaches its greatest width (more than 200 km) in the northwestern part of the sea, which is entirely located within the shelf zone. Almost over the entire extension of the eastern and southern coasts of the sea, the shelf is very narrow (only a few kilometers wide); in the western part of the sea, it is wider (a few tens of kilometers).

The continental slope includes up to 40% of the seafloor area; it descends down to depths of 2000 m. It is steep and cut by submarine valleys and canyons. Its lower part, located at depths of 1500–2000 m, is referred to as the continental foot; in selected cases, the boundary between the slope and the foot is poorly expressed. The floor of the basin (35% of the total area) represents a flat accumulative plain with a surface that gradually declines toward the center of the sea.

The Black Sea region has always attracted attention, thanks to its unique natural features, diversity of natural resources, and great economic and geopolitical importance of the basin of the Black Sea.

The particularity of the natural conditions of the Black Sea lies in the fact that it the largest basin in the world with a permanent halocline and a two-layered structure of the waters. The intensive pycno- and halocline prevents the waters from vertical mixing and oxygen penetration to deeper layers even in the period of the development of the wintertime vertical convection. Therefore, the entire water column below a depth of 100–200 m represents an inanimate hydrogen sulphide zone, in which only anoxic processes take place. About 90% of the water volume does not participate in the processes of self-purification of the sea.

Owing to the isolated inland position of the Black Sea, the formation of its hydrological regime proceeds under the control of external factors, such as heat and moisture fluxes, momentum across the sea surface, and riverine runoff. Therefore, the sea is distinguished by a high degree of variability in the hydrological and hydrochemical conditions, especially in shallow-water shelf areas. This, in turn, affects the biocenoses and, finally, leads to general changes in the ecosystem of the sea. Its pelagic ecosystem features a low resistance and is very sensitive to climatic changes and anthropogenic impacts, hiding manifestations of natural factors. During the past few decades, the technogenous influence on the marine environment has noticeably increased. It includes chemical pollution, impact of sea transport, mass development of the predator ctenophore *Mnemiopsis leidyi*, etc. This resulted in the aggravation of the ecological situation in the sea and led to the development of negative tendencies.

The natural regime of the Black Sea is relatively well studied. The peak of intensity of the studies occurred in the 1960s–1980s, when a great number of expeditions was performed. Starting from the 1990s, after the disintegration of the Soviet Union, which carried out the bulk of the regular observations, their number has sharply decreased. Meanwhile, during the last two decades, international activity at sea has intensified and a significant number of expedition have been performed under the auspices of special programs (CoMS-Black, NATO TU–Black Sea, Black Sea Environmental Programme, ARENA, IASON, ASCABOS, SESAME, ALTICORE, MOPED, a series of other projects of the INTAS, INCO-Copernicus, NATO, International Atomic Energy Agency, UNESCO IOC, as well as those of the Russian Academy of Sciences, the Ministry of Education and Science of the Russian Federation, and the Russian Foundation for Basic Research) with the use of modern instruments and techniques for research.

The economic significance of the Black Sea region is determined, first of all, by its transport and pipeline potentialities and recreation resources. Fishery is also retained in the region's economy, although, under the present-day conditions, its significance has strongly decreased. The role of hydrocarbon resources, to date, has been restricted to oil and gas prospecting in the shelf areas of the sea.

The high degree of urbanization of the coastal zone of the Black Sea represents a permanent menace of marine environment pollution. The principal pollution sources include industrial wastes of large cities and their sanitary condition. Rivers also influence the pollution of marine environment. This is especially manifested in shelf regions such as the northwestern part of the Black Sea. Here, the principal manifestations of the influence of the riverine runoff are represented by eutrophication and formation of a near-bottom hypoxy over vast areas.

One of the most important economic trends of the region is the development of transport. Intensive international shipping takes place and a significant number of large ports are located here. During these activities, a large part of the traffic is related to the transport of oil and oil products, which represents a permanent source of pollution of the sea. Oil enters the environment as a result of illegal, accidental, and operational discharges from vessels and oil terminals, as well as from land-based sources. Almost half of the inputs of oil from land-based activities are brought to the Black Sea via the Danube River. The most serious aftereffects are related to accidents with ships carrying dangerous cargo, first of all, with tankers. These accidents have damaged the fishery, mariculture (mussel and shrimp industry), and recreation zones. A special problem is the navigation of the Turkish straits, where the transport, legal, and ecological issues are closely interwoven.

Navigation also affects the migration of organisms, which is often undesirable. In the autumn of 1988, the predator ctenophore *Mnemiopsis leidyi* started its mass development in the Black Sea. Because of the activity of the Mnemiopsis population, the planktonic fodder base of planktivorous fishes, such as Black Sea anchovy has reduced by 3–5 times; the biomass of these fishes proper, whose juveniles may be directly consumed by this ctenophore, has also decreased. This book provides an up-to-date overview of the situation in the Black Sea and the Sea of Azov with regard to the jelly invasions that have been perturbing the local pelagic ecosystems since the 1980s.

One more effect of marine navigation is the operation of large ports and terminals. The prospecting and extraction of hydrocarbons and construction of underwater gas and oil pipelines represent large-scale invasions into the functioning of marine biota. During the past few decades, owing to these and other forms of anthropogenic impact, negative changes in the Black Sea ecosystem have occurred. Here, one observes a sharp fall in the fishery, a significant reduction in the stocks of the principal, valuable commercial fishes, as well of anchovies, and a suppression of the Black Sea population of dolphins. Pollution of beaches and bights by municipal waste, oil products, and abundant, and often pathogenous microflora, take place. The near-shore pollutants are supplied to the open sea area with the mesoscale vortical currents and now changes in the ecosystem may be traced over the entire area of the Black Sea. Owing to different effects, the changes in the physicochemical regime of the sea proceed, in particular, in the zone of interaction between oxic and anoxic waters.

At the end of the preceding century, an intensification of the studies of the Black Sea ecosystem occurred. Its necessity was mainly defined by three principal reasons: the influence of the regional climate changes during the last decade of the last century on the entire Black Sea ecosystem; the strongest impact of species–invaders on the pelagic and bottom biocenoses of the basin and catastrophic reduction in their commercial potential; and, finally, the large-scale construction and plans of construction of object of oil and gas complex in the sea area such as the oil terminal of the the Caspian Pipeline Consortium (CPC) on the Russian shelf (2001), and the "Blue Flow" (2003) and "Blue Flow – 2" (nearest future) underwater gas pipelines.

On 15 March 2007, in Athens, an agreement was signed between the governments of the Russian Federation, Republic of Bulgaria and Greek Republic outlining the cooperation for construction and use of the Burgas–Alexandroupolis oil pipeline. This project implies an extension of the CPC system and a significant increase in the volumes of the oil transported by tankers between the CPC terminal near Novorossiisk and Burgas, since the throughput capacity of the pipeline should range from 35 to 50 Mt per year.

On June 23, 2007, the "Gazprom" Company and the Italian oil firm "Eni" unveiled a plan for a large new pipeline to take Russian gas under the Black Sea to Europe. The 900-km "South Stream" pipeline would submerge on the Russian coast, come ashore in Bulgaria, and then branch to Austria and Slovenia in one spur and to southern Italy in another. All these projects require an adequate estimation of the aftereffects of the increasing anthropogenic stress on the sea environment.

The countries of the Black Sea basin try to protect the nature of the sea. They formulated international rules for the cleaning water areas from oil and waist and scientifically justified regulations of fishery. Special attention is paid to the most vulnerable shelf areas. The issues of the protection of the natural environment of the Black Sea are discussed in detail in one of the chapters of this book.

As to the international legislation, there are selected unsolved issues such as those about the delimitaion of near-shore aquatic areas between the Black Sea countries and the demarcation of the boundaries in the Black Sea and the Sea of Azov, for example in the regions of Tuzla and Zmeinyi islands.

The results obtained during the past decades are published in numerous articles and monographs; many of them are listed in the reference section [1–45]. The attention that great to the problems of the Black Sea is evidently a positive fact. Meanwhile, the quickly changing environment of the basin and the active introduction of new observation techniques and methods for analysis and calculations require further monographic generalizations. Therefore, without any doubt, a new book concerning the present-day features of the nature of the basin of the Black Sea and their variability is well-timed and urgent.

This monograph is characterized by the following features. First of all, it is multidisciplinary, dealing with the principal processes that form the structure of the sea; it assesses the fundamental particular features of its hydrology, hydrochemistry, and biology and also considers ecological and socioeconomic issues. The book consists of 17 chapters, which may be conventionally joined into the following sections. The first part contains the chapters devoted to the history of the studies of the Black Sea, the description of the bottom and coastal topography, the quaternary paleogeography of the basin, the detailed information about the riverine runoff, and the estimation of the hydrometeorological factors. The second section reflects the character of the hydrological structure and water circulation in the sea, including an analysis of its mesoscale vortical dynamics based on satellite observations. Subsequently, the hydrochemical part follows, which concerns both general hydrochemical structures of the sea and one of its fundamental issues – the character of the processes within the hydrogen sulphide zone and in the boundary layer at its upper interface. The biological chapters assess the biodiversity of the Black Sea and consider in detail the role of invaders in the ecosystem of the sea. The monograph is completed by a discussion of ecological aspects as well as social and political issues. An individual chapter of the book is devoted to the Sea of Azov. It should be emphasized that the authors of all the chapters are specialists who directly participate in the solution of the issues under consideration.

Another particular feature of the monograph lies in the combination of different methods used for analysis and calculations. They include addressing archived materials, analysis of "classical" observational data on hydrology and hydrochemistry, use of the satellite data, generalization of the results obtained in different calculations and numerical modeling. In so doing, the quantitative estimates obtained with different methods were converted to a compatible form. Only this kind of an analysis might allow us to adequately estimate the condition of the natural regime of the Black Sea and the changes occurring in it.

Thus, this book presents a systematization and description of the knowledge accumulated today on the physical oceanography, marine chemistry and pollution, and marine biology of the Black Sea. Special attention is paid to socioeconomic, legal, and political problems in the Black Sea region. Meanwhile, it is not a collection of individual papers; it is a monograph written by a team of scientists joined by a common understanding of the complicated phenomena and processes that are occurring in the Black Sea. The publication is based on numerous observational data, collected by the authors of the chapters during sea and shore expeditions, on the archive data of Moscow State University, State Oceanographic Institute, P.P. Shirshov Institute of Oceanology (Russia), and Kovalevskii Institute for Biology of the Southern Seas, Marine Branch of Ukrainian Hydrometeorological Institute (Sevastopol, Ukraine), and others, as well as on a wide range of scientific literature, mainly published in Russian editions. These data are complemented by the results of a series of Russian and Ukrainian national and international projects listed below, where an extensive research was carried out over the past decades.

This book addresses the specialists working in various fields of physical oceanography, marine chemistry, pollution studies, and biology and studying a cascade of problems: from regional climate to mesoscale processes and from remote sensing of the sea to numerical and laboratory modeling. It may also be useful to students and post-graduates specializing in oceanographic research of the seas. The editors and authors expect that this monograph would help the readers to complement the information on the nature of the Black Sea and the Sea of Azov, especially on the present-day condition of this extremely interesting basin. More information on special issues may be derived using the reference lists contained in each chapter.

The studies of the authors of this book were supported by the Russian World Ocean Federal Research Program (project 7), the "World Ocean" Project 17.4.3 of the Russian Academy of Sciences, the "Black Sea" Project of the Russian Ministry of Science, Science School Grant N 4376.2006.5 to IIV, the Program N 17 of the Presidium of the Russian Academy of Sciences (Project 5.4), the INTAS "ALTICORE" and "MOPED" projects, the "SESAME" and "ASCABOS" EC projects, CRDF Grant RUG1-2828-KS06, the Bilateral Russian–Greek project "Long-Term Variability of the Hydrophysical Processes and Zooplankton Key Species in the Black and Aegean seas: Interrelations and Dependencies upon Climate Changes", and by the Russian Foundation for Basic Research (projects NN 04-05-65149, 05-05-64927, 05-05-65092, 05-05-65110, 06-05-96676-Yug, 07-05-00141, 07-05-01024, and 07-05-00406).

The editors are grateful to the colleagues from the P.P. Shirshov Institute of Oceanology, Moscow State University, and Marine Hydrophysical Institute (Sevastopol, Ukraine) for their long-term fruitful cooperation on the Black Sea studies.

This book may be regarded as a follow-up volume to our first book in the Handbook of Environmental Chemistry series published by Springer-Verlag entitled "The Caspian Sea Environment" (2005) [46]. On behalf of the authors, we would like to thank Springer-Verlag Publishers for the timely interest to environments of the Black and Caspian seas and the support of the publication presented.

References

1. Filippov DM (1968) Water Circulation and Structure of the Black Sea. Nauka, Moscow (in Russian)
2. Bogdanova AK, Dobrzhanskaya MA, Lebedeva MN (1969) Water Exchange Through the Bosporus Strait and its Influence on Hydrology and Biology of the Black Sea. Naukova Dumka, Kiev (in Russian)
3. Goncharov VP, Neprochnov YuP, Neprochnova AF (1972) Topography and Deep Structure of the Black Sea Basin. Nauka, Moscow (in Russian)
4. Mordukhai-Boltovskoi FD (1972) Guide of the Black and Azov Seas Fauna. Naukova Dumka, Kiev (in Russian)
5. Degens ET, Ross DA (eds) (1974) The Black Sea: its Geology, Chemistry and Biology. Am Asso. Petrol Geol Mem 20, Tulsa, OK, USA
6. Skopintsev BA (1975) Formation of Present Chemical Composition of the Black Sea Waters. Gidrometeoizdat, Leningrad (in Russian)
7. Greze VN (1979) Bases of Biological Productivity of the Black Sea. Naukova Dumka, Kiev (in Russian)
8. Sorokin YuI (1982) The Black Sea: Nature, Resources. Nauka, Moscow (in Russian)
9. Blatov AS, Bulgakov NP, Ivanov VA, Kosarev AN, Tuzhilkin VS (1984) Variability of Hydrophysical Fileds in the Black Sea. Gidrometeoizdat, Leningrad (in Russian)
10. Altman EN, Gertman IF, Golubeva ZA (1987) Climatic fields of the Black Sea water salinity and temperature. Gosudarstvennyi Okeanograficheskii Institut. Sevastopolskoye Otdeleniye, Sevastopol (in Russian)
11. Zats VI, Finenko ZZ (eds) (1988) Dynamics of Waters and Productivity of the Black Sea. Nauka, Moscow (in Russian)
12. Ryabinin AI, Kravets VN (1989) Present State of the Hydrogene Sulphide Zone in the Black Sea (1960–1986). Gidrometeoizdat, Moscow (in Russian)
13. Keondjyan VP, Kudin AM, Terekhin YuV (eds) (1990) Practical Ecology of Marine Regions. The Black Sea. Naukova Dumka, Kiev (in Russian)
14. Simonov AI, Altman EN (eds) (1991) Hydrometeorology and Hydrochemistry of the USSR Seas. Vol IV: The Black Sea. Issue 1: Hydrometeorological conditions. Gidrometeoizdat, St.-Petersburg (in Russian)
15. Goptarev NP, Simonov AI, Zatuchnaya BM, Gershanovich DE (eds) (1991) Hydrology and Hydrochemistry of the Seas. vol V, the Azov Sea. Gidrometeoizdat, St.-Petersburg (in Russian)
16. Vinogradov ME (ed) (1991) Variability of the Black Sea Ecosystem: Natural and Anthropogenic Factors. Nauka, Moscow (in Russian)
17. Vinogradov ME, Sapozhnikov VV, Shushkina EA (1992) The Black Sea Ecosystem. Nauka, Moscow (in Russian)
18. Simonov AI, Ryabinin AI, Gershanovich DE (eds) (1992) Hydrometeorology and Hydrochemistry of the USSR Seas. Vol IV: The Black Sea. Issue 2: Hydrochemical conditions and oceanographic bases of formation of biological productivity. Gidrometeoizdat, St.-Petersburg (in Russian)
19. Izdar E, Murray JW (eds) (1991) Black Sea Oceanography. NATO Sci Ser C. Kluwer Academic Publishers, Amsterdam

20. Kovalev AV, Finenko ZZ (eds) (1993) Plankton of the Black Sea. Naukova Dumka, Kiev (in Russian)
21. Eremeev VN, Chudinovskikh TV, Batrakov GF (1993) Artificial Radioactivity of the Black Sea. UNESCO Rep Mar Sci 59, UNESCO
22. Bezborodov AA, Eremeev VN (1993) Interaction Zone between Oxic and Anoxic Waters. Marine Hydrophysical Institute, Sevastopol (in Russian)
23. Simonov AI, Ryabinin AI (eds) (1996) Hydrometeorology and Hydrochemistry of the Seas. Vol IV: The Black Sea. Issue 3: Modern state of the Black Sea pollution. ECOSEA-Gidrofizika, Sevastopol (in Russian)
24. Eremeev VN (1996) In: Griffiths PC (ed) Hydrochemistry and Dynamics of the Hydrogene-Sulphide Zone in the Black Sea. UNESCO Rep Mar Sci 69, UNESCO
25. Ozsoy E, Mikaelyan A (eds) (1997) Sensitivity to change: Black Sea, Baltic Sea and Northern Sea. Kluwer, Dordrecht
26. Zaitsev Yu, Mamaev V (1997) Biological Diversity in the Black Sea. A Study of Change and Decline. UN Publications, New York
27. Ivanov LI, Oguz T (eds) (1998) Ecosystem Modeling as a Management Tool for the Black Sea. Kluwer Academic Publishers, Netherlands
28. Besiktepe ST, Ünlüata Ü, Bologa AS (eds) (1999) Environmental Degradation of the Black Sea: Challenges and Remedies. NATO Science Partnership Sub-Series: 2, Vol 56. Kluwer Academic Publishers, Dordrecht
29. Sorokin Yu (2002) The Black Sea Ecology and Oceanography. Backhuys Publishers, Leiden
30. Zatsepin AG, Flint MV (eds) (2002) Multi-Disciplinary Investigations of the North-East Part of the Black Sea. Nauka, Moscow (in Russian)
31. Jaoshvili Sh (2002) The Rivers of the Black Sea. European Environmental Agency. Technical report N:71
32. Yilmaz A (ed) (2003) Oceanography of the Eastern Mediterranean and Black Sea. Tubitak Publishers, Ankara
33. Dumont HJ, Shiganova TA, Niermann U (eds) (2004) Aquatic Invasions in the Black, Caspian, and Mediterranean Seas. Nato Science Series: IV: Earth and Environmental Sciences, Vol 35. Kluwer Academic Publishers, Dordrecht
34. Mikhailov VN (ed) (2004) Hydrology of the Danube Delta. Geos, Moscow (in Russian)
35. Kosarev AN, Tuzhilkin VS, Daniyalova ZH, Arkhipkin VS (2004) Hydrology and Ecology of the Black and Caspian Seas. In: Geography, Society and Environment. Vol VI. Dynamics and Interaction of the Atmosphere and the Hydrosphere. Gorodetz, Moscow (in Russian)
36. Babii MV, Bukatov AE, Stanichny SV (2005) Atlas of the Black Sea Surface Temperature Based on Satellite Data (1986–2002). Marine Hydrophysical Institute, Sevastopol, Ukraine (in Russian)
37. Matishov G, Gargopa Yu, Berdnikov S, Dzhenyuk S (2006) The Regularities of Ecosystem Processes in the Sea of Azov. Nauka, Moscow (in Russian)
38. Matishov G, Matishov D, Gargopa G, Dashkevich L, Berdnikov S, Baranova O, Smolyar I (2006) Climatic Atlas of the Sea of Azov 2006. In: Matishov G, Levitus S (eds) NOAA Atlas NESDIS 59. US Government Printing Office, Washington DC (http://www.nodc.noaa.gov/OC5/AZOV2006/start.html)
39. Goryachkin YuN, Ivanov VA (2006) The Black Sea Level: Past, Present and Future. MHI NASU, Sevastopol (in Russian)
40. Zaitsev YuP, Aleksandrov BG, Minicheva GG (eds) (2006) North-Western Part of the Black Sea: Biology and Ecology. Naukova Dumka, Kiev (in Russian)

41. Zaitsev YuP (2006) An Introduction into the Black Sea Ecology. Aven, Odessa (in Russian)
42. Atlas of the Black Sea and Sea of Azov Nature Protection (2006) GUNIO MO RF, St.-Petersburg (in Russian)
43. Neretin LN (ed) (2006) Past and Present Water Column Anoxia. NATO Sciences Series. Springer, Dordrecht
44. Grinevetsky SF, Zonn IS, Zhiltsov SS (2006) In: Kosarev AN, Kostianoy AG (eds) The Black Sea Encyclopedia. Mezhdunarodnye Otnosheniya, Moscow (in Russian)
45. Yanko-Hombach V, Gilbert AS, Panin N, Dolukhanov PM (eds) (2007) The Black Sea Flood Question: Changes in Coastline, Climate and Human Settlement. Springer, Dordrecht
46. Kostianoy AG, Kosarev AN (eds) (2005) The Caspian Sea Environment. The Handbook of Environmental Chemistry. Springer-Verlag, Berlin, Heidelberg, New York, Tokyo

Hdb Env Chem Vol. 5, Part Q (2008): 11–30
DOI 10.1007/698_5_065

Published online: 13 February 2007

Brief History of the Black Sea Exploration and Oceanographic Investigation

Aleksey N. Kosarev[1] · Igor S. Zonn[2] (✉)

[1]Geographic Department, Lomonosov Moscow State University, Vorobievy Gory, 119992 Moscow, Russia

[2]Engineering Research Production Center on Water Management, Land Reclamation and Ecology, 43/1, Baumanskaya ul., 105005 Moscow, Russia
igorzonn@mtu-net.ru

Abstract Oceanographic investigations of the Black Sea have been conducted for more than 300 years. During this time they have passed from an antique-virtual notion about this water body, through descriptive geography and various cartographic images, to wide-scale studies of the hydrological, geochemical, biological, and other problems of the Black Sea region. Today the Black Sea is considered one of the best-studied regions. There are more than 3500 publications containing results of oceanographic investigations of this Sea. Knowledge of the history of investigations enables a deeper insight into the role of this most interesting water body in development of the Earth sciences, understanding better the dimensions of the socioeconomic significance of this vast region located at the juncture of Europe and Asia.

Keywords Black Sea · Cartography · History · Oceanography · Scientific investigations

1 Introduction

In studies of the Black Sea the greatest credit may be given to the scientists of the Circum-Black Sea countries, where the investigations of the Soviet and Russian scientists are most prominent. The Black Sea demands all-round attention not only in terms of the fundamental science, but also its applied aspects pertaining to utilization of the natural resources of the sea and its coastal zone. As a result of the breakdown of the Soviet Union two new independent states have appeared on the Black Sea – Ukraine and Georgia. This has given new shading to regional problems. The disintegration of the scientific-information domain had a negative effect on the Black Sea investigations. However, in the new geopolitical situation researchers are making their best to restore the scientific potential necessary for further cognition of this water area. The issues of the rational integrated utilization of marine resources, control of the observance of the respective international treaties, especially those related to the environment condition of this water body, are coming to the fore.

The history of the hydrographic and oceanographic investigations in the Black Sea has been conventionally divided by the authors into several basic stages: the Antique period, the Middle Ages, the Peter period, the post-Peter period, a wider approach to the Black Sea investigations, the period after World War II, and recent investigations.

2 Antique Period

Due to its geographical location the Black Sea was well known in antiquity. Initial data about the Black Sea may be met in the works of the geographers and writers of the Antique period, mostly in the epoch of Ancient Greece, and are very informative although rather abrupt. All they contain are the names that had been used by various peoples, data on the sea geography, shoreline configuration and nature, and about the population living on the coast. This enables better apprehension and reconstruction of the sea evolution in that historical period and, thus, a better understanding of the course of its further development. This was facilitated, to a great extent, by the surviving cartographic materials showing the accumulated information about the sea "condition".

Homer (seventh century B.C.) said that the Manych River was a strait via which one could get from Pontus Euxine (Black Sea) as far as the Caspian Sea. Greek Hecate Milesian (c. 546–480 B.C.) marked many new things on his map of the Earth. He described in detail the coasts of the Black Sea, likening its configuration to the "Scythian bow" whose bowstring corresponds to the southern coast, and the curved bowstaff to the northern coast with the Taurus peninsula. Such presentation of the contours of the Black Sea coast can quite often be found in many Antique works [1]. The Greek historian Herodotus (c. 484–425 B.C.) in his book "Histories" named eight rivers: Istr (Danube), Tiras (Dnestr), Hipanis (Southern Bug), Borisphen (Dnieper), Hipacirus (possibly Kalanchak), Gerus and Tanais (Don). All rivers were known to Herodotus in their lower reaches at their inflow into the Pontus. Strabon, a renowned geographer of the Ancient times, traveled through the whole southern coast of the Black Sea as far as Armenia. In his *Geography* he called Pontus the eastern bay of the Mediterranean Sea and estimated its circumference as 25 000 *stadia* (Roman unit of length equaling 185 m). At the same time he wondered why the sea had so much water: "Whether Pontus flowed into the Mediterranean Sea or the Mediterranean Sea flowed into Pontus?" Quite remarkable are the speculations of Strabon about the sea bed (Book 1, Chapter 3). Refuting the Eratosthenes opinion that Pontus overfilled with river waters and broke a channel nearby Byzantium through which water flowed into Propontis (Sea of Marmara) and the Mediterranean Sea, Strabon thought that this happened when the Pontus bed became higher than the bed of the Mediterranean Sea due to river sediments. The most ancient sailing book *Periple of Pontus Euxine* (fourth century B.C.) of the well-known Roman writer and statesman Flavy Arian, made on the basis of earlier sources and his voyage from Trapezund (Trabizond) to Dioscuriada (Sevastopol), contained names of many settlements. Arian gave a detailed description of the Achilles Island, presently Zmeinyi. Claudius Ptolemy (second century B.C.) in his *Geography* (Book VIII, Chapters 8–10) presented the eighth Europe map; the second Asia map covered the European and Asian Sarmatia. The hydrographic network here comprised three rivers: Borisphen, Tanais, and Ra (Volga). On the map were shown the Pontus (Black) and Meotian (Azov) Seas as well as Chersonese Tauric and Colchis, between the Black and Caspian Seas. Ptolemy had a great influence on the map-makers of the Roman Empire. Pliny the Senior (23–79 B.C.) paid much attention to the Black and Caspian Seas and nearby territories.

Paytinger on his map (allegedly dated 500 B.C.) located Pontus Euxine with great distortion as well as the Meotian Lake (Azov Sea). He also marked Agalinus (Dnestr), Celliani which flowed into a river running from the "Hipanis bog" (Southern Bug), and Nisanus (Dnieper).

A bronze Greek shield was found on the Euphrates with the engraved fragment of a traveler's map showing some parts of the Black Sea – Odessos, Bibona, Callatis, Chersonese and others.

Studies of the Black Sea began in the period of appearance of Greek settlements on its coast. At first the Black Sea was called by the Greek voyagers the Pontus Auxine, inhospitable sea, but when they settled down on its shores and formed their colonies there they started calling it Pontus Euxine, hospitable sea. The Black Sea was known to the Romans who in the first century B.C. extended their reign over the whole sea coast.

The Antique period gave initial geographical and cartographic notions about the Black Sea within the framework of the Earth description in general.

3
Middle Ages

In the Middle Ages the interest to the Black Sea was stirred mainly due to its favorable geographical location on the main merchant ways from Europe to the south and east. Here the well-known path "from Varangians to Greeks" appeared. In the ninth century A.D. the merchants from Kiev, Novgorod, and other Russian cities started using this way to get from the Baltic to the Black Sea, and from that time in some sources this sea was referred to as the Russia Sea (on some maps this name was used from the eighth to fifteenth centuries). By that time, Byzantine had already detailed descriptions of its coastline; maps (portolans) of some parts of the Black Sea coast were made. The outlines of the Black Sea on them are very close to the present ones.

In 1021 Al-Biruni prepared a map of seven seas. Inside a ring there was drawn a smaller-sized disc with five bays or seas jutting into it (one of the bays include two seas – Mediterranean and Bar Bontis (Black Sea)).

In 1154 Idrici prepared a map on 17 sheets and a round general map of the Earth. It is interesting to note that in its part oriented southwards, where the Russian lands located, the Black Sea was found in the upper part.

In the late twelfth to early thirteenth centuries, the Ottoman Empire became established on the Pontus coast and until the end of the seventeenth century the sea was often called the "Turkish Lake", as all other Circum-Black Sea territories belonged to the Ottoman Empire.

In 1318 Pyotr Vesconte made up the *Atlas of navigator charts* that included the map of the Black Sea. The outlines of the Black and Azov Seas in it were similar to their contours on ordinary sea maps.

The Medici Atlas of 1351 comprised the Black Sea map that was no different from other known maps of this sea. The Black Sea as well as the Azov Sea had its ordinary outlines although the Black Sea was linked with the Caspian by a river called "Asso". Three rivers flow along straight lines from the northwest to the Black Sea; the easternmost river (Dnieper) had a tributary.

With time, the ancient periples were modified into Medieval navigator charts. Many of them depicted the coast of the Black and Azov Seas. The Italian navigator charts usually presented only the northern and eastern coast of the Black Sea. In 1448, the Benedictine monk Andreas Valeperger in Constance made the World Map. The outlines of the Black and Azov Seas only vaguely resembled their contours on the Ptolemy maps or navigator charts. The Black Sea ports marked were Suastopolis (Fazis) on the Caucasian coast and Album Castorum ("White Camp", now Akkerman) in the mouth of the Nester (Dnestr) River.

In the Middle Ages the initial scientific–geographical notions about the sea were being shaped.

4 Peter-the-Great's Black Sea

The first Russian maps and descriptions of the Black and Azov Seas date back to the sixteenth century. With opening of an access for Russia to these seas the period of their detailed and systemic studies began.

The great drawing book (1627) provided brief data about the Black Sea with the description of its southern coast.

Simultaneously with creation of the Azov fleet, Russia organized hydrographic works. In 1695 a Colonel of the Preobrazhensky regiment, Yu.A. Mengden, conducted topographic-geodetic surveys in the area of Russian troop movement to the Azov Sea. After seizure of Azov in 1696, Ya.V. Bryus used the results of these surveys and other cartographic materials to prepare the *Map of Southern Russia* on which the Black and Azov Seas, and the basins of the Dnieper and Don Rivers were shown. In the same year by the order of Peter I the first nautical inventory of the Taganrog Bay of the Azov Sea from the Krivaya bar to the mouth of the Don River was made. This inventory was used for the preparation of a plan that became the first Peter's navigator chart created on the basis of direct measurements. In 1699 during the voyage of the "Krepost" ("Castle") ship from Azov to Kerch and further on to Constantinople sea measurements were carried out en-route. From the results obtained Admiral K. Kryuis "under direct supervision of Peter I" made up a map of the eastern part of the Azov Sea. The next year H. Otto, a senior navigation officer of the "Krepost" ship, prepared a Mercator's map of the Black and Azov Seas at a scale of about 1 : 1 800 000 on the basis of measurements made during the 1699 voyage. The map showed depths along the southern coast of the Crimea Peninsula, near the entrance and in the Bosphorus Strait. The map was published in Amsterdam. Using the results of hydrographic works the Dutch carver A. Schonebek engraved the map *Eastern part of the Palus Meotis Sea, now called the Azov Sea.* It showed a sea coastline, a grid of parallels and meridians, depths, anchorage places, and cities. The map scale was 1 : 700 000. It was published in Moscow.

In 1701–1702 Dutch naval officer Peter Bergman made the *Map of the Azov Sea and Don River from Korotoyak to Azov and Taganrog and Dolgaya bar.* In 1702 in Voronezh he drew the *New map of the Azov Sea.*

In 1702–1704 P. Pikar published in Moscow the map *Direct drawing of the Black Sea from Kerch city to Tsar Grad.* It showed such cities as Bendery, Ochakov, Taman, Trapezund, Tsargrad and had an insert showing the Bosphorus Strait with depths marked along the fairway.

In 1703–1704 in Amsterdam the *Atlas of the Don River, Azov and Black Seas* prepared by Admiral K. Kruis was published. It comprised descriptions and 17 maps. The maps were made by the cartographic materials collected during sea voyages of the ship "Krepost" in 1699. The maps showed depths, shoals, inflowing rivers, harbors, and cities. All maps were hand-painted. This atlas is considered the first significant work of Russia cartography.

In 1706 the *Description of the Black Sea* was published. It was prepared by P.A. Tolstoy, the first permanent ambassador in Russian history in the Ottoman Empire [2].

5
Post-Peter's Black Sea

After signing of the Karlovits Peace Treaty (1699), which attached Azov to Russia, the Russian society started showing a growing interest in the Black Sea basin. At the same time this stirred anxiety among the political and military elite in the Ottoman Empire. The early eighteenth century in Turkey was the time of appearance of descriptions of strategic points and of the modern state of Ottoman ports and fortifications on the Black Sea, especially on its northern shores, with a view to drawing attention to the position of Black Sea fortresses and their significance for the Empire.

In 1728 the Academy of Sciences started publishing calendars (from 1768 they were called "menologies"). Many maps included in them represented original works that stirred great interest. Thus, Ya. Schmidt prepared the *Map of the Black Sea* and *Map of regions on the Black and Caspian Sea coast.*

In 1734 the *General map of the Russian Empire* prepared by I.K. Kirilov was published. The sheet of the European part of Russia showed the Black and Azov Seas. In 1739 the map of the Black Sea made by Vitsen was included into the special navigation atlas of L. Renare.

The Russian–Turkish war and actions of the Russian fleet on the Black Sea demanded new truthful maps, which is why in 1755 Lieutenants I. Bersenev and L. Pustoshkin started surveys of its coast. In 1785 they prepared the *Atlas of the Black Sea* comprising 11 handmade maps. After the creation in 1768 of the Azov fleet, the surveys of the Azov Sea were started guided by Admiral A.N. Senyavin and in 1771 his new map appeared. Continuing with this work, Lieutenants Yelchaninov and Zaostrovsky made a description of the coast near Taganrog. In 1778 the Kerch Strait was described by Lieutenant Karyakin, and in 1785 by Rostovsev.

However, none of these maps came out and foreign seamen used French maps prepared by Bellen in 1772 and Delmarche in 1785. It should be noted that the Bellen map contained serious mistakes: the shores of the Black Sea were marked with an error of up to 120 km, and 75 km for the Azov Sea. The Don mouth was shown 185 km eastward of its actual location.

In 1783 after attachment of the Crimean Peninsula and Northern Circum-Black Sea region to the Russian state the powerful Black Sea fleet was created; new ports, military bases, and lighthouses were built. Beginning from this time, regular hydrographic works were conducted in the Black Sea, primarily for preparation of navigation charts to meet the needs of the Black Sea navy. The information about the Black Sea was updated to a great extent. New atlases and maps were prepared and published, including abroad, that verified the coastline, showed the sea depths to more than 500 m, and showed bottom sediments. In 1771 Admiral A.N. Senyavin published a map of the Kerch Bay and the Azov Sea. In 1783–1786 naval officer Bersenev described the shores of a 50 km long stretch from Tarkhankut Cape to the Kerch Strait. In 1795 the map of the northern coast of the Black Sea on three sheets appeared that was made by order of Vice-Admiral De Ribas. Captain I.I. Billings prepared a handmade atlas on 45 sheets with views of the ports and coast of the Black Sea.

In 1797–1798 the reconnaissance description of the northern coast of the Black Sea (from the Dnestr to the Kuban River) was made under supervision of I.I. Billings and with the participation of I.M. Budischev, A.Ye. Vlito and others. These works were the basis for the *Atlas of maps and layouts of the Black Sea from Dnestr to Kuban River* (1799).

In 1799, as a result of studies of the mounts of the Dnieper, Dnestr, and Danube Rivers, I.I. Billings prepared the first *Atlas of the Northern Black Sea.* On the basis of these works, in 1807 I. Budischev prepared the manuscript *Sailing chart or guidelines over the Black and Azov Seas.* As a result, in 1817 the *Atlas of Maps* appeared, whose quality was superior to all available foreign maps of the Black Sea.

6
A Wider Approach to Black Sea Investigations

With the development of the scientific geographical idea and formation of some "sectoral" sciences, the issue of organization of regular works for the study of specific physical, hydrological, and hydrobiological features of the Black Sea and its water regime came to the fore. A network of hydrometeostations was created on the sea coasts and regular observations of the hydrometeorological regime were conducted.

In 1801–1802 the inventory of I.I. Billings was continued by Lieutenant I.M. Budischev and Warrant Officer N.D. Kritsky, who covered the western coast of the Black Sea from Odessa to the Bosphorus Strait, while Lieutenant-Commanders A.Ye. Vlito and P.A. Adamopulo covered the Anatolian coast from Constantinople to the Samsun Cape. A.Ye. Vlito and N.D. Kritsky also prepared the inventory of the whole Azov Sea and Kerch Strait and in 1804 the *Azov Sea map* was published. In 1807 on the basis of these works and I.I. Billing's inventory, maps of the Black and Azov Seas were published.

In 1804 A. Vilbrecht prepared a general map of the Black Sea on two sheets, which was called *Sea map of the Black, Azov and Marmara Seas based on most recent surveys and astronomical observations of the Russians and French.*

Beginning from 1807 and through 1825 the reconnaissance sea inventories of some areas of the Black Sea were prepared under the supervision of I.M. Budischev, F.F. Bellinsgausen, N.D. Kritsky and others.

In 1808 the first part of the navigation chart or *Sea guide of the Azov and Black Seas* of I.M. Budischev came out. Its contents were close to the present-day navigation charts.

In 1811 the third volume of the works of Academician P.S. Palass *Zoographia Rosso – Asiatica*, devoted to the Black Sea ichthyofauna, came out. Later fauna and flora studies were conducted by other biologists who visited the Black Sea in the early nineteenth century: M.G. Ratke, A.D. Nordman, and A.F. Middendorf.

The hydrographic surveys for the Budischev's navigation chart were added to the results of the Caucasian coast inventory conducted in 1816 by Captain F.F. Bellinsgausen and enabled publication in 1817 of the *General map of the Black and Azov Seas.*

In 1820, by agreement between the Russian and French governments, a description of the Black Sea was made on the ship "La-Chevrette" under the command of Captain Gautier, and two years later the sea map came out in France.

At that time in 1832 at the Russian Black Sea Navy headquarters, the Hydrographic Department was organized such that it comprised a hydrographic depot and an instrumental chamber. In this way the Hydrographic Department of the Black Sea Navy started its activity.

In 1833 during a visit of a squadron commanded by M.P. Lazarev to Turkey the description of the Bosphorus and Dardanelles was made by Lieutenants Ye.V. Putyatin and V.A. Kornilov. In 1842 the *Atlas of ices of the Black and Azov Seas* was prepared.

Similar mapping of the Black and Azov Sea coasts was conducted during expeditions of Ye.P. Manganari (1825–1836), as a result of which in 1936 he prepared the new *General map of the Black Sea* and in 1842 the detailed *Atlas of the Black and Azov Seas coasts* showing bottom sediments, currents, and depths to 300 Russian fathoms (1 Russian fathom = 1.852 m).

In 1841–1842 the Depot of Maps of the Black Sea Navy produced the first copies of the *Atlas of the Black and Azov Sea maps.* This publication was completed in 1844. For the first time in the Russian sea cartography the underwater relief on maps of the atlas was depicted with the help of isobaths.

Coastal studies and measurements of depths of the Black Sea were continued on instructions from famous Russian seafarer and naval commander M.P. Lazarev.

In 1847 Lieutenant-Commanders G.I. Butakov and I.A. Shestakov (cutters "Pospeshny" and "Skoryi") conducted sea surveys of the Caucasian and Crimean coasts, Bug River, Dnestr lagoon and a coastal part between Ochakov and the Danube River mouth (1848), the Anatolian coast of Turkey (1849), and the shore of the Constantinople Strait to the Danube River mouth that added to the data of Ye.P. Manganari. In 1851 on the basis of these materials the *Navigation chart of the Black Sea* was published, which for a long time was the basic aid during voyages over the Black Sea.

In 1854 the *Navigation chart of the Azov Sea* was published. It was prepared by A. Sukhomlin on the basis of his works of 1850–1851.

Systematic observations of the sea level by tide gauges and also current studies with the help of "bottle mail" were started in 1855. In 1868, deep-water measurements of water temperature and density were carried out from the corvette "Lvitsa" during installation of a telegraph cable from Feodosia to Adler. They showed that in the Black Sea the water density increases with the depth.

In the same period, stationary hydrometeorological observations were initiated. The first hydrometeorological station on the Black Sea, which later became an observatory, was founded in Nikolayev in 1801. Stations were then founded in Kherson (1808), Odessa (1821) and Sevastopol (1824). Systematic observations were conducted by stations in Nikita (1826), Karadag (1831), Kerch (1863), Yevpatoria (1866), Yalta (1869) and others. In some years hydrometeorological observations were interrupted for some reasons and then resumed again. By the end of the century the Hydrographic Department had already several dozens of hydrometeorological stations, including those on the shores of Turkey (Sinop, Trabzon).

A new stage in detailed and systematic investigations of the basin began with organization in 1871 of the hydrographic expedition to the Black and Azov Seas under command of Captain V.I. Zarudny. The result of this expedition was a publication of finely designed maps of the Black Sea coast, both of the whole sea and detailed maps of its separate parts. Within the framework of this Hydrographic Expedition a special hydrological team was functioning led by F.F. Vrangel that carried out significant oceanographic investigations (measurements of currents, temperature and salinity in surface and deep waters) in the northwestern part of the sea, Kerch Strait, near Crimean and Caucasian coasts.

A significant event in the study of the Black Sea was an organization in 1871 of the first sea biological station in Odessa, which several years later was moved to Sevastopol. Its first director was Academician A.O. Kovalevsky. Now, this is a widely known Institute of Biology of the Southern Seas (IBSS) of the National Academy of Sciences of the Ukraine, named after its founder. With the appearance of the biological station in Sevastopol the studies of flora and fauna of the Black Sea acquired a systematic nature. Thus, in the early twentieth century the hydrobiological investigations led by S.A. Zernov covered an extensive offshore zone. In 1909 in the northwestern part of the sea there was found a great community of red algae phyllophore, known in the literature as "Zernov's phyllophore field". In 1913 S.A. Zernov published the results of his investigations in the monograph *Study of the life in the Black Sea.* Fauna studies of the Novorossiysk Biological Station of the Odessa University and the Karadag Biological Station (established in 1914) deserve high credit.

In 1881–1882 the outstanding Russian oceanographer and Naval Commander (later Admiral) S.O. Makarov using the Istanbul-based Russian ship "Taman" carried out detailed hydrological observations in the Bosphorus Strait that included measurements of water temperature, water salinity, current velocity and direction. On the basis of these observations he found that there were differently oriented currents: the upper current in the Strait went from the Black Sea to the Sea of Marmara and the lower one in the backward direction. In this way the remarkable phenomenon of the Black Sea was discovered that explained the specific features of its hydrological structure. In 1885 the results of works in the Bosphorus Strait were published by S.O. Makarov in his famous paper *About water exchange of the Black and Mediterranean Seas* [3], which got the award of the Russian Academy of Sciences.

The late nineteenth century is characterized by further systematic pooling of knowledge on the hydrology of the Black and Azov Seas. Special expeditions were organized for carrying out hydrological investigations, regular observations of currents were conducted on floating lighthouses, and new hydrometeorological stations were opened.

Of great significance for the extension of knowledge on the general hydrology of the Black Sea was the organization in 1890–1891 of the first complex oceanographic Black Sea expedition on the gunboats "Chernomorets" (1890), "Donets" and "Zaporozhets" (1891). The expedition was led by I.B. Shpindler. Hydrographer F.F. Vrangel, geologist N.I. Andrusov, chemist A.A. Lebedintsev, and biologist A.A. Ostroumov [4] also participated. During this expedition approximately 200 deep oceanographic stations were carried out, and a cold intermediate water layer with a temperature lower than 8 °C was discovered. Quite unexpectedly, it was found that the whole water column below 200 m was contaminated with hydrogen sulfide. Its presence was discernable even by the smell of deep water samples. Chemist N.D. Zelinsky (later, an Academician), who also worked in this expedition, attributed the presence of hydrogen sulfur to the action of a specific group of anaerobic bacteria. In 1891 samples were taken with a special bathometer with a gilded internal surface. The deep-sea waters were devoid of any living organisms. Near the entrance into the Bosphorus was found water with a salinity of approximately 34‰, which came with the deep current from the Sea of Marmara. This finally proved that the deep waters of the Black Sea are formed by mixing of local water with water from the Sea of Marmara. Therefore, this expedition brought outstanding oceanographic discoveries. The work of this expedition was later continued by the sea observatory of the Black Sea Navy.

Generalized data on the hydrometeorological regime of the Black Sea were placed in navigation charts and in special publications. In 1908 the *Atlas of winds and mists of the Black Sea* was published, and in 1915 the next issue of the navigation chart of the Black Sea. Publication of monthly weather reviews was initiated, and some climatic data on the Turkish coasts of the Black Sea were published. In 1908–1914 hydrographic surveys of the coasts were conducted and the maps of the Black Sea verified.

The works performed till 1914 provided the first detailed, triangulation-based description of the Black and Azov Seas. By the beginning of the First World War Russia could boast the best knowledge of hydrography of these seas.

7
USSR Investigations

The Soviet period was a qualitatively new stage in the investigations of the Black Sea: more planned and regular, more active, and covered all aspects of ocean science.

In 1921 a decree was passed that established a hydrometeorological service of the Soviet Union. By this time the Azov–Black Sea basin already had a network of hydrometeorological stations that conducted standard coastal observations. Offshore observations were conducted by fishing vessels and hydrographic ships, often incidentally while passing from one port to another. After the Civil War in Russia (1917–1922) the sea scientific investigations became systematic. In 1922 these researches were conducted by the Azov–Black Sea research-fishing expedition led by N.M. Knipovich, one of the top sea researchers. The investigations covered a strip of the Black Sea along the USSR coast more than 100 km wide. The expedition functioned successfully until 1928. During this time it studied in detail the physical and chemical conditions of sea water influencing the formation of a fishing base and evaluated the fish resources. The data obtained helped to verify the distribution of oxygen and hydrogen sulfide in the Black Sea. The results of extensive expedition works in 1922–1928 were included in the first generalizing summary on Black Sea hydrology prepared by N.M. Knipovich [5]. In particular, it gave a scheme of surface currents, the main features of which are recognized at present (large-scale cyclonic vortices in the eastern and western parts of the sea – "Knipovich goggles").

Initial oceanographic investigations of the Black Sea were finalized, in general, by the expedition under the command of Yu.M. Shokalsky [6] and later V.A. Snezhinsky. The expedition took place in the period from 1928 to 1935. Observations were conducted on vessels "Ingul", "Dunai", and "Gidrograf" in all seasons of a year. Sea works were assigned to the sea observatory and biological station in Sevastopol. During this expedition there were organized 53 cruises in which approximately 1600 oceanographic stations were made. These measured sea water temperature and salinity at all depths down to the seabed, and collected over 2000 biological and geological samples. The materials of this expedition were the basis for the fundamental work of A.D. Arkhangelsky and N.M. Strakhov on the geological structure and the history of the Black Sea depression development [7]. Detailed investigation of the water vertical hydrological structure had shown that mixing of the upper oxygen and the lower hydrogen sulfide layers of sea water did occur, although rather slowly. Chemical analyses helped to find that nearly all hydrogen sulfide contained in deep waters is formed as a result of reduction of sea water sulfates by the carbon of organic matter with the participation of bacteria. Biological investigations revealed seasonal variations of plankton, and by-depth distribution of benthos. The Black Sea oceanographic expedition was one of the most successful ventures in the study of this water body.

During 1928–1938 regular oceanographic investigations in the Black Sea were conducted by the Sevastopol sea observatory. At that time, monthly synchronous 50-mile profiles normal to the coast were made. In 1929, on the initiative of V.V. Shuleikin, a sea hydrophysical station (now a branch of the Marine Hydrophysical Institute, National Academy of Sciences of the Ukraine, MHI NASU) was established in Katsiveli near Yalta. In 1932 the expedition of the Hydrometeorological Service of the Black and Azov Seas, together with the Sevastopol Biological Station led by V.V. Shuleikin, conducted important oceanographic investigations in the eastern part of the Black Sea during wintertime. The results of these works were the basis for plotting dynamic maps of currents and making descriptions of cold water distribution in the sea.

In the late 1940s classical works of the famous biologist V.A. Vodyanitsky were published where he, proceeding from the results of biological and hydrological observations, validated the logical model of a vertical structure and general circulation of waters in the Black Sea [8]. Its main idea was that the whole sea water column represented a unity subject to a system of vertical and horizontal movements from the surface to the bottom. Here V.A. Vodyanitsky admitted that exchange between the surface and deep water masses went on slowly, while evaluations of its rate were rather approximate [8].

In the Soviet Union, the research institutes and organizations of Sevastopol had at their disposal well-equipped research vessels and played a leading role in the studies of the Black Sea. These were, first of all, the Marine Hydrophysical Institute (R/V "Mikhail Lomonosov", R/V "Akademik Vernadsky", R/V "Professor Kolesnikovç" and the Institute of Biology of the Southern Seas (R/V "Akademik Kovalevsky", R/V "Professor Vodyanitsky") of the USSR Academy of Sciences, now the National Academy of Sciences of the Ukraine. In addition, there were the bases of research vessels of the Sevastopol Branch of the State Oceanographic Institute (SB SOI) and the Hydrographic Service of the Black Sea Navy. Sevastopol was also a port of registration for research vessels of the M.V. Lomonosov Moscow University ("Akademik Petrovsky", "Moskovsky Universitet") whose summer cruises were combined with the student's practical course. For nearly three decades, (1961–1989) R/V "Miklukho-Maklai" was used by the Odessa Branch of IBSS for detailed investigation of the environmental conditions in the northwestern part of the sea, including its deterioration from the early 1970s.

The 1950s were characterized by in-depth studies of the sea hydrology. By the joint efforts of the hydrographic organizations, and establishment of the Hydrometeorological Service and of research institutes of the USSR Academy of Sciences, Ministry of Fishery and other departments, the Black Sea was covered by a dense network of synchronous complex oceanographic surveys. During a year up to 1000 oceanographic stations were made.

In Gelendzhik, the Southern Branch of the Institute of Oceanology of the USSR Academy of Sciences was actively functioning. The main subject of the Institute research was study of specific features of water mass and currents. Observations were carried out in open sea, from vessels, and on autonomous (the so-called buoy) stations in the coastal zone. Here in the late 1950s the expeditions were organized on research vessels "Akademik S. Vavilov", "Akademik Shirshov", particularly for study of the shelf area and geological structure of the Black Sea depression. From the results of these expeditions, more accurate bathymetric, geophysical, and geomorphological sea maps were prepared. Novorossiysk was the base of large vessels of the Institute of Oceanology: R/V "Vityaz" and R/V "Professor Shtokman". In the 1950s–1980s the most detailed and diversified investigations of the Black Sea were conducted during many of the aforementioned expeditions. Other research vessels were also used for these purposes. The expeditions were conducted by institutes and organizations of the Soviet Union and later on of Russia, Ukraine, and other Black Sea countries.

International cooperation was widely developed. In 1957–1959 a sizable contribution to a collection of on-field data was made by interdepartmental expeditions under the program of the International Geophysical Year. Such large organizations as the USSR Hydrometeorological Service, the USSR Academy of Sciences, and others took part in it. For the first time, the observations and measurements enabled tracing of the mesoscale variability of a thermohaline structure of waters and currents. From 1961 the USSR Hydrometeorological Service initiated regular seasonal observations on several "century" sections. Six of them were selected in the Black Sea: Bolshoy Fontan Cape – Tarkhankut Cape, Tarkhankut Cape – Zmeinyi Island, Khersones – Bosphorus Strait, Sarych Cape – Inebolu Cape, Kadosh Cape – Unye, and Yalta – Batumi. On each section there were from nine to 20 oceanographic stations. The materials received from these sections helped to reveal an interannual and seasonal variability of the sea regime.

In 1950–1970, using the materials of multi-year observations of the Basin Hydrometeorological Service, the following hydrometeorological handbooks very important for the national economy were published: *Climatic and hydrological atlas of the Black and Azov Seas* (1956), *Atlas of ices of the Black and Azov Seas* (1962), *Catalog of level observations in the Black Sea* (1969), and *Handbook on hydrological regime of seas and river mouths* (1970). They were arranged in four volumes and summed up the basic hydrological characteristics of the coastal zone of the Black and Azov Seas. There was also the *Handbook on the Black Sea climate* (1974) and other works.

The results of oceanographic studies of the Black Sea accumulated by that time were published in the summary work of A.K. Leonov *Regional oceanography* [9] and in the monograph of D.M. Filippov *Circulation and structure of the Black Sea waters* [10]. In 1976–1978 the joint program of integrated oceanological investigations of the Black Sea ("SKOIC") was implemented with participation of the main Black Sea organizations (MHI, IBSS, SB SOI, Hydrographic Service of the Black Sea Navy). The works under this program were targeted to assessment of hydrological changes in the Black Sea with regard to anthropogenic effects. The results were summarized in collections of articles. The materials received by the 1980s made it possible to move from characterization of the average multi-year regime of the Black Sea to analysis of its dynamics. This was reflected in the monograph prepared by A.S. Blatov, N.P. Bulgakov et al. *Variability of hydrophysical fields in the Black Sea* [11], which was the first to provide assessments of a wide range of variability from short-term to year-by-year.

The specific features of these investigations in the period in question were an integrated approach, participation of several organizations, and also application of the newest autonomous and remote-sensing devices, non-contact methods of measurements (from aircrafts and satellites) and others. Some observations were conducted in accordance with international cooperation plans. Investigation results were also summarized in collective monographs of MHI edited by B.A. Nelepo *Integrated investigations in the Black Sea* (1979) and *Integrated oceanographic investigations of the Black Sea* (1980).

In the 1980s the P.P. Shirshov Institute of Oceanology of the USSR Academy of Sciences started intensive study of the present state and changes in the ecosystem of the Black Sea pelagic zone, led by Academician M.Ye. Vinogradov. Within the framework of this project there were organized expeditions on the R/V "Vityaz" (March–April 1988) and R/V "Dmitry Mendeleev" (July–September 1989). Special attention was drawn to the effect on the Black Sea ecosystem of the invader comb-jellyfish (*Mnemiopsis leidyi*), which spread extensively in the sea in the late 1980s. In 1991 the investigations of the Black Sea ecosystem on the R/V "Vityaz" were continued in the winter season (February–April), which had been less studied. The results of this program were published in the book *Black Sea ecosystem variability* [13]. At the same time, the USSR Hydrometeorological Committee within the framework of the project "USSR Seas" published the summarized results and estimates of many parameters characterizing the Black Sea regime [14].

Scientific investigations of the Black Sea were characterized by a considerably increased volume of works, especially targeted systematic works to meet the needs of the national economy for multipurpose utilization and transformation of natural (biological, mineral, and recreational) resources of the sea. These investigations were aimed, first of all, at receiving information about natural–ecological conditions of the sea, its regime, main processes, interactions among some components of the water environment, and regularities contributing to its formation; study of the genesis of natural events, their variability in time, and their forecast. Thus, there was a transition from general description to study of life processes in the water environment and their casual links.

8
Investigations Made by Bulgaria, Romania, Turkey, and USA

In Bulgaria, the Black Sea studies were concentrated in Varna, where the Fish Resources Institute founded in 1954 (in 1932 – Marine Biological Station) and the Institute of Oceanology of the Bulgarian Academy of Sciences (BAS), formed in 1973, are located. Bulgarian researchers paid much attention to the oceanography of the western part of the Black Sea, including such issues as the effect of the Danube outflow on the regime of this water area, and oceanography of the Circum-Bosphorus region. The international experiment "Kamchia" deserves special mention. It was conducted in the late 1970s under the program of the CMEA member countries (Council of Mutual Economic Assistance) using the experimental base of the Institute of Oceanology, BAS near Varna. Specialists from Bulgaria, Soviet Union, German Democratic Republic, Poland, and Romania took part in these research efforts. The experiment was targeted to study the interaction and exchange processes in the atmosphere–hydrosphere–lithosphere system in the coastal zone of the sea. The results of the international experiment "Kamchia-77" entitled *Interaction of the atmosphere, hydrosphere and lithosphere in the sea coastal zone* were published in 1980. A team of Bulgarian scientists also prepared the review *Black Sea* describing the main specific features of this water body [12].

In Romania, oceanographic studies of the Black Sea are conducted by the Marine Research Institute in Constanta. Their target is the shelf zone. For a long time the Romanian–Ukrainian team had been engaged in studies of the Danube River delta, which is under the jurisdiction of both countries. In 1979–1999 environmental monitoring in the Danube mouth area was conducted.

Turkish oceanographers study the vast zone along the southern Anatolian coast and also water exchange in the Bosphorus Strait. They have at their disposal the research vessels "Piri Reis" of the Institute of Marine Sciences and Technology in Izmir, "Bilim" of the Institute of Marine Sciences, Middle East

Technical University in Erdemli, and other vessels. Turkish specialists take an active part in voyages over the Black Sea organized by western countries and international organizations.

Scientists from the USA, in particular the Woods Hole Oceanographic Institute, are seriously engaged in Black Sea studies. In 1969 the voyage of the R/V "Atlantis II" was organized, which lasted for seven weeks. Determination of the age of the upper sedimentary layers in the sea depression helped to reveal that the modern sedimentation processes developed here much more intensively than on the Atlantic seabed. In 1975 there was organized a short voyage on the R/V "Chain". Among the obtained results was a chronology of geological sediments determined on the basis of ground samples analyses. In summer 1975 a special drilling ship "Glomar Challenger" was working in the Black Sea and drilled three wells that enabled more accurate timing of the sea depression formation. In 1988 there was organized an international geological expedition on R/V "Knorr" that included five cruises. Oceanographers from the USA, Turkey, and several European countries worked on it. In 1974 Woods Hole Oceanographic Institution published *The Black Sea (it's geology, chemistry, biology), a bibliography* by Phyllis N. Laking.

9
Recent Studies

Present-day Black Sea investigations are characterized by application of probing devices, autonomous units, and remote sensing (satellite-based) techniques for receiving information, in combination with in-situ observations. Satellite data (optical and infrared radiometric, altimetric, radar, and drifter tracking) are used mostly by scientists in the P.P. Shirshov Institute of Oceanology, Russian Academy of Sciences (SIO RAS) and MHI NASU, which work in close collaboration. These enable tracing of the large scale circulation, mesoscale water dynamics, vortex formation, dynamics of upwelling fronts and zones, areas of higher biological productivity, oil pollution, etc.

Very important for study of the condition and variability of the Black Sea ecosystems is implementation of international regional agreements and programs supported by targeted financing. Thus, in April 1991 a Cooperative Marine Science Program for the Black Sea (ComsBlack) designed for 5 years was approved. In 1992 the Convention on the Control of the Black Sea Pollution was signed. It was ratified by all countries of the region.

In 1993 the Global Environmental Facility (GEF) provided financial support to the Black Sea Environmental Program (BSEP). Its task is to concentrate the efforts of scientists and specialists from the Black Sea countries around the main concept of the program, i.e., definition of the present condition and variations in the Black Sea ecosystem, elaboration of actions on preservation, and development of its biodiversity.

The International Atomic Energy Agency (IAEA) makes a considerable contribution into environmental programs of the Black Sea. The Black Sea countries consider the monitoring of radioactive pollution of the sea to be the priority issue. The IAEA Program is aimed at study of the presence of radionuclides in the Black Sea and tracing variations of its radioactivity.

NATO is also involved in studies of the Black Sea. Within the framework of the program "Science for Peace", in 1993 this organization provided financial support to the project "Modeling of the ecosystem as a means of the Black Sea management" designed for 5 years. The USA took part in this project as well as the Black Sea countries. One of the significant results of this activity was creation of an integrated database of hydrological, hydrochemical, and hydrobiological observations conducted in the Black Sea.

On the request of UNESCO IOC the scientists from MHI NASU prepared and published in the series "UNESCO reports in marine sciences" two summing-up publications *Artificial radioactivity of the Black Sea* and *Hydrochemistry and dynamics of the hydrogen sulfide zone in the Black Sea.* INTAS and INCO-Copernicus of the European Union also funded several international projects on the Black Sea investigations.

The results of multi-year oceanographic observations on the Black Sea are taken together in several databases including, with some minor differences, about 100 000 hydrographic stations. The main bases of oceanographic information on the Black Sea are available at the All-Russia Research Institute of Hydrometeorological Information in Obninsk, MHI NASU in Sevastopol, SOI, and at the Department of Oceanology of the Moscow State University.

Many publications on oceanography of the Black Sea, numbering several thousands of sources, are systematized in reviews, monographs and special bibliographies, in particular in *Black Sea bibliography* (1974–1994).

In recent years, due to development of the offshore resources, investigations in the coastal zone have been conducted more actively. These included local dynamic experiments in the shelf zone of the Southern Coast of the Crimea (SCC) – LDExp involving synchronous mesoscale CTD surveys and installation of automatic buoy stations (ABS) in different seasons of a year, and organization of marine special measurements and hydrographic and hydrometeorological stations in the regions of intensive shipping (Experimental Department of MHI NASU in Katsiveli, Kerch Port, and others).

For more than two decades the seasonal observations have been conducted on the 20-mile profile from the Danube mouth along latitude 45°20′N on nine stations; mesoscale surveys with ABS installation are carried out in the Zmeinyi Island area, 10-mile "century" section and coastal mesoscale surveys in the SCC region. Every year approximately 200 oceanographic casts are made in this region.

For almost three decades a stationary oceanographic platform has been functioning in the SCC shelf nearby Katsiveli. It conducts, in a semi-automatic regime, a full complex of meteorological observations, measurements of the sea level, waves, vertical distribution of water temperature and salinity, and velocities of currents at different depths.

From 1999 to the present SIO RAS has undertaken regular complex investigations of the Black Sea that have helped trace the process of degradation and partial restoration of its ecosystem with regard to appearance of foreign invaders (*Mnemiopsis* and *Beroe*). Regular in-situ observations enabled a detailed study of the dynamics in the population structure of comb jellyfish, which invaded into the Black Sea ecosystem in the eastern part of the basin. It was shown that the new invader *Beroe* became "established" in the ecosystem and in 2000 maintained its dominating position in a peak of its seasonal development from late September to late November, which enabled speaking about a long-time invasion. Less pressing of *Mnemiopsis*, which is eaten by *Beroe*, would lead to increased populations of plankton-eating fish. For the first time it was demonstrated how the mesoscale circulation in the Black Sea influences the composition, distribution, and productivity of plankton communities. The received data formed a basis for forecast and adequate analysis of modern anthropogenic trends in the basin.

The results of the Black Sea drifter experiment conducted in 1999–2003 were interpreted and analyzed. The drifter trajectories confirmed the existence of intensive mesoscale vortices, both in the coastal and deep-water parts of the Black Sea. The coefficient of horizontal vortex diffusion was estimated and the time scale of the exchange between the central and coastal zones of the sea was determined. It was found that intensification of the Rim Current due to wind forcing is accompanied by weakening of the mesoscale vortex dynamics and related horizontal (cross-shelf) water exchange, while the Rim Current weakening in lower wind forcing leads to the opposite effect. Such a conclusion was made on the basis of the results obtained by analysis of the NCEP wind field characteristics over the Azov–Black Sea region, satellite IR images, shipboard hydrographic observations, and the results of laboratory modeling of physical mechanisms of variability of macro- and mesoscale water dynamics in the Black Sea. The results of this complex study were published in 2002 in a book *Multidisciplinary investigations of the northeastern Black Sea* edited by A.G. Zatsepin and M.V. Flint [15].

On the basis of research conducted by the Department of Oceanology of the Moscow State University, the oceanographic and environmental conditions of the Black Sea were summarized in the paper by A.N. Kosarev et al. [16]. The fundamental book of Yu. Sorokin *Black Sea* [17] published in the Netherlands in 2002 sums up many years of "traditional" investigations of the Black Sea. It pools together the knowledge in oceanography, chemistry, biology, and microbiology of the sea.

By 2000 the epoch of mostly occasional observations over some of the manifestations of natural processes in the Black Sea was over. It was replaced with systemic (remote satellite and contact autonomous) observations of the spatial and temporal structure of marine processes, which should be interpreted with regard to their compliance with fundamental laws [18]. Here wide possibilities were opened for addressing the earlier inaccessible problems and for development of a new outlook on traditional Black Sea oceanographic issues, supported by more reliable, numerous, and adequate facts received from observations.

The European Union continues to support international collaboration devoted to complex investigation of the Black Sea. This is done, for example, within the framework of the following projects: "A Regional Capacity Building and Networking Programme to Upgrade Monitoring and Forecasting Activity in the Black Sea Basin, ARENA" (2003–2006), "International Action for Sustainability of the Mediterranean and Black Sea Environment, IASON" (2005–2006), "ASCABOS" (2005–2008), "SESAME" (2006–2010), "ALTICORE" (2006–2008), "MOPED" (2007–2009), etc.

References

1. Agbunov MV (1987) Antique navigation chart of the Black Sea. Nauka, Moscow (in Russian)
2. Tolstoy PA (2006) Description of the Black Sea, Aegean Archipelago and Ottoman Fleet. Vostochnaya Literatura, Moscow (in Russian)
3. Makarov SO (1990) On water exchange of the Black and Mediterranean Sea. In: Oceanographic papers. Gidrometeoizdat, Moscow (in Russian)
4. Shpindler IYe (1893) Results of hydrological investigations of the Black Sea in 1890–1891. In: Marine collection. (in Russian)
5. Knipovich AM (1932) Hydrological investigations in the Black Sea. In: Transactions of the Azov-Black Sea scientific-fishing expedition, Issue 10. (in Russian)
6. Leonov AK (ed) (1950) To the memory of Yu.M. Shokalsky. In: Collection of papers, Chap 2. AN SSSR, Moscow (in Russian)
7. Arkhangelsky AD, Strakhov NM (1938) Geological structure and history of the Black Sea development. AN SSSR, Moscow (in Russian)
8. Vodyanitsky BA (1948) Main water exchange and history of salinity formation in the Black Sea. In: Transactions of the Sevastopol biological station, vol 6. (in Russian)
9. Leonov AK (1960) Regional oceanography, Chap 1. Gidrometeoizdat, Leningrad (in Russian)
10. Filippov DM (1968) Water circulation and structure of the Black Sea. Nauka, Moscow (in Russian)
11. Blatov AS, Bulgakov NP, Ivanov VA, Kosarev AN, Tuzhilkin VS (1984) Variability of hydrophysical fields of the Black Sea. Gidrometeoizdat, Leningrad (in Russian)
12. Vylkanov A, Danov Kh, Marinov Kh, Vladev P (1983) Black Sea Collection. Gidrometeoizdat, Leningrad (in Russian, translated from Bulgarian)
13. Vinogradov MYe (ed) (1991) Black Sea ecosystem variability. Nauka, Moscow (in Russian)
14. Simonov AI, Altman EN (eds) (1991) Hydrometeorology and hydrochemistry of seas in the USSR, vol IV. Black Sea. Issue 1. Hydrometeorological conditions. Gidrometeoizdat, Leningrad (in Russian)
15. Zatsepin AG, Flint MV (eds) (2002) Multidisciplinary investigations of the northeastern Black Sea. Nauka, Moscow (in Russian)

16. Kosarev AN, Tuzhilkin VS, Daniyalova ZH, Arkhipkin VS (2004) Hydrology and ecology of the Black and Caspian Seas. In: Geography, society and environment, vol VI. Dynamics and interaction of the atmosphere and the hydrosphere. Gorodetz, Moscow (in Russian)
17. Sorokin YuI (2002) Black Sea ecology and oceanography. Backhuys, Leiden, The Netherlands
18. Ivanov VA, Kosarev AN, Tuzhilkin VS (2004) To the history of expedition oceanographic investigations of the Black Sea. In: Collection of scientific works of the Marine Hydrophysical Institute NASU, Issue 10. Sevastopol (in Russian)

Hdb Env Chem Vol. 5, Part Q (2008): 31–46
DOI 10.1007/698_5_079

Published online: 26 May 2007

Quaternary Paleogeography of the Azov–Black Sea Basin

Alexander A. Svitoch

Department of Geography, Moscow State University, 119992 Moscow, Russia
geography@inbox.ru

Abstract During the Pleistocene history of the Black Sea, a series of basins may be traced noticeably different in their hydrological and faunistic characteristics. The formation and evolution of these basins was defined by many factors; among them, the principal factors were the availability of the connection with the Caspian and the Mediterranean seas, the character of the water exchange with them, and the proportions between the principal components of the water balance (freshwater supply and precipitation). The transgressive basins almost did not differ in their area and sea level height; on the whole, the latter was close to its present-day position. The maximum of the sea level rise was determined by the depth of the Bosporus threshold and the level of the Mediterranean Sea; it is most probable that this rise never exceeded a value of +6–8 m. The main differences between the transgressive basins were related to the salinity changes, which defined the character and composition of the fauna dwelling in them.

Keywords Black Sea · Sea of Azov · Stratigraphy · Paleogeography · Quaternary system

1 Introduction

The Black Sea and the Sea of Azov are elements of the system of inland basins, which represent relics of the formerly vast ocean basin of the Paratethys (Fig. 1). This enormous basin existed at the end of the Paleogene and the beginning of the Neogene and was connected to the Atlantic and Indian oceans. During its geological history, it has passed a complicated evolution from large sea basins to individual isolated brackish-water and freshwater basins.

The entire history of the Azov–Black Sea basin is closely related to the geological structure of the region, which is located within the Alpian geosynclinal belt. Here, starting from the Oligocene–Miocene, orogenic processes proceed actively, accompanied by fractures of the earth's crust, block displacements,

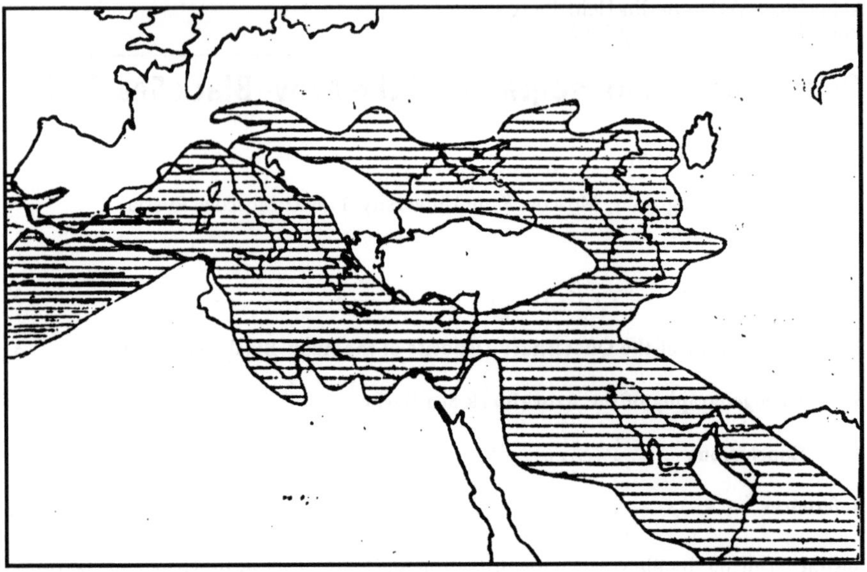

Fig. 1 Paratethys at the end of the Oligocene (after Rogel et al. [22])

and manifestations of volcanic activity. The final (Quaternary) stage of this epoch, despite its small duration (about 1.6 million years), is extremely important for the entire history of the region. During this time, the basins have acquired their present-day outlines, seaside landscapes have been formed over the surrounding coasts, and the systems of historical civilizations have emerged and developed.

Beginning with N.I. Andrusov, the issues of the stratigraphy and paleogeography of the Azov–Black Sea basin have been considered by numerous researchers such as A.D. Arkhangel'skii, N.M. Strakhov, P.V. Fedorov, L.A. Nevesskaya, A.F. Shnyukov, F.A. Shcherbakov, D.A. Ross, and many others.

This study is based upon abundant materials obtained by many scientists in reference cross-sections on the coasts of the Black Sea and the Sea of Azov as well as on numerous data referring to the shelves of these basins. The information about the Quaternary deposits of the Black Sea basins is poor and was obtained owing to the deep-sea drilling within the frameworks of the program of cruise 42 of D/V *Glomar Challenger* [1–3].

2
Stratigraphy of the Quaternary Deposits of the Azov–Black Sea Basin

In this paper we have used the stratigraphic scale of the IUGS [4], subdividing the Quaternary system into the Lower (1.6–0.8 million years), Middle

(0.8–0.12 million years), and Upper (0.12–0.01 million years) Pleistocene and the Holocene (< 0.01 million years). The stratigraphic subdivision of the marine Pleistocene of the Azov–Black Sea basins is based on the distribution of the mollusk fauna over the section of marine deposits. During the Pleistocene, the ancient Pontian, which was an intermediate basin between the Caspian and the Mediterranean seas, has undergone repeated invasions of the fauna from adjacent basins. Together with the evolution of the endemic Black Sea fauna, this became the reason for the diverse composition of the fossil mollusk fauna. Usually, the stratigraphic consideration of the marine Pleistocene involves an analysis of the ratios between the representatives of the Mediterranean, brackish-water (Black Sea–Caspian), Azov (freshened Caspian and Black Sea forms), and fresh-water fauna (Fig. 2). In so doing, the account for the ratio between the Mediterranean and brackish-water species in the mollusk assemblages is the most important. For the Mediterranean fauna, the abundances of stenohaline (*Cardium tuberculatum* and others) and euryhaline (*Cerastoderma glaucum*, *Abra ovata*, and others) species are determined. The brackish-water (Black Sea–Caspian) fauna of the Black Sea consists of two groups – those of the Pliocene relics and of didacnas of the Black Sea and Caspian origins.

Within the Azov–Black Sea Quaternary sediments, with respect to the combinations of the mollusk fauna, the following stratigraphic units are recognized: the Chaudian, the Bakunian, the Old Euxinian, the Uzunlarian, the Karangatian, the Tarkhankutian, the New Euxinian, and the Black Sea formations.

The Chaudian Formation (Lower Pleistocene – beginning of the Middle Pleistocene) consists of the Gurian, Lower, and Upper Chaudian sequences.

The Gurian Chaudian layers were reliably registered only on the coast of Georgia; they are represented by shallow-water deposits of the initial stage of the Chaudian transgression and contain abundant shells of Pliocene relics (*Tschaudia tschaudae*) and Black Sea species (*Didacna pseudocrassa* and others).

The Lower Chaudian Formation, similar to the Gurian one, is spread only in restricted areas. These deposits are localized in the freshened part of the Chaudian basin, such as the Kerch–Taman region and the south of Moldova. In the stratotypical section described near Cape Chauda, they are represented by the sediments of a strongly freshened basin – sands with shells of *Monodacna subcolorala, Didacna baeri-crassa, Dreissena polymorpha*, and others [5].

The Upper Chaudian deposits are the most developed; different facies of them are spread over almost the entire Black Sea coast. In the reference section on Cape Chauda they are represented by shallow-water carbonate sediments with abundant fauna consisting of Pliocene relics (*Tschaudia tschaudae*), Black Sea endemic species (*Didacna pseudocrassa, D. olla*, and others), and fresh-water elements (*Dreissena polymorpha*). In other sections of the

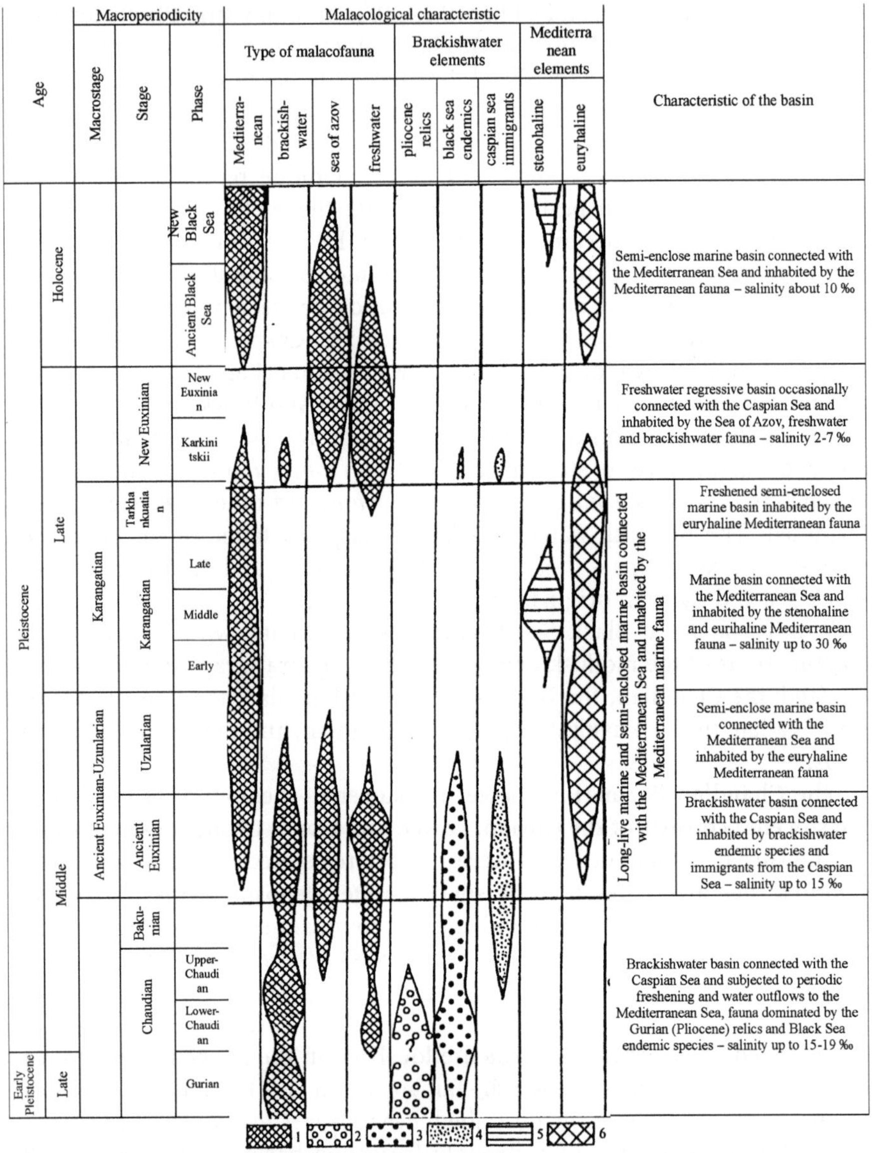

Fig. 2 Pleistocene basins of the Black Sea. *1–4* Types of the mollusk fauna: *1* brackish-water fauna; *2* Pliocene relics; *3* Black Sea endemic species; *4* Caspian immigrants. *5–6* Mediterranean fauna: *5* stenohaline; *6* euryhaline

Upper Chaudian Formation, rare shells of Mediterranean mollusks (*Cerastoderma glaucum*, *Mytilus edulis*, and others) were noted.

On the Caucasian coast, the Upper Chaudian deposits form an accumulative cover of the lower (50–60 m) Chaudian terrace; they are usually represented by coarse-grained coastal–marine sediments. In the region of the Rioni Lowland, the Upper Chaudian deposits include a thick layer of clayey sands containing *Didacna pseudocrassa, D. baericrassa, Tschaudia tschaudae*, and *Dreissena tschaudae* [6]. A similar mollusk composition is also characteristic of the shallow-water silty-clayey sediments of the Bulgarian shelf.

In the Black Sea basins, the Chaudian deposits are not subdivided; here, they are represented by uniform clays and silts with sapropel interlayers in the upper part, that contain assemblages of fresh-water and brackish-water cryophilic diatoms [2].

The Bakunian Formation (middle Middle Pleistocene) is represented by the deposits of the final stage of sedimentation in the Chaudian basin. They are reliably registered only on the Taman Peninsula [5, 7], where they are represented by shallow-water deposits of a freshened basin.

The Old Euxinian Formation corresponds to the initial stage of the evolution of a vast and long-existing marine brackish-water basin. Within the Crimean–Taman coast, its deposits are represented by sands and silts of a deepened sublittoral zone with numerous shells of the governing species *Didacna pontocaspia.*

On the Caucasian coast, the Old Euxinian deposits form a thin accumulative cover of one or two marine terraces. The deposits are coarse-grained and sandy–pebbly; they refer to a dynamic near-shore sedimentation environment. On the Bulgarian shelf and in estuarine parts of river valleys, the Old Euxinian deposits are characterized by a fine-grained lithological composition and, in addition to the Black Sea and Caspian brackish-water fauna, contain numerous shells of the fresh-water species *Dreissena polymorpha.*

Up the section, the Old Euxinian deposits are gradually replaced by the sediments of **the Uzunlarian Formation** (end of the Middle Pleistocene). The boundary between these formations is rather conventional; it is traced by the reduction in the number of brackish-water mollusk species and the increase in the abundance of the representatives of euryhaline Mediterranean fauna. In the stratotypical section on the coast of Lake Uzunlarian, this formation is represented by two layers. The lower layer is formed by clayey sands and silts, which, along with brackish-water and fresh-water mollusks and benthic foraminifers, contain numerous shells of euryhaline Mediterranean mollusks (*Cerastoderma glaucum, Abra ovata*, and others). Above, one finds gray-green clays with interlayers of coquina matter mostly formed by marine Mediterranean species.

The two-layered structure of the Uzunlarian deposits is also registered on the Caucasian coast. Here, they form a coarse-grained terrigenous (conglomerates and calcareous sands) accumulative cover of the fifth (55 m high) terrace.

In the estuarine parts of the rivers of the Bulgarian coast and on the adjacent shelf, the Uzunlarian deposits are represented by sands and silts with plant remains; they lie over the Old Euxinian sediments sand contain shells of *Cerastoderma glaucum*, *Monodacna caspia*, and *Dreissena polymorpha.*

In the south of Moldova, the Uzunlarian deposits include the lagoonal sediments of the third terrace of the Prut and Danube rivers [8]. They do not contain Mediterranean mollusk species, while Caspian (*Didacna nalivkini*) and local (*D. poratica* and *D. raricostata*) forms with a large admixture of representatives of Azov (*Hypanis plicatus* and *Monodacna* sp.) and fresh-water (*Dreissena polymorpha, Corbicula fluminalis, Viviparus diluva,* and others) fauna are abundant.

In the deep-water Black Sea basins, the sediments of the Middle Pleistocene are represented by clayey diatomaceous oozes and silts with three diatom assemblages; the upper and lower assemblages are marine and brackish-water, while the middle assemblage contains numerous freshwater cryophilic species [7].

The Karangatian Formation (beginning of the Upper Pleistocene) is the best stratified sequence of the Black Sea Pleistocene deposits; it reflects an important paleogeographic event – the epoch of the existence of a Mediterranean-type basin with a characteristic mollusk assemblage containing mass stenohaline Mediterranean species. The stratotypical sections of the Karangatian Formation are located on the Kerch Peninsula (Cape Karangat, El'tingen, and Lake Chokrak); they are subdivided into a series of layers that represent the sequence of the basin evolution. Most of the scientists [9–11] note the three-layered structure of the Karangatian Formation: the lower (Tobechik) layer is represented by freshened and marine silts and clays with the mollusks *Cerastoderma glaucum, Abra ovata, Mytilaster lineatus, Paphia senescens*, and *Chione gallina*. The middle (El'tingenian or Karangatian) layer is composed of shallow-water sands and pebbles with coquina of rich Mediterranean fauna (*Acantocardia (Cardium) tuberculatum, Paphia senescens*, and others). The upper (Late Karangatian) layer consists of shallow-water sandy deposits with impoverished Karangatian fauna (*Ostrea edulis, Mytilaster lineatus*, and others).

On the Caucasian coast, the Karangatian deposits compose accumulative covers of two terraces. On the Bulgarian coast, one can also trace two Karangatian terraces, whose deposits are characterized by the *Ostrea–Cardium–Mytilus* and *Corbula gibba* faunistic assemblages. On the Bulgarian shelf, the sediments feature fine-grained clayey-silty composition and contain rich Mediterranean fauna.

In the south of Moldova, the Karangatian epoch corresponds [8] to the lagoonal sediments of the second terrace of the Prut and Danube rivers; they contain one or two layers with shells of Caspian didacnas separated by a layer of lacustrine deposits.

To date, many absolute age determinations of the Karangatian deposits with the use of radiocarbon, uranium-ionium, and thermoluminiscent methods have been performed. The uranium-ionium data seem to be the most reliable. They show a rather wide time range of 33–139 thousand years B.P., though most of the datings fall within the intervals 110–80 and 70–60 thousand years B.P. This allows one to suggest that, during the Karangatian transgression, two peaks of sedimentation occurred.

The Karangatian deposits, similar to the underlying Uzunlarian sediments, contain Mediterranean mollusks, though their fauna is richer and includes a greater number of stenohaline species.

With respect to their composition, the deep-water Karangatian deposits are represented by clayey diatomaceous oozes with a rich assemblage of thermophilic marine diatoms dominated by the characteristic tropical species *Thalassiosira oestrupii* [1].

The Tarkhankutian Formation. The Tarkhankutian deposits were separated [5] in the bottom sediments cores from Karkinit Bay; they were regarded as the post-Karangatian marine sediments with *Cerastoderma glaucum, Abra ovata, Dreissena polymorpha,* and rare *Mytilaster lineatus.* They represent the epoch of low sea level standing and were confidently established only in the near-shore region of the present-day Black Sea area at sea depths of 20 m and greater. The most complete sections were described using drilling materials from the Kerch Strait. The sediments are represented by the facies of coastal pebbles and sands, by more fine-grained deposits of the deepened sublittoral zone, and by lagoonal loams and clays. They contain euryhaline Mediterranean (50%) and Caspian (40%) mollusks with rare fresh-water species.

The New Euxinian Formation (the end of Upper Pleistocene), composed by the deposits of a strongly freshened basin, represents a very important though short-term stage of the Pleistocene history of the Black Sea. These sediments were first recognized by N.I. Andrusov [12], while A.D. Arkhangel'skii and N.M. Strakhov [13] suggested that they be referred to as New Euxinian deposits. These sediments are developed on the Black Sea floor at sea depths from 20 m and lower; they are represented by diverse sandy-clayey deposits with the Azov (*Monodacna caspia, Adacna vitrea, Hypanis plicatus,* and others) and fresh-water (*Dreissena polymorpha, Viviparus,* and others) types of fauna. They partly cover the basin of the Sea of Azov and the Kerch Strait, where they feature a fine-grained composition with diverse Azov–Caspian and fresh-water fauna.

The Black Sea Formation corresponds to the sediments of the last-Holocene-transgression of the Black Sea; it is distinguished by the appearance of Mediterranean mollusk species. The deposits feature a universal development in the bottom sediments of the sea area and compose lower terrace levels on its coasts. They are very diverse in the facial respect, being represented by beach sands and pebbles, coquinas of the shallow-water shelf areas, lagoonal oozes, and even deep-water clays contaminated by hydrogen sulfide. A series of layers may be recognized with regard to the changes in the composition of Mediterranean mollusks [9]. A mass dating of the Holocene sediments of the Black Sea showed that the chronological range of the Black Sea Formation falls into the interval from 10 000 years B.P. and before [14–16].

3
Paleogeography of the Azov–Black Sea Basin

In the recent history of the Black Sea, a series of transgressive stages with particular types of hydrological conditions, faunistic assemblages, and coastal settings are recognized, such as the Chaudian, the Old Euxinian, the Uzunlarian, the Karangatian, the Tarkhankutian, the New Euxinian, and the Black Sea stages (see Fig. 2).

The Chaudian transgressive stage is related to the most long-lasting transgression in the Black Sea during its Quaternary history; it began in the Early Pleistocene and terminated at the beginning of the Middle Pleistocene. At that time, the sea represented a vast brackish-water basin of a mesohaline type with long-term sea level oscillations; it was inhabited by a mollusk community dominated by brackish-water species, among which relic Pliocene forms and endemic Black Sea didacnas prevailed.

Judging from the character of the changes in the mollusk fauna, one can distinguish the following substages in the evolution history of the basin: the Gurian, the Lower and Upper Chaudian, and the Bakunian substages.

The Gurian substage refers to the epoch of the Early Pleistocene, when a small brackish-water basin existed; it seems to have been absolutely isolated from adjacent basins and represents a relic lake-sea inhabited by a particular mollusk fauna dominated by Pliocene relics (*Tschaudia tchaudae*) and a number of representatives of typical Pleistocene fauna (*Didacna pseudocrassa*).

On the Caucasian coast, where the Gurian sediments were established, the climate at that time was warm and wet. The mean monthly temperatures in January and June were 5 °C and 25 °C, respectively; beech and coniferous forests were characteristic [17].

The Early Chaudian substage is the epoch in the middle of the Early Pleistocene, when a freshened brackish-water basin existed; it featured an impoverished fauna dominated by endemic brackish-water (*Didacna baeri-crassa*) and fresh-water (*Dreissena polymorha*) species. Judging from the mollusk assemblages, the salinity in this basin never exceeded 8–10‰.

With respect to its area and the sea level height, the Early Chaudian sea did not exceed the present-day Black Sea basin. It was definitely not connected with the Mediterranean Sea and, probably, not with the Caspian Sea; the latter suggestion may be inferred from the absence of Caspian mollusks in the Lower Chaudian deposits. Presumably, the climate of this epoch, which coincided with the first Pleistocene cooling over the Russian Plain, was cold; open steppe landscapes existed in coastal lowlands and cold steppes were developed in the northern regions.

The Late Chaudian substage (the first half of the Middle Pleistocene) refers to the maximum of the Early Pleistocene transgression with the largest territories flooded and the highest sea level standing, which probably reached and even exceeded (by 2–5 m) the present-day position of the Black Sea level. Among the bottom sediments, facies of organogenic sediments of warm shallow-water areas prevail; they are dominated by the relics of the Gurian brackish-water fauna and Black Sea endemic species. The occurrence of Chaudian mollusks on the coasts of the Dardanelles (Galliopoli sections) and the presence of Caspian immigrants (*Didacna rudis* and others) in faunistic assemblages point to an inflow of Black Sea waters to the Mediterranean Sea and of Caspian waters to the Chaudian basin. Judging from the mollusk assemblages, at the beginning of the transgression, the salinity of the Late Chaudian sea was probably either equal to the present-day Caspian Sea salinity (about 13‰) or was somewhat higher than it (about 15‰). At the peak of the transgression, the sea deeply penetrated into river valleys. The coasts were dominated by forest–steppe landscapes with prevalent broad-leaved species.

The Bakunian substage marks the final epoch of the Chaudian transgression; it coincided with the major glaciation of the Russian Plain and the Bakunian transgression of the Caspian Sea, whose waters were delivered to the Black Sea via the Manych depression. Together with the probable decrease in the evaporation caused by the cooling, this stimulated a significant desalination of the residual Chaudian basin (down to 10–13‰), the disappearance of Gurian fauna in it, and the abundance of Caspian and fresh-water fauna. The sea level position seems to have been either close to the present-day one or slightly lower than it.

Thus, the Chaudian transgressive stage represents a durable (about one million years) epoch of the existence of a vast brackish-water basin inhabited by a specific mollusk assemblage dominated by Late Pliocene relics and Black Sea endemic species. This basin passed different stages of evolution with two epochs of a relative desalination (Early Chaudian and Bakunian), when the salinity in the basin was close to that of the present-day Caspian Sea. During the transgression maximum, the salinity of the basin probably corresponded to the present-day Black Sea value, while the sea level was significantly higher. Except for the initial stage, the Chaudian basin was dumping its waters to the Mediterranean Sea and was being filled by the waters of the Bakunian transgression of the Caspian Sea.

The Old Euxinian transgressive stage represents the initial phase of a very long-existing sea basin that appeared at the middle of the Middle Pleistocene (about 0.4 million years). It was separated from the Chaudian basin by a profound regression, during which the sea level fell by 40–60 m and the Chaudian brackish-water fauna was replaced by the assemblages of Old Euxinian mollusks. The level of the Old Euxinian basin was seemingly lower than its present-day position and its area was smaller than the present-day sea area.

Via the Manych depression, the Old Euxinian basin was connected with the Caspian Sea, where the Khazar transgression developed at that time. A mass migration of brackish-water fauna from the Caspian to the Black Sea occurred. At the end of this epoch, the connection with the Mediterranean Sea via the Bosporus Strait was re-established and euryhaline species most resistant to salinity variations (*Cerastoderma glaucum* and others) penetrated to the Black Sea. At that time, the salt content in the seawater increased from 10–13‰ to 15–17‰ and the adjacent regions of the Russian Plain featured a warm moderately humid climate; later, it was followed by a strong cooling (Dnieper glaciation).

The Uzunlarian transgressive stage represents the further evolution of the Old Euxinian basin and its transition from the brackish-water basin of Caspian type to a freshened marine basin with a salinity close to the present-day Black Sea value (about 18‰).

The Uzunlarian transgression, which proceeded in the second half of the Middle Pleistocene (0.25–0.1 million years B.P.), provided a one-way and permanently strengthening connection with the Mediterranean Sea. In terms of its area and sea level height, the basin seem to have been not greater than the present-day Black Sea. The near-mouth parts of rivers were flooded and deep estuaries existed. The climate of the Uzunlarian age was moderately warm with a tendency to increasing dryness and contrasts.

The Karangatian transgressive stage resulted in a basin largest during the Quaternary history of the Black Sea. At the maximum of the transgression, the level of the basin seems to have been located 6–8 m higher than at present.

Great amounts of warm saline Mediterranean waters were supplied to the Black Sea basin and a one-way migration of euryhaline and stenohaline Mediterranean fauna occurred. A long (more than 200 km) bay extended over the Manych depression toward the Caspian Sea; episodically, it transformed into a strait providing a fauna exchange with the Late Khazarian sea. However, no noticeable amounts of Caspian mollusks penetrated into the Karangatian sea. At the beginning of the transgression, impoverished Mediterranean fauna (*Cerastoderma glaucum*, *Abra ovata*, and *Paphia senescens*) appeared in it; at the maximum of the transgression, its species diversity increased (Mediterranean species *Cardium tuberculatum*, *Ensis ensis*, and others). The abundance of Mediterranean fauna in the waters of the Karangatian transgression seems to be related to its relatively high salinity rather than to the high water temperature. This suggestion is supported by the fact that, among the Mediterranean fauna that dwelled in the Karangatian epoch, along with relatively thermophilic (Mediterranean–Lusitanian and Mediterranean–Canarian) species, one also encounters numerous representatives of cryophilic fauna (*Cerastoderma glaucum*, *Ostrea edulis*, *Corbula gibba*, and others). At the maximum of the transgression, the salinity of the open part of the Karangatian sea is estimated at 30‰ [9]. In the near-shore parts of the basin, at the beginning and at the end of the transgression, the salinity was somewhat lower and closer to the present Black Sea salinity.

The Karangatian transgression coincided with the major (Riss–Wurmian) interglacial period on the Russian Plain ($\sim$ 0.12–0.07 million years B.P.). Judging from the uranium-ionium datings, its duration may be estimated at 70–50 thousand years. The climate of this epoch was warm and moderately dry; coastal lowlands were occupied by steppe and forest-steppe landscapes. During the second half of the Karangatian epoch, the climatic conditions became worse and the coastal lagoons transformed into lakes inhabited by fresh-water mollusks (Planorbis, Limnea, and others).

The Tarkhankutian transgressive stage. The Tarkhankutian epoch represents the final stage of the existence of a Mediterranean-type basin in the second half of the Late Pleistocene. Conventionally, this phase may be regarded as the final stage of the Karangatian basin evolution. On the northern shelf of the Black Sea, the submerged accumulative coastal features with radiocarbon ages of 40–25 thousand years B.P. that refer to the Tarkhankutian transgression are located at depth marks of about – 20 to – 30 m. The Tarkhankutian basin was small; it was located inside the outlines of the present-day Black Sea. It was inhabited by impoverished Mediterranean fauna and featured a salinity of 3–5 to 8‰ [9].

One can infer that, during the period from the middle of the Middle Pleistocene to the second half of the Late Pleistocene, the area of the Black Sea was occupied by a brackish-water sea basin inhabited by euryhaline and stenohaline Mediterranean fauna. At the maximum of its development, its salinity was almost twice as great as the salinity of the present-day sea and its level seems to have reached absolute marks of + 6–8 m; sea bays and estuaries deeply penetrated into the land over coastal depressions. With respect to selected characteristics, this basin strongly differed from the preceding (Chaudian) and the subsequent (New Euxinian) basins. First of all, this refers to the large-scale one-way connection with the Mediterranean Sea, which defined the relatively high salinity of the basin and the character of its dwellers.

The long-term evolution history (about 0.5 million years) of this basin, which is commonly referred to as the Pantikapean basin [7], consists of a series of stages that reflect the sequence of its evolution from the brackish-water basin of a Caspian type to a freshened marine basin (Black Sea type), normal marine basin (Mediterranean type), and, finally, to a strongly freshened basin (Azov type).

The New Euxinian stage. The fall of the level of the Pantikapean basin that started in the middle of the Late Pleistocene finally resulted in a deep regression with level drops to marks of – 80, – 100 m, and lower [15], in a break of its connection with the Sea of Marmara, and in its transformation into a fresh-water basin with a volume of about 500 000 km^3 and an area of approximately 320 000 km^2 [16]. It was inhabited by fresh-water mollusks and featured a salinity that never exceeded 1.5–3.0‰. At those times, the Black Sea shelf, the Kerch Strait, and the Sea of Azov were drained. The floor of the Sea of Azov represented a flat lowland crossed by the channel of the pra-Don River. Its mouth was located 50 km south of the Kerch Strait, while the mouths of the Danube and Dnieper rivers were located 200 km away from their present-day position. In the valleys of the Caucasian coast, river mouths were deepened and coarse-grained alluvial facies were formed.

About 15 000 years B.P., the waters of the Khvalynian transgression of the Caspian Sea overcame the Manych threshold (with an absolute mark of about 45 m) and started to enter the Black Sea basin. At the beginning of the transgression, the Kerch Strait and the Sea of Azov were dried; here, drilling revealed [10, 12, 18] a system of major channels, via which the Caspian waters were dumped into the Black Sea.

The New Euxinian transgression developed in a stadial mode. During the period from 15.0 to 12 500 years B.P. (Enikal stage), the level in the basin rose by 20 m, later on it underwent a short-term fall down to a mark of – 45 m and then (11 000 years B.P.) sharply rose up to its maximum at about – 20 m (the New Euxinian stage proper). Judging from the mollusk composition, it was a strongly freshened brackish-water basin with no Caspian didacnas. At the maximum of the New Euxinian transgression, its waters flooded the floor of the Kerch Strait and a part of the basin of the Sea of Azov. The floor of the Sea of Azov was never fully filled with transgressive waters. Starting from 13 000 years B.P., the dumping of the Black Sea waters to the Sea of Marmara via the Bosporus was steadily resumed; at this time, the Bosporus threshold, which at present lies at marks of approximately – 40 m, was flooded. The water discharge via the strait at the maximum of the New Euxinian transgression is estimated at 60 km^3/year [16].

The end of the Caspian water supply to the Azov–Black Sea basin and the termination of the New Euxinian transgression occurred after 11 000 years B.P., when the level of the regressing Khvalynian sea fell below the position of the Manych threshold. During the Holocene, there was no connection between the Caspian Sea and the Azov–Black Sea basin and the basins have been evolving independently.

The New Euxinian epoch corresponded to the cold climate of the postglacial age; landscapes close to those of cold steppes were developed over the drained areas of the Sea of Azov shelf and on the low Black Sea coasts. At the end of this epoch, an aridization of the climate occurred and semidesert landscapes became widespread.

The Black Sea (Flandrian) transgressive stage. The Holocene transgression of the Black Sea represents the terminal stage of its Quaternary history and its transformation into a modern freshened marine basin inhabited by euryhaline Mediterranean fauna; its salinity is about 18‰ in the open part of the sea, 7–12‰ in semi-enclosed bays and lagoons, and up to 20–22‰ in abyssal layers.

The transgression was caused by the breach of Mediterranean waters, which filled the basin, via the Bosporus. According to some estimates, it occurred 7–8 thousand years B.P. [10, 16], while other estimates yield dates of 12–15 thousand years B.P. [2, 19]. The composition of the Mediterranean fauna that penetrated into the basin and the sequence of its appearance show that, during the Holocene transgression, the salinity in the sea has been gradually and continuously increasing. The transgression proceeded at different rates; it was most intensive at the initial stage, when the level of the Black Sea rose by 20 m over 3500 years; this growth was especially accelerated in the period 5–3.5 thousand years B.P. [7]. The supply of large masses of heavy saline sulfate-rich Mediterranean waters resulted in the suppression of the vertical water exchange in the Black Sea basin and in the formation of a vast layer of hydrogen sulfide contamination. It is assumed [19] that the origin of the present-day hydrogen sulfide contamination of the Black Sea abyssal waters is related to the activity of halophilic anoxic bacteria, mostly those of the *Desulfovibrio* genus. The reduction of the sulfate contained in seawater by these bacteria leads to the formation of reduced forms of sulfur.

The Holocene transgression coincides with the postglacial epoch of the climate improvement. At its beginning, the low northern coasts of the Black Sea were dominated by semidesert and steppe landscapes; later on, the role of forests increased [20].

In the Middle Holocene, steppe landscapes also dominated the coasts of the Black Sea and the Sea of Azov, though with a significant proportion of mixed forests: there, formation of black soils began.

In the Late Holocene, the arid and continental properties of the climate enhanced; steppe landscapes dominated again, broad-leaved species disappeared from the forests of river valleys, and xerophytic frutescent plants spread widely [20].

On the Caucasian coast, intensive recent tectonic movements went on. The deformation of the Holocene terrace in the Sochi region over the past 10 000 years is estimated at 10–13 m at a rising rate of up to 1.3 mm/year [21]. The Holocene stage of the Azov–Black Sea evolution history, despite its short duration, represents an important epoch of formation of the main features of the natural environment of these basins.

4 Conclusions

During the Pleistocene history of the Black Sea, a series of basins may be traced that are noticeably different in their hydrological and faunistic characteristics. The formation and evolution of these basins was defined by many factors, the principal factors being the availability of the connection with the Caspian and the Mediterranean seas, the character of the water exchange with them (Fig. 3), and the proportions between the principal components of the water balance (freshwater supply and evaporation).

A flow-through regime characterized by the water inflow from the Caspian Sea via the Manych depression and water outflow via the Bosporus was typical of the Late Chaudian epoch (beginning of the Middle Pleistocene) and the Karangatian stage (beginning of the Late Pleistocene). Water inflow to the Black Sea via the Bosporus proceeded at the end of the Chaudian stage, at the end of the Middle Pleistocene (the Uzunlarian stage), and in the Holocene (the Black Sea epoch). Drainless basins existed in the New Euxinian epoch and during the maximums of regressions.

The transgressive basins almost did not differ in their area and sea level height (Fig. 4); on the whole, the latter was close to its present-day pos-

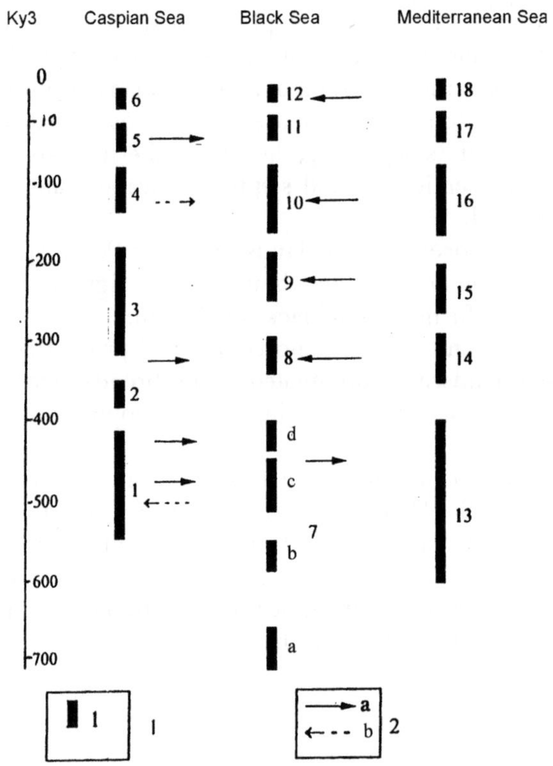

Fig. 3 Hydrological connections between the Black Sea and adjacent basins in the Pleistocene. *Inset 1* transgressive basins: *1* Bakunian; *2* Urundzhik; *3* Early Khazarian; *4* Late Khazarian; *5* Early Khvalynian; *6* New Caspian; *7* Chaudian and its stages *a* Gurian, *b* Early Chaudian, *c* Late Chaudian, *d* Bakunian; *8* Old Euxinian; *9* Uzunlarian; *10* Karangatian; *11* New Euxinian; *12* Black Sea; *13* Sicilian; *14* Milacian; *15* Paleotyrrhenian; *16* Eutyrrhenian; *17* Neotyrrhenian; *18* Verzilian. *Inset 2* direction of the water outflow and fauna migration: *a* established, *b* inferred

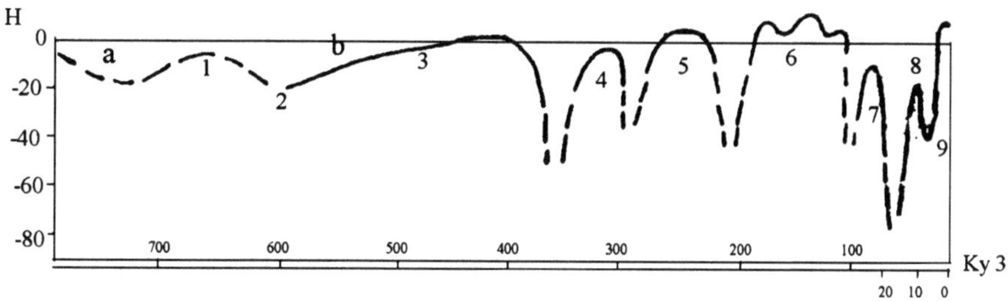

Fig. 4 Oscillations of the Black Sea level during the Pleistocene. Basins: *1* Gurian, *2* Early Chaudian, *3* Late Chaudian, *4* Old Euxinian, *5* Uzunlarian, *6* Karangatian, *7* Tarkhankutian, *8* New Euxinian, *9* Black Sea (Flandrian)

ition. The maximum of the sea level rise was determined by the depth of the Bosporus threshold and the level of the Mediterranean Sea; it is most probable that this rise never exceeded a value of + 6–8 m. The main differences between the transgressive basins were related to the salinity changes, which defined the character and composition of the fauna dwelling in them.

The changes in water salinity in the Pleistocene Azov–Black Sea basins were rather significant and have been well established from the mollusk assemblages. Salinity ranged from 2–4‰ in the New Euxinian time to 30‰ in the Karangatian epoch. The sea passed through the stages of a normal marine basin (Karangatian), freshened marine basin (Uzunlarian, Tarkhankutian, Flandrian), brackish-water Caspian-type basin (Chaudian, Old Euxinian), and fresh-water basin (New Euxinian); inhabited by stenohaline and euryhaline fauna, by brackish-water (Caspian and Pontian), Azov, and fresh-water species, respectively.

The transgressive basins were separated by regressive epochs. These were short-term but characterized by significant sea-level oscillations – from a few tens to a few hundreds of meters. The regressions proper seem to have mainly been related to the breaks in the connections with adjacent basins and to the transformation of the regressing basins into drainless fresh-water basins or strongly freshened lakes – lagoons.

References

1. Neprochnov YP, Trimonis ES, Shimkus KM et al. (1980) Geological history of the Black Sea from deep-sea drilling data. Nauka, Moscow (in Russian)
2. Dagens ET, Ross DA (1972) Chem Geol 1:1
3. Ross DA (1974) In: Black Sea, geology, chemistry and biology. Tulsa, Oklahoma
4. Cowie TW, Bassett MG (1989) Episodes 12:2
5. Nevesskaya LA (1963) Identifier for bivalve mollusks of the Quaternary deposits of the Black Sea. AN SSSR, Moscow (in Russian)
6. Tsereteli DV (1966) Pleistocene deposits of Georgia. Metsniereba, Tbilisi (in Russian)

7. Svitoch AA, Selivanov AO, Yanina TA (1998) Pleistocene Paleogeographic events in the Ponto-Caspian and Mediterranean. Nauka, Moscow (in Russian)
8. Mikhailesku KD, Markova AK (1992) Paleogeographic stages of the fauna evolution in the south of Moldova during the Anthropogene. Stinitsa, Kisinau (in Russian)
9. Nevesskaya LA (1965) Tr PIN AN SSSR 105:387 (in Russian)
10. Fedorov PV (1978) The Pleistocene of the Ponto-Caspian. Nauka, Moscow (in Russian)
11. Chepalyga AL, Mikhailesku KD, Izmailov YA et al.(1989) Problems of stratigraphy and paleogeography of the Pleistocene Black Sea. In: Chetv Per Stratigr. Nauka, Moscow, p 113 (in Russian)
12. Andrusov NI (1948) Izv AN SSR 12:23 (in Russian)
13. Arkhangel'skii AD, Strakhov NM (1938) Geological structure and evolution history of the Black Sea. AN SSSR, Moscow (in Russian)
14. Arslanov KhA, Gerasimova SA, Izmailov YA (1975) Bull Kom Izuch Chetv Per 44:107 (in Russian)
15. Balabanov IP, Izmailov YA (1989) In: Alekseev MN (ed) Quaternary geochronology. Nauka, Moscow, p 42 (in Russian)
16. Kvasov DD (1975) Late Quaternary history of major lakes and inland seas of Eastern Europe. Nauka, Leningrad (in Russian)
17. Dolukhanov PM (1988) History of Mediterranean seas. Nauka, Moscow (in Russian)
18. Popov GI (1983) The Pleistocene of the Black Sea–Caspian straits. Nauka, Moscow (in Russian)
19. Nikolayev SD (1995) Isotope paleogeography of inland seas. VNIRO, Moscow (in Russian)
20. Vronskii VA (1988) Holocene Paleoclimates of Europe and the territory of the USSR. Nauka, Moscow, p 150 (in Russian)
21. Nesmeyanov SA, Izmailov YA (1995) Tectonic deformations of the Black Sea terraces of the Caucasian coast of Russia. Nauka, Moscow (in Russian)
22. Rogel F, Atreininger FF, Müller C (1978) Init Rep DSDP 42:985

Hdb Env Chem Vol. 5, Part Q (2008): 47–62
DOI 10.1007/698_5_078

Published online: 12 May 2007

Coastal and Bottom Topography

Evgenii I. Ignatov

Department of Geography, Lomonosov Moscow State University, Vorobievy Gory, 119992 Moscow, Russia
ign38@mail.ru

Abstract This chapter is devoted to a description of the present-day bottom topography and types of coasts of the Black Sea, as well as to the general character of the bottom sediments. Two maps of topography and sediments illustrate the morphology of the Black Sea basin and the particular features of the evolution of its coasts.

Keywords Coasts · Shelf · Continental slope · Continental footstep · Underwater canyons · Deep-sea floor · Bottom sediments

1 Introduction

The Black Sea represents an inland basin. Its coasts, located within different continents, were the birthplace of ancient civilizations. At present, they are developed industrially and socioeconomically. They host promising fields of oil and gas and other mineral resources. Here, the problems of distribution of the population, development of transport communications, and geopolitical relations are urgent and require quick solution.

The Black Sea basin, which now is connected with the Sea of Azov and the Mediterranean Sea and was in the geological past also connected to the Caspian Sea, underwent a complicated history of geological evolution. The latest stage of the post-glacial epoch was especially important for the processes of formation of the topography and of the sedimentation. During the past 20 000 years, the Black Sea basin has evolved from a desalinated lake with a level located about 100 m below the present-day marks to to a sea basin within its modern boundaries.

In this chapter, we present a description of the present-day topography of the floor and the coasts of the Black Sea and of the distribution of the bottom sediments over its floor surface.

2
Coastal Topography

The coasts of the Black Sea are distinguished by their favorable natural conditions and landscape diversity. Their geographical position and warm climate favored their active mastering and population from antique times up to present. One can still find traces of the penetration of ancient civilizations into the coastal areas in the form of ruins of old Greek temples and Genoese fortresses. By the third millennium A.D., the Black Sea coasts have suffered a strong anthropogenic stress. This refers to the near-shore areas off the most densely populated regions near the major ports of Burgas, Varna, Constanta, Odessa, Nikolaev, Sevastopol, Novorossiisk, Sochi, Sukhumi, Batumi, Samsun, Sinop, Trabzon, and the navigable Bosporus and Kerch straits.

On the coasts of the Black Sea there are located large international recreation zones such as Albena, Slnchev Bryag, Zlatny Pyastsy, Magnalia, Costinesti, Odessa, Yalta, Gelendzhik, Anapa, Sochi, Gagra, Pitsunda, and others.

The Black Sea washes the coasts of Europe and Asia Minor. The length of the coastline reaches 4125 km; of these, the lengths of the Bulgarian, Rumanian, Ukrainian, Russian, Georgian, and Turkish coasts equal 380, 240, 1330, 410, 315, and 1450 km, respectively [1].

With respect to the features of the topography, the northern coast of the sea significantly differs from the eastern and southern coasts. Its particular features include the generally plain character (except for the Crimean Mountains) and the presence of estuaries of the major rivers of the Black Sea basin.

The eastern and southern coasts are mountainous. The coastal ridges of the folded structures of the Caucasus and Anatolia extend parallel to the coastline, forming a longitudinal type of mountainous topography. At places, the slopes of the mountains descend directly into the sea in the form of steep escarpments. In the southeastern part of the Black Sea, within the Kolkhida lowland, mostly accumulative coasts are developed.

The general outline of the coast is slightly irregular (Fig. 1). The largest bays are located in the western part of the sea: off the coasts of Bulgaria (Burgas and Varna bays), in the northwest (Odessa and Karkinit bays), and in the Crimea (Kalamit and Feodosia bays). In the eastern part of the sea, one finds Novorossiisk and Gelendzhik bays; in the south, Sinop Bay and Samsun Bight should be noted. The Crimean Peninsula is the largest one; others are the Taman', Indzheburun, and Yasun peninsulas.

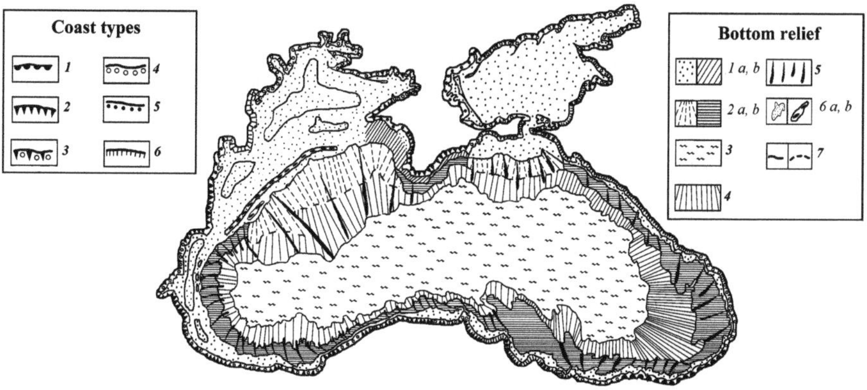

Fig. 1 Topography of the coasts and floor of the Black Sea. *Bottom relief*: *1* shelf *a* accumulative, *b* abrasive; *2* continental slope *a* accumulative, *b* stepwise; *3* floor of the basin; *4* continental footstep; *5* underwater canyons; *6* bars *a* sandy, *b* marginal; *7* morphological boundaries *a* distinct, *b* fuzzy. *Coast types*: *1* landslide; *2* abrasive; *3* abrasive-accumulative; *4* accumulative; *5* lagoonal; *6* deltaic

The character of the coastal zone defines the morphology and type of the coasts. In mountainous areas, abrasive coasts dominate. In many cases, they are complicated owing to the development of intensive landslide and caving processes and thus may be referred to as the abrasive–denudational type. In plain and low areas, the coasts are mostly accumulative. Lagoonal and deltaic coasts are confined to the areas near river mouths.

The northwestern part of the Black Sea is rimmed by the low plains of the steppe zones of Bulgaria, Rumania, and the Ukraine. Here, the major rivers of the regions fall into the Black Sea such as the Danube, Dnieper, Dniester, and Yuzhnyi Bug rivers. Their estuaries and lagoons are superimposed upon the coastline. One can encounter lagoons separated from the sea by sandy spits and lagoons that deeply penetrate into land, such as the Dniester and Dnieper–Bug lagoons [2].

The coasts of Odessa Bay are presently eroded and their condition is defined by landslide and caving processes widely developed on the steep coastal slopes. Here, the heights of the cliffs sometimes exceed 30 m. Intensive engineering coastal protection is performed.

The marginal zone of the delta of the Danube River is rimmed by lagoonal sandy bars and low marine terraces (about 2 m high). This coastal region is of great economic significance. One of the projects is related to the construction of a Danube–Dnieper irrigation canal and operation of marine navigation canals.

Coastal bars (sandy and sandy–coquina) are low and feature widths of 40–100 m. Numerous traces of minor erosional channels suggest that stormy waves override the bars and seawater penetrates into lagoons. At present, accumulative coasts are gradually receding.

Abrasive coasts are mostly composed of clayey deposits, poorly-cemented rocks, or limestones. The cliff heights range from 15 to 35 m. The abrasion rate is controlled by the composition of the coastal rocks and the cliff heights. The recession of the cliffs is also caused by the development of landslide and caving processes, especially under the conditions of strong storms. The length of individual sliding blocks reaches 500 m at a width of up to 15 m. Deep surf niches are formed in the lower parts of the cliffs.

The coasts of Odessa Bay represent an abrasive–accumulative arc with beaches in its top part and abrasive areas of capes North Odessa and Langeron at its edges.

Of special interest are the elongated accumulative features of the bar of Tendrovskaya Spit (65 km) and Dzharylgach Island (40 km), which are extended over a single line.

The picturesque coasts of the Crimea are among the most beautiful areas of the Black Sea. Along the southern coast of the Crimea, spurs of the Crimean Mountains extend crowned by Roman Kosh (1545 m), Chatyr Dag (1527 m), and Ai Petri (1234 m) mountains. Coastal slopes are complicated by ancient landslide formations terminating in bights and beautiful wide bays such as Laspi, Yalta, Feodosiya, and other bays. They are separated by capes composed of crystalline rocks (Ayu Dag, Kara Dag, and others).

Cape Kara Dag is especially picturesque when seen from the sea. This volcanic remnant creates exclusive shore topography and a particular coastal landscape. Between capes Kara Dag and Kiik Atlama is located one of the most beautiful bights of the Crimea, Koktebel' Bight.

The steep slopes of the Southern Crimea advance into the sea in the form of capes, between which open bays with sandy and pebbly beaches are located. In the amphitheaters formed by the coastal slopes of the bays, one finds resort towns such as Yalta, Gurzuf, Alupka, Alushta, Simeiz, Sudak, and Feodosiya. Here, parks, recreation complexes, and palaces remarkable in their beauty and landscape features are situated. Cypresses, palms, magnolias, and other subtropical plants decorate the Nikita Botanical Gardens, the Alupka (Vorontsov), Livadia, Miskhor, Gurzuf, and other parks [3].

The coastal topography is controlled by the geological structure of the Crimea. Capricious features such as abrasive remnants and cliffs are formed depending on the rock strength.

The high coastal escarpments composed of Jurassic limestones recover complicated geological folds broken by tectonic faults and overthrusts. At the sites where intrusive rocks are developed, remarkable capes are formed. The largest of them are capes Ayu Dag, Medved' Gora, Ai Todor, Nikita, Foros, Laspi, Sarych, and Aia.

The western coast of the Crimea from Cape Aia to Tarkhankut Peninsula is characterized by a great variety of coast types, from vertical escarpments to the low lagoonal coasts of Kalamit Bay. The coasts of the Geraklei Peninsula with Cape Khersones are dissected by narrow deeply penetrating Riassic bights such as Balaklava, Sevastopol, Karantin, and others. Wide sandy bars separate Saka, Sasyk, and Donuzlav lagoons from the sea. Here, sea ports and resorts such as Sevastopol and Evpatoriya are located.

The coasts of Saka Lagoon host institutions for mud-cures. Beaches and bars are subjected to erosion because of the non-regulated economic activity on the coasts and related deficiency of the alongshore sediment transport. At places, the rate of the coastline recession reaches 5 m/year. The coastal erosion may result in a complete destruction of the Evpatoriyan beaches.

The Tarkhankut Peninsula features abrasive coasts composed of coquina-limestone rocks with high (20–40 m) cliffs, surf niches, and deepened bench. Numerous grottos, abrasive niches, and underwater rocks are attractive for submarine excursions and diving [2].

In the east of the Crimea, the Kerch Strait separates it from the low Taman' Peninsula. In the north, on the Sea of Azov side, the width of the strait between Capes Khroni and Akhilleon reaches 15 km at maximum depths of 9.5–10 m. In the south, on the Black Sea side, between Capes Takyl and Panagiya, the width of the strait is maximum, reaching 21.8 km at a depth of 19 m. The narrowest place of the Kerch Strait is located off the cape on the northern termination of the Tuzla spit, where the strait is only 3.5 km wide. The length of the ferry line between the ports of the Crimea and Caucasus is 4.6 km.

Surges represent a serious hazard for the low coasts of the Kerch Strait. For example, during the hurricane of 28–29 October 1969, a surge wave fully flooded a coastal band a few kilometers wide. The surge height exceeded 3 m. The flood was accompanied by the destruction of near-shore constructions and communications and human losses.

The Tuzla Spit is an accumulative coastal topographic feature located on the eastern side of the Kerch Strait. Before 1925, it represented a uniform body attached to the Taman' Peninsula and dammed the strait on the Black Sea side. At present, it consists of two parts – the island portion and that strengthened by an artificial dam. The island part of the Tuzla Spit is a narrow accumulative feature up to 500 m wide formed by two series of coastal bars.

The Caucasian coasts between Anapa and Gagra are mountainous. The Greater Caucasian Ridge extends subparallel to the coastline. In the nearshore area, its spurs feature terraces and, at places, steeply descend into the sea in the form of escarpments. Southward of Gagra, the Kolkhida lowland separates the Caucasian Ridge from the sea.

Rivers form deltas representing large accumulative salients in the regions of Pitsunda, Adler, and Sukhumi. The Caucasian coasts are of great importance for their economic and recreation use. With respect to their beauty and diversity, they are equal to the Crimean coasts. Here, woody terraced slopes with flourishing subtropical plants are developed along with steep rocky escarpments. The warm sea attracts numerous tourists and vacationists. Along the entire Caucasian coast, recreation institutions, hotels, and summer camps are located. The major towns and ports of this portion of the coast are Anapa, Novorossiisk, Gelendzhik, Tuapse, Sochi, Sukhumi, and Batumi. At present, the coasts of the Caucasus are subjected to erosion and this fact requires urgent operations for their protection and preservation. Many parts of the coastal zone are strengthened with various hydraulic constructions, wave breakers, and artificial beaches.

The Turkish coasts of the Black Sea are mountainous and steep; over almost their entire extension, they are referred to as the abrasive type. The East and West Pontian mountains, which approach the Black Sea from the south, feature maximum marks of 3937 m (Kachkar Dag), 3439 m (Karchkhal Dag), and 3711 m (Verchennik Tepe). On relatively smooth rectilinear coasts, one can observe an alternation of rocky capes and wide bays that do not penetrate deeply into land.

Accumulative coasts are confined to the near-mouth river areas. They mostly represent narrow local beaches. Only in the middle part of the Turkish coast do the major rivers Eshil Irmak and Kyzyl Irmak form accumulative deltaic plains. The erection of ports and protecting hydraulic constructions prevents the accumulative features from erosion.

In the west, the coasts of Bulgaria are mountainous in their southern part, while in the north, closer to the Rumanian boundary, they give place to the lowlands of the Danube Plain. Nevertheless, the abrasive type dominates over the coasts of this region. The height of the cliff increases up to 60 m near Cape Kaliakra and to 220 m north of the Batov River. The abrasive coastal slopes feature numerous landslides; the abrasion rate of these coasts is up to 0.5 m/year [4].

Summarizing the review of the Black Sea coasts, we should note that, during the recent decades, the process of their erosion has been significantly intensified. This was strongly favored by the anthropogenic impact on the coastal zone and, probably, by the global sea level rise against the background of the global climate warming. This means that it is necessary to significantly change the strategy of human behavior in the coastal zone and to modify the operation of natural protection in order to provide sustainable development of the environment.

3
Bottom Topography

The studies of the topography and geological structure of the floor of the Black Sea are assessed in publications of many prominent scientists such as N.I. Andrusov, A.D. Arkhangel'skii, N.M. Strakhov, V.P. Goncharov, Yu.P. Neprochnov, A.F. Neprochnova, V.P. Zenkovich, N.A. Aibulatov, P.N. Kuprin, E.F. Shnyukov, G.A. Saf'yanov, and others.

Their concepts and the results of the author's observations allow one to outline the general pattern of the structure of the coastal topography and the relief of the floor of the Black Sea basin.

The Black Sea depression represents a remnant of the ancient Tethys Ocean. Its topography has undergone a complicated evolution history, which may be subdivided into two principal stages. The earlier stage (from the Mesocenozoic to Quaternary, 160–200 million years B.P.) was when the principal geological structures of the Black Sea depression and its mountain rimming were formed. The later stage (the Pleistocene–Holocene period, of about one million years), covered the formation of the Black Sea level, accompanied by its oscillations from deep regressive falls and formation of a lacustrine basin in the glacial periods to sharp transgressive rises in the post-glacial times with reconnection to the Mediterranean Sea via the Bosporus Strait [5, 6]. At this stage, the pattern of the coastal and bottom topography was formed, which was later inherited by the present-day features.

The level of the Black Sea seems to be still confined to its optimal marks characteristic of the existing climatic conditions in its watershed. It is not the highest level position over the past 18 000 years. In the Holocene (6000 years B.P.), it was 3–4 m higher than at present. Now, the sea level position is also changing and responds not only to the seasonal variations in the water budget (approximately 15–30 cm) but also to the global warming. According to different estimates, this may lead to a significant sea level rise as early as the current century.

This scenario should probably be taken into account when planning economic activity on the mastering of the coastal zone in order to provide environmental security and sustainable development of the cis-Black Sea region.

The submarine topography of the Black Sea can be naturally subdivided into the zones of the shelf, continental slope, continental footstep, and the floor of the deep-sea depression (Fig. 1).

The Black Sea shelf is an inclined abrasive–accumulative surface formed owing to the large-scale sea level oscillations in the Pleistocene–Holocene. At selected places, this surface is complicated by relics of ancient coastlines and buried fragments of river valleys recognized at depths down to 40–60 m. On the shelf edge, one may encounter relics of ancient accumulative bars with a relative height up to 4 m.

The outer shelf edge is limited by the border of an escarpment; it is located at depths of 100–160 m. In selected shelf areas, there is an older terrace step that reaches depths of 200 m.

With respect to the particular features of the topography, the Black Sea shelf may be subdivided into the following regions: Bulgarian–near-Bosporus, Northwestern, Crimean, Kerch–Taman', Caucasian, East Anatolian, and West Anatolian.

It should be especially noted that the division of the Anatolian (Turkish) shelf into two parts – eastern and western – is caused, mainly, by the fact that the parts of the shelf adjacent to the Pontian Mountains differ from one another in morphological features. The boundary between these parts is confined to the Sinop salient.

The Bulgarian–Near-Bosporus shelf. Its inner zone is identified as an underwater coastal slope descending to the lower boundary of the impact zone of the waves with a 1% recurrence at depths of 25–30 m. Off capes, the slopes are steep, while in bays they are gentle. The central zone is occupied by a slightly dissected hilly accumulative plain; from the inner zone it is separated by underwater depressions well manifested at depths from 15 to 40–70 m.

The outer zone of the shelf is occupied by an accumulative plain with relic features of the coastal topography such as bars. These coastal bars were probably formed at regressive stages, when the Black Sea level was located at marks of 80–100 m.

The Northwestern shelf occupies the area limited by the coastline in the north and by the shelf edge in the south; the latter is located at sea depths of 130–200 m and runs from Cape Kaliakra in the west to Cape Khersones in the east. The maximum width of the shelf is 220 km. The present-day morphology of the shelf is defined by the major rivers that fall into the northwestern part of the sea. Their near-mouth features were formed in the Quaternary over the paleorelief flooded during regressive stages of the sea level history. The present-day surface of the shelf retains fragments of relic topographic features in the form of paleodetltas and ancient coastlines. On the whole, the shelf represents a stepwise alluvial–marine plain with superimposed underwater bars, relics of river valleys, and abrasive escarpments. In the south, it is rimmed by marginal bars extended along the shelf edge at depths of 100–130 m. The inner zone of the northwestern shelf consists of the underwater coastal slope with signs of intensive wave action extending down to depths of 30–40 m. The width of this zone sharply changes from 10 km in the west to 110 km at the meridian of Berezanskii Lagoon. Here, the manifestation of the sediment-forming activity of the rivers is especially strong.

The central zone of the shelf is dominated by accumulation processes, which results in smoothing and burying of relic topographic features. This zone is 35–90 km wide and occupies areas with sea depths from 40 to 60 m.

The outer shelf zone is located at depths greater than 60 m; its marginal part is characterized by steeper slopes than those in the inner and central zones. The greatest depths here reach 60–100 m; its smallest width (down to 10 km) is observed in the east, while in the west, off the Danube River mouth, it reaches 60 km. Here, the shelf edge is confined to 130–150 m depth contours.

The portion of the shelf between Cape Tarkhankut and Cape Khersones represents a slightly inclined abrasive–accumulative plain with traces of relic coastal topographic features down to sea depths of 40–50 m.

The Crimean shelf extends from Cape Khersones in the west to Cape Meganom in the east. It is widest off Cape Sarych (35–40 km) and narrowest off Cape Ayu Dag (5 km) [1, 2]. This region is subjected to intensive wave action because it is exposed to all the southerly winds. The boundary of the underwater coastal slope is located at depths of 30–40 m. The near-shore zone is the area of alongshore sediment transport and smoothing of the bottom topography. Underwater and dried abrasive remnants are common; the largest of them are confined to the capes composed of strong volcanic rocks [7, 8].

The central and outer zones of the shelf are mostly represented by inclined accumulative plains in bay areas (for example, off Yalta) and by abrasive surfaces off capes Sarych, Meganom, and others. At selected places of the shelf, one can recognize manifestations of tectonic activity in the form of faults and block displacements.

The Kerch–Taman' shelf is located south of the Kerch Strait from Cape Meganom in the Crimea to Cape Utrish on the Caucasian coast. South of the Tuzla Spit of the Taman' Peninsula, the shelf width up to sea depths of 200 m exceeds 60 km. The edge of the shelf is distinctly manifested in the western and eastern parts of the region at depths of 90–130 m; on the seaward side, it is limited by abrasive escarpments with heights from 4–5 to 10 m. The shelf features a stepwise surface. Its inner zone down to depths of 30–40 m occupies the wide shallow-water shoal, forming a gently sloping accumulative plain. Its surface is complicated by flooded ancient coastal features, numerous underwater ridges, reefs, and hollows. Below, down to depths of 80 m, is located the central shelf zone, where processes of non-wave-action accumulation mostly proceed. The outer zone is defined by ancient abrasive escarpments and underwater accumulative bars. In the region of the Kerch Strait, the shelf edge forms an arc-shaped line convex toward the sea. Here, no distinct marginal topographic features are observed and the transition to the continental slope occurs gradually owing to the increase in the floor sloping angles. One may suggest that, with respect to the general topographic pattern, this area represents an accumulative fan of the paleodeltas of the Don and Kuban' rivers.

The Caucasian shelf extends southward along the coastal ridges of the Greater Caucasus from Cape Utrish to Cape Pitsunda and farther to the south up to Batumi. Its width does not exceed 10–12 km (off Sochi and Dzhugba) and reaches a value of 20–30 km between Cape Pitsunda and the Gumista River and in the area between Capes Iskuriya and Anakliya. These two wide bumps off Gudauta and Ochamchira are formed by the paleodeltas of the Bzyb' and Kodori rivers. The narrowest part of the shelf is observed off Leselidze, where steep escarpments of the underwater slope start directly from the coastline, and off Sukhumi, where it is not greater than 3–4 km. Nevertheless, within the Caucasian shelf, one can also distinguish the inner and outer zones.

The inner zone represents an abrasive–accumulative coastal slope down to depths of 25–30 m, with a characteristic ridge–stepwise topography and marks of active wave impact. Farther, a slightly inclined surface descends to depths of 60–70 m, below which the outer shelf zone begins. Its edge is complicated by abrasive–landslide escarpments with numerous cuts related to the top parts of underwater canyons. The depth of the outer edge of the shelf never exceeds 110–130 m.

The East Anatolian (Pontian) shelf extends in the form of a narrow band in the southeast of the Black Sea along the East Pontian mountains from Batumi to Sinop. Usually, its width does not exceed 3–4 km and increases up to 12 km off Cape Yason and to 16–20 km off Cape Shatly; in Sinop Bay, it reaches even 25–30 km. Its outer edge lays at sea depths of 110–130 m. The margin of the shelf is dissected by the tops of submarine canyons that capture alongshore sediment fluxes. The narrow portions of the shelf are characterized by steeply inclined abrasive surfaces; slightly inclined accumulative plains with underwater ridges and hollows are spread in wide arc-shaped bays between capes.

The West Anatolian shelf, from Sinop to the Bosporus Strait, features a maximum width of 25–30 km. The narrowest shelf is observed off Zonguldak and Karasu, where it is less than 3–4 km wide. The edge of the shelf is confined to depths of 100–110 m. The shelf is mostly represented by a stepwise abrasive or abrasive–accumulative plain with superimposed ridges, escarpments, and hollows.

The continental slope of the Black Sea basin is located below the outer edge of the shelf. It has a complicated heterogeneous structure caused by the particular features of the tectonics of the adjacent plains and mountain ridges – the Crimean, Caucasian, Stara Planina, and West and East Pontian ridges. The depth of the edge of the continental slope ranges from 100 to 200 m. Its lower boundary is marked by a topographic bend at sea depths of 1100–1500 m.

In tectonically active areas of the continental slope, structural topographic features dominate and, in the transverse profiles of the slope, relatively gentle (1–3°) accumulative surfaces are sharply replaced by steep almost vertical (10–30°) escarpments, often featuring a stepwise profile and cut by systems of faults. Over the steep slopes, landslide processes develop [9].

For example, these kinds of processes actively proceed on the Caucasian continental slope off Dzhugba and Arkhipo-Osipovka. Here, the underwater relief is characterized by extreme complicacy and irregularity. Landslide formations are encountered at depths of about 850 m at a distance of 6 km from the coast. The thickness of the sliding units is 20–25 m at a length of 350–400 m. The landslides descend to depths of 1200–1500 m at the foot of the continental slope.

The steepest and narrowest portions of the continental slope are confined to the Crimean coast, the Adler segment of the Caucasian coast, and to the regions off Trabzon and Zonguldak of the Anatolian coast. These parts of the continental slope are dissected by series of underwater canyons.

On the southern coast of the Black Sea, off capes Bafra and Yason, the continental slope is complicated by a series of rises. They extend over distances of 40–80 km almost parallel to the coastline; the largest of them is the Arkhangel'skii Rise.

Among the types of the continental slope of the Black Sea, with respect to their morphology and origin, one can clearly distinguish the northwestern and the northeastern areas adjacent to the near-mouth regions of the Danube, Dniester, and Dnieper rivers and to the Kerch Strait. In these areas, the continental slope was formed owing to the growth of paleodeltas of the rivers cited in the west and those of the Kuban' and Don rivers in the east. Therefore, here, the continental slope is significantly advanced into the sea by 70–90 km in the form of a gentle accumulative plain and is coupled with the continental footstep.

Virtually over its entire extension, the continental slope of the Black Sea is dissected by numerous faults and underwater canyons. These canyons, confined to tectonic dislocations (fracture zones or grabens), are later transformed by turbidity flows, which use them as channels for the transport of mineral particulate matter from the near-shore zone to the foot of the continental slope. At the places of discharge of turbidity flows, alluvial fans are formed, which may be cut by runoff channels [10].

The tops of underwater canyons are confined to river mouths and form a complicated branched pattern of tributaries, which cross the shelf edge and join the main channel of the canyon within the continental slope.

The largest of the canyons known in the Black Sea is the Danube Canyon; it has a length of 220 km and eight tributaries up to 58 km long. The greatest number of tributaries (69) is distinguished in the underwater canyon system off the Bosporus Strait.

The Danube underwater canyon system is crowned by a thick alluvial fan. The relative height of this topographic feature reaches 500 m at a width changing from 40 km in its upper part to 60 km at its base. The alluvial fan is advanced by 100 km in the southeastern direction into the western depression of the sea. At the center of the fan, one observes an underwater valley rimmed by high (up to 300–400 m) near-channel bars [11].

A similar pattern is also observed in the Kerch–Taman' portion of the continental slope, where the system of underwater canyons features a length of the main valleys up to 280 km and a great number of secondary channels (52).

The highest degree of dissection by canyons is observed on the continental slopes of the Caucasian, West and East Pontian, near-Bosporus, and Bulgarian regions. In these areas, the tops of underwater canyons almost approach the coastline, terminating at a depth of 5–7 m, and play their part in capturing the alongshore sediment fluxes. The main channels of the canyons cross the continental slope and are 15–20 to 100–150 km in length.

The morphologies of the different underwater canyons have much in common. They feature V-shaped or U-shaped transverse profiles. With depth, the width of the canyons increase from 150–200 m at the shelf edge to to

300–500 m in the middle part of the continental slope and to 1.5–5 km at its base. Alluvial fans of the underwater canyons often merge, forming a particular hilly topography at the foot of the continental slope.

The continental footstep of the Black Sea occupies an intermediate position between the continental slope and the floor of the central depression at depths from 1100–1200 to 1800–2000 m. Morphologically, it is represented by a slightly inclined plain that borders the base of the continental slope. It is a kind of accumulative tail formed owing to the merging of numerous alluvial fans near the mouths of underwater canyons and to the sedimentation matter supplied from the shelf and continental slope due to the sediment runoff and landslide processes.

The continental footstep has a greater area than the continental slope. Its smooth topography is complicated by hilly features, underwater valleys, and channels of turbidity flows that serve for the sediment discharge. At depths of 1800–2000 m, the lower boundary of the continental footstep is indistinct. The sloping plain gradually gives place to the floor of the deep-water depression. The greatest widths of the continental footstep (up to 90–100 km) are observed in the northwestern, northeastern, and southeastern parts of the Black Sea.

The floor of the deep-water depression occupies the central part of the Black Sea basin. Its area inside the 2000-m depth contour comprises about 34% of the total area of the sea.

The Andrusov Rise, which is poorly manifested in the bottom topography, together with the Arkhangel'skii Rise, which is its southern continuation, divides the central Black Sea depression into its western and eastern parts. Their sublatitudinal extensions comprise 450 and 300 km, respectively. The western depression is deeper; here, sea depths greater than 2200 m have been registered. In the eastern depression, sea depths range from 2000 to 2160 m. Over the entire history of the geological evolution of the Black Sea basin, these depressions have been sites of accumulation of sedimentary matter, whose thickness can reach 10–15 km.

The floor of the depressions is significantly smoothed and slightly inclined, featuring a gradual depth increase from the margins toward the center of the sea. In terms of morphology and origin, its topography may be referred to as underwater plains of terminal marine accumulation. Precisely here, the decay of all the hydrogenous and gravity matter fluxes delivered from the shelf over underwater canyons, continental slope, and continental footstep to the enclosed depressions, where they are captured and accumulated, occurs.

4 Bottom Sediments

The distribution of the recent bottom sediments in the Black Sea is distinguished by its complicacy. Their composition and origin depend on the provenance areas, hydrodynamic and lithodynamic activity in the contact zone of the sea, and on the morphology of the bottom topography (Fig. 2).

The sedimentation is strongly controlled by the inherited character of the relief-forming processes that proceeded against the background of the Holocene history of the Black Sea. Sediment formation is also influenced by the solid riverine runoff and coastal abrasion, slope-derived supply, and biogenic and chemogenic matter.

Wave action, currents, and gravity processes define the particular features of the redistribution of the bottom sediments, their zonation, and the existence of coarse-grained matter in the near-shore zone subjected to wave action, and of the fine-grained fraction beyond this zone at greater depths. Unusual features in the bottom sediment distribution may be caused by the activity of turbidity flows and landslide processes, which distort the general regularities of the lithological zonation.

The first schematic of the bottom sediments of the Black Sea was compiled by A.D. Arkhangel'skii and N.M. Strakhov in 1938. Later, these data have been refined by different scientists. Recent sediments of the Black Sea are characterized by a significant spatial variability and lithological diversity with respect to their mineral and grain-size compositions [12, 13].

Over plain areas, one observes no sharp contrasts and boundaries between different facial types of sediments. With respect to the composition and distribution of the sediments, the Black Sea can be subdivided into the same regions as the shelf, continental slope, continental footstep, and the floor of the deep-water depression.

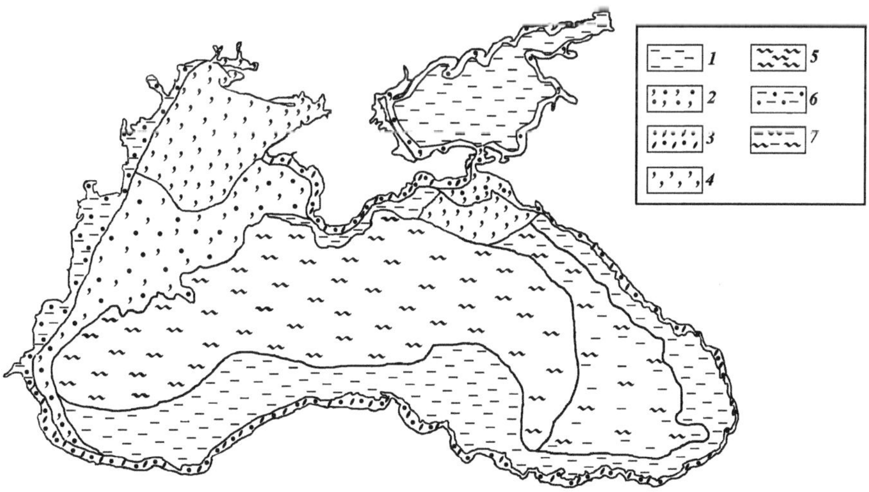

Fig. 2 The bottom sediments of the Black Sea: *1* terrigenous oozes, *2* sandy–coquina, *3* sandy–pebbly, *4* coquina, *5* coccolith oozes, *6* sandy–silty, *7* terrigenous–coccolith oozes

The northwestern region is characterized by large provenance areas, from which alluvial matter is supplied via the deltas of the Danube, Dniester, and Dnieper rivers. The Danube River, which features a wide delta with numerous channels and branches provides the greatest solid runoff to the shelf zone. The delta is composed of terrigenous sediments and is advanced toward the sea by 7–10 km.

In the solid runoff of other rivers, regulated owing to the construction of reservoirs, biogenic supply has a greater significance. While in the near-shore zone at low depths the proportion of the biogenic matter ranges from 30 to 50% of the total sediment, the proportion of coquina and detritus in the sediments of the central zone of the northwestern part of the sea reaches 80% and more.

At the outer edge of the shelf and in the upper part of the continental slope, silty and clayey coquina and lowly-carbonate oozes are mostly developed. The sediments of the continental slope (as well as those of the major part of the deep-water basin of the Black Sea) are represented by clayey-calcareous (coccolith) oozes.

The continental slope represents a transit zone of the sediment fluxes supplied in the form of detrital matter from the rivers and the products of abrasion, as well as the sediments carried by turbidity flows. The continental slope is covered with compacted clayey oozes (grain-size fractions of 0.001–0.01 mm). At selected places on steeper parts of the slope, remains of mollusk fauna such as shells of *Dreissena rostriformis* have been encountered. At sites with gentle sloping, they are overlain by the Holocene and recent sediments [11].

A special type of sediment is formed at the foot of the continental slope of the Black Sea in the zone of discharge of turbidity flows. In the western part of the sea and in the Crimean sector, the sediments of the base of the continental slope are represented by finely stratified biogenic or organogenic oozes. In the eastern and southern parts of the basin, a terrigenous component of clayey sediments becomes more significant [6, 14].

The sediments of the abyssal plain of the central Black Sea region are mostly biogenic and are enriched with organic matter. The floor of the deep-water depression is covered with coccolith oozes. In peripheral zones, in addition, terrigenous lowly calcareous oozes and carbonate-free silts are observed.

The Crimean region is characterized by alongshore variations in the sediments of the underwater slope. West of the Tarkhankut Peninsula up to Evpatoriya, biogenic coquina deposits dominate; they cover the limestone bedrock. On the Crimean shelf, terrigenous sediments are also observed represented by boulders and pebbles in the near-shore zone, sands at depths down to 7–10 m, and fine sands and silty oozes at greater depths. Meanwhile, at depths of about 30 m, there exists a sandy–pebbly bar formed by extreme waves.

In the shallow-water Kerch–Taman' region, in the zone of wave action, coquina sediments are developed; at greater depths, terrigenous oozes with an admixture of coquina matter are observed [15].

The sediments of the Caucasian region are formed under the influence of the solid runoff of mountain rivers and due to the intensive development of the processes of abrasion and denudation. Waves and coastal currents significantly affect the distribution of the terrigenous–detrital matter over the underwater slope, concentrating largest particles of the matter of boulders, pebbles, and sands close to the coastline and on the beach. Beyond the zone of the wave action, fine-grained sands and silty oozes are accumulated. Often, bedrocks are exposed at abrasive surfaces of the underwater slope at depths down to 60 m.

In the West and East Pontian regions, the sediment formation is strongly influenced by mountain rivers and numerous water channels that supply terrigenous matter as well as by coastal abrasion.

The sediments of the underwater coastal slope of the Bulgarian region from the shore to the depth of near 30 m consist of differently-grained sands, silts, and coquina; bedrock exposures are abundant.

Analyzing the particular features of the sediment distribution in the nearshore zone and on the floor of the Black Sea, one should note the extremely high sedimentation rates, reaching 20 cm in 100 years.

5 Conclusions

The history of the formation of the Black Sea basin is usually referred to the Upper Paleozoic epoch, when its basaltic basement was formed and the foundation was laid to the depression that later developed in the Mesozoic–Cenozoic time [1, 7]. By the beginning of the Mesozoic, a series of major rises at the margins of the depression had been formed, while the adjacent areas were covered with sea.

During the Lower and Upper Cretaceous, geosynclines started to grow at the margins of the depression, while in the south, the Arkhangel'skii and Inebolu Rises were formed.

The intensive Upper Cretaceous transgression covered the entire area and existed until the Eocene. Against its background, the prototype of the mountain rim of the Black Sea depression that was formed at that time looked like a system of individual island massifs. This pattern sharply changed in the Oligocene. The mountain ranges of the Crimea, Caucasus, and Pontides began growing, which resulted in the isolation of the Black Sea basin, activation of exogenous processes, and accumulation of a thick sedimentary sequence.

During the subsequent Miocene–Pliocene and Quaternary periods, the thickness of the sediments increased.

In the Pleistocene–Holocene, the Black Sea underwent a series of intensive transgressions and regularities. At this stage, the topographic pattern inherited by recent marine–coastal features was formed, accompanied by the appearance of the shelf, marine terraces, and systems of underwater canyons on the continental slope.

At the present-day stage of the basin development, evolution of the topography and sedimentation in the Black Sea proceed against the background of the global climate warming and sea level rise. This predicted process may lead to extreme and, probably, catastrophic events in the near-shore zone. In the practice of the use of the Black Sea as a natural resource, it is necessary to revise the measures for provision of the ecological security of the environment.

References

1. Jaoshvili SV (2002) The rivers of the Black Sea. Technical report 71, European Environment Agency, Copenhagen, Denmark
2. Zenkovich VP (1958) Coasts of the Black Sea and the Sea of Azov. Geografgiz, Moscow (in Russian)
3. Kaplin PA, Leontiev OK, Lukyanova SA, Nikiforov LG (1991) Sea coasts - nature of the world. Mysl, Moscow, p 479 (in Russian)
4. Popov VL, Mishev K (1974) Geomorphology of the Bulgarian Black Sea coasts and shelf. BAK, Sofia (in Bulgarian)
5. Muratov MV (1972) Geotektonika 5:362 (in Russian)
6. Shimkus KM, Emelyanov EM, Trimonis ES (1975) Earth's crust and evolution history of the Black Sea basin. Nauka, Moscow, p 138 (in Russian)
7. Solov'ev VF (1971) In: Solov'ev VF (ed) Geological structure of the shelves of the the Caspian Sea, the Sea of Azov, and the Black Sea with respect to their oil and gas bearing properties. Nauka, Moscow (in Russian)
8. Goncharov VP, Neprochnov YuP, Neprochnova AF (1972) Topography and deep structure of the Black Sea basin. Nauka, Moscow (in Russian)
9. Kuprin PN (1988) In: Kuprin PN (ed) History of geological evolution of the continental margin of the western part of the Black Sea. MGU, Moscow (in Russian)
10. Leont'ev OK, Saf'yanov GA (1973) Canyons beneath the sea. Mysl', Moscow (in Russian)
11. Scherbakov FA, Kuprin PN, Polyakov AS (1975) In: Kuprin PN (ed) Multidisciplinary studies of the nature of the ocean. MGU, Moscow, p 141 (in Russian)
12. Tugolesov DA, Gorshkov AS, Meisner LB et al. (1983) Dokl AN SSSR 269:440 (in Russian)
13. Tugolesov DA, Gorshkov AS, Meisner LB et al. (1985) Tectonics of the Mesocenozoic deposits of the Black Sea basin. Nedra, Moscow (in Russian)
14. Moskalenko VN (2000) Science Kuban 4(11):33 (in Russian)
15. Shimkus KM, Kuprin PN, Sorokin VM (1996) Unconsolidated bottom surface sediments. Scale 1 : 2 000 000. St. Petersburg, Russia

Hdb Env Chem Vol. 5, Part Q (2008): 63–89
DOI 10.1007/698_5_091

Published online: 9 October 2007

The Sea of Azov

Aleksey N. Kosarev[1] · Andrey G. Kostianoy[2] (✉) · Tamara A. Shiganova[2]

[1]Geographic Department, Lomonosov Moscow State University,
Vorobievy Gory, 119992 Moscow, Russia

[2]P.P. Shirshov Institute of Oceanology, Russian Academy of Sciences,
36 Nakhimovsky Pr., 117997 Moscow, Russia
kostianoy@online.ru

Abstract Based on the data of long-term observations, principal hydrological and hydrochemical characteristics of the Sea of Azov such as the water temperature, salinity, and oxygen and nutrient contents are assessed. Features of the seasonal and interannual variations under the action of natural and anthropogenic factors are shown. The biodiversity of the sea and the influence of the invader species on the state of the sea ecosystem are analyzed in detail.

Keywords Biodiversity · Invaders · Nutrients · Oxygen content · Sea of Azov · Water salinity · Water temperature

1 Introduction

The first studies of the oceanographic and biological features of the Sea of Azov started in the middle of the nineteenth century [1]. Regular surveying of the hydrological and meteorological regime of the sea began with the development of a network of coastal hydrometeorological stations, as well as with sea expeditions on board research vessels conducted at the end of the nineteenth and at the beginning of the twentieth century by Wrangel (1873), Shpindler (1890), Antonov (1913), and others [2–4]. From 1922 to 1928, Knipovich headed sea expeditions to the Sea of Azov and the Black Sea with the aim of oceanographic and fishery research [5–7]. From 1928 to 1932, sea expeditions were continued by a special fishery station, which was later reorganized into the Azov and Black Sea Fisheries Research Institute (AzCherNIRO). In 1936, the USSR State Hydrometeorological Service set up a network of hydrometeorological stations and standard hydrographic sections in the Sea of Azov; later, it was used by the State Oceanographic Institute (SOI) for the studies of the hydrological regime and regional climate. After World War II, the AzCherNIRO restarted research activities in the Sea of Azov. The results of sea expeditions have been regularly published in Marine Hydrometeorological Yearbooks. Since 1952, the Azov Institute for Fishery (AzNIIRKH) has been carrying out comprehensive research of hydrological, chemical, biological parameters and fishery in the sea.

In the late 1980s, more than 20 coastal hydrometeorological stations were providing daily hydrological and meteorological information. Six standard hydrographic sections across the Sea of Azov were used to collect physical, chemical, and biological data. This set of data was used for the description of the state of the sea, its seasonal and interannual variability, as well as for the assessment of biological resources [8–10]. Since 1997, Murmansk Marine Biological Institute (MMBI) and its Azov Branch (since 1999) have conducted more than 40 scientific expeditions in the Sea of Azov [11–16]. The Southern Scientific Center of the Russian Academy of Sciences established in Rostov-on-Don in 2002, made a decision to rescue the historical oceanographic data and to make them available to the international scientific community in order to stimulate studies of the Sea of Azov. In 2006, this resulted in the NOAA publication *Climatic Atlas of the Sea of Azov 2006*, edited by Matishov and Levitus [17]. This atlas and the accompanying CD-ROM contain oceanographic data collected by specialists of the USSR and Russian Academy of Sciences, Ministry of Fisheries, and the Hydrometeorological Service of the USSR and Russia in the Sea of Azov and the adjacent part of the Black Sea from 1913 to 2004. The atlas contains monthly climatic maps of temperature and salinity at the sea surface and depth levels of 5 and 10 meters. The interannual variability of temperature and salinity of the Sea of Azov is also discussed.

The interest to the Sea of Azov was always related to its large fish stocks, which are inferior only to that of the Caspian Sea. Annual fish hauls (sturgeons, pike-perchs, breams, and sea roaches) in this small sea reached 300 kt. This triumph of fishery was confined to the period of the natural harmony between the processes in the sea, when it was characterized by a high quality of the environment.

Previous to the early 1950s, under the natural water regime, the Sea of Azov was distinguished by its extremely high biological productivity. The riverine runoff delivered great amounts of nutrients, 70–80% of which were supplied during the spring flood period. This provided abundant development of phytoplankton, zooplankton, and benthos. The area of the spawning zones related to flooded regions and lagoons in the lower reaches of the Don and Kuban' rivers reached 40 000–50 000 km^2. Along with the good heating, low salinity, sufficient saturation with oxygen, long vegetation period, and rapid cycling of nutrients, these factors provided conditions favorable for ichthyofauna that included up to 80 species [18].

Today, the basin of the Sea of Azov represents a well-developed industrial and agricultural region. The formation of the industrial and agricultural complex in the basin of the Sea of Azov is related to the regulation of the riverine runoff, to the partial use of the runoff, to the intensive industrial and civil construction, creation of irrigation systems in the sea watershed, and to the development of road–transport hubs, etc. This resulted in significant changes in the sea owing to the decrease in the volume of the freshwater supplied. The ecological changes resulted in a sharp drop in the biological productivity of the sea. The trophic base for fish decreased by several times and the total hauls reduced, mostly at the expense of valuable fish species.

2 Physico-Geographical Conditions

During its rich history, the Sea of Azov had many different names. Ancient Greeks called it *Maeotian Lagoon* (*Maeotian Lake*), while Romans referred to it as *Palus Meotis* (*Maeotian Marsh*) after the tribe *Maeotae* that dwelled on its coasts. In the antique epoch, locals called it *Temerinds*. In medieval times, Russian name for it was the *Surozh Sea* after the name of the Crimean town of Surozh (now Sudak).

The Sea of Azov is the most shallow-water and one of the smallest seas of the world. Its area is 39 000 km^2 at a volume of 290 km^3; the average depth is 7 m with a maximum value of 14 m. It is connected with the Black Sea by the narrow (up to 4 km), and shallow-water (up to 15 m) Kerch Strait. The maximum length of the sea is 360 km at a maximum width of 180 km. The first sailing directions for the Sea of Azov (1854) were compiled by Sukhomlin, who spent two years studying the coasts of the sea.

The sea features rather simple outlines. The northern coast is even and steep with accumulative sandy spits. In the northeast, the largest of the sea bays – Taganrog Bay – penetrates into the land; its top coincides with the delta of the Don River. In the west, the Arabatskaya Strelka Spit separates Sivash Bay from the sea. The bay is connected with the sea by the Genichesk Strait. Sivash Bay (or the Gniloye Sea) represents a system of shallow-water bays with a total area of 2560 km^2. Their depths are 0.5–1.5 m, with a maximum value of 3 m. Annually, Sivash accepts up to 1.5 km^3 of the water from the Sea of Azov. Owing to the strong evaporation, the Sivash water transforms into saturated salt solution (brine, or rapa) with a salinity reaching 170 psu. Similar to Kara Bogaz Gol of the Caspian Sea, Sivash Bay provides various chemical resources. It contains millions of tons of salt, magnesium sulfate, sodium sulfate, bromine, and other ingredients. For a long time, table salt works existed in Sivash Bay. Mirabilite is also extracted from the Sivash brines through salt precipitation.

In the southeast, the delta of the Kuban' River with vast flooded plains and numerous channels extends over about 100 km. The Kuban' River enters the top part of the open Temryuk Bay. Low seacoasts gradually descend to a flat sandy bottom. The depths smoothly increase with the distance from the coast. The largest depths are observed in the central part of the sea; in Taganrog Bay, they range from 2 to 9 m. In Temryuk Bay, mud volcanoes are known. The main sources for the supply of the terrigenous matter that forms the bottom sediments of the Sea of Azov are represented by the products of coastal abrasion and the riverine alluvium. The bottom sediments are mostly represented by clayey and silty oozes and sands.

Essentially, the Sea of Azov is a vast zone of mixing between the riverine and Black Sea waters. Almost the entire riverine runoff to the sea (more than 90%) is provided by the Don and Kuban' rivers and its major part is confined to the spring–summer season. The principal exchange between the waters of the Sea of Azov and those of the Black Sea is implemented via the Kerch Strait.

The climate of the Sea of Azov, which deeply penetrates into land, is continental. It is characterized by cold winters, and dry and hot summers. In the autumn–winter period, the weather is determined by the influence of a spur of the Siberian anticyclone with a domination of easterly and northeasterly winds with a speed of 4–7 m/s. Enhancements of the impact of this spur cause strong winds (up to 15 m/s) and are accompanied by invasions of cold air masses. The mean monthly temperature in January ranges from – 1 to – 5 °C; during northeasterly storms, it may fall down to – 25 to – 27 °C.

In the spring–summer period, warm and fair weather with weak winds prevails. In July, the mean monthly temperature over the entire sea equals 23–25 °C, while its maximum values reach more than 30 °C. In this season, especially in the spring, Mediterranean cyclones often pass over the sea; they are accompanied by westerly and southwesterly winds with speeds of 4–6 m/s, and sometimes by gusts.

The water balance of the Sea of Azov consists of the following components: the incoming part contains the riverine runoff and the atmospheric precipitation, while the outgoing part includes evaporation. The water exchange via the Kerch Strait should also be taken into account. According to the data averaged over 1923 to 1985, the riverine runoff, precipitation, and evaporation comprised 38.6, 15.5, and 34.6 km^3/year, respectively. The annual inflow of the Black Sea waters via the Kerch Strait was 36–38 km^3/year, while the outflow of the Azov waters comprised 53–55 km^3/year; this provided a value of the resulting water removal from the Sea of Azov of about 17 km^3/year.

An analysis of the water balance of the Sea of Azov over the years cited shows that the components of the balance changed in the period before the regulation of the riverine runoff in1952 and after it. On average, the riverine runoff to the sea reduced by 5.7 km^3/year, the supply of the Black Sea waters increased by 1.5 km^3/year, and the outflow of the waters of the Sea of Azov to the Black Sea increased by 1.9 km^3/year. Meanwhile, on the whole, the water balance remained almost the same.

The winds that dominate over the sea induce significant surge (onset) sea level oscillations. The highest sea level rises were registered in Taganrog, where they reached 6 m. At other sites, rises ranging from 2 to 4 m are possible (Genichesk, Eisk, and Mariupol'); in the Kerch Strait, they reach a height of about 1 m. The limited sizes and small depths of the sea favor rapid development of wind waves. The waves are short and step; in the open sea, they are up to 1–2 m, sometimes 3 m high.

Sharp changes in the atmospheric pressure and winds over the Sea of Azov may also induce seiches – freestanding oscillations of the sea level. In port areas, seiches with periods from a few minutes to a few hours are generated. In the open sea, seiches with a diurnal period up to 20–50 cm high are noted.

Seasonal changes in the sea level mainly depend on the regime of the riverine runoff. The annual sea level change is characterized by its rise in the spring–summer months and a fall in the autumn and winter with average total range of 20 cm.

The currents in the sea are mostly induced by the wind. Under the forcing by westerly and southwesterly winds, an anticlockwise water circulation in the sea is formed. The cyclonic water movement is also characteristic under easterly and northeasterly winds as well when they are stronger in the eastern part of the sea. If these winds are stronger in the southern part of the total abundance, the circulation has an anticyclonic character. At weak winds and calm, insignificant currents of intermittent directions are observed. Since weak and moderate winds dominate above the sea surface, currents with velocities lower than 10 cm/s feature the highest recurrence rates. Under strong winds up to 15–20 m/s, current velocities increase up to 60–70 cm/s.

In Taganrog Bay, the resulting water transport is controlled by the runoff of the Don River and is directed from the bay toward the sea. In the Kerch Strait, under northerly winds, the current flows from the Sea of Azov to the Black

Sea; winds with a southerly component provide the supply of the Black Sea waters to the Sea of Azov. The dominating current velocities in the strait grow from average values of 10–20 cm/s to 30–40 cm/s in its narrowest part. After strong winds, compensatory currents are generated in the strait.

3 Ice Conditions

In the Sea of Azov, ice is formed every year; in so doing, the ice coverage (sea area covered with ice) strongly depends on the character of the winter (severe, moderate, or mild). In moderate winters, ice is formed in Taganrog Bay by the beginning of December. During December, fast ice is formed along the northern coast of the sea and somewhat later along its other coasts. The width of the fast ice band ranges from 1.5 km in the south to 6–7 km in the north. In the central part of the sea, floating ice is formed only at the end of January or the beginning of February; subsequently, it freezes together and forms ice fields with high ice concentration numbers (9–10). The ice cover is most developed at the beginning of February, when its thickness reaches 30–40 cm (60–80 cm in Taganrog Bay).

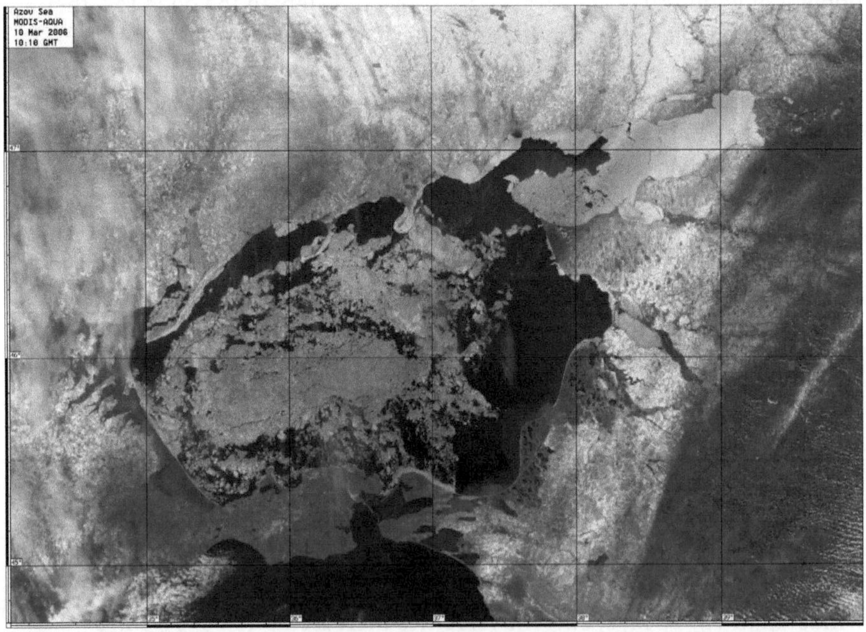

Fig. 1 Ice in the Sea of Azov revealed from a *MODIS-Aqua* satellite image on March 10, 2006. Image courtesy of D.M. Soloviev, Marine Hydrophysical Institute, Sevastopol, Ukraine

Throughout the winter, the ice conditions feature instability. The mutual replacements of the cold and warm air masses and wind fields over the sea caused repeated breaking and drifting of ice fields and their hummocking. In the open sea, the heights of hummocks never exceeds 1 m, while off the Arabatskaya Spit, hummocks may reach a height of 5 m. As a rule, during mild winters, the central part of the sea is free from ice; it may be observed only in bays and lagoons along the coasts.

In mild winters, the release from ice occurs during March first in the southern regions and in river mouths, then in the north, and, finally, in Taganrog Bay (Fig. 1). The average duration of the ice period is 4.5 months. In anomalously warm or severe winters, the times of ice formation and thawing may be shifted by 1–2 months or even greater.

4 Thermohaline Structure

Due to the small sizes and small water depths of the sea, the principal characteristics of the hydrological and hydrochemical regime are subjected to significant natural and anthropogenic variations.

In the shallow-water Sea of Azov seasonal changes in the water temperature are very strongly manifested. In the winter (January to February), over the greater part of the sea area, the sea surface temperature equals 0–1 °C; only in the region of the Kerch Strait, it grows up to 2–3 °C. In the summer (July to August), the temperature is homogeneous over the entire sea area being equal to 24–25 °C (Fig. 2). The maximum values in the open sea reach 28 °C, while near the coasts they may exceed 30 °C. In the near-bottom layer of the sea, the temperature distribution is generally close to the values registered at the surface of the basin.

The shallow-water character of the sea provides rapid propagation of wind and convective mixing down to the bottom, which leads to equalizing the vertical temperature distribution; in most cases, the temperature difference is less than 1 °C. Meanwhile, during summertime calm periods, the thermocline is formed which prevents the near-bottom layer from water exchange.

Under the conditions of natural riverine runoff, the salinity distribution in the sea was rather homogeneous; horizontal gradients were observed only in Taganrog Bay, at the exit from which, salinity values of 6–8 psu dominated (Fig. 3). This bay is filled with desalinated waters with a salinity of about 2–7 psu. In the open sea the salinity ranged from 10 to 12 psu; in almost all of the regions, gradients were episodically observed and they were mainly related to the supply of the Black Sea waters. The seasonal salinity changes never exceeded 1 psu, except for Taganrog Bay, where they enhanced under the influence of the intraannual runoff distribution. Most frequently, high vertical salinity gradients are formed in Temryuk Bay, where the waters of the

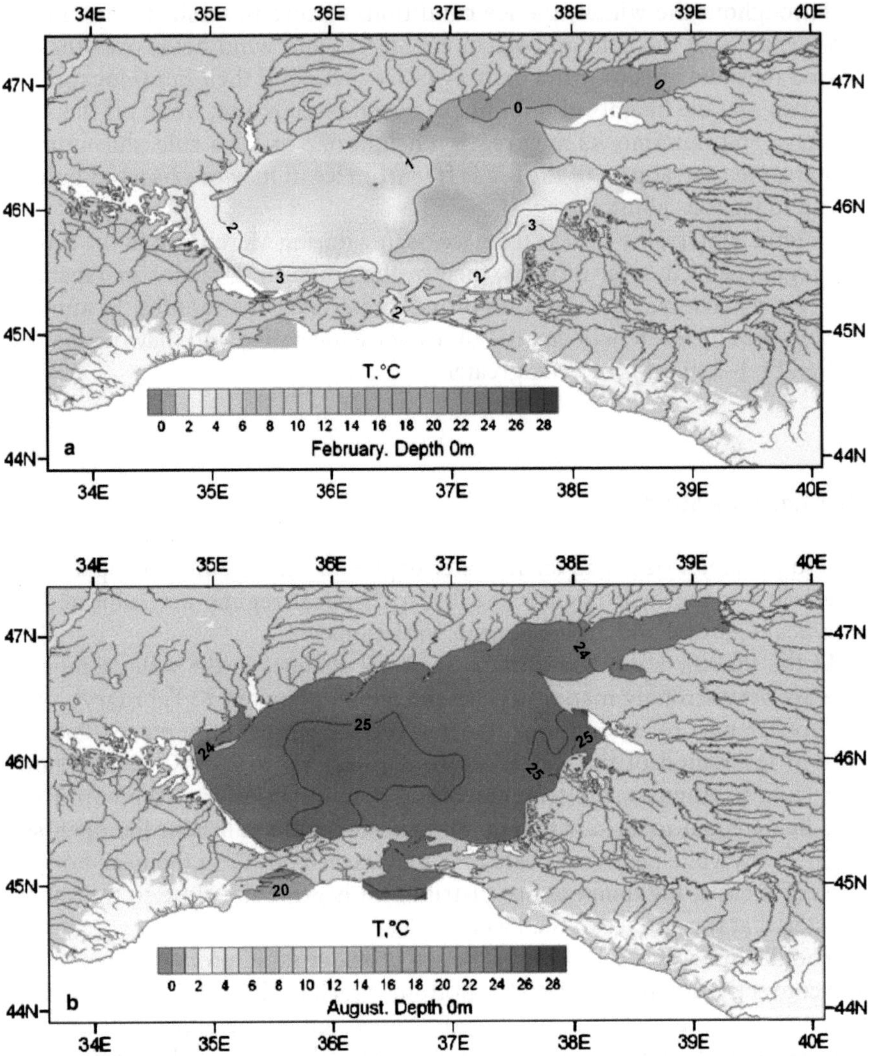

Fig. 2 Mean multiannual water temperature (°C) in the Sea of Azov at a level of 0 m in **a** February and **b** August [17]

Kuban' River are delivered to a relatively deep-water near-mouth sea area. In the spring of 1980, the salinity difference between the surface and the bottom here reached 9 psu.

The multiannual changes in the salinity of the Sea of Azov are closely related to the variability in the overall humidity in its watershed. For example, during the stage of enhanced humidity in 1924–1932, the average salinity of the sea has decreased from 10.5 to 9.6 psu. During the period of reduced humidity in 1945–1951, the salinity rose from 10.7 to 12.7 psu. The regulation of

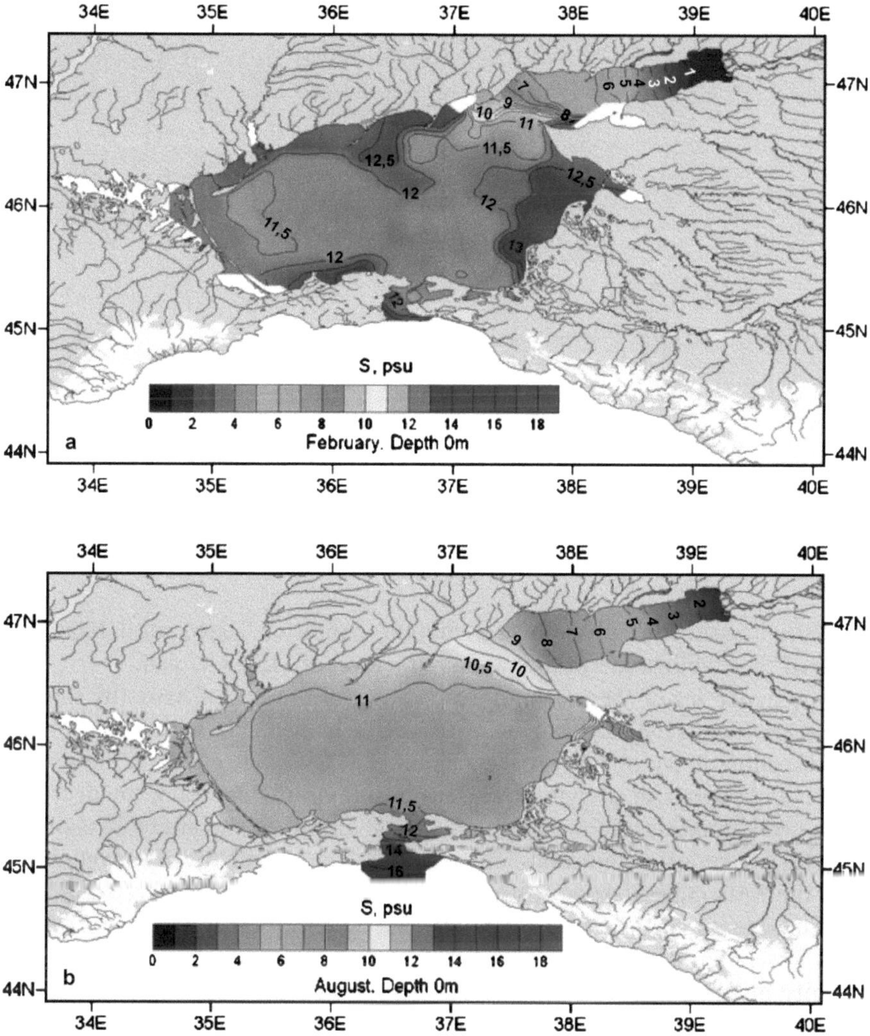

Fig. 3 Mean multiannual water salinity (psu) in the Sea of Azov at a level of 0 m in **a** February and **b** August [17]

the Don River runoff in 1952 and the one-time removal of about 25 km^3 of the Don River waters in order to fill in the Tsimlyansk Reservoir provided the rapid growth of the salinity of the Sea of Azov (Fig. 4).

In 1953–1955, the average salinity in the sea reached 12.6–12.7 psu. A growth that great was caused not only by the irreversible withdrawal of the runoff but also by the fact that it was preceded by a depressive phase of the total humidity of the sea basin with a maximum at the beginning of the 1950s. Later (starting from 1956), the humidity in the basin of the Sea of Azov

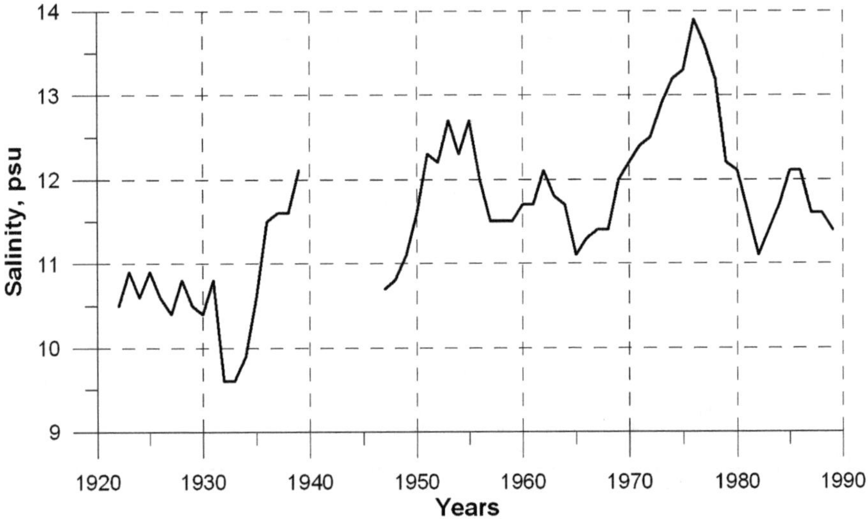

Fig. 4 Long-term (1922–1989) changes in the average salinity of the Sea of Azov (psu)

increased again; this phase of the climatic condition lasted until 1968 and favored the stabilization of the salinity at a level of 11.3–11.7 psu.

The dynamics of the mean annual salinity of the Sea of Azov (by the example of 1956–1968) shows that, even active irreversible withdrawal of the waters, climatic factors may make a significant favorable effect on the results of the anthropogenic activity in the sea basin. On the contrary, the decrease in the total humidity in the basin after 1968 amplified the aftereffects of the irreversible runoff withdrawals when, in 1973, the Kuban' River was regulated and the Krasnodar Reservoir was filled. In the 1970s, the integrated annual riverine runoff to the Sea of Azov was 22–27 km^3/year, a value more than 40% lower than the natural norm. As a result, the tendency to increase the salinity of the sea was enhanced. The strongest salination was observed in 1975–1977, when the salinity in the sea comprised 13.3–13.9 psu, while in Taganrog Bay it was 9.5–11.1 psu.

On the whole, as a result of the coupled effect of climatic and anthropogenic impacts, the salinity maximum in the Sea of Azov in 1975–1977 exceeded the natural norm by up to 3.0 psu and, in Taganrog Bay, even more. Beginning from 1978, the regime of the sea reached its new water-rich phase, the mean annual salinity of the sea acquired a tendency to fall, and, in 1980s, it comprised 11–12 psu. By 2000 salinity lowered even to 10–11 psu.

In 1950–1970, the increased withdrawals of freshwaters for municipal purposes resulted in a decrease in the riverine runoff to the sea and a corresponding increase in the delivery of the Black Sea waters. The spatial inhomogeneity of the salinity became noticeable; in the near-Kerch region, especially in low-water years, its values grew to 15–18 psu, i.e., up to values that have never been observed from the beginning of the century.

At present, the Sea of Azov is characterized by the existence of salinity frontal zones in the regions of the riverine water transformation in the near-mouth areas of the Don and Kuban' rivers and in the zone of mixing between the waters of the Sea of Azov and the Black Sea. The salt exchange with Sivash Bay is insignificant and its influence involves only a small region near the Tonkii Strait. The central part of the sea is occupied by a homogeneous water mass with a salinity of 11–12 psu (see Fig. 3).

The enhancement of the propagation of the Black Sea waters in the near-bottom layers resulted in a growth in the vertical salinity and density gradients and deterioration of the conditions of mixing and ventilation of the near-bottom waters. Also increased the probability of the formation of the oxygen deficiency (hypoxy) and conditions lethal for hydrobionts.

In accordance with the distributions of the water temperature and salinity, the vertical density gradients reach their maximal values in frontal regions – Taganrog Bay and the near-mouth area of the Kuban' River, and in the near-Kerch region.

The regulation of the riverine runoff, its reduction by 13–15 km^3/year, the creation of reservoirs, and other aftereffects of the anthropogenic activity in the basin caused serious negative changes in the sea ecosystem. The 30% drop in the annual runoff of the Don River and the significant decrease in the flooding volumes resulted in the reduction of the spawning areas and violated the conditions of reproduction of freshwater fish species [10, 18].

5 Hydrochemical Conditions

The oxygen regime of the Sea of Azov is mainly rather favorable; water saturation with oxygen is sufficient and oxygen distributions over area and depth are rather uniform. According to the generalized data of multiyear observations, in the winter, the average absolute oxygen content in the surface layer is 330–370 μM in the open sea and 420–470 μM in Taganrog Bay. In the spring, the oxygen content equals 360–380 μM both in the open sea and in Taganrog Bay and the water is well aerated from the surface to the bottom (the relative oxygen content is about 100% or even slightly higher). In the summertime, when the water temperature grows, the absolute oxygen content decreases. In the open sea, its average value in the warm season is about 260 μM (100%) in the surface layer and 200 μM (about 80%) in the near-bottom layer. In Taganrog Bay, the oxygen contents in respective layers equal 270 μM (100%) and 220 μM (85%). Meanwhile, in the central part of the sea, the oxygen content in the near-bottom layer may locally drop down to 110 μM. In the autumn, the water temperature fall causes a uniform oxygen distribution: 280–290 μM in the Sea of Azov and 230–310 μM (90–95%) in Taganrog Bay.

In the summer, the weakening in the vertical mixing in the sea results in the formation of oxygen-deficient zones in the near-bottom layers. These conditions lead to the appearance of suffocation zones often accompanied by extinction of bottom fauna. The principal reasons for the summertime oxygen deficiency are related to the high contents of the easily mineralizable organic matter in the water and bottom sediments, the enhancement of the vertical temperature stratification owing to the sea heating, and the salinity gradient increase caused by the changes in the riverine runoff. The frequency of the formation of oxygen-deficient zones as well as their area and intensity are closely related to the general character of the hydrological conditions in the sea (wind activity, salinity regime, amount and composition of the nutrients supplied to the sea, etc.). Therefore, the summertime near-bottom oxygen deficiency in the Sea of Azov is subjected to a significant interannual variability. This phenomenon was long known in the Sea of Azov; however, it was observed only episodically. In the 1960s and 1970s, an active development of the reduction processes in the near-bottom layer occurred related to the beginning of the process of anthropogenic salination of the sea accompanied by a weakening of the wind activity over its area. In July 1987, the presence of hydrogen sulphide was registered in the Sea of Azov for the first time in history; its content in Temryuk Bay was 20–35 μM.

The significant anthropogenic impact that the Sea of Azov suffers is manifested in the delivery of great amounts of organic matter and nutrients as well as of technogenous pollutants. Under these conditions, the dwelling environment of young sturgeon fish was reduced to the area of Taganrog Bay (about 12% of the total sea area). In 1989–1990, the presence of hydrogen sulphide was detected in the central part of the sea and in Berdyansk and Temryuk bays.

The present-day hydrochemical conditions in the Sea of Azov, including the distribution of nutrients, are described using the materials of the cruise of R/V *Akvanavt* that was carried out in July to August 2001 [19]. According to the observations, the Sea of Azov features a two-layered structure. In so doing, the upper layer, 7–10 m thick, consisted of three water masses. Transformed Black Sea waters with a salinity higher than 11.5 psu, oxygen content of 170–190 μM (less than 85% of saturation), phosphate content of 0.7–0.9 μM, silicate content of 8–12 μM, nitrate content of 0.4–0.6 μM, nitrite content of 0.05–0.1 μM, and ammonium content of 1.0–2.0 μM, occupied the eastern part of the sea. Fresher waters of riverine origin with a salinity lower than 10 psu were supplied from Taganrog Bay and propagated in the form of a tongue extended along the northern coast of the sea. They were characterized by enhanced contents of oxygen (250–300 μM, or more than 120% of saturation), silicate (28–32 μM), and nitrites (0.15–0.25 μM), by an insignificant growth in phosphate contents (1.0–1.3 μM), and by a decrease in the contents of nitrate (0.3–0.5 μM) and ammonium (lower than 1 μM).

The surface water of the central and western parts of the sea was characterized by intermediate values of these parameters. Over the entire sea area, the concentrations of the main forms of nutrients were sufficiently high and did not restrict the phytoplankton development, which points to the significant trophicity of the sea.

Among the most interesting results of the cruise was that suffocation zones were found no thicker than 1.5 m with high hydrogen sulphide contents revealed in the near-bottom layer. The anoxic layer was separated from the overlying layers by a pycnocline 0.5–1.5 m thick. The high density gradients prevented the layers of the sea from the vertical water exchange. The near-bottom layer was characterized by an increase in the contents of hydrogen sulphide (80–90 μM), ammonium, and phosphates. In the Sea of Azov, the concentrations of all the reduced compounds near the boundary of the hydrogen sulphide layer corresponded to those in the Black Sea at depths 50–100 m below this boundary. This suggests a higher intensity of the processes of mineralization of organic matter and a stability of the stratification in the waters of the Sea of Azov.

The principal reason for the formation of anoxic conditions in the Sea of Azov (as well as in other inland seas) lies in the imbalance between the organic matter supply and the income of dissolved oxygen required for its oxidation. In the formation of the suffocation conditions in the Sea of Azov, certain roles belong both to the allochtonous organic matter delivered with the riverine runoff and to the autochtonous matter generated in the sea proper. The supply of nutrients is implemented with the runoff of the Don and Kuban' rivers and via the Kerch Strait, while their removal is related to the outflow to the Black Sea, extraction in the course of fishery activity, and burying in the bottom sediments. The results obtained in the cruise are evidence of the strong eutrophication of the entire area of the Sea of Azov and of its critical ecological state.

A characteristic trend in the present-day nutrient dynamics in the waters of the rivers of the Sea of Azov basin lies in the decrease in the content of phosphorus (P_{tot}) and the increase in the nitrogen content. After regulation of the runoff, the internal structure of the nutrient runoff – nutrient to phosphorus concentration ratio – has been sharply distorted. Under the natural conditions, the values of the N/P ratio were 3.9 and 4.5 for the Don and Kuban' rivers, respectively, and later they became 17.6 and 11.5, respectively. The differently directed trend of the changes in the nitrogen and phosphorus contents in the riverine waters of the Sea of Azov basin are related to the fact that in the Tsimlyansk and Krasnodar reservoirs, accumulation of particulate mineral and organism phosphorus takes place. In the Don River waters, a strong growth in the nitrogen concentration occurred. Balance estimates for nitrogen and phosphorus in the Sea of Azov showed that the most variable component of the balance is related to the burying of nutrients in the bottom sediments.

It seems impossible to calculate an exact nutrient balance for the Sea of Azov. According to the estimates available for the period from 1952 to 1976, the input/output values are evaluated in the ranges 75–122 and 8–17 kt for the total nitrogen and for the total phosphorus, respectively.

The hydroeconomic situation in the Sea of Azov basin is very tense. At present, about 30–40 km^3 of riverine waters are annually supplied to the sea. Under this runoff intensity, there is a possibility of retaining the seawater salinity in the range up to 11–12 psu. The further growth in the water consumption is prohibitive since it should cause a salinity growth up to the Black Sea values thus deteriorating the dwelling conditions for most valuable marine organisms.

6 Biodiversity

After the Caspian Sea, the Sea of Azov is the second most significant inland basin of the former USSR with respect to its fish stocks. Recently, every hectare of its area provided 80 kg of fish, half of which was represented by valuable and very valuable species (sturgeons, pike-perches, breams, sea roaches, and others). The annual fish production in this smallest sea reached 300 kt. This triumph of the Sea of Azov fishery was related to the times of complete harmony between the natural processes, when the sea was characterized by a high quality dwelling environment.

This harmony was provided by three principal factors:

- The sufficient delivery of riverine waters rich in nutrients to the sea and the high rate of their cycling.
- The low salinity of the waters of the sea.
- The high provision of the reproduction of fish stocks: for migratory and semi migratory species alone, the total spawning area exceeded 600 000 hectares, while the habitat of marine fish covered the entire sea area.

The positive effect of the interaction between these factors was enhanced owing to the shallow-water character of the sea and its geographical position. Since the middle of the last century the environmental state of the Sea of Azov has been under great anthropogenic pressure, resulting in negative changes in the sea biota.

6.1 Phytoplankton

In the Sea of Azov and Taganrog Bay, based on multiyear studies of phytoplankton, 605 species, varieties, and forms of purely or optional planktonic algae were discovered. With respect to the species number, diatoms and green algae dominate. Blue-green algae and pyrophytes also feature a high species diversity; euglene and yellow-green algae comprise about 5% of the total species number [20].

The principal alga representatives in the Sea of Azov are planktonic algae. The low water transparency suppresses the development of bottom plants. Essentially, the production formed by phytoplankton serves as a source for life of the entire heterotrophic population of the Sea of Azov. Owing to the particular features of the hydrological regime, the phytoplankton of the sea has certain special features that are mostly typical of lagoons. The shallow-water character of the sea and its good response to heating allow the algae to inhabit almost the entire water column. The quantitative development and species composition of phytoplankton in the open part of the sea are almost similar to those in the near-shore zone. The Sea of Azov is characterized by intensive and rather long-term "blooming" periods, high concentrations of particulate organic matter in the water, and frequent events of oxygen deficiency. Meanwhile, with respect to the salt composition, the relations between the total salt content and chlorinity, the domination of pyrophytes and diatomaceous algae (marine species) in the sea, the Sea of Azov water is close to oceanic water.

The great volumes of fresh and Black Sea waters delivered to the Sea of Azov supply assemblages of algae and animals dwelling in these basins. However, the extremely variable salinity of the seawaters makes dwelling and developing possible for euryhaline species only. The tolerance of algae to salinity changes precisely defines the boundaries of their distribution in the Sea of Azov and Taganrog Bay.

The region of the Sea of Azov proper with an average salinity of 11–12 psu is mostly inhabited by three ecological alga assemblages: the marine, brackish-water–marine, and brackish-water assemblages. The flora of Taganrog Bay, whose salinity varies from 3–4 to 9–10 psu, usually consisted of species of the freshwater–brackish-water assemblage from the groups of blue-green, green, and diatomaceous algae [20].

The increase in the salinity of the sea caused essential changes in the structure of phytoplankton assemblages. Significant changes were also noted in the phytoplankton productivity, which is clearly manifested in the distribution of phytoplankton and in the general a decrease in its biomass [20].

6.2 Zooplankton

The planktonic fauna of the Sea of Azov consists of representatives of different origins. Species of the freshwater, relic brackish-water, Pontian–Caspian, and marine assemblages are encountered [21].

Each of the components inhabits this or that zone of the sea depending on its dwelling and reproduction range and features the highest density under the conditions of its optimal salinity.

The eastern part of Taganrog Bay, where the salinity changes within the range of 0.5–4 psu, is inhabited by the freshwater and brackish-water Cladocera and Copepoda. Rotifers are represented by mass amounts of the armored rotifers *Brachionus plicatilus*, *Keratella curdata*, *Asplanchna*, which dwell in fresh and brackish waters.

In the central part of the bay, the salinity changes within the limits 3–7 psu. Here, the composition of plankton features a mixed character. Along with brackish-water and freshwater forms, marine forms are also encountered.

In the western part of the bay, which directly faces the open sea, marine fauna are almost completely represented by plankton; it contains all the three forms of *Acartia clausi* (the Azov, the small Black Sea, and the large Black Sea forms), *Centropages ponticus*, meroplankton, larvae of the balanus *B. improvisus*, and larvae of Gastropoda, Bivalvia, and Polychaeta.

The salinity variations in Taganrog Bay cause migration of the assemblages. For example, under salination, the marine assemblage penetrates into the eastern assemblage domain, while at desalination, brackish-water and even freshwater assemblages penetrate into the western part of the bay.

In the Sea of Azov proper, zooplankton are represented by a small number of marine species mostly consisting of copepods. The dominating forms are *A. clausi* (the Azov, the small Black Sea, and, sometimes, the large Black Sea forms), *C. ponticus*, meroplankton, etc., as well as the rotifers of the *Synchaeta* genus. Earlier, small amounts of *A. latisetosa* were encountered.

In the periods of salination of the Sea of Azov, the invasion of a group of planktonic species from the Black Sea occurred [22].

6.3 Zoobenthos

The bottom biocoenoses of the Sea of Azov are characterized by low species diversity and a rather high level of domination [23]. The principal components of the bottom fauna are represented by worms, crustaceans, bottom protists, coelenterates, and mollusks. The latter comprise up to 60–98% of the total biomass of bottom invertebrates.

The structure and biomass of the biocoenoses change over a wide range, which are defined by the combination of biotic and abiotic factors. Among the latter, most important are the salinity, the gas regime, and the properties of the sediment. The biotic factors are the zoobenthos grazing by fish and the competition for the dwelling environment between the benthos representatives.

In Taganrog Bay, zoobenthos representatives referring to the freshwater, relic brackish-water, and marine assemblages are observed. Before the runoff regulation, the Pontian–Caspian species *Hypanis colorata*, *Dreissena polymorpha*, and *Hyraniola kowalevskyi* densely inhabited the bay and some of them distributed beyond its limits to the northeastern part of the Sea of Azov. The marine species *Nereis succinea* and *Cerastoderma lamarcki* were encountered at the interface with the Sea of Azov and, partly, in the western part of the bay.

During the periods of salination, marine species widely inhabited the western and central parts of the bay, while the Pontian–Caspian species were encountered only in more desalinated regions [23].

6.4 Ichthyofauna

In the Sea of Azov proper (including Taganrog Bay) and in the northeastern part of the Black Sea, ichthyofauna are comprised of 183 species and subspecies of fish referring to 112 genera, 55 families, and 22 divisions. Of them, 50 species may be regarded as rare, 19 species are vulnerable and subjected to the hazard of disappearance, and the sturgeon *Acipenser nudiventris* has most probably become extinct. In all, 39 marine species, 8 freshwater species, 14 species of anadromous and catadromous migrants, and 42 species of inhabitants of brackish-water regions were registered [24] (Fig. 5).

In 1975–1977, when the salinity in the Sea of Azov was extremely high (in particular, in its southern part, values up to 15 psu were often noted), this region was visited, in addition to usual seasonal invaders (the anchovy *Engraulis encrasicolus maeoticus*, *E. encrasicholus ponticus*, the garfish *Belone belone euxini*, the mullet *Liza* (*Mugil*) *cephalus*, *L.*(*M.*) *auratus*, *L.*(*M.*) *saliens*, the friar *Atherina mochon pontica*, the whiting *Merlangus merlangus euxinus*, the pickarel *Spicara smaris*, and others), by the species that were extremely rare or those that have never been reported in the Sea of Azov. The first group consists of the bluefish *Pomatomus saltatrix*, the Black Sea plaices *Scophathalmus rhombus* and *Psetta maxima maiotica*, the blue stingray *Dasyatis pastinaca*, the spurdog *Squalus acanthias*, the kingfish *Sciaena umbra* and *Umbrina cirrosa*, the Black Sea salmon *Salmo trutta*, the mackerel *Scomber scombrus*, the wrasse *Crenilabrus ocelitus*, the blenny *Blenius zvonimiri*, the blanket bullhead *Aphya minuta*, the sea smelt *Atherina hepsetus*, the thick pipe-fish *Syngnathus variegatus*, and others. For the first time in the Sea of Azov, the corkwing *Crenilabrus griseus*, the rock hoppers *Blenius ponticus* and *B. sanquinolentus*, the bullheads *Pomatoschistus minutus* and *Gobius niger*, the puntazzo *Puntazzo puntazzo*, and the Mediterranean sea eelpout *Gaidropsarus mediterraneus* were encountered. Meanwhile, all the above-listed species (rare and first encountered) were met only in small amounts mostly in the southern part of the sea. Only *Gobius niger*, after penetration into a new basin, in two years became a rather common fish not only in the southern areas but also in the northern regions (off Obitochnaya and Berdyansk spits) [25]. These facts suggest that, in the years of salination of the waters of the Sea of Azov, its ichthyofauna may be naturally supplemented by Black Sea immigrants that use to dwell in the northeastern and northwestern parts of the Black Sea and can resist water temperatures lower than 3–5 °C.

Fig. 5 Commercial species of the Sea of Azov. **a** Sprat *Clupeonella delicatula*. **b** Azov Sea anchovy *Engraulis encrasicolus maeoticus*. **c** Turbot *Scophtalmus maeoticus torosus*. **d** Sturgeon *Acipenser guldenstadti*

The fauna of Taganrog Bay is twice as poor as that of the Sea of Azov proper. It includes 55 species referring to 36 genera and 16 families, and mainly consists of freshwater and brackish-water forms and migrants. Among the latter, three species are rare and six species are vulnerable and subjected to extinction [24].

The ichthyofauna of the lower reaches of the Don River, Kuban' lagoons, and other basins of the region (in addition to the Sea of Azov and the Black Sea proper) including waterlogged areas, sea bays and lagoons is represented by 134 species and subspecies referring to 90 genera and 42 families; they are joined into 17 divisions of three classes (one species of the lamprey Cephalaspidomorphi, one species of the chondral Chondrichthys and 132 species of the bony fish Osteichthyes) [24]. With respect to their ecology, 33.6% of the forms are freshwater, 26.1% are marine, 17.9% are brackish-water, 10.4% are anadromous, and 0.7% are catadromous species. Most of the fish (77%) dwell in the near-bottom layer, while the rest are pelagic species [24].

7 Introduced Species

All the alien species entered the Sea of Azov from the Black Sea via the Kerch Strait with currents or with ships. Selected species negatively affected the ecosystem, while the others enrich its flora and fauna [22, 26, 27]. Selected Black Sea fish make their seasonal migrations to the Sea of Azov for spawning and fatting. In the years with enhanced supply of the Black Sea waters, purely marine species of Mediterranean and Atlantic origins invaded the Sea of Azov. Some of them spread over the entire sea and even featured outbursts of their abundance.

7.1 Phytoplankton

Owing to the continuous water exchange, the Black Sea serves as a supplier of phytoplankton species to the Sea of Azov. Among them, one finds *Melosira moniliformis* (O.Mull), *Ag. v. moniliformis*, *Cerataulina bergonii* Perag., *Nitzschia seriata* Cl., *Coscinodiscus radiatus* Ehr., *Chaetoceros rigidus* Ostf., *Ch. affinis Laud. v. affinis*, *Pseudosolenia fragilissima* Bergon, and others. Meanwhile, the distance of their penetration and the intensity of their development are limited by the salinity of the waters of the Sea of Azov. Therefore, marine stenohaline species are encountered within a narrow band near the Kerch Strait and represent temporary components of the phytoplankton community of the sea, while the most euryhaline species extend over the entire area of the Sea of Azov [20]. In the warm seasons, phytoplankton of the freshwater assemblage from the Don and Kuban' rivers and from the lagoons of the Kuban' seaside also penetrate to the Sea of Azov. These seasonal invaders make the greatest contribution to the local species communities in the desalinated regions of Taganrog Bay and in river deltas.

Among the phytoplankton species, the first real alien species is the diatom alga *Pseudosolenia calcar-avis* (= *Rhizosolenia calcar-avis*); in 1908–1920, it penetrated into the Black Sea and, in 1924, it invaded the Sea of Azov. As early as 1924 to 1928, its abundance in the Sea of Azov reached 2.4–4.5 million cells; a similar outburst was also registered in the 1950s. At present, pseudosolenia also often dominate the phytoplankton community of the Sea of Azov in warm seasons [20]. *Pseudosolenia* appears in phytoplankton in April–March and develops throughout the entire summer; it is capable of dwelling in the Sea of Azov at a salinity of 10.4–12.9 psu and a temperature of 5.6–26 °C. Because of its large size, pseudosolenia can hardly be consumed by fish or by invertebrates, and, at its mass development, it replaces more valuable aboriginal phytoplankton species. Among other alien phytoplankton species one can mention two diatom species – *Bacteriastrum hyalinum* Laud. and *B. delicatulum* Shad – discovered by Gubina in the 1980s and episodically encountered in the Sea of Azov. Three planktonic diatom species more alien for the Sea of Azov penetrated in from the northwestern part of the Black Sea: *Cerataulina pelagica* (= *C. bergoii*), *Chaetocerus socialis*, and *Ch. tortissimus*. All of them are mass species and *Cerataulina pelagica* often dominates during the periods of increased salinity values in the Sea of Azov [20].

Of all the phytoplankton species introduced to the Sea of Azov, only *Pseudosolenia calcar-avis* may cause negative effects on the sea ecosystem during its bloom.

7.2 Zooplankton

In the period of salination of the Sea of Azov, selected plankton species entered from the Black Sea. Some of them such as *Penilia avirostris*, *Sagitta setosa*, *Paracalanus parvus*, and *Rhisosthoma pulmo* dwelled only in the regions of the maximal influence of the Black Sea waters and did not penetrate beyond the southern part of the Sea of Azov. Other species such as *Oithona nana*, *O. similis*, *Labidocera brunescens*, and the medusa *Aurelia aurita* actively developed and soon covered the entire area of the basin [22]. In the 1970s, a mass development of the Black Sea medusa *Aurelia aurita* in the Sea of Azov was observed. At present, the appearance of these kinds of species may take place only at intensive advection of the Black Sea waters and they may be regarded only as temporary invaders.

At the beginning of the 2000s, a new species that came from the Black Sea – *Acartia tonsa* – was encountered. At present, it is present in all the zooplankton samples collected in the warm season of the year both in the Sea of Azov proper and in Taganrog Bay; thus, it has formed its own reproductive population.

In August 1988, a new Black Sea invader – the ctenophore *Mnemiopsis leidyi* – appeared in the Sea of Azov; it was first encountered near the Kerch Strait in the southern and eastern parts of the sea. From that time, every spring or summer, *Mnemiopsis leidyi* has penetrated into the Sea of Azov from the Black Sea with currents to provide an outburst in its development in the summer or early autumn. Then, it became extinct at the temperature drop below 4 °C at the end of October to November [28]. *Mnemiopsis leidyi* negatively affected the ecosystem of the Sea of Azov and undermined its fish stocks.

The other ctenophore *Beroe ovata*, which spontaneously appeared in the Black Sea in 1997, invaded the Sea of Azov in September to October 1999 [29, 30] during its first bloom in the Black Sea [30]. The penetration of *Beroe ovata* into the Sea of Azov and its migration are identical to those of *Mnemiopsis leidyi*: it appears in the southern part of the sea and then expands over other areas. The main prerequisites that control the formation of the habitat of *Beroe ovata*, as well as in the case of *Mnemiopsis leidyi*, are the intensity of the Black Sea water advection, the character of the winds, and the *Beroe ovata* abundance in the prestrait area of the Black Sea. The *Beroe ovata* population ends at the onset of low seawater temperatures.

7.3 Benthos

Among the representatives of benthos, the first alien species in the Sea of Azov was *Balanus improvisus*, which appeared in the Black Sea as early as the nineteenth century. At present, this is a typical representative of the benthos of the Sea of Azov. Its maximal biomass registered in the near-shore waters of Taganrog Bay equals 7 kg/m^2 [31]. Being a fouling species, balanus causes damage to the ecology; meanwhile, its mass larvae serve as a food base for planktivorous fish. In addition, empty valves of balanus provide shelter for 18 bottom invertebrate species; some of them such as the amphipods *Gammarus locusta*, *Stenothoe monoculoides*, and *Jassa ocia* reproduce inside the valves [32]. Balanuses participate in the cleaning of the environment as filtrators.

In 1956, the rapana *Rapana venosa* entered the Sea of Azov from the Black Sea. It was encountered only in the southern regions of the Sea of Azov adjacent to the Kerch Strait, where the salinity is the highest. The low salinity of the sea seems to prevent rapana from expanding over the bulk of its area.

Among bivalve mollusks, shipworms *Teredo navalis*, were the first encountered; they invaded from the Black Sea in 1953–1955 during the increase in the salinity of the Sea of Azov. Under low salinity values, their abundance is low, but when the salinity grows, in the warm seasons, outbursts of the mass development of shipworms are possible; sometimes, they result in a rapid damage or even destruction of wooden constructions. Among other bivalve mollusk species, in 1966–1967, the brackish-water species *Mya arenaria* was encountered; it distributed over almost the entire Sea of Azov. At present, this species represents an important component of benthos, mainly in the near-shore zone and, especially, in the regions with low oxygen contents, where the biomass of mya in the biocoenosis varied from 110 to 1700 g/m^2 [33].

In April 1989, one more species alien for the Black Sea, penetrated into the Sea of Azov – *Anadara inaequivalvis*. *A. inaequivalvis* widely spread over the Sea of Azov; it is regularly encountered in the *Cerastoderma lamarki* biocoenoses and, under the decrease in the oxygen content, forms an independent biocoenosis in the near-bottom layers with a high biomass (up to 600 g/m^2) and a high species diversity (up to 30 species) over significant areas of the seafloor. At the present-day condition of the gas regime in the Sea of Azov, the development of *A. inaequivalvis* as well as that of mya is a positive phenomenon, because these mollusks populate biotopes with low oxygen content not available for other species. Their assimilation provides an enhancement of the productivity through enrichment of the fodder base for pelagic and benthivorous fish. These species represent promising objects for commercial use because they contain up to 40% of delicate meat [33].

The Black Sea mussel *Mytilus galloprovincialis* is one more bivalve mollusk species that invaded the Sea of Azov at the end of the 1950s at the salinity increase. Before the Don River runoff was regulated, only single mussel specimens were encountered; later, when the salinity increased, mussels obtained optimal conditions for their development and started to spread over the entire area of the basin [34]. Presently, mussels also play an important role in the benthic biocoenoses of the Sea of Azov.

The Dutch crab *Rhithropanopeus harisii*, which inhabited the basin including Taganrog Bay in the 1960s, refers to the old invaders to the Sea of Azov; it is encountered even in the freshwater regions of the lower reaches of the Don River. This species dwells over sandy and clayey sediments in seagrass thickets and serves as an additional food resource for benthivorous fish such as bullheads, plaices, turbots, and sturgeons. Their invasion did not hurt the ecosystem and was rather useful for it.

7.4 Fish

The habitats of selected Black Sea fish are adjacent to the Sea of Azov; however, usually, they were not encountered in it. For example, Taman' Bay and the northern part of the Kerch Strait represent the northern boundary for 11 species and 36 species more permanently dwell in these areas (the region near Feodosiya–Kerch–Novorossiisk). In the years of intensive advection of the Black Sea waters to the Sea of Azov, these species may penetrate farther replenishing the fauna of the basin, especially off the Crimean coasts [24]. In 1975–1976, in addition to the usual seasonal migrants (anchovy, garfish, mullets, friar, Black Sea whiting, pickerel, and others) species rarely encountered here (bluefish, turbot, chuco, spurdog, Black Sea salmon, mackerel, and others) penetrated into this region. For the first time in the Sea of Azov, corkwings, rock hoppers, bullheads, and eelpouts were observed. However, almost all of the above-listed species are mostly characteristic of the southern half of the sea and are observed in small amounts [25]. These facts suggest that the ichthyofauna of the Sea of Azov during the period of its salinity increase may be in a natural way significantly replenished by Black Sea immigrants that use to dwell in the northeastern part of the Black Sea and are capable of resisting water temperatures lower than 3–5 °C. Therefore, the ichthyofauna of the Sea of Azov is subjected to similar processes of the temporary replenishment by Black Sea species, and may even be tidal to a greater degree than in the case of the plant and invertebrate communities.

In the Sea of Azov, measures for intentional introduction of commercial fish species from other basins have been taken. Among them, 15 introduced species (11%) originate from fresh and brackish waters of the Far East and North America. Most of the fish introduced refer to the near-bottom species (77%), the rest (23%) are pelagic species [24]. However, the percentage of the established species is low. The most successful intentional introduction action was the introduction of the mullet *Liza haematochila* (*Mugil soiuy*) (Fig. 6). The principal prerequisite for its introduction into the Azov–Black Sea basin was its high resistance to a wide range of salinity and dissolved oxygen changes. This fish is a typical detritofage and it was believed that it would not compete with local fish species. In addition, it had to utilize organic matter, whose amount strongly increased after the runoff regulation causing suffocation phenomena [35]. The conditions in the Sea of Azov were very favorable for this mullet. It spread over almost the entire area of the sea, in lagoons, channels, and river mouths. Every year, a part of the spawning population leaves the Sea of Azov for the Black Sea. At present, the hauls of mullets are increasing. It has become an important commercial fish both in the Black Sea and in the Sea of Azov.

Among other intentionally introduced fish introduced in line with the development of the pond and lagoon-lake fish culture, one should note three buffalo species, the motleys *Aristichthys nobilis*, the white *Hypophthalmichthys molytrix* (Valenciennes, 1844), the silver carps, and the white amur *Ctenopharingodon idella*. Silver carps and amurs spread in the lower reaches of rivers and lagoons; now, they are commercial fish of the Azov

Fig. 6 Haarder *Liza haematochila* (*Mugil soiuy*)

basin [24]. The Caspian kilka *Clupeonella cultriventris ñaspia*, which is widely spread in the Volga basin, penetrated also into the lower and upper Don River and the reservoirs of the Manych system, and the Tsimlyansk Reservoir.

8 Conclusions

The regulation of the Don (1952) and Kuban' rivers (1973) and the withdrawal of the riverine runoff for reservoir filling caused negative qualitative and quantitative aftereffects in the runoff to the sea, in particular, reduced flooded and spawning areas. In the sea proper, one observes a growth in the vertical temperature and salinity gradients and an increase in the formation of oxygen-deficient zones in the near-bottom layer. In 1987, the presence of hydrogen sulphide was first registered in the lower layers of the sea.

Under present-day conditions, the amount and composition of the nutrients supplied to the sea radically changed as well as their distribution throughout the year. The major part of the particulate matter precipitates in the Tsimlyansk Reservoir, while its amount delivered to the sea in the spring and at the beginning of the summer significantly decreased; simultaneously, the supply of mineral forms of phosphorus and nitrogen reduced, while the amounts of their organic forms that are hardly assimilated by organisms sharply increased.

Meanwhile, the pollution of riverine and sea waters by different hazardous chemicals such as pesticides, phenols, and, at selected places, oil products also increased. The highest pollution degree is observed in the near-mouth regions of the Don and Kuban' rivers and in the areas adjacent to major ports. These ecological changes resulted in a sharp drop in the biological productivity of the sea. The trophic base for fish multifold reduced and the total fish hauls, especially those of valuable fish species, also decreased.

Summing up the composition of alien species in the Sea of Azov, it is important to mention that the species which could establish themselves in the sea belong to euryhaline, eurytherm, euryoxygen and stenobathno-shallow-water species. Total numbers of aliens comprised 46 species. When analyzing the ecological role of species-invaders in the Sea of Azov, one should first mention the enormous negative effect at all levels of its ecosystem, fish resources included, caused by the invasion of the predator ctenophore *Mnemiopsis leidyi*. The *Pseudosolenia calcar-avis* diatom alga, at its mass development, supplants more valuable aboriginal species of trophic phytoplankton.

The introduction of other organisms may be regarded as positive events. Benthic species such as mya and anadara widely spread over the regions with low oxygen contents unfavorable for other benthos representatives; they provided valuable food resources for benthofagous fish, while their larvae are consumed by small pelagic fish. The role of the fouling species *Balanus im-*

provisus is negative; meanwhile, its larvae are consumed by small pelagic fish. The crab *Rhithropanopeus harisii* also became an additional food source for benthofagous fish.

The ctenophore *Beroe ovata* is, beyond doubt, a useful invader; unfortunately, according to its seasonal dynamics, it appears in the Sea of Azov too late, when mnemiopsis has already reproduced, widely spread, and undermined the stocks of trophic zooplankton. No positive role of *Beroe ovata* in reducing the *Mnemiopsis leidyi* population in the Sea of Azov was noted to date. Meanwhile, its development in the Black Sea influences the size of the *Mnemiopsis leidyi* population; therefore, after the *Beroe ovata* appearance, *Mnemiopsis leidyi* enters the Sea of Azov later and its abundance is significantly lower.

References

1. Danilevskiy NYa (1871) Investigation of Fishery in Russia. Description of Fishery in the Black Sea., vol. VIII. St. Petersburg (in Russian)
2. Klossovskiy A (1890) Level and Temperature Oscillation on the Sea Shores of Azov and Black Seas. Marine Ministry, St. Petersburg (in Russian)
3. Shpindler IB, Vrangel FF (1899) Data on Hydrology of Azov and Black Seas Collected During Expeditions 1890–1891. Emperor Acad Sci, St. Petersburg (in Russian)
4. Antonov L (1926) Notes Hydrograph 51:195 (in Russian)
5. Knipovich N (1926) The Essay on Works Conducted by the Scientific and Fishery Expedition in the Black Sea and the Sea of Azov in 1925. Moscow (in Russian)
6. Knipovich N (1932) The Hydrological Studies in the Sea of Azov. Papers of the Scientific and Fishery Expedition of the Black Sea and the Sea of Azov, Issue 5. , Moscow (in Russian)
7. Knipovich N (1938) The Hydrography of Seas and Brackwaters (in application to the fisheries). Pishchepromizdat, Moscow, Leningrad (in Russian)
8. (1962) Hydrometeorological Reference Book of the Sea of Azov. Gidrometeoizdat, Leningrad (in Russian)
9. Terziev FC (ed) (1986) Hydrometeorological Conditions of the Shelf Zone of the Seas of the USSR, Sea of Azov. Gidrometeoizdat, Leningrad (in Russian)
10. Goptarev NP, Simonov AI, Zatuchnaya BM, Gershanovich DE (eds) (1991) Hydrology and Hydrochemistry of the Seas, vol V, the Azov Sea. Gidrometeoizdat, St. Petersburg (in Russian)
11. Matishov GG (ed) (2001) Environment, Biota, and Modeling of the Ecological Processes in the Sea of Azov. Kola Scientific Center of the Russian Academy of Sciences, Apatity (in Russian)
12. Matishov GG (ed) (2002) Ecosystem Studies of the Sea of Azov and the Coastal Zone. Kola Scientific Center of the Russian Academy of Sciences, Apatity (in Russian)
13. Matishov G, Abramenko M, Gargopa Yu, Bufetova M (2003) The Latest Ecological Phenomena in the Sea of Azov (second half of the 20th century). Kola Scientific Center of the Russian Academy of Sciences, Apatity (in Russian)
14. Matishov GG (ed) (2004) Complex Monitoring of the Environment and Biota of the Azov Basin, vol 6. Kola Scientific Center of the Russian Academy of Sciences, Apatity (in Russian)

15. Matishov GG (ed) (2005) Ecosystem Studies of the Environment and Biota of the Azov Basin and Kerch Strait. Kola Scientific Center of the Russian Academy of Sciences, Apatity (in Russian)
16. Matishov G, Gargopa Yu, Berdnikov S, Dzhenyuk S (2006) The Regularities of Ecosystem Processes in the Sea of Azov. Nauka, Moscow (in Russian)
17. Matishov G, Matishov D, Gargopa G, Dashkevich L, Berdnikov S, Baranova O, Smolyar I (2006) Climatic Atlas of the Sea of Azov 2006. In: Matishov G and Levitus S (eds) NOAA Atlas NESDIS 59. US Government Printing Office, Washington DC. http://www.nodc.noaa.gov/OC5/AZOV2006/start.html.
18. Zalogin BS, Kosarev AN (1999) The Seas. Mysl, Moscow (in Russian)
19. Yakushev EV, Sukhinov AI, Lukashev YuF et al. (2003) Okeanologiya 1:44 (in Russian)
20. Studenikina EI, Aldakimova AYa, Gubina GS (1999) Phytoplankton of the Sea of Azov under Anthropogenic Impact. AzNIIRKH, Rostov-on-Don (in Russian)
21. Mordukhai-Boltovskoi FD (1972) Guide of the Black and Azov Seas Fauna. Naukova Dumka, Kiev (in Russian)
22. Mirzoyan ZA, Volovik SP, Kornienko GG, Dudkin SI, Logichevskaya TV (2000) Biology of ctenophore *Mnemiopsis leidyi* in the Sea of Azov. In: Volovik SP (ed) Ctenophore *Mnemiopsis leidyi* (A. Agassiz) in the Sea of Azov and the Black Sea and its effects on ecosystems. AzNIIRKH, Rostov-on-Don, p 101 (in Russian)
23. Studenikina EI, Volovik SP, Tolokonnikova I, Frolenko LN, Selivanova EV (1998) Current characteristics of benthic communities in the Sea of Azov. In: Makarov EV, Volovik SP, Tyutina YuE (eds) The main problems of the fishery and preservation of fish stocks of the Azov-Black Sea basin. AzNIIRKH, Rostov-on-Don, p 67 (in Russian)
24. Volovik SP, Chikharev AS (1998) Antropogenic changes of ichthyofauna of the Azov basin. In: Makarov EV, Volovik SP, Tyutina YuE (eds) The main problems of the fishery and preservation of fish stocks of the Azov-Black Sea basin. AzNIIRKH, Rostov-on-Don, p 7 (in Russian)
25. Volovik SP, Dakhno VD (1983) On species composition of the Azov ichthyofauna in conditions of salinity increasing. Abstract on results of research of AzNIIRKH for 25 years. AzNIIRKH, Rostov–on-Don (in Russian)
26. Bronfman AM, Khlebnikov EP (1985) The Azov Sea. The basic reconstruction. Gidrometeoizdat, Leningrad (in Russian)
27. Volovik SP (1985) Productivity and Management of Conservation of the Azov Sea Ecosystem. Dissertation. Rostov-on-Don (in Russian)
28. Studenikina EI, Volovik SP, Mirzoyan ZA, Luts GI (1991) Okeanologiya 3:722 (in Russian)
29. Mirzoyan ZA, Volovik SP, Martynyuk ML (2002) Caspian Floating University 3:105 (in Russian)
30. Shiganova TA, Bulgakova YuV, Volovik SP, Mirzoyan ZA, Dudkin SI (2000) A new alien species *Beroe ovata* and its effect on ecosystem Azov-Black Sea basin in August–September 1999. In: Volovik (ed) Ctenophore *Mnemiopsis leidyi* (A. Agassiz) in the Sea of Azov and the Black Sea and its effects on ecosystems. AzNIIRKH, Rostov-on-Don, p 432 (in Russian)
31. Partalii EM (1980) Seasonal Changes in Biocenoces of Foaling Animals in the Sea of Azov (Taganrog Bay). AzNIIRKH, Mariupol (in Russian)
32. Zakutskiy VP (1965) Zool J 7:1092 (in Russian)
33. Studenikina EI, Frolenko LN (2003) In: Proceedings of conference on evolution of marine ecosystem under impact of invaders and artificial mortality of fauna. RAS, Min. Prom. Sci. Tekhnology RF, Rostov-on-Don, p 133 (in Russian)

34. Nekrasova MYa, Zakutskiy VP, Nekrasov SN (1980) Stocks and distribution of the mussel in the Sea of Azov. In: Biological Productivity of the Caspian and Azov Seas. Rostov-on-Don, p 101 (in Russian)
35. Volovik SP, Pryakhin YuV (1997) Azov Haarder population. In: Makarov EV, Volovik SP, Tyutina YuE (eds) The main problems of the fishery and preservation of fish stocks of the Azov-Black Sea basin. AzNIIRKH, Rostov-on-Don, p 210 (in Russian)

Hdb Env Chem Vol. 5, Part Q (2008): 91–133
DOI 10.1007/698_5_061

Published online: 30 January 2007

River Mouths

Vadim N. Mikhailov[1] (✉) · Maria V. Mikhailova[2]

[1]Faculty of Geography, Moscow State University, Vorobievy Gory, 119992 Moscow, Russia
vmikh@hydro.geogr.msu.su

[2]Water Problems Institute, Russian Academy of Sciences, Gubkina, 3, 119991 Moscow, Russia

Abstract In this paper, we consider the main peculiarities of river mouths as geographical objects forming as a result of river and sea interaction, presenting the principles of classification of river mouths and their subdivision into the parts are presented. Furthermore, structure and regime of mouths of the rivers flowing into the Black Sea and the Sea of Azov and natural and human governing factors including changes in river water runoff and sediment load, and sea level rise are considered. Results of the calculation of the present-day river water runoff and suspended sediment load of the rivers flowing into the Black Sea within six sections of the coasts (northeastern, eastern, southern, southwestern, northwestern, Crimean) are presented. Besides, we estimate total water and sediment input to the Black Sea and the Sea of Azov. Special attention is focused on the major river mouths of the region – the Danube, Dniester, Dnieper and Southern Bug, Rioni, Don and Kuban as well as natural and human-induced changes in their structure and regime in the 20th century. Lastly, problems of the influence of hydrological processes at river mouths on the environmental state of the seas are also discussed.

Keywords River · Mouth · Delta · Water runoff · Sediment load

1
Introduction

Morphological features, hydrological regime and landscape of the present-day river mouths in the coastal zones of the Black Sea and the Sea of Azov, as well as of other river mouths of the world, are formed as a result of an interaction between rivers and seas [1,2]. River mouths are very vulnerable and sensitive to changes in the riverine (natural and human-induced alternation of water runoff and sediment load) and marine factors (mean sea level variations, tides, storm surges, waves). Therefore, river mouths can be considered as efficient indicators of the large-scale changes in river and sea regime [3].

Growing scientific and practical interest in investigations of river mouths in the 20th century was related to the fact that these geographical objects became very important for human activity. At this time, activities such as agriculture, fishing and navigation are having a considerable impact on the the natural resources of mouths of numerous rivers of the world, including the region of the Black Sea and the Sea of Azov.

In addition to this, it has been found that the impact of flow regulation and water withdrawal in river basins is the most evident at mouths as terminate elements of river networks.

History of use and investigations of river mouths of the Black Sea and the Sea of Azov is very long, complex and interesting. The first descriptions of river mouths in this region belong to antique geographers and historians. An overview of these and more recent investigations is presented in [4,5].

Some information on evolution, structure and regime of the Danube, Dniester, Dnieper, Rioni, Don and Kuban river mouths at the period up to the first part of the 20th century is given in [4]. More complete and present-day characteristics of the above-mentioned river mouths are considered in [5].

Results of large-scale investigations of the Danube River mouth are discussed in [6–9]. The most comprehensive analysis of the present-day peculiarities of hydrological and morphological processes in the Danube delta is made in [8].

Main characteristics of the mouth area of the Dniester River are described in [10–12]. Natural and human-induced changes in regime of the combined Dnieper and Southern Bug rivers mouth area are studied in [13,14]. Very rapid and drastic changes of the Rioni delta due to engineering works are considered in [15].

Extensive studies were carried out at the Don and Kuban river mouths (the Sea of Azov) during the second part of the 20th century [16–19]. The most complete information on present-day environmental situation at the Don and Kuban river mouths is presented in [19].

In spite of published results of the extensive studies of the considered river mouths, a systematic overview of these investigations and analysis of present-day environmental state of these objects did not perform up to now. An attempt to draw up such overview will be made in this chapter.

2 River Mouths as Peculiar Geographical Objects

The term *mouth* as the point, where a river empties into the ocean, inland or marginal sea, lake, found wide use in geography long ago. At the same time, the term *river mouth* features a considerable uncertainty, because it is not clear without further specification whether it means a point (site) of river inflow into another water body or a certain object elongated in space and covering a certain area. As applied to the inflow of a large river to a sea or a large lake, I.V. Samoilov [4] introduced the term *river mouth area* instead of the uncertain term *river mouth*. The river mouth area covers a complex geographical object including a part of the river and a part of the sea. The term introduced by Samoilov turned out to be appropriate and found wide use among scientists.

The following definition of the river mouth area can be considered to be most optimal [1, 2]: it is a peculiar geographical object covering the region of river inflow to a receiving basin (ocean, sea, lake) and having a transitional (from fluvial to marine) hydrological regime. The river mouth area is formed under the influence of specific mouth processes, the principal of which are dynamic interaction and mixing of river water with water of the receiving basin, deposition and redeposition of fluvial and partially marine sediments resulting in the formation of a mouth fan and often a delta.

The term *mouth* is recommended to be used either in a wide sense as a reduced version and absolute synonym if the term *mouth area* (in this case, the use of the term *river mouth* is quite necessary) or in a narrow sense as applied to the point of inflow of any watercourse to another watercourse (*tributary mouth*) or to a receiving basin (*delta branch mouth*) as well as to the outlet to the open sea from a semi-enclosed coastal water body (*liman mouth, lagoon mouth, estuary mouth*).

The main features of the river mouth area (or, in a shorthand form, river mouth) as a geographical object are as follows [1, 2]:

It covers a part of the lower reach of a river (mouth reach if the river, including a delta if it is available) and a part of the coastal zone of a receiving basin (nearshore zone of the river mouth) usually with a complex and very changeable hydrographical system inherent in them. This hydrographical system is represented by a combination of *watercourses* (river within the boundaries of the mouth reach, delta branches, distributary channels and outlets, man-made canals, etc.) and *water bodies* (deltaic lakes and lakes adjoining the deltas, bogs, plavs, salt marshes, mouth lagoons, limans and estuaries, outlets, inlets, open nearshore zone, etc.).

Two types of water, i.e., river and sea water, quite different in their physical (including dynamic), chemical, and biological properties interact in the river mouth area. The river hydrological regime dominates in the mouth region of the river; however, it is under an intense impact of the receiving basin (mean sea level long-term changes, tides, storm surges, sea water intrusion). To the contrary, the hydrological regime typical of the receiving basin dominates in the nearshore zone of the river mouth; however, it is under an intense impact of the river (river flow currents, propagation of river plume with fresh and turbid waters into the sea).

The river mouth area is characterized by accumulative land forms flooded with river water and sometimes with water of the receiving basin, low-lying lands composed of interpenetrating layers of fluvial, marine, and lacustrine deposits.

The river mouth area usually has a specific soil and vegetation cover with predominately boggy and meadow soils, aquatic or hygrophilous plants; it has peculiar and rich fauna (fish, fowl, fur-bearing animals, etc). The landscape of land patches within the river mouth area sometimes differs radically from the surrounding area landscape and is azonal, particularly in geographical zones insufficiently wet (steppes, semideserts, and deserts).

River mouth areas on the coasts of oceans, seas, and large lakes are very diverse in structure and hydrological regime. This diversity depends, firstly, on the morphological peculiarities of the lower reach of the river, nearshore zone of the receiving basin and coastal zone as a whole, and, secondly, on peculiarities of the hydrological regime of the river and nearshore zone of the sea.

Therefore, it is possible to classify the river mouth areas only with the use of a complex of classification characteristics related to the structure and regime of both the river mouth parts – mouth reach of the river and the nearshore zone [1, 2].

According to the morphological characteristics, all the mouth reaches of the river can be subdivided into mouth reaches without deltas (single-branch) and deltaic mouth reaches. The latter can be multi-branch (the number of delta branches exceed 5) or with few branches (the number of delta branches does not exceed 5).

The open nearshore zone of the mouth area can be subdivided into wide or narrow, deep or shallow types. The nearshore zone is considered to be deep, when the river streamflow entering the receiving basin separates from the bottom by the seawater layer; if this streamflow occupies the whole water column, the nearshore zone is considered to be shallow.

Sometimes semi-enclosed coastal water bodies are situated between a mouth reach of the river and open nearshore zone [20]. These intermediate parts of the river mouth areas can be presented as narrow sea bays, lagoons, limans, and estuaries. These coastal water bodies are often separated from the open nearshore by coastal bars, spits, barrier islands, etc. and connected with it through relatively narrow outlets. These semi-enclosed coastal water bodies are characterized by active interaction and mixing of river and seawater.

Two morphological types of the deltas can be distinguished: filling (or bayhead) deltas, which are formed in the semi-enclosed coastal water bodies (limans, lagoons, estuaries) and protruding deltas, which are developed in the open nearshore zone.

Therefore all the river mouth areas are subdivided into the following types regarding their structure [1, 2]:

I Simple – with an open nearshore zone and without deltas;
II Semi-enclosed – with semi-enclosed coastal water bodies and without deltas;
III Semi-enclosed deltaic – with semi-enclosed coastal water bodies and with filling (or bayhead) deltas;
IV Open deltaic – with an open nearshore zone and protruding deltas.

The proposed morphological classification of river mouth areas also reveals the scheme of river mouth evolution. Only two genetic series of this evolution are possible under the condition of a relatively steady water level in the receiving basin: I → IV and II → III → IV.

Mouth areas of different types and subdivision of river mouth into the parts are shown in Fig. 1.

According to the hydrological classification [1, 2], the following indices can be used for the mouth reach of the river: water regime and river flow recharge patterns, mean water turbidity, and thermal and ice regime patterns. As for a semi-enclosed part of the river mouth and an open nearshore zone, the following hydrological characteristics can be taken into consideration: the pattern of mean sea level changes, the rate of tides, and storm surges, the dominating current, the mode of waves, water salinity, and the peculiarities of thermal and ice regime.

The boundaries of a river mouth area are defined by the intense manifestation of mouth processes (Fig. 1). The river boundary of a mouth area or the head of a mouth area are defined either by a maximum propagation distance of water level fluctuations of marine origin (tides, storm surges) into a river during low-flow period or by the point, where the river channel is divided into delta branches (a delta head or a delta apex), if water level fluctuations of marine origin do not reach this point. In the first case, there is *a part of the river mouth reach above the delta head* (or *a near-delta reach*) between the river boundary of the mouth area and the delta head. In the second case, these two components coincide. The first principle of defining the upper boundary of a river mouth area is usually applicable to river mouths with small deltas or

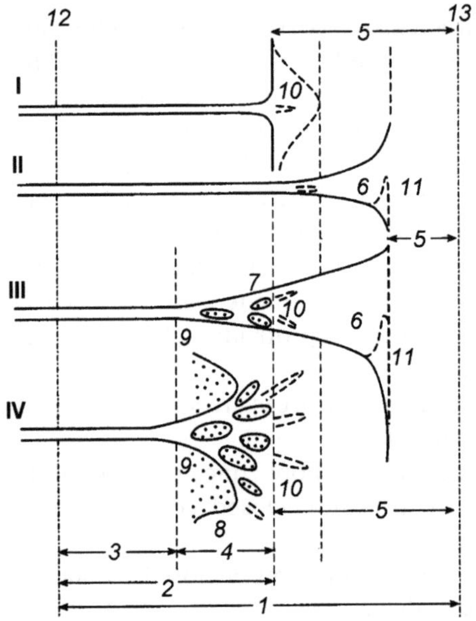

Fig. 1 Scheme of river mouth areas of different types and their subdivisions into the parts. I – simple river mouth area, II – semienclosed river mouth area without delta; III – semienclosed deltaic river mouth area with filling (or bayhead) delta; IV – open deltaic river mouth area with protruding delta. *1* – river mouth area, *2* – mouth reach of the river, *3* – part of the river mouth reach above the delta head (near-delta reach), *4* – delta, *5* – open nearshore zone, *6* – semienclosed coastal water body (narrow sea bay, liman, lagoon, estuary), *7* – filling (bayhead) delta, *8* – protruding delta, *9* – delta head (delta apex), *10* – subaquatic channel, *11* – blocking spit, *12* – river (*upper*) boundary of the river mouth area (head of the river mouth), *13* – sea boundary of the river mouth area

to mouths without any deltas, while the second principle is applicable to river mouths with large deltas.

The definition of the sea boundary of the mouth area is related to the term *mouth-mixing zone.* Water salinity within this zone increases from the salinity inherent in river water (usually 0.2–0.5‰) to the salinity of seawater (usually 10–40‰ in different seas). The salt composition of water radically changes within the mixing zone: river water of hydrocarbonate class and calcium group transforms into seawater of chloride class and sodium group.

The location of the mixing zone depends on the river water runoff, the pattern of the nearshore zone (deep, shallow, wide, narrow), the range and phase of tide, wind direction and force, and sea level fluctuations. Therefore, this zone undergoes both seasonal and short-term variations.

The mixing zone usually includes an area of the most intense interaction and mixing of river and sea water, where horizontal and vertical gradients of hydrological and hydrochemical characteristic, and primarily water salinity, are maximum. This area is called *frontal zone.*

The sea boundary of a river mouth area or the outer boundary of an open nearshore zone of a river mouth is defined by a maximum propagation distance of the outer (marine) part of the frontal zone into the sea, when river and sea waters are mixed in the surface layer. This boundary is arbitrarily defined by the location of the isohaline equalling about 90% of water salinity in the adjacent part of the sea at the river high-flow period.

3 Mouths of Rivers Flowing into the Black Sea and the Sea of Azov and Their Types and Main Features

The drainage areas of the Black Sea and the Sea of Azov equal about 2.5×10^6 km^2 and 0.56×10^6 km^2 respectively. More than 500 rivers flow into the Black Sea [21, 22]. In the first half of the 20th century, about 20 rivers emptied into the Sea of Azov [19].

All the rivers can be divided into four categories according to their drainage area (F):

- Very small rivers with a drainage area from 50 to 200 km^2;
- Small rivers with a drainage area between 200 and 2000 km^2;
- Middle-size rivers with a drainage area from 2000 to 50 000 km^2;
- Large rivers with a drainage area more than 50 000 km^2

There are only four rivers, which belong to the third category (F from 2000 to 50 000 km^2): the Caucasian rivers Kodori, Inguri, Rioni and Chorokhi. Six rivers (the Danube, Dniester, Southern Bug, Dnieper, Don and Kuban) can be classified as large rivers ($F > 50\,000$ km^2).

The majority of very small and small rivers have similar type mouths. They have, as a rule, a single channel, which can be blocked during low-flow period by coastal bars or spits composed of sand or pebble. Only several rivers of these categories have in their mouths small coastal water bodies similar to lagoons or limans and belong to the II type (semi-enclosed mouths without deltas). Mouths of this kind can be found in very low and flat coastal plains along the southwestern and northwestern parts of the Black Sea and the northern part of the Sea of Azov.

Large in size and the most important river mouths including deltas are typical of only middle-size and large rivers with sufficient water runoff and sediment load. These rivers enter the eastern and northwestern coasts of the Black Sea and the eastern coast of the Sea of Azov.

Mouth areas of the Danube, Dniester, Dnieper and Southern Bug, Don, Kuban and Rioni rivers are the most important in the region of the Black Sea and the Sea of Azov. Mouth areas of these rivers differ from other mouths not because of only their large size and a great diversity of landscape. These river mouths play the most important role in environmental state and hydrological and hydrochemical regime of the seas.

Table 1 Types of the major river mouths of the region of the Black Sea and the Sea of Azov

River mouth	Country	Type of river mouth	Type of the delta	Type of semienclosed coastal water body	Type of open nearshore zone
Danube	Romania, Ukraine	Open deltaic	Protruding, multi-branch	*	Deep
Dniester	Ukraine	Semienclosed deltaic	Filling (bayhead), with few branches	Liman	Deep
Dnieper and Southern Bug	Ukraine	Semienclosed deltaic, complex	Filling (bayhead), multi-branch	Liman	Deep
Rioni	Georgia	Open deltaic	Protruding, with few branches	*	Deep with canyon
Don	Russia, Ukraine**	Semienclosed deltaic	Filling (bayhead), multi-branch	Narrow sea bay	Shallow
Kuban	Russia	Open deltaic	Protruding, multi-branch	* <	Deep

* semi-enclosed coastal water body is absent
** only the northwestern part of the Taganrogskiy Bay

Table 2 Morphometrical characteristics of the major river mouths of the region of the Black Sea and the Sea of Azov

River mouth	Delta area, km^2	Number of delta branch mouths	Length of the main delta branch, km	Length of delta coastline, km	Area of the semi-enclosed coastal water body, km^2	Area of open near-shore zone km^2 **	Refs.
Danube	4200	16	116	190	*	1360	[5, 8]
Dniester	49	2	13	22	360	–	[5, 12]
Dnieper and Southern Bug	350	12	47	15	1000	–	[5]
Rioni	20	3	7	10	*	60	[15]
Don	540	22	38	55	5240	–	[5, 19]
Kuban	4190	7	116	150	*	500	[5, 19]

* semienclosed coastal water body is absent
** dash means lack of data

Besides, these river mouths play a great role in development of different branches of economy in Romania, Ukraine, Russia and Georgia: agriculture, fishery, navigation, etc. Main characteristics of these six river mouth areas are presented in Tables 1 and 2. More comprehensive information on these river mouths will be considered in Sects. 6 and 7.

4
Natural and Human Factors Influencing the Evolution and Regime of the River Mouths

The following are the main factors, affected by natural and human-induced changes, influencing the structure and regime of the river mouths in the region of the Black Sea and the Sea of Azov: river water runoff and sediment load and their variations, sea level changes (both the eustatic sea level changes and the relative sea level changes, which accounts for the effect of land subsidence), and sea waves.

River water runoff is responsible for many hydrological features of river mouths and their parts – deltas, semi-enclosed coastal water bodies and open nearshore zone. Water runoff influences water levels in deltas, delta inundation, water salinity at the mouths, etc. Water runoff depresses action of the storm surges on the deltas. River sediments load determines channel processes, sedimentation in a delta and nearshore, delta formation processes, the rate of delta progradation into the sea.

Changes in the water runoff and sediment load that could have impact on the river mouths under consideration during historical times (except for the last 60–100 years) could by caused by the following factors [3]:

- The natural increase in the water runoff and sediment load during cold, wet periods, which occurred in Europe in 1400–1300 and 900–300 B.C. and in 400–750, 1150–1300, and 1550–1850 A.D. The latter interval is called the Little Ice Age.
- The natural decrease in the water runoff and sediment load during warm, dry periods. An example of such periods in Europe is the period of 900–1100 called the Medieval Climatic Optimum.
- The human-induced increase in the sediment runoff caused by enhanced erosion that resulted from extensive deforestation and land ploughing in the river basins of Europe. Two main such periods are distinguished: the 5th century B.C.–the 5th century A.D. and the 16th–19th centuries.
- The human-induced decrease in the sediment load caused by the reduction of erosion that resulted from the decrease in arable lands and implementation of the program of reforestation in the river basins of Europe. One of such periods is the first half of the 20th century.

Different causes can coincide in time and strengthen their effect on the rivers. For example, in the 16th–19th centuries, the human-induced erosion caused by deforestation and land ploughing coincided with the Little Ice Age. Under these circumstances, the sediment load of rivers noticeably increased.

During the last 60–100 years [5, 21, 22], the majority of rivers under consideration were subject to the human-induced decrease in the water runoff and sediment load. The decrease in the water runoff was mainly due to the water withdrawal for economical needs and losses through evaporation from the free-water surface of reservoirs. At the Danube River mouth, however, the climate-induced increase in the water runoff over the recent 30 years has exceeded the human-induced decrease in it. Human-induced decrease of water runoff among large rivers was the greatest in the Dnieper, Don and Kuban rivers and comprised 1.2–1.3 times.

Due to accumulation of sediments in reservoirs, the sediment load of the majority of rivers decreased much more than their water runoff. A maximum decrease in sediment load has occurred in the Kuban, Don, Rioni, and Danube rivers (by a factor of 5.6, 2.8, 2.4 and 1.5, respectively).

Drastic changes in river water and sediment distribution between delta branches in the Danube and Rioni deltas took place after the artificial canalization, deepening and straightening of some channels.

Changes in the river water runoff and sediment load at the major river mouths of the region will be considered in detail in Sects. 5–7.

In the 20th century, an appreciable rise in the water levels of the Black Sea was recorded. At the gauging stations, the sea level rise is measured with respect to the land surface; hence, it is called the relative sea level rise (RSLR).

The RSLR substantially affects the river mouths. It results in flooding in coastal areas, increase in coastal water bodies, backwater in delta branches, and a certain intensification of the wave erosion of delta coastline.

The RSLR consists of two components: the eustatic rise caused by an increase in the water volume of the sea (due to changes in the ratio between the input and output components of the water balance of the sea) and the land subsidence. In many river mouths, the effect of the second factor prevails.

The eustatic rise in the levels of the Black Sea and the Sea of Azov is the consequence of the ocean level rise and the positive water balance of these seas.

Land subsidence in deltas can be associated with two factors: (1) tectonic subsidence of the seabed and its peripheral parts and (2) gradual thickening (and dehydration) of young deltaic deposits under the pressure of their own weight. Extraction of groundwater and gas can also cause the subsidence of deltaic deposits. Worldwide, river deltas are the main areas of land subsidence. It can reach 10 mm $year^{-1}$ (the Mississippi River mouth) and more [3].

The current rate of the eustatic rise in the world ocean level is estimated at 1–2 mm $year^{-1}$ (about 1.5 mm $year^{-1}$ on average). According to [23], in the

20th century, the average rate of the level rise in the world ocean as a whole was equal to that in the Atlantic Ocean, 1.7 mm $year^{-1}$.

The mechanism of the "transfer" of the eustatic component of the world ocean level rise into the Black Sea and the Sea of Azov is an intricate and poorly understood process. It is clear that in these seas the eustatic rise differs from that in the ocean as a whole. The main cause is relative isolation of these seas from the ocean. Thus, the volumetric changes in the sea water levels are likely to be more susceptible to changes in the ratio between the components of the water balance of the sea (river runoff, precipitation, and evaporation) than to the eustatic rise in the ocean level.

In 1923–1985, the Black Sea was characterized by an increase in its water volume, the rate of which attained + 1.5 km^3 $year^{-1}$ [3]. If one converts this value into the sea level change for the same period (with the sea area being 423×10^3 km^2), then the rate of the eustatic sea level rise will be 3.5 mm $year^{-1}$. By different assessments, the eustatic (volumetric) increase in the Black Sea level comprises 2.0–4.0 mm $year^{-1}$ [3, 8, 24].

The above data evidence that in the 20th century the eustatic rise in the Black Sea level markedly exceeded the eustatic rise of the level in the Atlantic Ocean (1.7 mm $year^{-1}$ [23]) and in the Mediterranean Sea (1.1–1.3 mm $year^{-1}$ [3]). In most publications, the rise in the Black Sea level is explained by some specific features of its water balance in this period: an increased river water runoff into the sea (first of all, that of the Danube River), abundant precipitation over the Black Sea's water surface, and decreased evaporation.

The eustatic rise of the level in the Sea of Azov is less than in the Black Sea and equals 2.0–2.5 mm $year^{-1}$ [25, 26].

It is impossible to accurately assess the contribution of the subsidence of deltaic deposits to the RSLR, because the amount of subsidence greatly varies (both in space and time) within the same delta. The only way to assess this contribution is to compare the actual level rise (RSLR) with the average rates of the eustatic level rise of the sea.

The rate of the RSLR in the Black Sea and the Sea of Azov and at their river mouths (Table 3) varies over a wide range [3, 8, 24–26]. These data show that actual eustatic sea level rise increased with time and by the beginning of the 21st century reached in the Black Sea 4 mm $year^{-1}$. Besides, these data confirm the assumption that the land subsidence plays very important role in the RSLR in river mouth areas and can reach 3–6 and even 12 mm $year^{-1}$. The most values of the RSLR and subsidence are typical of the coastal parts of the Danube and Rioni deltas (Table 4).

The open nearshores of the majority of the rivers under study are deep. Mean slope of the open nearshore bottom at the Rioni mouth is 6.5‰ ("old" delta) and 14.3‰ ("new" delta) [3]. Less deep are the open nearshores of the Danube and Kuban rivers [3, 5]. The open nearshores of this kind are typical of the Dniester, Dnieper and Don mouths [5].

Table 3 Approximate data on the relative sea level rise (RSLR) in the Black Sea and the Sea of Azov

Water object	Gauging station	Period	Average RSLR, mm $year^{-1}$	Refs.
Black Sea	Odessa	1950–1973	4.5	[24]
		1974–1995	6.6	[24]
	Batumi	1958–1996	7.5	This study
Sea of Azov	Taganrog	1882–1957	0.1	[25]
		1958–1998	2.1	[25]
	Eisk	1958–1998	2.2	[25]
Danube mouth	Primorskoye	1958–1984	4.0	[8]
		1958–1997	5.7	[3]
		1971–2003	3.3	[8]
		1985–1997	7.2	[8]
	Ust'-Dunaysk	1985–2003	16.2	[8]
	Prorva	1971–2003	4.0	[8]
		1985–2003	10.2	[8]
Dnieper and Southern Bug mouth	Kherson	1945–1997	7.8	This study
	Novaya Odessa	1945–1988	4.9	This study
	Nikolaev	1945–1997	3.3	This study
	Ochakov	1945–1997	4.0	This study
Rioni mouth	Poti	1958–1996	7.6	[3]
Don mouth	Azov	1881–1957	0.3	[25]
		1958–1998	3.5	[25]
Kuban mouth	Temryuk-port	1910–1973	3.6	[26]
		1974–1998	5.4	[26]
	Slobodka	1962–1973	5.5	[26]
		1974–1998	10.4	[26]

The relative steepness of these open nearshores can result in a decelerated progradation of deltas into the seas (even at constant sediment load) and the marked effect of sea waves.

The effect of waves on the seashores, including the delta coasts, depends on the energy of waves. Waves are capable of destroying the deltaic deposits in the periods of low sediment load and to slow down the progradation of deltas into the sea, as well as to transport sandy sediments along the delta coastlines and to form beaches and sand coastal bars. In some cases, separate deltaic lobes were completely destroyed after the water and sediment input to the deltas had ceased (the deltas of the Danube, Rioni rivers).

No reliable data on the regime of waves in the nearshores under study are available in publications. Rough assessments of the rate of sea waves can be made using [3, 7, 24]. On the whole, moderate waves prevail in the nearshores under study. Mean wave height in the Danube and Rioni nearshore zones equal 0.6 and 0.2 m, respectively [3]. In 1960–1988, decrease of the wind speed

Table 4 Average water runoff and suspended sediment load of the rivers entering the Black Sea along the northeastern coast within Russia [21, 22]

River	Drainage area, km^2	Annual water discharge, $m^3\,s^{-1}$	Specific water discharge, $l\,s^{-1}\,km^{-2}$	Water runoff, $km^3\,year^{-1}$	Suspended sediment load, $10^3\,t\,year^{-1}$
Sukko	89.2	0.69	7.7	0.022	12.3
Dyurso	53.7	0.45	8.4	0.014	1.32
Ozereika	52.5	0.35	6.6	0.011	7.40
Tsemes	82.6	0.51	6.2	0.016	11.3
Mezyb	194	3.86	19.9	0.122	67.2
Dzhankhot	49.0	1.14	23.3	0.036	20.5
Pshada	358	9.82	27.4	0.310	56.8
Vulan	278	6.36	22.9	0.200	59.0
Dzhubga	100	1.52	15.2	0.048	30.5
Shapsukho	303	7.03	23.2	0.222	113
Nechepsukho	225	4.59	20.4	0.145	87.7
Tu	59.1	1.36	23.0	0.043	24.8
Nebug	73.3	2.53	34.5	0.080	42.3
Agoi	91.8	3.39	36.9	0.107	56.0
Tuapse	352	12.8	36.3	0.404	111
Shepsi	57.5	1.93	33.5	0.061	31.4
Ashe	282	12.4	43.9	0.390	57.0
Psezuapse	290	15.4	53.7	0.486	91.5
Shakhe	553	36.8	66.5	1.161	211
Dagomys	103	2.06	20.0	0.065	44.5
Sochi	296	16.1	54.3	0.508	101
Matsesta	67.5	2.28	33.8	0.072	31.3
Khosta	93.5	4.90	52.4	0.155	31.5
Kudepsta	87.1	3.39	38.9	0.107	38.2
Mzymta	885	49.5	55.9	1.562	258
Total	5076	201	39.6	6.35	1596

and correspondingly of the wave energy in the northwestern nearshore zones were observed [24]. Waves tend to strengthen in autumn and winter.

5 River Water and Sediment Input to the Black Sea and the Sea of Azov

5.1 General Considerations

Values and variations in river water runoff and suspended sediment load of the rivers flowing into the Black Sea and the Sea of Azov (especially of small and very small rivers) have not been estimated satisfactorily [5, 19, 21, 22].

Systematic and reliable hydrological observations in lower parts of many rivers began only recently.

Water runoff and sediment load of many rivers in the 20th century were subject to strong human impact because of water withdrawal, dam and reservoir construction, and river flow regulation.

Present-day average data on river water runoff and suspended sediment load will be presented below. This information is based mainly on [19, 21, 22] and own investigations of the authors of this chapter.

Available data on characteristics of river water runoff and sediment suspended load before river flow regulation will be also presented.

5.2
River Water and Sediment Input to the Black Sea

Coastal zone of the Black Sea, in the context of the river input, can be divided into the following six sectors: northeastern (coast of Russia), eastern (coast of Georgia), southern (coast of Turkey), southwestern (Bulgarian coast), northwestern (coast of Romania and Ukraine), coast of Crimea (Ukraine).

Along the Russian coast, the majority of rivers flowing into the Black Sea are small ones (Table 4); the largest among them is the Mzymta River, which has the drainage area of 885 km^2, average annual water runoff of 1.56 km^3 and suspended sediment load of 0.26×10^6 t. The annual water runoff of the most of the rivers is less than 0.15 km^3. Only the Pshada, Tuapse, Ashe, Psezuapse, Shakhe and Sochi rivers carry out to the sea more than 0.3 km^3 $year^{-1}$ of water.

The rivers in the northeastern part of the Black Sea are slightly affected by human-induced changes.

The overall amounts of water runoff and suspended sediment load into the Black Sea from rivers within Russia are about 6.4 km^3 $year^{-1}$ and 1.6×10^6 t $year^{-1}$ respectively.

More water abundant rivers flow into the Black Sea in its eastern part (Table 5). The overall annual volume of river water inflow to the Black Sea from Georgia (including small rivers) is 45.7 km^3. Of this, almost three quarters comes from the major rivers: Bzyb' (3.79 km^3), Kodori (4.17 km^3), Eritskali Canal (the conduit for the canalized and controlled the Inguri River) (3.15 km^3), Rioni (13.38 km^3) and Chorokhi (8.71 km^3).

Several rivers in West Georgia are regulated by reservoirs and are used in the production of hydroelectricity. However, this control has insignificant impact on the Rioni or Gumista rivers, and their total average annual water runoff is practically unchanged. At the Inguri mouth, however, the average water discharge decreased since 1976 from 165 to 39.5 m^3 s^{-1} due to water diversion to the Eritskali Canal with annual water discharge of 100 m^3 s^{-1}.

The overall amount of the suspended sediment load of the considered rivers is significant and equal to 18.6×10^6 t $year^{-1}$ (Table 5). At present the

Table 5 Average water runoff and suspended sediment load of the rivers entering the Black Sea along the eastern coast within Georgia [21, 22]

River	Drainage area, km^2	Annual water discharge, $m^3\,s^{-1}$	Specific water discharge, $l\,s^{-1}\,km^{-2}$	Water runoff, $km^3\,year^{-1}$	Suspended sediment load, $10^3\,t\,year^{-1}$
Psou	421	19.2	45.6	0.606	158
Khashupse	200	9.5	47.5	0.300	80.5
Zhove-Kvara	72	6.11	84.8	0.193	53.7
Bzyb	1510	120	79.5	3.79	767
Mchishta	169	7.71	45.6	0.243	20.2
Khipsta	166	9.76	58.8	0.308	34.4
Aapsta	243	10.8	44.4	0.341	37.7
Gumista	576	33.3	57.8	1.051	264
Besleti	81.5	3.53	43.3	0.111	12.0
Kelasuri	220	13.2	60.0	0.416	84.2
Majarka	114	5.1	44.7	0.161	15.9
Kodori	2030	132	65.0	4.166	1295
Tumish	62.2	1.64	26.3	0.052	3.35
Dgamysh	120	4.32	36.0	0.136	9.0
Tskhenistskali	61	1.61	26.4	0.051	3.35
Mokva	336	18.1	53.9	0.571	46.8
Galidzga	483	29	60.9	0.928	94.7
Okumi	265	14.5	54.7	0.458	34.5
Eristskali Canal	–	100	–	3.15	–
Inguri	4060	39.5 165*	– 40.6*	1.247 5.207*	450 2700*
Khobi	1340	60.1	44.8	1.898	221
Rioni	13 400	424	31.6	13.38	6020 7590**
Supsa	1130	50.1	44.3	1.581	246
Natanebi	657	24.5	37.3	0.773	146
Kintrishi	291	16.7	57.4	0.527	22.3
Chakvistskali	172.6	12.5	72.4	0.394	19.0
Korolistskali	55	3.8	69.1	0.200	8.30
Chorokhi	22 100	276	12.5	8.71	8440
Total	50 335	1447	28.7	45.7	18 587

* before the Eristskali Canal construction
* before river flow regulation

most part of this sediment load falls on the Kodori ($1.29 \times 10^6\,t\,year^{-1}$), Rioni ($6.02 \times 10^6\,t\,year^{-1}$) and Chorokhi ($8.44 \times 10^6\,t\,year^{-1}$) rivers. Suspended sediment load of the Rioni River due to deposition of sediments in reservoirs slightly decreased (see Sect. 6.4).

The water runoff of the rivers in Turkey was computed using a method of water balance [21, 22] (Table 6). The total volume of water contribution to the

Black Sea from Turkish rivers amounts to 37.7 km^3 $year^{-1}$ (not including the border Chorokhi and Veleka rivers). More than one-half of this inflow falls on major rivers, the Yesil-Irmak, Kizil-Irmak, Filyos and Sakarya.

Many rivers in Turkey are used for irrigation and other water needs, and the Yesil-Irmak, Kizil-Irmak, Riva, Karasu, Gulluk and Abdal rivers are subject to significant irreversible water losses. The annual volume of irreversible water losses can be up to 3–5 km^3. In natural conditions, the volume of water runoff would be around 42 $km^3$$year^{-1}$. Presently, the total average suspended sediment load of the rivers along this sector of the coast is about 13.6×10^6 t $year^{-1}$. More than one-half of this value (55%) falls on the Filyos and Sakarya rivers.

On the coast of Bulgaria the majority of the rivers are small or very small. The largest river is the Kamchea with present-day average annual water runoff of 0.6 km^3 and suspended sediment load of 4.6×10^6 t $year^{-1}$.

Annual river water inflow of all the rivers directly to the sea is 1.2 km^3; if the water drains from rivers flowing into the coastal water bodies is included,

Table 6 Average water runoff and suspended sediment load of the rivers entering the Black Sea along the southern coast within Turkey [21, 22]

River and parts of the coast	Drainage area, 10^3 km^2	Specific water discharge, l s^{-1} km^{-2}	Water runoff, km^3 $year^{-1}$	Suspended sediment load, 10^3 t $year^{-1}$
From the Chorokhi to the Harsit	9.5	19.0	5.70	1270
Harsit	3.5	10.0	1.10	510
From the Harsit to the Yesil-Irmak	11.0	13.0	4.51	1440
Yesil-Irmak	36.1	4.65	5.30	330 / 12 500*
From the Yesil-Irmak to the Kizil-Irmak	2.5	12.0	0.95	298
Kizil-Irmak	78.6	2.38	5.90	440 / 16 700*
From the Kizil-Irmak to the Filyos	9.9	10.0	3.12	1020
Filyos	13.1	7.0	2.90	3700
From the Filyos to the Sakarya	3.6	10.0	1.14	370
Sakarya	56.5	3.15	5.60	3800 / 4600*
From the Sakarya to the Rezovska	4.8	9.60	1.45	420
Total	229.1	5.21	37.67	13 598

* before river flow regulation

the total volume will be 1.8 km^3. In sum, annual suspended sediment load of considered rivers is close to 0.75×10^6 t year^{-1} (Table 7).

The most water abundant rivers (the Danube, Dniester, Southern Bug, Dnieper) empty into the Black Sea within its northwestern part (Table 8). Their present-day total average annual water runoff and suspended sediment load are equal to about 263 km^3 and 41.5×10^6 t respectively. The Danube is the second river in Europe in length, drainage area and water runoff after the Volga River and the first in sediment load. The water runoff of the Danube River is markedly subject to climatic changes. In spite of the water withdrawal and flow regulation the Danube water runoff in the second part of the 20th century increased (Table 8) due to positive changes in precipitation over the river watershed. In contrast, the sediment load of the Danube River strongly decreased after construction of several large reservoirs, including the Iron Gate-I in 1971 (see Sect. 6.1).

Water runoff and suspended sediment load of the Dniester and Dnieper rivers also decreased after construction of reservoirs (Table 8) (see Sects 6.2 and 6.3).

Table 7 Average water runoff and suspended sediment load of the rivers entering the Black Sea along the southwestern coast within Bulgaria [21, 22]

River	Drainage area, km^2	Annual water discharge, m^{3}s^{-1}	Specific water discharge, l s^{-1}km^{-2}	Water runoff, km^3 year^{-1}	Suspended sediment load, 10^3 t year^{-1}
Rezovska	183.4	0.79	4.30	0.025	17.4
Veleka	995	8.76 9.41*	8.80 9.45*	0.276 0.297*	65 78*
Karaagach	224.3	0.96	4.28	0.033	21.3
Dyavolska	133.2	0.57	4.28	0.018	12.7
Ropotamo	248.7	1.17	4.70	0.037	23.6
Akheloi	141.0	0.61	4.33	0.019	13.4
Khadzhiiska	355.8	1.53	4.30	0.048	33.8 46.0*
Dvoinitsa	478.8	2.06	4.30	0.065	45.5
Perperidere	58.2	0.25	4.29	0.008	5.5
Shkorpilovska	78.7	0.34	4.32	0.011	7.5
Kamchea	5358	19.2 27.7*	3.58 5.17*	0.606 0.874*	462 1122*
Kranevska	84.5	0.36	4.26	0.011	8.0
Batova	338.8	0.73	2.15	0.023	35.4 40.8*
Total	8678	37.3	4.30	1.18	751.1

* before river flow regulation

Table 8 Average water runoff and suspended sediment load of the rivers entering the Black Sea along the northwestern coast within Romania and Ukraine

River	Drainage area, 10^3 km^2	Annual water discharge, m^3 s^{-1}	Specific water discharge, l s^{-1} km^{-2}	Water runoff, km^{3}year^{-1}	Suspended sediment load, 10^6 t year^{-1}	Refs.
Danube	817	6590 6320*	8.1 7.7*	208 199*	36.3 52.4*	[8]
Dniester	72.1	288 320*	4.0 4.4*	9.1 10.1*	4.1 5.5*	[5,11, 12]
Southern Bug	63.7	69	1.1	2.18	0.20	[5]
Ingul	9.7	18.5	1.9	0.58	0.126	[21, 22]
Dnieper	503	1375 1683*	2.7 3.3*	43.4 53.1*	0.80 2.10*	[5, 21, 22]
Total	1465	8340 8410*	5.69 5.74*	263.2 265.0*	41.5 60.3*	

* before river flow regulation

Before the river flow regulation in the first part of the 20th century total water runoff and suspended sediment load of all the rivers in this coastal sector equalled about 265 km^3 year^{-1} and 60.3×10^6 t year^{-1} correspondingly. Combined decrease in the average water runoff comprised only about 2 km^3 year^{-1}, which is less than 1%. Decrease in the suspended sediment load of the rivers is more significant – 19×10^6 t year^{-1} that forms 31%.

The total water runoff of the small mountain rivers of Crimea (Ukraine) does not exceed 0.3 km^3 year^{-1}. Their suspended sediment load is about 132×10^6 t year^{-1} (Table 9).

Results of the calculation of the river water and sediment contribution to the Black Sea are given in Table 10. Total drainage area of all the considered rivers approximately equals 1.8×10^6 km^2. At present, values of the water runoff and suspended sediment load of all the rivers are equal to 354.5 km^3 year^{-1} and 76.2×10^6 t year^{-1} correspondingly.

Some amounts of sand riverine sediments are transported in channels in the form of sand ripples, waves and dunes formed on the river bed under the influence of the near-bottom currents. This sediment load (called bedload) usually comprises about 10% of the suspended sediment load. Taking into account this fact, we find that the total river sediment input to the Black Sea may range up to 84×10^6 t year^{-1}.

About 80% of the total water runoff of all the rivers flowing into the Black Sea falls on five the most water abounded rivers: the Danube (208 km^3 year^{-1} or 59%), Dnieper (43.4 km^3 year^{-1}, 12%), Rioni (13.38 km^3 year^{-1}, 3.8%), Dniester (9.1 km^3 year^{-1}, 2.6%) and Chorokhi (8.71 km^3 year^{-1}, 2.4%). Only three rivers are responsible for the most part (67%) of the combined river suspended

Table 9 Average water runoff and suspended sediment load of the rivers entering the Black Sea along the Crimean coast (Ukraine) [21, 22]

River	Drainage area, km^2	Annual water discharge, $m^3 s^{-1}$	Specific water discharge, $l s^{-1} km^{-2}$	Water runoff, $km^3 year^{-1}$	Suspended sediment load, $10^3 t year^{-1}$
Al'ma	633	1.40	2.2	0.044	44.3
Kacha	110	1.32	12.0	0.042	12.1
Kokozka	836	1.17	1.4	0.037	25.9
Bel'bek	270	2.16	8.0	0.068	32.4
Chernaya	47.6	1.47	30.9	0.046	0.57
Derekoika	49.7	0.48	9.7	0.015	2.78
Ulu-Azen'	64.8	0.56	8.6	0.018	6.48
Demerdzhi	53	0.13	2.4	0.004	4.66
Taraktash	153	0.06	0.4	0.002	2.65
Total	2217	8.75	3.95	0.276	132

Table 10 Present-day average water runoff and suspended sediment load of the rivers flowing into the Black Sea

Sector of the coast	Drainage area, $10^3 km^2$	Annual water discharge, $m^3 s^{-1}$	Water runoff, $km^3 year^{-1}$	Suspended sediment load, $10^6 t year^{-1}$
Northeastern (Russia)	5.1	201	6.4	1.6
Eastern (Georgia)	50.3	1450	45.7	18.6
Southern (Turkey)	229	1190	37.7	13.6
Southwestern (Bulgaria)	8.7	37.3	1.2	0.75
Northwestern (Romania and Ukraine)	1465	8340	263.2	41.5
Crimea(Ukraine)	2.2	8.8	0.28	0.13
Total	1760	11 230	354.5	76.2

sediment contribution to the Black Sea: the Danube (36.3×10^6 t year^{-1} or 48%), Chorokhi (8.44×10^6 t year^{-1}, 11%) and Rioni (6.02×10^6 t year^{-1}, 8%).

5.3 River Water and Sediment Input to the Sea of Azov

At the present time, water and sediments only of two rivers (the Don and Kuban) run into the Sea of Azov (Table 11). The region around these rivers

Table 11 Average water runoff and suspended sediment load of the Don and Kuban rivers

River	Drainage area, 10^3 km^2	Annual water discharge, m^3 s^{-1}	Specific water discharge, l s^{-1} km^{-2}	Water runoff, km^3 year^{-1}	Suspended sediment load, 10^6 t year^{-1}
Don	422	684	1.62	21.6	1.60
		873*	2.07*	27.6*	4.54*
Kuban	57.9	353	6.10	11.1	1.52
		411*	7.10*	13.0*	8.49*
Total	479.9	1037	2.16	32.7	3.12
		1284*	2.68*	40.6*	13.03*

* before river flow regulation

has experienced strong human impact (see Sect. 7). Until the middle of the 20th century, many small rivers entered to the Sea of Azov along its northern coast and between deltas of the Don and Kuban rivers [19]. For example, average annual discharges of the small Obitochnaya, Berda, Kalmius and Mius rivers (northern coast) were equal to 1.63, 2.53, 12.4 and 11.8 m^3 s^{-1} respectively [19]. Water discharges of the small Kagalnik, Eya, Chelbasa, Beisug, Kirpili rivers (eastern coast) were estimated as 1.19, 2.45, 3.41, 4.81 and 2.06 m^3 s^{-1} correspondingly [19]. During the middle of the 20th century, most of the small rivers were regulated; many small pools and reservoirs were constructed. Presently, all above-mentioned small rivers practically do not empty into the sea [19].

In the middle of the 20th century, the amount of water and sediment input of the Don and Kuban rivers drastically decreased after river flow regulation (Table 11). At present time, average combined water inflow of the Don and Kuban rivers to the Sea of Azov amounts to about 32.7 km^3 year^{-1}. The total suspended sediment load of two rivers at the heads of their deltas now equals about 3.1×10^6 t year^{-1}.

6 Major River Mouths of the Black Sea

6.1 Danube River Mouth

The Danube River mouth area is the largest among other river mouths in the region of the Black Sea and the Sea of Azov. This mouth belongs to the open deltaic type. The Danube mouth area includes the near-delta river reach, the delta itself, the head of which is coincident with the river bifurcation into two largest delta branches – the Chilia (left) and the Tulcea (right), and the open and deep nearshore zone (Fig. 2).

The Chilia branch (116 km in length) is a continuation of the Danube River and the major branch of the delta. It forms two internal and one external or marine (Chilia) deltas. The majority of side secondary branches of the internal deltas have died out. The main branches in the Chilia delta are the Ochakovskiy (left) and Starostambul'skiy (right).

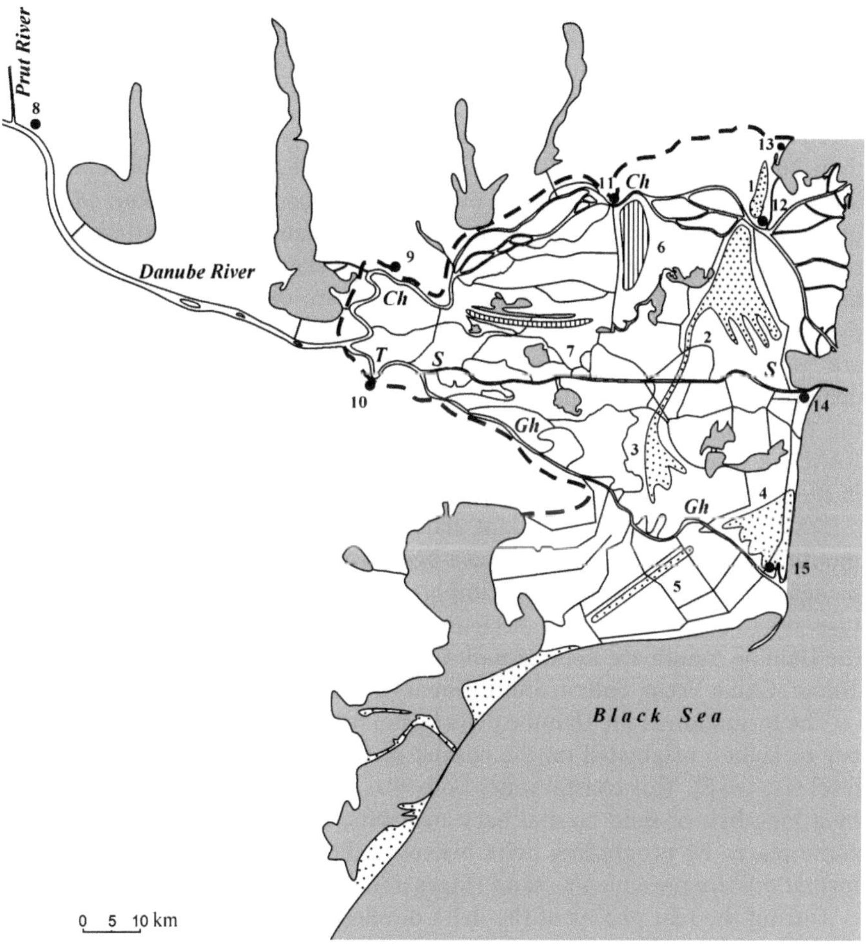

Fig. 2 Scheme of the Danube River mouth. Main delta branches: *Ch* – Chilia, *T* – Tulcea, *S* – Sulina, *Gh* – Gheorghe. Sand ridges (ancient coastal bars): 1 – Jibrieni, 2 – Letea, 3 – Caraorman, 4 – Saraturile, 5 – Crasnicol. Clay ridges: 6 – Chilia, 7 – Stipoc. Settlements: 8 – Reni, 9 – Izmail, 10 – Tulcea, 11 – Chilia, 12 – Vilkovo, 13 – Primorskoye, 14 – Sulina, 15 – Sfintu Gheorghe. *Dash line* is land boundary of the delta

The Tulcea branch (17 km in length) divides into the Sulina branch (76 km) and Gheorghe branch (77 km).

The present-day area of the Danube delta equals 4200 km^2 [8]; this delta is the third in Europe in size after the deltas of the Volga and Terek rivers.

The Danube delta is located in the territory of two countries, Romania in the south and Ukraine in the north. Romanian and Ukrainian parts of the delta comprise about 80% and 20% of the total delta area respectively.

The multi-branch Danube delta represents a unique geographical object with a large variety of landscapes including wide periodically inundated floodplain, swamps, and plavs, numerous large and small lakes, sand ridges inside the delta, sand beaches along the delta coastline, forests, gardens, cultivated lands, etc. Dense channel network consists of main delta branches, numerous natural distributary channels, artificial navigation, irrigation and drainage canals, etc.

The Danube delta is known because of its very high biodiversity. More than 1600 species of plants have been recorded here. In the Danube delta there is the largest compact zone of reed, which occupies 1560 km^2 [8]. Part of the reed cover is represented by so-called "plaur", or a floating layer of reed roots on the surface of several lakes. The Danube delta is a habitat for more than 3,500 species of fauna, including rare or valuable ones: pelican, pygmy cormorant, swan, red-breasted goose, migratory sturgeons, etc. There are two large and well-known Biosphere Reserves on the territories of the delta belonging to Romania and Ukraine. The Danube delta was included in the World Natural Heritage list under the World Heritage Convention, in the RAMSAR Convention List as a wetland zone of international importance and in the "Man and Biosphere" UNESCO Program.

Since ancient times, abundant natural resources of the Danube River mouth (water, land, biological) have been used for fishing, agriculture, and navigation. At present, the most important intensive land-users are agriculture, port construction, and pisciculture. Now the main towns and ports at the Danube mouth are Reni, Izmail, Chilia, Vilkovo, Ust'-Dunaisk (Ukraine), Tulcea, Chilia Vechi, Sulina, Sfintu Gheorghe (Romania).

The formation of the Danube delta began 6,000 to 7,000 years ago in a large bay or lagoon originated on the coastal plain after the postglacial Black Sea level rise [6–8]. This coastal water body was partially separated from the sea by a long belt of sand coastal bars and spits. Later this belt was broken in some places by prograding delta branches. Presently, these ancient coastal formations are presented as sand ridges and dunes inside the delta (Fig. 2).

During the first phases of the delta development, the main river flow ran to the sea along the southern boundary of the delta (the Gheorghe branch at present). Then the ancient Sulina branch was formed. Only by the 18th century, after filling up the shallow inland parts of the former bay by river sediments, formation of the "marine" protruding Chilia delta in the open nearshore zone started. This event took place around 1740s. The most active progradation of this delta into the Black Sea was observed at the end of the 19th and at the beginning of the 20th century (Fig. 3).

Over the 19th century, the area and mean length of the Chilia delta increased from 80 to 220 km^2 and from 8 to 15 km respectively [8]. By the beginning of the 21st century, its area and mean length reached 360 km^2 and 20 km correspondingly. The most intensive growth of the Chilia delta was in 1872–1883 (4.3 $km^2\ year^{-1}$), 1884–1894 (3.6 $km^2\ year^{-1}$), and 1894–1922 (2.5 $km^2\ year^{-1}$) [8, 27]. Recently, rate of the progradation of the Chilia delta into the sea decreased (up to 1.5 $km^2\ year^{-1}$ in 1958–1972 and 0.36 $km^2\ year^{-1}$ in 1992–2002) [8]. This process was connected with progradation of the delta to large marine depths, reduction in the sediment load of the Danube River, some redistribution of river water and sediments from the Chilia branch into the adjacent Tulcea branch at the delta head. Besides, retreat of the Danube delta coastline as a whole was strengthened by abrasion [8, 28]. Abrasion was especially marked along the coast in the Romanian part of the Danube delta.

During the progradation into the sea, the Chilia delta has gone through a number of phases of development [8]. In the first phase (up to the beginning of the 19th century), it had less than 10 branch mouths and very intended coastline. From 1830 to 1922, the number of branch mouths increased to 50–60. After that, the number of branch mouths steadily decreased to 19 in 1957, 15 in 1980, and 12 at present time. Simultaneously the delta coastline has become smoother (Fig. 3). Along this coastline, sand beaches were formed and their length increased from 4 km in 1930 to 15 km in 1957 and 36 km in 2002 [8].

The Danube River contributes to its delta about 200–210 km^3 water per year. New estimates of long-term and seasonal changes in the river water runoff are presented in [8].

Over the period of 1840–2002, the water runoff was subject to periodical changes without any marked trends (Fig. 4). Long-term average water dis-

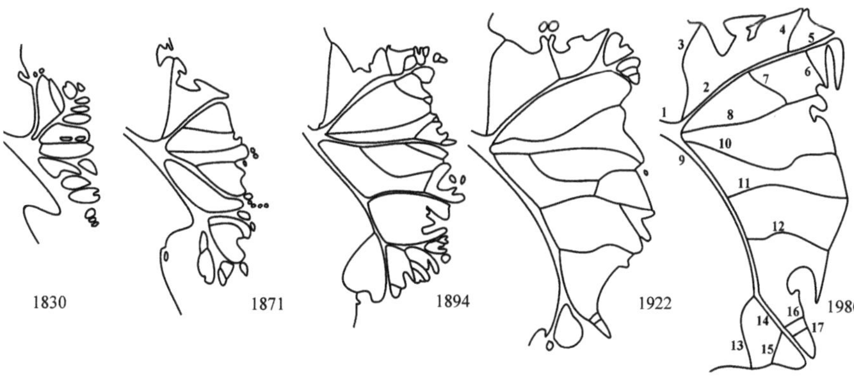

Fig. 3 Scheme of the evolution of the Chilia delta. The delta branches and their ordinal numbers are listed in Table 13

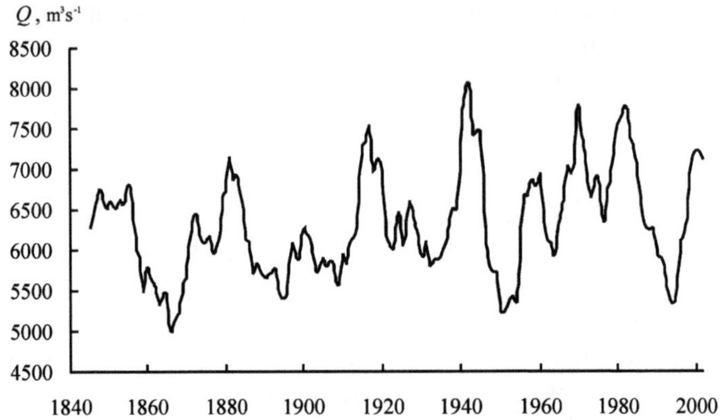

Fig. 4 Long-term changes in the annual water discharges of the Danube River at the delta head over the period 1840–2002 (six year moving mean was used)

charges of the Danube River at the delta head are given in Table 12. The last 30–40 years have been under the strong influence of climate factors and have seen more water abundant than previous ones.

By contrast, the sediment load of the Danube River has decreased because of reservoir construction, which was started in the Danube basin in the middle of the 20th century (Table 12). These data show marked human-induced decrease in the suspended sediment discharge and concentration with time. Reduction of the suspended sediment load of the Danube River was especially significant after 1971 and 1984, when the Iron-Gate-I and Iron Gate-II reservoirs were put into operation (Fig. 5).

The following periods in seasonal regime of the Danube River can be distinguished [8]: high-flow period (spring-summer rainfall and snow melt

Table 12 Water runoff and suspended sediment load at the delta head of the Danube River over different periods

Period	Water discharge, $m^3\ s^{-1}$	runoff, $km^3\ year^{-1}$	Suspended sediment discharge, $kg\ s^{-1}$	load, $10^6\ t\ year^{-1}$	mean concentration, $g\ m^{-3}$
1840–1920	6140	194	1990	62.8	324
1921–1960	6320	199	1660	52.4	263
1961–1970	7020	222	1520	48.0	217
1971–1984	6890	217	1450	45.8	210
1985–2002	6370	201	906	28.6	142
1961–2002	6700	211	1230	38.8	184
1971–2002	6590	208	1150	36.3	174

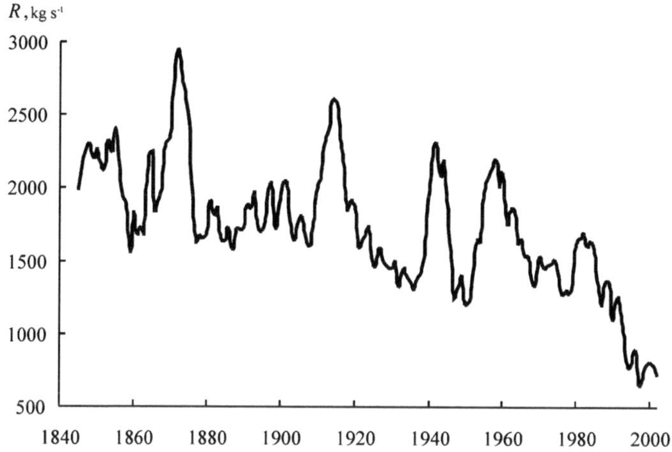

Fig. 5 Long-term changes in the annual suspended sediment discharges of the Danube River at the delta head over the period 1840–2002 (six year moving mean was used)

flood) from March to July (about 51% of the annual water runoff); low-flow period from August to October; autumn and winter period of rainfall floods from November to February.

Seasonal distribution of water runoff and suspended sediment load was practically unchanged in spite of reservoir construction on the Danube River.

Water runoff of the Danube River at the delta head (delta apex) is distributed between the Chilia and Tulcea branches. Downstream the water runoff of the Tulcea branch is distributed between the Sulina and Gheorghe branches. This water distribution in the Danube delta was not steady (Fig. 6). At the beginning of the 20th century, the share of the water runoff of the Chilia branch began to decrease due to rapid progradation of the Chilia branch into the Black Sea and essentially as a consequence of an artificial deepening and canalization of the Sulina branch during 1868–1902. This share changed from 72% of the Danube runoff in 1890s–1910s to 62–63% in the middle of the 20th century. To the contrary, the share of the water runoff of the Tulcea and Sulina branches increased. New changes of the water runoff distribution within the delta occurred in 1980s–1990s as a result of large-scale engineering works on straightening the meanders in the Gheorghe branch in 1981–1992 [8]. These works led to reduction of the Gheorghe branch length from 110 to 77 km.

At present, the average shares of the river water input to the Chilia, Tulcea, Sulina, and Gheorghe branches are approximately equal to 52, 48, 21 and 27%, respectively. The tendency in redistribution of the Danube water runoff in favor of the Tulcea and Gheorghe branches is retained.

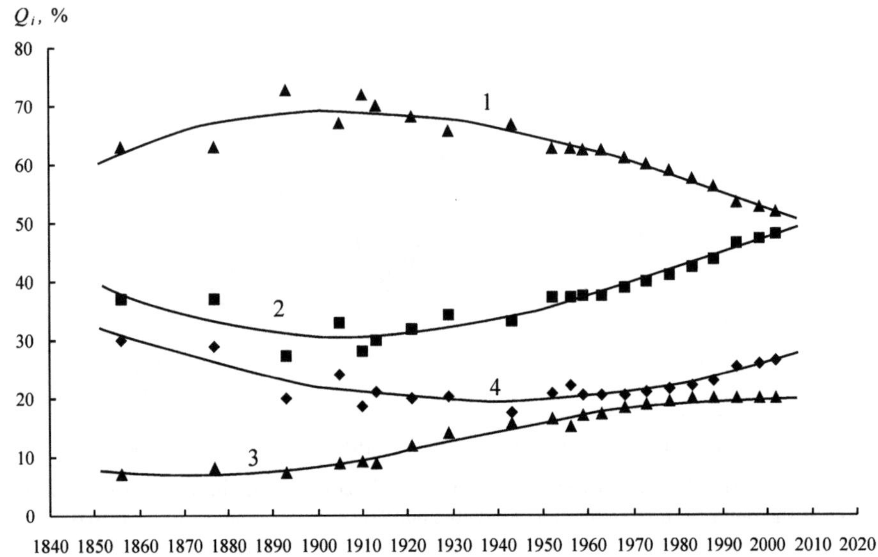

Fig. 6 Long-term redistribution of the river water runoff between main branches of the Danube delta (in % of the average water runoff of the river at the delta head). Branches: 1 – Chilia, 2 – Tulcea, 3 – Sulina, 4 – Gheorghe. *Marks* show observation data

Distribution of water runoff between branches of the Chilia delta is very changeable (Table 13). During the second part of the 20th century, some branches (the Bystry, Tsyganskiy) were subject to erosion and increased water runoff; certain of the branches periodically changed their water runoff (the Prorva, Gneoushev, Vostochny, Starostambul'skiy); some branches were undergone strong sedimentation and decreased their water runoff (the Ochakovskiy, Potapovskiy branches). In individual cases, some branches were filled up by sediments and died out (the Sredniy and Zavodninskiy branches).

Field observations [8] show that distributions of the sediment load between delta branches is proximately in proportion to distribution of the water runoff.

The Danube River mouth, including the delta, is subject to action of storm surges. In the nearshore zone, during strong northeastern and eastern winds magnitude of the storm surges reaches 1.0–1.2 m. At the low-flow periods, the propagation of surge-induced water level variations into the Danube River can consist of 200–250 km [8].

During the low-flow periods and under the influence of the eastern winds, seawater can penetrate into the deep delta channels [8]. The maximum length of seawater intrusion recorded during recent decades was 16.5 km in the Sulina branch [29] and 14.0 km in the Ochakovskiy branch [8]. As a rule, seawater penetrates into the delta branches in the form of "salt wedge".

Table 13 Redistribution of water runoff between branches of the Chilia delta (% of the Danube River average water runoff)

N	Branch	1958–1960	1961–1965	1966–1970	1971–1975	1976–1980	1981–1985	1986–1990	1991–1995	1996–2000	2001–2003
1	Chilia, head of the Chilia delta	62.5	62.5	61.2	60.0	58.5	56.2	54.6	51.9	51.1	50.4
2	Ochakovskiy	25.1	23.[illegible]	21.5	19.7	18.0	17.4	16.9	14.4	13.3	12.7
3	Belgorodskiy	0.2	0.[illegible]	0.1	0.2	0.2	0.1	0.1	0.1	0.1	0.1
4	Prorva	5.8	6.[illegible]	7.1	7.5	7.6	7.7	7.9	6.6	5.1	4.8
5	Potapovskiy	14.6	11.7	9.1	6.3	4.0	3.6	3.2	2.8	3.3	3.2
6	Gneoushev	1.5	1.7	2.1	2.4	2.2	2.4	2.2	1.6	1.3	1.2
7	Poludenny	1.4	1.2	1.6	1.8	2.0	2.0	2.0	2.0	2.0	1.9
8	Ankudinov	1.3	1.2	1.3	1.3	1.4	1.3	1.3	1.2	1.2	1.1
9	Starostambul'skiy, head	37.4	39.2	39.7	40.3	40.5	38.8	37.7	37.5	37.8	37.7
10	Sredniy	0.7	0.4	0.2	0.1	0.1	0	0	0	0	0
11	Bystry	10.0	10.8	12.3	13.5	14.2	15.3	16.2	17.1	18.2	18.5
12	Vostochny	1.5	1.5	1.7	2.1	2.3	2.5	2.5	2.3	2.0	2.0
13	Limba	0.8	0.8	0.7	0.6	0.6	0.5	0.4	0.3	0.3	0.3
14	Starostambul'skiy, mouth	24.1	25.4	24.6	23.7	23.1	20.3	18.5	17.5	17.1	16.6
15	Kuril'skiy	1.0	0.3	0.8	0.7	0.7	0.6	0.5	0.4	0.4	0.4
16	Zavodninskiy	0.2	0.2	0.1	0.1	0.1	0	0	0	0	0
17	Tsyganskiy	0.7	0.5	1.0	1.1	1.4	1.7	1.9	2.0	2.1	2.2

During the second part of the 20th century, river flux of organic matter, nutrients and pollutants markedly increased. For example, increase in concentration of ammonia nitrogen is 2.5 times, of nitrites and nitrates 4 and 5 times respectively, of phosphate 2 times. Since the beginning of 1970s, the concentration of heavy metals and oil in river water has also increased [9]. Over the period 1996–2000, the input of the contaminants to the Black Sea with the Danube waters comprises: oil 53×10^{12} t, Cu 1.2×10^{12} t, and Zn 3.3×10^{12} t [9].

Salinity field in the mouth nearshore zone of the Danube depends on water runoff and winds [7, 9, 30]. During high-flow period, water salinity near the Zmeiny Island can decrease to 3‰ in comparison with a normal value of 15‰ [7].

The mixing zone of river fresh water with salinity less than 0.5‰ and sea brackish water with salinity up to 18‰ occupies a bend up to 20–30 km in width during high-flow period and western winds and of 3–5 km during the low-flow period and eastern winds [7, 30]. The Danube water spreads mainly on the surface with layer from 1–3 to 5 m [9]. The Danube River water runoff and its distribution between delta branches play a very important role not only in hydrological and hydrochemical regime of the delta and mouth nearshore zone but also in formation of ecological conditions in the northwestern part of the Black Sea as a whole [7, 9, 30].

During the high-flow periods, waters with small salinity can reach the Zmeiny Island in the east, Bulgarian coast in the south and the Dniester mouth in the north. During very significant spring-summer river floods, the area of the Danube influence occupies 70% of the northwestern part of the Black Sea. The total area of this direct river influence, defined according to the freshwater phytoplankton presence, is not less than 10^5 km^2 [9].

Increase in the input of the organic matter and nutrients into the Black Sea causes increase of total phytoplankton biomass. In summer due to formation of the temperature, salinity and density stratification and algal blooms, decay of dead phytoplankton leads, in turn, to the oxygen lack and near-bottom hypoxia [9]. These processes have to consider as the consequence of anthropogenic eutrophication of the sea [9].

Observations show that the area of hypoxia directly depends on volume of the Danube water runoff during spring-summer flood. Besides, this area depends on the time of the flood peak [9, 31]. If the flood peak takes place in April, the river fresh waters are driven out of the northwestern part of the Black Sea in the south direction under the influence of predominated northern winds in this time, and hypoxia is absent. Other situation takes place if the flood peak falls on May or June, when under the impact of the southern winds, main mass of river fresh water remains in the northwestern part of the Black Sea. In this case, hypoxia forms later and in the area between the Danube and Dniester mouths [9, 31].

In 1973–1990, the zone of hypoxia occupied from 3.5×10^3 to 40×10^3 km^2 in the northwestern part of the Black Sea and caused the mass mortality of 60×10^6 t of bottom animals and 5×10^3 t of fish, especially juveniles [9].

6.2 Dniester River Mouth

The Dniester River mouth area is of the semi-enclosed deltaic type and includes the Dniester delta, semi-enclosed Dnestrovskiy Liman, Tsaregradskoye outlet, and the open nearshore zone (Fig. 7). Small filling (or bayhead) delta of the Dniester River has an area of 48.6 km^2 [12]. There are only two small branches in the Dniester delta: the Glubokiy Turunchuk (right) and the Dniester (left) 5 and 12 km in length respectively. The first and second branches derive around 60 and 40% of the river water runoff correspondingly.

The Dnestrovskiy Liman has an area of 508 km^2, a length of 42.5 km, a width from 4.2 to 12.0 km, and an average depth of 1.5 m [5, 11].

Deep Tsaregradskoye outlet connects the semi-enclosed liman and the open nearshore zone. Inlet's width in the narrowest place equals 280 m, its depths are 10–12 m [11].

The liman and open nearshore zone are separated by a spit (or coastal bar) 0.5 km in width and 9.0 km in length [11]. At present, this spit is subject to abrasion and retreats in the northwestern direction.

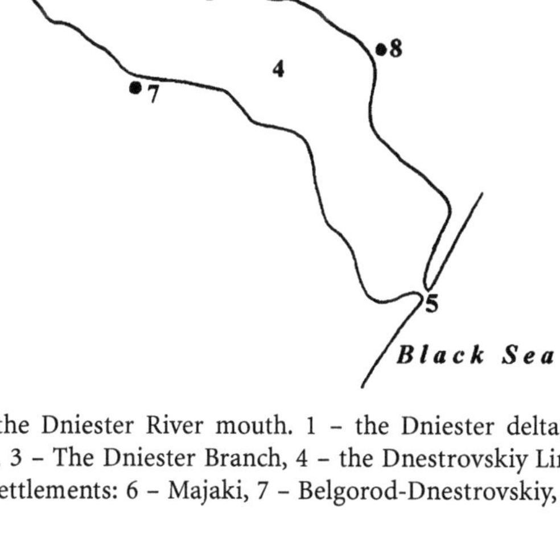

Fig. 7 Scheme of the Dniester River mouth. 1 – the Dniester delta, 2 – the Glubokiy Turunchuk branch, 3 – The Dniester Branch, 4 – the Dnestrovskiy Liman, 5 – the Tsaregradskoye outlet. Settlements: 6 – Majaki, 7 – Belgorod-Dnestrovskiy, 8 – Ovidiopol'

Over the period of 1830–1993, the Dniester delta was markedly prograded into the shallow Dnestrovskiy Liman. Delta area and its coastline length increased by 14.1 km^2 (41%) and by 6.9 km (33%) correspondingly [12]. During 163 years, average growth of the delta was relatively large: about 0.09 $km^2\ year^{-1}$.

Over the above-mentioned period, average value of the deposition of riverine sediments in the delta and liman was equal to 2.3×10^6 t $year^{-1}$ or close to one-half of the river sediment load [12]. Thickness of layer of deposition during the same period comprised 0.55 m [12], and average rate of accumulation exceeded 0.3 mm $year^{-1}$.

Hydrological and hydrochemical regimes of the delta and especially of the liman are mainly determined by the river fresh water inflow, local climatic conditions, including wind regime, and water exchange between the liman and the sea through the outlet.

In 1954 and 1987, in the middle part of the Dniester River the Dubossarskoye and Dnestrovskoye reservoirs were constructed. Total storage of these reservoirs is 0.48 and 3.0 km^3 respectively [11]. After putting of the reservoirs into operation, water runoff of the Dniester River decreased from 10.1 to 9.1 $km^3\ year^{-1}$ and the suspended sediment load decreased from 5.5×10^6 to 4.1×10^6 t $year^{-1}$ (Table 8).

Water flow regulation had only a slight influence on regime of the delta and liman. Water runoff in spring and summer (high-flow period) somewhat decreased, but in autumn and winter (low-flow period) slightly increased.

Waters in the liman are fresh near the delta and brackish in other parts. Mean water salinity in the central part of the liman is nearly 1.5‰, but in the vicinity of the outlet can change from 2 to 16‰ [5]. Currents in the liman are induced by river water runoff and winds. Because of the influence of waves (on the average 0.5 m in height [12]) turbidity of liman's waters is usually greater than that of river waters. Seaward currents and water discharges in the outlet during the strong wind-induced negative surges can reach 3 m s^{-1} and 3000 $m^3 s^{-1}$, respectively.

In the middle of the 20th century, average annual components of the water balance of the Dnestrovskiy Liman were computed as follows [5, 10]:

Total water input (14.19 km^3) consists of river water runoff, equaled 10.16 km^3 (71.6%); water inflow from the sea (3.75 km^3, 26.4%); precipitation (0.238 km^3, 1.7%); and ground water inflow (0.038 km, 0.3%).

Output components of the water balance (total output equals 14.19 km^3) are the flow through the outlet into the sea (13.77 km^3 or 97.1%) and evaporation (0.418 km^3, 2.9%). One can show that evaporation exceeds precipitation, but both of these components are very small.

Ecological conditions in the liman (distribution of water temperature and salinity, concentration of dissolved oxygen, and phytoplankton, etc.) depend on the fresh and saline waters inflow, nutrient input and river- and wind-induced water circulation in the liman [10, 11]. The above-mentioned hydrological, hydrochemical and hydrobiological characteristics should be taken into account with the use of waters for irrigation, fishery, etc.

6.3 Dnieper and Southern Bug Rivers Mouth

The mouth of the Dnieper and Southern Bug rivers has a complex structure and as a whole belongs to the semi-enclosed deltaic type. It consists of the following parts (Fig. 8):

- Mouth reaches of three rivers – the Dnieper, including its delta, Southern Bug and Ingul;
- Large semi-enclosed coastal water body named the Dneprovsko-Bugskiy Liman;
- The Kinburnskiy outlet connecting the liman and the Black Sea;
- Open nearshore zone of the sea in the vicinity of the outlet.

The upper boundaries of the mouth reaches of the rivers are determined by propagation of the water level variations induced by storm surges during low-flow periods. The lengths of these reaches for considered rivers equal 160, 100 and 20 km respectively. For the Dnieper River, the Kakhovskaya dam limits a distance of storm surge propagation.

The Dnieper delta is of the filling or bayhead and multi-branch type (Fig. 9). An area and a length of the delta equal to 350 km^2 and 47 km cor-

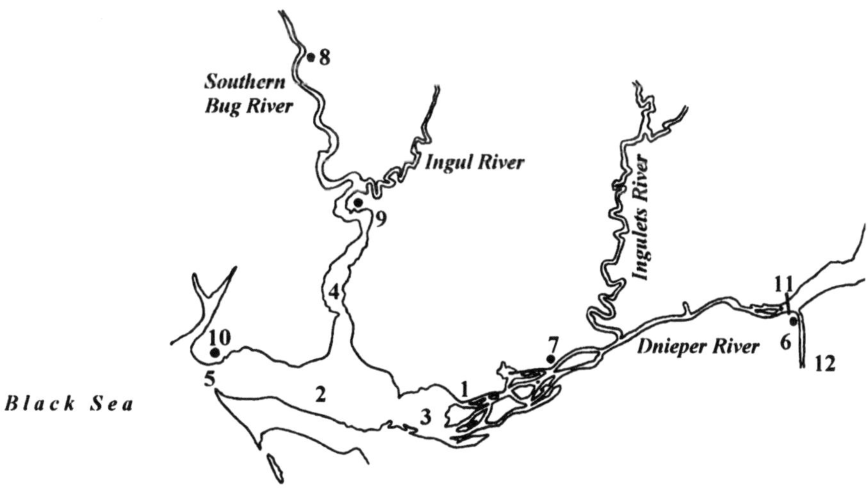

Fig. 8 Scheme of the Dnieper and Southern Bug rivers mouth. Parts of the mouth area: 1 – Dnieper delta, 2 – Dneprovsko-Bugskiy Liman, 3 – Dneprovskiy Liman, 4 – Bugskiy Liman, 5 – Kihburnskiy outlet. Settlements: 6 – Novaya Kakhovka, 7 – Kherson, 8 – Novaya Odessa, 9 – Nikolaev, 10 – Ochakov. 11 – Kakhovskaya dam, 12 – North-Crimean canal

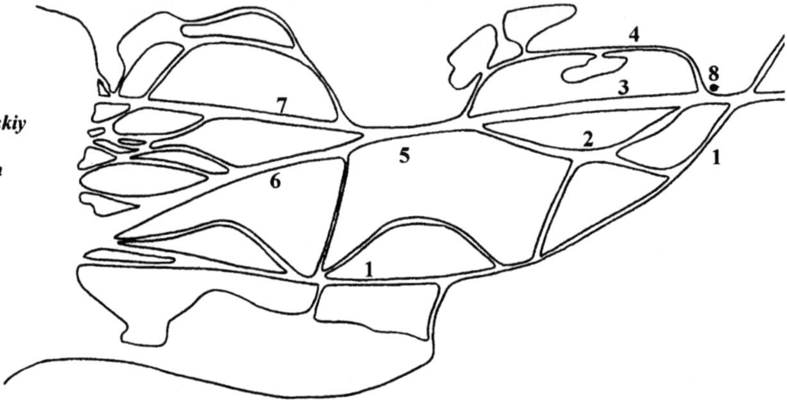

Fig. 9 Scheme of the Dnieper delta. Main branches: 1 – Konka, 2 – Stary Dnieper, 3 – Novy Dnieper, 4 – Koshevaya, 5 – Zabich, 6 – Rvatch. 8 – Town of Kherson

respondingly [5]. The delta head is located in the place where small delta branches begin to form. Near the town of Kherson (28 km from the liman) main delta channel is divided into four large branches (from left to right): the Konka, Stary Dnieper, Novy Dnieper, and Koshevaya. Downstream these branches join again. Seaward of this place there are the next main delta branches: the Konka (left, 15% of the total river water runoff) and Zabich (right, 85%). The Zabich branch after that is divided into the Bokay (left, 50% of the total river water runoff) and the Rvatch (right, 35%) branches (Fig. 9).

The Dneprovsko-Bugskiy Liman represents flooded river valleys widened in the seaward directions. The liman involves three parts: the Dneprovskiy Liman, the Bugskiy Liman and their combined part. An area of the Dneprovsko-Bugskiy Liman equals about 1000 km^2, its length from the Dnieper delta to the Kinburnskiy outlet is equal to 63 km, its width is up to 15 km, and its mean and maximum depths are 4.4 and 12 m [5, 14]. A length and a maximum width of the Bugskiy Liman are equal to 42 and 2 km, respectively. The Kinburnskiy outlet has a width in the narrowest place of 3.7 km; its mean depths are 4.3–4.5 m, and a maximum depth within navigation channel reaches 20 m [5].

The considered river mouth area plays a very important role in economy and environmental management of this part of Ukraine. The Dnieper and Southern Bug rivers are important water ways. Many large industrial centers and ports such as Kherson, Nikolaev, and Ochakov are situated here. River waters are widely used for irrigation and water supply. In this region, agriculture and fishery are also developed.

Because of the complex structure of the object under study and strong human-induced changes in governing factors, regime of the mouth area is very complicated [5, 10, 13, 14]. It is subject to combined influence, firstly, of natural and human-induced changes of the river water runoff, secondly, of marine factors, including storm surges and water and salt exchange with the sea through the outlet, and, thirdly, of local winds.

Present-day average values of the water runoff of the Dnieper, Southern Bug and Ingul rivers are amounted as 43.4, 2.2 and 0.58 km^3 $year^{-1}$ correspondingly (Table 8).

The water runoff of the Dnieper River was highly regulated as a result of the reservoirs cascade construction in 1950s – 1960s. The strongest impact of the river regulation on the mouth area occurred after the construction in 1958 of the Kakhovskoye reservoir with a total storage of 18.2 km^3 (Table 8).

As a result of the river flow regulation and water withdrawal for irrigation, an average annual water runoff has been reduced by 22%. Besides, water runoff during high-flow period from April to August decreased and during low-flow period from October to March (especially in winter months) increased.

These long-term and seasonal changes in river water runoff led to marked alternation of mouth processes [5, 10, 14]. Inundation of the delta decreased, mineralization of the waters in the delta and liman increased. Water salinity in the lower part of the Dneprovsko-Bugskiy Liman increased during summer – autumn period up to 5–10‰. Besides, input of the riverine organic matter, nutrients and pollutants into the liman increased. Its ecological state was impaired. Algal blooms have become more frequent. Anoxic events became to be occurred in the deep part of the liman and navigation channels.

Analysis of changes in the water balance of the Dneprovsko-Bugskiy liman is of special scientific interest [5, 10, 14]. Among main input components of the water balance of the Dneprovsko-Bugskiy liman are inflows of river and sea waters. These values are estimated as 56.7 and 37.4 km^3 $year^{-1}$ (or 60.0 and 39.6% of the total water input) under the natural conditions and 46.0 and 40.74 km^3 $year^{-1}$ (or 52.8 and 46.7% of the total water input) after river flow regulation. It is evident that influence of the river water runoff decreased, and the impact of the inflow of waters from the sea increased. Among the output components of the water balance amount of the contribution of water outflow to the sea through the outlet as a result of the river flow regulation decreased from 93.6 to 86.3 km^3 $year^{-1}$. Nevertheless, its share in the total water outflow was practically unchanged (99.2 and 99.1%). Contribution of precipitation and evaporation to the water balance of the liman are very small, 0.368 and 0.802 km^3 $year^{-1}$ respectively, and the second value markedly exceeds the first one.

Mixed with the waters of the Dnieper and Southern Bug rivers, a large amount of human-made nutrients, including nitrogen and phosphorus compounds, contaminants including heavy metals and hydrocarbons, as well as organic matter flows into the northwestern part of the Black Sea. The impact on the sea has become a very serious ecological problem.

6.4 Rioni River Mouth

The Rioni River mouth area belongs to the open deltaic type (Fig. 10) and comprises a protruding delta with a few branches and a deep nearshore zone. The delta head is near the water division sluice constructed at 7 km from the sea in 1959 for distributing river flow between the two main branches of the delta. The channel network of the Rioni mouth includes delta branches, the City Canal (left) and the Rioni-spillway (right). The City Canal, 7.5 km long, is divided into Northern and Southern branches separated by the Bolshoy Island to form the "old" delta. The Rioni-spillway, 7 km long, is also divided into Northern and Southern branches separated by an island to form the "new" delta. The total area of the present-day delta of the Rioni River is about 20 km^2.

The Rioni River (Phasis) was mentioned for the first time in the works of the ancient scientists Herodotus and Hippocrates. Strabo was the first to present data on the hydrography of the Rioni delta: "At the Phasis River, there

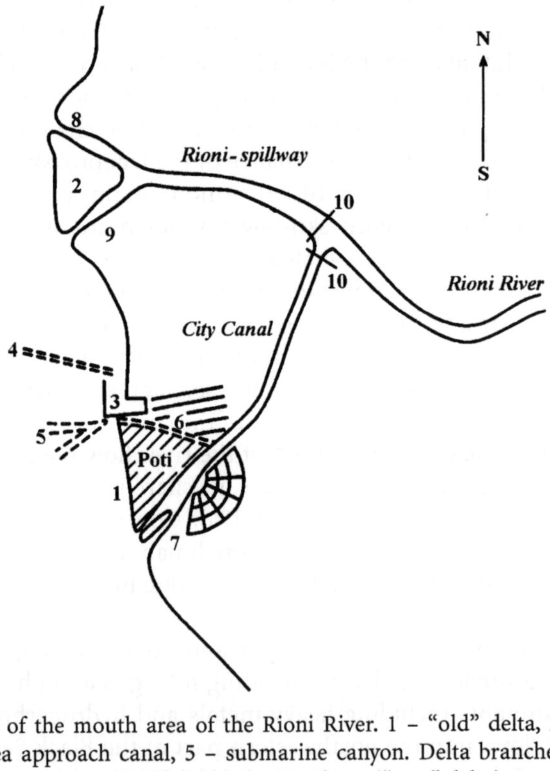

Fig. 10 Scheme of the mouth area of the Rioni River. 1 – "old" delta, 2 – "new" delta, 3 – port, 4 – sea approach canal, 5 – submarine canyon. Delta branches: 6 – Northern ("old" delta); 7 – Southern ("old" delta); 8 – Northern ("new" delta); 9 – Southern ("new" delta). 10 – Water division sluice

is a city of the same name, Colchis' commercial port, facing a river on one side, a lake on the other side, and a sea on the third side" [32]. Pliny the Elder called the Phasis "the most famous of the Pontic rivers" [32].

An analysis of the literary, cartographic, and paleogeographic data [4, 5, 15, 32] leads to the inference that at the ancient times the Phasis flowed into the sea through two branches, and a city of the same name was located on the delta island between the river branches.

In modern times, the development of the protruding delta of the Rioni River has been as follows [15, 28]: the "old" delta was formed at the Rioni mouth before the diversion of river flow in 1939 (the data on the progradation of the "old" delta are very scarce) and a "new" delta was formed at the mouth of the Rioni-spillway after 1939; the "old" delta virtually inherited the distinctive hydrographic features of the ancient times (two branches, Northern and Southern), divided by the Bolshoy Island.

The surveys of the "old" protruding delta of the Rioni River [15] make it possible to describe its evolution for the last 190 years (Fig. 11a). The "old" delta prograded rapidly up to 1926. The average progradation rate was 9–12 m $year^{-1}$ during this period. The construction of the seaport Poti that began in 1860 resulted in the obstruction of the alongshore drift of sediments from the south by the port structures. It is possible that this contributed to the more active supply of river sediments to the submarine canyon near the northern branch and to the canyon's further growth. The shift of the head of the canyon to the delta coastline retarded the progradation of the delta and favored its subsequent abrasion.

In 1939–1986, the extent of the average abrasion of the "old" delta amounted to 250 m, and the largest washout was 540 m (Fig. 11b); the abrasion rate was 6–7 m $year^{-1}$. The eroded urban area of Poti amounted to

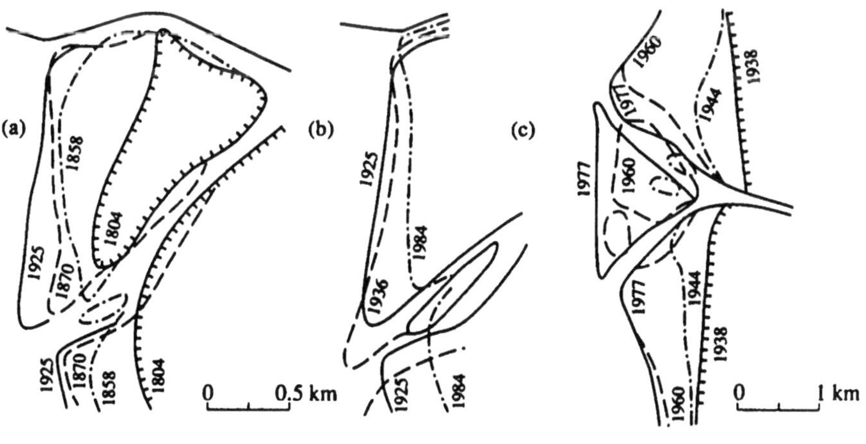

Fig. 11 Schematic maps of the development of the "old" delta (**a** and **b**) and the "new" delta (**c**) of the Rioni River

2.8 km^2 despite of coast protection measures. The abrasion was intensified, firstly, by the diversion of a part of the Rioni River water to the north in 1939, when a "new" delta began to form at the mouth of the new channel, and secondly, by a considerable excavation of alluvium from the river channel for construction needs.

In the early 1980s, the northern branch of the City Canal was drained.

The "new" protruding Rioni delta (Fig. 11c) began to form after the water runoff and sediment load was redistributed in 1939 to prevent Poti from being flooded [15]. Most of the river water was directed along the artificial Rioni-spillway.

The average annual water runoff of the Rioni River (from 1928 to 1991) amounts to 13.4 km^3 $year^{-1}$ (Table 6).

The water regime of the river is characterized by a spring–summer flood caused by snow melt and rainfall and by rainfall floods during the year. The highest water runoff is observed in May. The seasonal water runoff distribution is fairly uniform. The river flow regulation practically did not influence the water runoff.

The average suspended sediment load of the Rioni River for the 1934–1991 period amounted to 6.24×10^6 t $year^{-1}$. The regulation if the Rioni River by the Rioni, Ladzhanuri, and Vartsikhe reservoirs had a small effect on the sediment load of the river. However, the construction of the Gumati hydroelectric plant produced a substantial effect on the sediment load of the Rioni River: the sediment load was close to half of what it was in 1958. Before the river flow regulation, the sediment load amounted to 7.59×10^6 t $year^{-1}$. After the river flow regulation, the sediment load decreased to 4.10×10^6 t $year^{-1}$ in 1958–1967, and 3.72×10^6 t $year^{-1}$ in 1968–1977. The excavation of sand from the river for construction needs also contributed to the decrease in total sediment load. In 1981–1991, the amount of sediment load was recovered and nearly returned to what it was before the river flow regulation. This may be explained by the silting of the Gumati reservoir, by the resumption of the transit of sediments through it, and by the erosion in the lower reaches of the Gumati and Vartsikhe reservoirs. At present, sediment load of the river is equal to 6.02×10^6 t $year^{-1}$ (Table 6).

In 1971–1991, 70% of the water runoff went through the Rioni-spillway and 30% flowed through the City Canal. An average of 55% of the river sediments were delivered to the Rioni-spillway and to the "new" delta in 1971–1991.

The flow bifurcated into the Northern and Southern branches of the "new" delta and, later, an island was formed (Fig. 11c). By its dimensions and outlines, this island closely resembles Bolshoy Island, which was formed at the mouth of the main channel of the Rioni River in the "old" delta.

The mean annual increment of the "new" delta amounted to 0.27 km^2 $year^{-1}$ for 20 years (from 1939 to 1958). In 1959–1967, the rate of the delta growth slowed down to 0.1 km^2 $year^{-1}$ because of the decrease in sediment load after the construction of the Gumati Hydroelectric Plant, the partial diversion of the river water to the City Canal, and the development of the sand quarry in the "new" delta. Later, in 1968–1980, the delta increment averaged 0.16–0.17 km^2 $year^{-1}$ [15].

7 Major River Mouths of the Sea of Azov

7.1 Don River Mouth

The mouth area of the Don River is of the semi-enclosed deltaic type [5, 16, 19]. It consists of three parts (Fig. 12): the near-delta river reach subjected to water level fluctuations due to influence of storm surges, small delta and the relatively narrow Taganrogskiy Bay. The near-delta reach is about 110 km long. The delta has an area of 540 km^2 and a length of 38 km. The delta coastline is 55 km long. The bay has an area of 5240 km^2, a width from 26 to 52 km, a length of 140 km, and a mean depth about 5 m. It is limited in western part by the Dolgaya and Belosarayskaya spits.

The Don delta belongs to the filling (or bayhead) and multi-branch type (Fig. 13).

In spite of the small size of the Don delta, its channel network is relatively dense. Two main directions of the river flow into the sea are the following: firstly, the Don River → the Don branch → the Stary Don branch → the Peschany branch, artificially deepened for navigation, and secondly, the Don

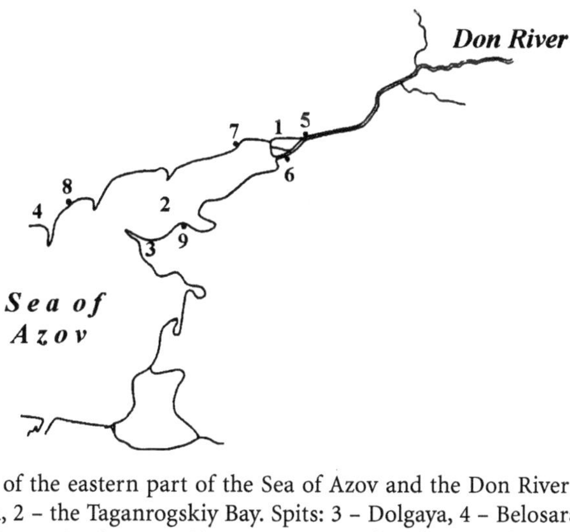

Fig. 12 Scheme of the eastern part of the Sea of Azov and the Don River mouth. 1 – the Don River delta, 2 – the Taganrogskiy Bay. Spits: 3 – Dolgaya, 4 – Belosarayskaya. Towns: 5 – Rostov-na-Donu, 6 – Azov, 7 – Taganrog, 8 – Mariupol', 9 – Eisk

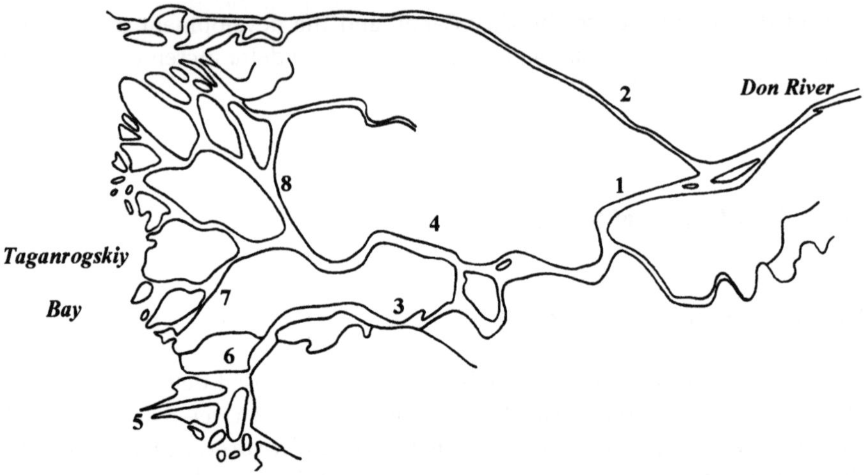

Fig. 13 Scheme of the Don River delta. Main branches: 1 – Don, 2 – Mertvy Donets, 3 – Stary Don, 4 – Bol'shaya Kalancha, 5 – Peschany, 6 – Merinovo, 7 – Mokraya Kalancha, 8 – Bol'shaya Kuterma

River → the Don branch → the Bol'shaya Kalancha branch → the Bol'shaya Kuterma branch. The Don branch as a direct extension of the river within the delta represents the major delta branch.

The Don River mouth region is highly economically developed. Its natural resources are widely used for agriculture, water transport, and fishery. There are many large towns in the mouth area: Rostov-an-the-Don, Azov (on the river and its branch), and Eisk, Taganrog (Russia), Mariupol' (Ukraine) on the bay coast.

Since 1952, the regime of the Don River was regulated by the Tsymlyanskoye reservoir with a total storage of 23.9 km^3 [5, 25]. As a result of the water flow regulation and withdrawal, an average river water runoff decreased from 27.6 $km^3\ year^{-1}$ in 1881–1951 to 21.6 $km^3\ year^{-1}$ in 1952–2000 (Table 11).

Reduction of the suspended sediment load was more significant: from 4.54×10^6 t $year^{-1}$ (before river flow regulation) to 1.6×10^6 t $year^{-1}$, or in 2.8 times (Table 11).

Since river flow regulation, seasonal water flow distribution has also changed: water runoff in high-flow period (April and May) is greatly decreased, but in other months – from August to February – increased.

At present time, river water runoff at the delta head divides between the Don branch and the right small Mertvy Donets branch (Fig. 13) in the proportion about 97 and 3% of the total river water runoff. Downstream water runoff of the Don branch distributes between the Stary Don (left) and Bol'shaya Kalancha (right) branches as 27 and 70% of the total river water runoff. Then water runoff of the Stary Don branch is divided between the Peschany (17% of the total river water runoff) and Merinovo (10%) branches. Bol'shaya Kalancha branch, in its turn, subdivides into the Mokraya Kalancha (24%) and Bol'shaya Kuterma (46%) [5].

The Don mouth is subject to significant influence of storm surges. In the Taganrogskiy Bay, heights of positive storm surges can reach 2.5 m and those of negative surge – 3 m [5]. Penetration of water level fluctuations induced by storm surges into the Don River depends on the height of level rise in the bay and river water discharge. Since 1952, regulation of the river flow has led to some reduction in propagation of surge-induced level variations into the river [5].

Water salinity in the Taganrogskiy Bay depends on changes in the river water runoff and a distance from the delta coastline. Along the bay, water salinity can change from 0.5–1‰ near the delta to 9–11‰ at western boundary of the bay. Seasonal changes of water salinity in the bay depend inversely on variations of river water discharges. Before river flow regulation, mean salinity in the bay was 5–6‰. After river flow regulation, it increased to 8–9‰. At present, mean water salinity ranges from 6 to 8‰ in accordance with changes in mean annual water runoff of the Don River.

7.2 Kuban River Mouth

The mouth area of the Kuban River (Fig. 14) belongs to the open deltaic type [5, 17–19]. It consists of the large delta 4190 km^2 in area and narrow and deep nearshore zone 5–10 km in width. The delta coastline is 150 km long.

At the delta head the Kuban River is divided into two main branches: the Kuban (left) and the Protoka (right), which are similar in size and wa-

Fig. 14 Scheme of the Kuban' River mouth. Main branches: 1 – Kuban, 2 – Protoka. Settlements: 3 – Slavyansk-na-Kubani, 4 – Temryuk, 5 – Temryuk-port, 6 – Slobodka

ter runoff. Lengths of these branches are 116 and 130 km, respectively. In the lower parts of these branches, there are several small distributary channels: the Kazachiy Erik, Pryamoy, Sredniy, Golinskiy (a system of the Kuban branch) and Pravy and Levy (a system of the Protoka branch).

The Kuban delta represents a well irrigated region and has a very large and complex system of offtake and spillway structures, irrigation, and drainage canals. Recently a water distribution structure was constructed at the delta apex; its function is to control of the water supply to main delta branches and heads of main irrigation canals.

There are a great number of large and small lakes in the delta. Some of them are connected with the sea and represent water bodies of the lagoon type. But here they are known as limans.

The Kuban delta widely used for agriculture and fishery. Towns of Temryuk, and Slavyansk-na-Kubani are situated here.

Water runoff of the Kuban River was alternated after water diversion into the Nevinomysskiy canal in 1949 and especially as a result of the Krasnodarskoye reservoir construction in 1972 [26, 33]. A total storage of this reservoir is 3.0 km^3. The average water runoff at the delta head of the Kuban delta decreased from 13.0 $km^3\,year^{-1}$ in 1929–1948 to 11.6 $km^3\,year^{-1}$ in 1949–1972 and to 11.1 $km^3\,year^{-1}$ in 1973–2000. Decrease in the suspended sediment load was more marked: from $8.5 \times 10^6\,t\,year^{-1}$ before 1949 to $6.8 \times 10^6\,t\,year^{-1}$ in 1949–1972 and to $1.5 \times 10^6\,t\,year^{-1}$ in 1973–2000. Reduction of suspended sediment load after the reservoir construction comprised 4.5 times (Table 11). Seasonal water runoff distribution was also alternated: a share of the total annual runoff in spring and summer (especially from May to July) decreased, but in autumn and winter (especially from September to January) increased [5, 26].

At present, river water runoff and sediment load are distributed between the Kuban and the Protoka branches in the proportions 49 : 51% and 47 : 53% respectively [5, 33]. Over the second part of the 20th century, the Kazachiy Erik branch decreased its share of the total river water runoff because of sediment accumulation from 17 to 3%.

Recent investigations showed that about 60% of the river sediment load is accumulated within the delta [33].

Sometimes the Kuban delta is subject to flooding due to storm surges and ice dams. Significant inundation induced by storm surge was recorded in October 1969, when water level rise at the town of Temryuk at a distance of 7 km from the sea exceeded 3 m [5]. The last large inundation due to ice dam was observed in winter 2003.

In the last 30–35 years, delta coast has experienced abrasion. The retreat of the delta coastline has consisted of 3–5 $m\,year^{-1}$. The main reasons of this process are reduction of the river sediment load after reservoir construction and sea level rise. Slight progradation of the delta is observed only near the mouths of the main branches.

8
Possible Changes of the River Mouths in the Future

Approximate prediction of hydrological, hydrochemical and morphological processes in the river mouth areas in the future can be made only taking into account expected natural and human-induced alternations of governing factors and the present-day tendency of changes in structure and regime of the objects under the consideration.

Possible changes of main external governing factors influencing structure and regime of the river mouths of the Black Sea and the Sea of Azov region are the following:

1. further human-induced decrease of the annual river water runoff and especially sediment load as a result of water withdrawal and reservoir construction;
2. seasonal water runoff redistribution because of river flow regulation: water runoff decrease during high-flow periods and increase in low-flow periods;
3. further sea level rise and attendant intensification of wave action on the delta coasts;
4. increase of the input of riverine nutrients, organic matter and pollutants.

Besides, large-scale engineering works, including deepening of the channels, construction of water diversion and distribution structures, construction of embankments, etc., are possible in the future within the deltas.

The above-mentioned external impact can reinforce the present-day tendencies of processes in the river mouths areas. Therefore expected changes of structure and regime of the river mouths of the Black Sea and the Sea of Azov region may be summarized as follows:

1. decrease of the areas of inundation in deltas during high-flow periods as a result of river flow regulation and dike construction;
2. decrease of the lengths of propagation of water level variations induced by storm surges during low-flow periods after river flow regulation;
3. intensification of sea water intrusion into artificially deepened delta branches;
4. redistribution of the water runoff into artificially deepened delta channels;
5. strengthening the abrasion processes along the coasts of the Danube, Rioni and Kuban deltas as a consequences of the river sediment input reduction and sea level rise and wave action. The progradation of the delta coastline into the open sea is possible only at the mouths of main delta branches (the Bystry and Starostambul'skiy in the Danube delta, the Kuban and Protoka in the Kuban delta). Slow progradation of the Dniester delta into the shallow Dneprovskiy liman will continue;

6. formation of complexes comprising of coastal bars and small lagoons along low-laying peripheral parts of the large Danube, Dnieper and Kuban deltas under the influence of the sea level rise and wave action;
7. deterioration of the water quality at the river mouths, including deltas, semi-enclosed limans and bays, and open nearshore zones, as a result of the delivery of river-born nutrients, organic matter and human-made pollutants; accumulation of these substances in water and bottom deposits in channels, lakes, bays, swamps, etc; processes of the resuspension and secondary pollutants.

It must be emphasized that river water and matter inputs and hydrological processes at river mouths would have a profound and increasing impact on vast adjacent sea areas and coasts.

It is possible that eutrophication processes and such adverse events as algal blooms and hypoxia will be more frequent in some sea areas in the future. Conditions like that can take place in the northwestern part of the Black Sea in the vicinity of the mouths of the Danube, Dnieper and Southern Bug rivers and in the Taganrogskiy Bay of the Sea of Azov.

Significant decrease of the river sediment input to the seas together with the sea level rise and wave action can disturb the sediment balance in the coastal zone and intensify the abrasion and retreat of the sea coasts.

The contribution of the river water runoff to the water balance of the Black Sea and the Sea of Azov in the next decade would be expected around the present-day state.

Acknowledgements This study was supported by the Russian Foundation for Basic Research Grants No 04-05-65149 and 05-05-65110.

References

1. Mikhailov VN (1998) Hydrology of river mouths. Moscow University Press, Moscow (In Russian)
2. Mikhailov VN (2004) Water Res 31:5
3. Mikhailov VN, Mikhailova MV (2003) Water Res 30:655
4. Samoilov IV (1952) River mouths. Geographgiz, Moscow (In Russian)
5. Mikhailov VN (1997) River mouths of Russia and adjacent countries: past, present and future. Geos, Moscow (In Russian)
6. Petrescu IGh (1957) Danube delta: genesis and evolution. Editura stiintifica, Bucurest (In Romanian)
7. Diaconu C, Nikiforov YaD (eds) (1963) Hydrology of the Danube mouth area. Gidrometeoizdat, Moscow (In Russian)
8. Mikhailov VN (ed) (2004) Hydrology of the Danube delta. Geos, Moscow (In Russian)
9. Berlinsky N, Bogatova Yu, Garkavaya G (2006) Estuary of the Danube. In: PJ Wangersky (ed) The Handbook of Environmental Chemistry, vol 5, Part H (Estuaries). Springer-Verlag, Berlin Heidelberg, p 233
10. Timchenko VM (1990) Ecologo-hydrological studies of the water bodies on the northwestern coast of the Black Sea. Naukova dumka, Kiev (In Russian)

11. Braginsky LP (ed) (1992) Hydrological regime of the Dniester and its water bodies. Naukova dumka, Kiev (In Russian)
12. Shuisky YuD (1995) Dopovedi natsionalnoi academii nauk Ukraini 5:76 (In Russian)
13. Kostianitsin MN (1964) Hydrology of the Dnieper and Southern Bug mouth area. Gidrometeoizdat, Moscow (In Russian)
14. Zaitsev YuP (ed) (1989) Dnieper-Bug estuarine system. Naukova dumka, Kiev (In Russian)
15. Mikhailova MV, Dzaoshvili ShV (1998) Water Resources 25:134
16. Rodionov NA (1958) Hydrology of the Don mouth area. Gidrometeoizdat, Moscow (In Russian)
17. Simonov AI (1958) Hydrology of the Kuban mouth area. Gidrometeoizdat, Moscow (In Russian)
18. Bogucharskov VT, Ivanov AA (1979) The Kuban delta. Rostov University Press, Rostov-na-Donu (In Russian)
19. Simov VG (1989) Hydrology of river mouths of the Sea of Azov. Gidrometeoizdat, Moscow (In Russian)
20. Schubel JR, Pritchard DW (1971) What is an estuary? In: Schubel JR (ed) Estuarine environment. American Geological Institute, Washington, p 1
21. Dzaoshvili ShV (1999) Water Resources 26:275
22. Jaoshvili Sh (2002) The rivers of the Black Sea. European Environmental Agency. Technical report N:71
23. Douglas BC (1991) J Geogr Res 96:6981
24. Shuisky YuD (1999) Geogr Fis Dinam Quat 22:87
25. Mikhailov VN, Povalishnikova ES, Zudilina SV, Tiguntsev LA (2001) Water Res 38:645
26. Mikhailov VN, Povalishnikova ES, Ivanov AA (2002) Water Res 29:133
27. Mikhailova MV (1995) Water Res 22:452
28. Mikhailova MV (1995) Formation of the Danube and Rioni deltas and their coasts. In: Proc 2nd Int Conf on the Mediterranean Coastal Environment. Spain, Tarragona, p 911
29. Bondar C (1972) Hydrol Studies 32:1 (In Romanian)
30. Mikhailova MV, Povalishnikova ES (1996) Mixing of river and sea waters at the Danube river nearshore and its dependence on water discharges and wind. In: Proc. XVIIIth Conf. of the Danube Countries on Hydrological Forecasting and Hydrological Bases of Water Management. Austria, Graz, p 159
31. Berlinsky NA, Kosarev AN, Garkavaya GP, Bogatova YuI (2004) Vestnik Moskovskogo universiteta Ser. 5 5:17 (In Russian)
32. Agbunov MV (1987) Antique saling directions for the Black Sea. Nauka, Moscow (In Russian)
33. Magritsky DV, Ivanov AA (2003) Vestnik Moscovskogo universiteta. Ser. 5 5:46 (In Russian)

Hdb Env Chem Vol. 5, Part Q (2008): 135–158
DOI 10.1007/698_5_086

Published online: 27 June 2007

Hydrometeorological Conditions

Aleksey N. Kosarev (✉) · Viktor S. Arkhipkin · Galina V. Surkova

Geographic Department, Lomonosov Moscow State University, Vorobievy Gory,
119992 Moscow, Russia
akosarev@mail.ru

Abstract Based on long-term and seasonal data, the basic hydrometeorological features that form the natural regime of the Black Sea are under consideration, which include climate (regional atmospheric circulation, winds, atmospheric pressure, air temperature, moisture content, precipitation), wind waves, water balance, sea level (multiannual and seasonal changes, storm surges, seishes, tidal oscillations), as well as sea ice.

Keywords Black Sea · Climate · Hydrometeorological conditions · Sea ice · Sea level · Water balance · Wind waves

1 Introduction

The main distinctive property of the Black Sea is its inland location and high isolation from the World Ocean. Because of this, formation of the sea hydrological regime and water structure is governed by outer factors: the fluxes of heat, moisture, and wind stress via the sea surface, as well as the river runoff. In this connection, the sea is characterized by a high level of environmental variability. At the same time, in different parts of the Black Sea, the influence of outer factors are very unequal. Therefore these factors exert a different impact on the formation of hydrological fields and vertical thermohaline structure in the sea. All this determines the necessity of more detailed and regular observations of hydrometereological parameters of the Black Sea.

2
Climate

The climate of the Black Sea and its coastal regions is defined by three principal factors, which depend on the latitude and topography of the area—the irradiance processes, the atmosphere circulation (both large-scale and local), and the character of the underlying surface. With regard to the type of the air masses that dominate throughout the year, the northern and the southern parts of the Black Sea may be referred to as the temperate and subtropical climatic zones, respectively [1].

Solar irradiance. The southern position of the region determines the supply of a great amount of solar irradiance. The low albedo of the underlying water surface (4–5% under calm weather at a sun height of $h = 60°$) and the adjacent land areas at the absence of a permanent snow cover leads to the high values of the irradiance balance (Fig. 1); it is positive over the major part of the Black Sea throughout the year except for selected sites off its northern coasts.

Atmospheric circulation. The atmospheric circulation represents the most important process that defines the movements of the air masses over the Black Sea. Owing to its particular features, the climates in the western and eastern parts located within the same latitudinal belt differ in their thermal regimes and moisture contents, and therefore, over the eastern part of the Black Sea, the winter is warmer than over its western part and the monthly sum of precipitation is correspondingly several times higher.

The recurrence of the principal types of synoptic processes over the sea area and the related dominating wind directions in the lower troposphere vary during the year (Fig. 2).

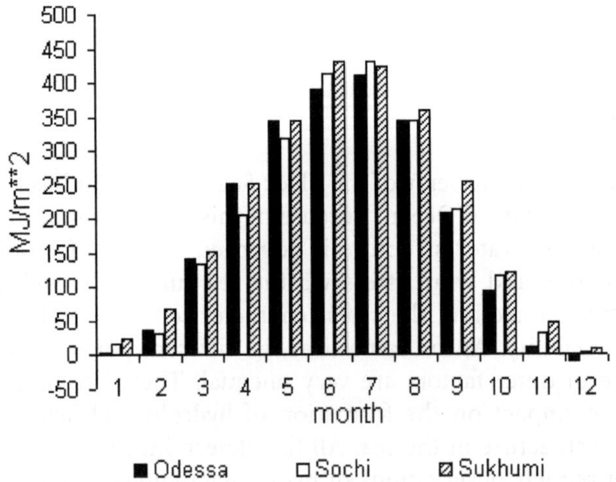

Fig. 1 Monthly values of the irradiance balance, MJ/m^2

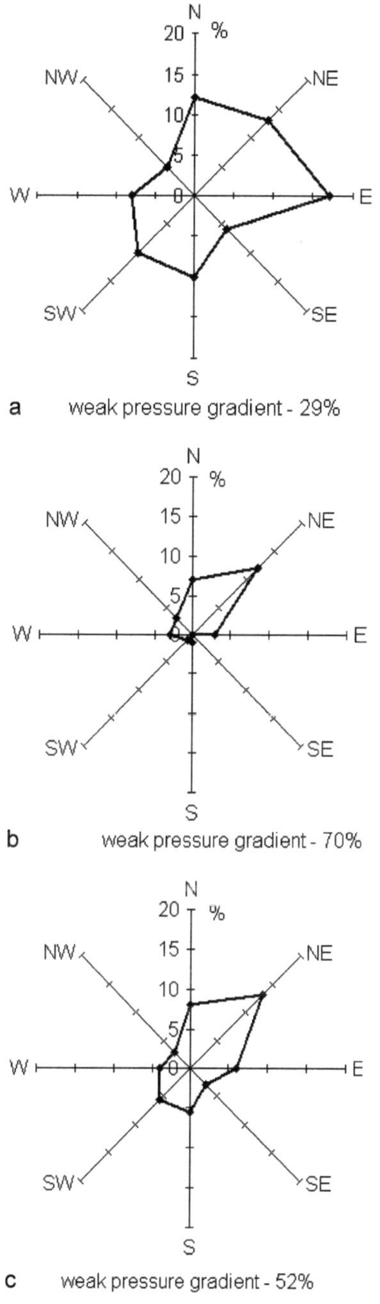

Fig. 2 Recurrence of synoptic processes, %, **a** in February, **b** in August, and **c** throughout the year. The names of the synoptic processes corresponding to the directions of the winds in the lower troposphere that dominate during the process. In the line below, the recurrences of the situations with low-gradient baric field are indicated

At the northerly processes, the Black Sea finds itself at the southeastern periphery of a vast anticyclone centered over Europe and Scandinavia. The strongest northerly winds accompany rapid displacement of the anticyclone from the region of the Balkan Peninsula in the course of the development of salinity activity over the Caucasus, the Caspian Sea, and, more rare, over the eastern part of the Black Sea. In these cases, the rate of atmosphere pressure growth in the western section of the Black Sea may reach 3–5 hPa during 3 h.

At the northeasterly synoptic process in the Black Sea region, the center of the anticyclone is located over the western regions of the European part of Russia. Owing to the advection of cold air from the north and northeast, the cyclonic activity over the southeastern part of the Black Sea is intensified. The passage of cyclones over the southern part of the Black Sea is accompanied by strong easterly and northeasterly winds, especially in the northeast of the Black Sea and off the western coast of the Crimea. At the same time, the southeastern part of the sea is usually dominated by weak and moderate winds of different directions.

For the easterly type of processes, it is characteristic of the anticyclone to be centered over the central regions of the European part of Russia. Meanwhile, the cyclonic activity also develops over the Mediterranean Sea and Turkey. In so doing, the Mediterranean cyclones tend to the southern regions of the Black Sea and result in a significant strengthening of easterly winds over the major part of its area.

The northerly, northeasterly, and easterly processes noticeably dominate in the wintertime and generally throughout the entire year.

The southeasterly processes are also observed mostly during the cold time of the year; they are related to the situation when the high-pressure area is located over the east of the European part of Russia and Kazakhstan, while its spur extends into the western regions of the European part of Russia. In so doing, the Balkan Peninsula and the Mediterranean Sea experience the influence of a low-pressure area while the Mediterranean cyclones displace to the southwest of the Black Sea and favor the strengthening of the southeasterly wind in its eastern part. While this process develops, the airflows mostly feature an easterly direction over the northwest of the sea and a southerly direction over its southwestern part.

The southwesterly processes develop in the situation when the depression of the air pressure in the lower troposphere is directed from the Baltic Sea toward the Balkan Peninsula. The cyclone development in this depression results in the strengthening of southerly and southwesterly winds over the Black Sea.

The cyclonic activity over the central part of the European part of Russia leads to the development of westerly winds over the Black Sea. In this case, the strongest winds are observed under the passages of deep Scandinavian cyclones over the south of the Ukraine and in the rear parts of the Mediterranean cyclones.

The northwesterly type of synoptic processes is related to the development of the cyclonic activity in the southeast of the European part of Russia and to an anticyclone over Europe with a spur toward the Balkan Peninsula. Similar to the westerly type of the process, the strongest winds are observed at the displacement of Scandinavian cyclones to the southeast of the European part of Russia across the southern part of the Ukraine in the rear of the Mediterranean cyclones.

In the case where a cyclone is located over the central part of the Black Sea, strong easterly and westerly winds dominate over its northern and southern parts, respectively. At this time, an anticyclone is usually located over the European part of Russia.

Local winds. Local (mesoscale) circulations and winds such as breezes, mountain–valley circulation, slope winds, foehns, boras, etc., along with the atmospheric processes on synoptic scale, play an important role in the formation of the climate on the coasts of the Black Sea.

The development of breezes, mountain–valley circulation, and slope winds is best favored by synoptic situations low-gradient baric field and low velocities of the main flow in the lower troposphere. In these cases, a distinctly manifested diurnal variation in all the meteorological parameters is observed [2, 3].

Breezes are induced by temperature contrasts between land and sea; they are characteristic of the entire Black Sea coast. The greatest recurrence of breezes universally fall in the period from March to October; on the southern coast of the Black Sea, they are probable throughout the entire year.

The greatest number of days with breezes (more than 50 days/year, at places up to 190 days/year) is noted on the southern coast of the Crimea, where the temperature contrast between land and sea is best expressed. Along the Caucasian coast, the recurrence of breezes increases from the north to the south from 18 to 50 days/year. Breezes are most rarely encountered on the western and northwestern coasts of the Black Sea and in the Kerch region.

The durations of the offshore and onshore breezes during the day are approximately equal (11–12 h). The morning and evening changes in the breeze direction occur very quickly—during 15–20 min. The onshore breeze is changed by the offshore one approximately in 2.5 h after the sunrise; the opposite change occurs almost simultaneously with the sunset [4].

The speed of breezes is relatively low. For the offshore breezes, its average value is 3–5 m/s; for the onshore breezes, it is 1–3 m/s. Speed increases may be observed in the regions where mountains approach the coastline and the orientation of mountain and river valleys coincides with the dominating direction of the breezes. In such cases, the breezes are enhanced by slope winds or due to the mountain–valley circulation.

Within the mountainous portions of the coast, foehns are often developed. They represent strong and gusty winds, which cause sharp changes in the temperature and moisture contents. They may last from a few hours to a few days. In the wintertime, they are more common than in the summer. The generation of foehns is related to the air masses that have overcome the mountain ridge. On the windward side of the ridge, the air upwells, the water vapor condenses, and clouds ("foehn bar") are formed; it reaches the crest of the ridge and sharply terminates on the leeward side. On the leeward side, an intensive downwelling airflow is induced, which, at the foot of the mountains, may reach characteristics similar to a hurricane. When warm foehns develop, the air descending downhill is adiabatically heated, which causes a sharp (sometimes up to 10–15 °C) temperature growth and a drop in the moisture content at the foot of the ridge. These kinds of foehns are characteristic of the Crimean and Caucasian mountains.

In the region of Novorossiisk, especially in the wintertime, bora (Nordost) is often observed, which represents a cold foehn accompanied by a sharp temperature drop and significant wind speeds (up to 40 m/s). During a bora, the cold air that has overridden the low (600–700 m) Markotkh Ridge extended along the shore flows downhill. The high wind speed values are caused by the large density differences between the still warm air over the sea surface and the cold air; the latter downwells from the ridge as a waterfall featuring strong acceleration. The topography of the ridge may provide an additional bora enhancement. Similar to a fluid, the air moves over the lines of least resistance; therefore, boras are stronger on the ridge passes.

The Novorossiisk boras most often emerge when the center of the European part of Russia is occupied by an intensive cold anticyclone, at the southern margin of which strong northeasterly winds develop. Often, boras are induced in the rear parts of the so-called "diving cyclones" that travel from Scandinavia and Karelia to the Lower Volga or Southern Urals. In all these cases, the boras on the Black Sea coasts are related to the penetration of cold air into the southern part of the European part of Russia. Usually, boras are observed in the wintertime, while the most intensive events are confined to the end of the fall to the beginning of the winter, when the sea is still warm as in the summer, while invasions of very cold Arctic air from the continent are already possible.

The wind speed during bora events may reach 25–30 m/s; in selected cases on mountain passes it is as high as 60 m/s. Boras result in icing-over of ships and port constructions since seawater splashes immediately freeze over their surfaces. Many ships couldn't withstand the attack of bora and went down under the ice load and hurricane winds.

Wind speed. Over the open sea, the wind speed is greater than that on the coasts throughout the year. In all the months, the highest speed values are noted in the northern part of the sea except for the southeastern coasts of the Crimean Peninsula. The least values are observed in the southeastern part of the sea. According to the data of meteorological stations, weak winds with speeds less than 5 m/s dominate throughout the year over the major part of the coasts. The number of days with strong winds (> 15 m/s) is the greatest on the northeastern and southwestern coasts (34–35 days per year). The least number of such days (20–22 days per year) are characteristic of the southern coast of the Crimea and the southeastern regions of the Caucasian coast. On average, the mean annual wind speed over the sea increases from the south to the north and comprises 4–6 m/s. The highest wind speed over the open sea probable once per 100 years makes 40 m/s. Almost everywhere, the annual trend of the wind speed is characterized by an increase in the cold period and a decrease in the warm period of the year (Fig. 3).

Atmospheric pressure. The regime of the atmospheric pressure over the Black Sea is defined by the influence of the Azores and Asian anticyclones, by the area of the wintertime cyclonic activity over the Mediterranean Sea, and by the summertime thermal depression over North Africa and Asia Anterior. Seasonal changes in the air and sea surface temperatures additionally affect the pressure field. In addition, cyclone passages may cause rather rapid and significant aperiodic changes it atmospheric pressure.

The annual trends of the atmosphere pressure feature a regularity common of the entire Black Sea and represented by a distinctly manifested low in the summertime (July) and an insignificant secondary low in the spring, in April (Fig. 4). During the winter half of the year, the general pressure background is elevated and is rather similar over the entire Black Sea. Pressure growth is most rapid in August and September and lasts until January; later, the pressure begins to irregularly decrease.

During the period from May to October, mean pressure values over the western part of the sea are higher than those over its eastern part. In other months, the pressure low is mostly observed over the central parts of the sea. The variability of the atmospheric pressure throughout the year is the highest in the northern part of the sea and the lowest in its southeastern part.

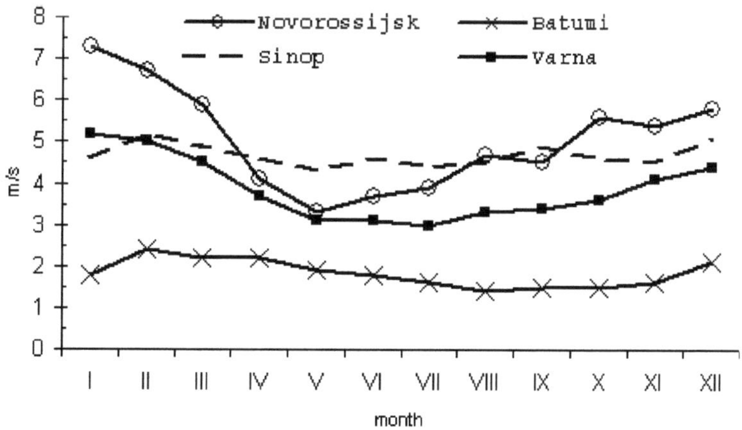

Fig. 3 Mean monthly wind speed, m/s

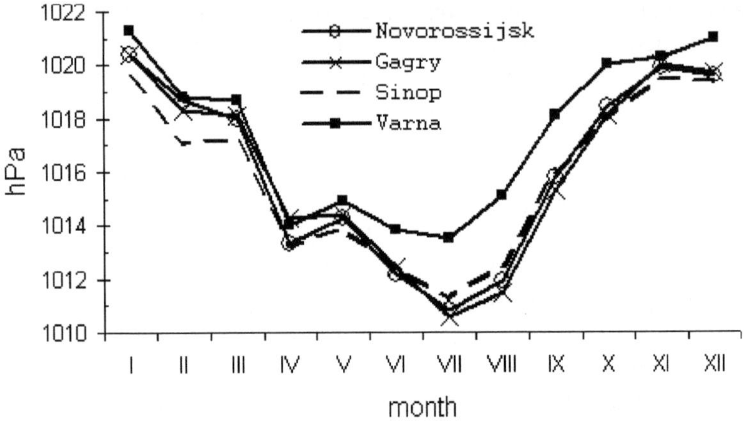

Fig. 4 Mean monthly atmosphere pressure at hydrometeorlogical stations, hPa

Air temperature. The mean annual air temperature over the Black Sea ranges from 10 °C in the northwest to 14–15 °C in the southeast. From August to March, the air temperatures over the open sea are higher than on the coasts. In the spring the situation changes. Owing to the increase in the solar irradiance flux, the land is rapidly heated and becomes warmer than the sea, which has greater heat capacity and, during the summer, accumulates a large amount of heat. Due to the effect of the sea, the variability of the air temperatures on the coasts is greater than that over the open sea. The greatest annual temperature variations are characteristic of the northwestern part of the sea, while the central and southeastern parts feature the least variations.

From September to March, the air temperature distribution over the sea is quasi-zonal. The highest temperatures are noted in the southeast and southwest of the sea in the regions with great sea depths. The lowest mean monthly temperature (down to negative average values of –1 to –2 °C), are observed in February in the northwest; the highest values (up to 24 °C) were registered in August off the Caucasian coast (Fig. 5b).

The differences in the annual temperature trends over the sea are best expressed in the wintertime, when the mean monthly values in February range from – 2 °C in the northwest to 7.5 °C in the southeast. During the warm season, the differences are not so great (Fig. 6) – the mean values for August vary from 21.5 °C in the northwest up to 24.0 °C in the southeast.

The low mean daily air temperatures (–23 to –25 °C) are noted under the northerly, northeasterly, and easterly synoptic processes in January and February. When these processes give place to the southeasterly, southerly, and southwesterly ones, the temperature growth may be rather rapid, reaching 8–10 °C per day. On the whole, negative temperatures occur over the entire sea area; they are mostly noted in January and February with highest recurrences over the northwestern and northeastern parts of the sea—in these

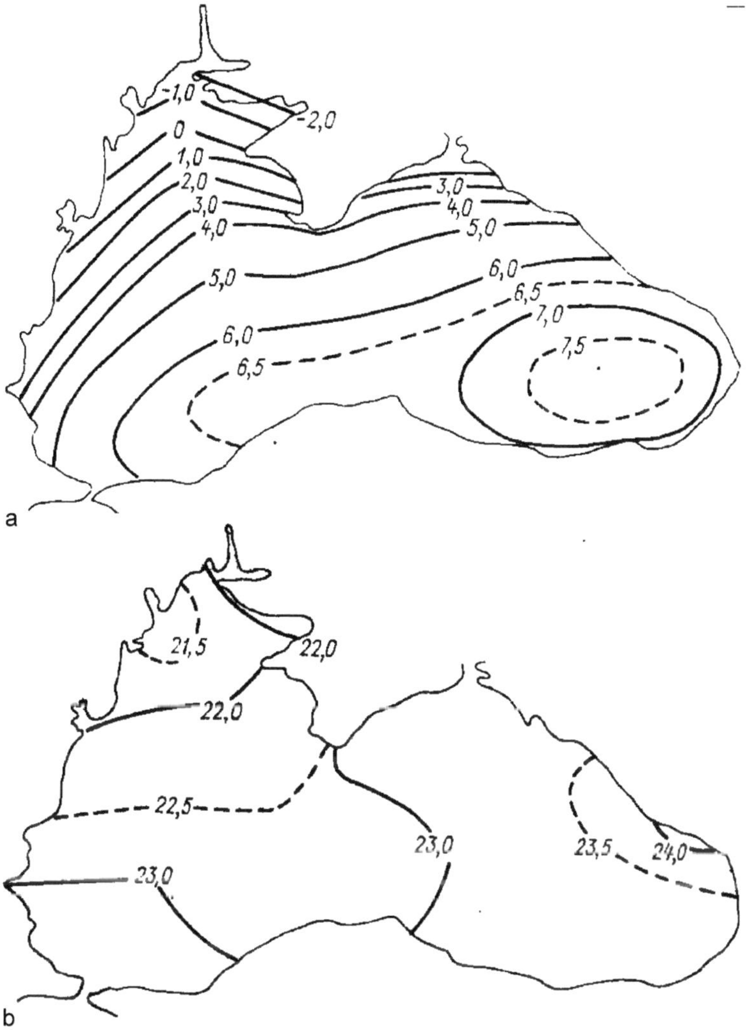

Fig. 5 Mean air temperature, °C, **a** in February and **b** in August

regions, from 6 to 10 days per month with temperatures from 0.0 to – 4.9 °C are registered. In selected years, the number of days with negative air temperatures may reach 22–26 in January and February and 13–15 in December and March.

The warmest regions are the Caucasian and Anatolian coasts of the Black Sea. There, air temperatures rarely drop below zero. In selected years, the number of such days may comprise 5–8 in January and February and 1–2 13–15 in December and March. In these regions, mean daily temperatures below -5 °C are noted once per 10–20 years, on the average. On the southern coast of

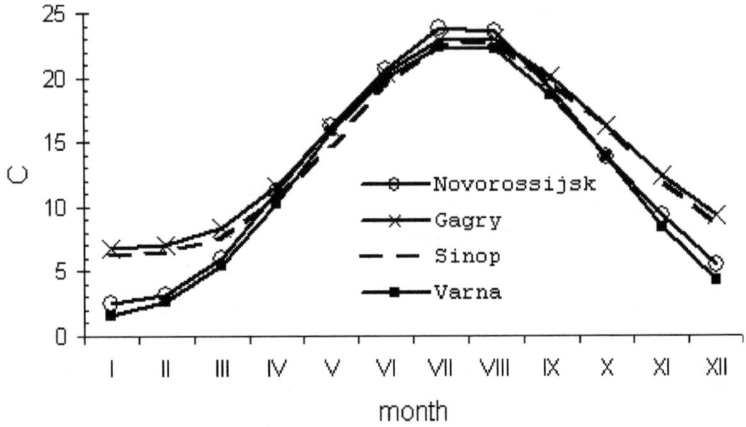

Fig. 6 Mean monthly temperature, °C

the Crimea, where mountains prevent cold from penetrating from the north, air temperatures below 0 °C are also rare: on the average, 3–4 days in January and February (up to 13–15 days in selected years) and 1–2 days in December and March.

On the coast, mean air temperatures below –10 °C are observed but not every year. Only in the northwest and northeast of the sea may these temperatures be observed during 7–14 days in January and February, 5–8 days in December, and 2–3 days in March. On the southern coast of the Crimea, over the past 75 years, only two cases were registered when the mean daily air temperature dropped below –10 °C.

Against the background of temperatures below 10 °C, sharp warmings are possible,which are caused by the warm air advection from the south in the process of cyclonic activity. In mountainous parts of the coast, warmings related to the foehn events are especially sharp. In these cases, the temperature growth rate may reach 15 °C per day and more. On the contrary, during the bora events, one observes a sharp temperature drop by 10–15 °C as compared to the temperature before the beginning of this wind.

The period with a stable mean daily air temperature of 20 °C and higher is the shortest in the northwest of the sea; here, lasts it from the end of June to the beginning of September. Its average duration in this region comprises 70–80 day; toward the southeast of the sea, it grows up to 100–110 days per year. Over the major part of the coasts, mean daily temperatures higher than 30 °C are possible in the summertime. Meanwhile, the mean daily temperature threshold of 35 °C was never overcome. The number of moderately hot days with a mean daily temperature from 20 to 25 °C is especially great in July and August and, on average, comprise 20 days per month (up to 25 days per month on the Anatolian and the southern part of the Caucasian coasts). Hot weather with a mean daily temperature higher than 25 °C is observed over

3–9 days per month in the north of the Black Sea coast, 10–11 days on the southern coast of the Crimea, and 2–7 days in the southern part of the Caucasian coast. Over the entire coast, the greatest recurrences of hot days are confined to July and August. At mean daily temperatures of 30 °C and higher, the maximum temperature values may reach 35–40 °C.

Daily trends of coastal air temperatures feature maximum values at 13–16 h; in the summertime, the peak is observed later than in the winter. The temperature minimum is observed early in the morning at sunrise. The extreme temperature values in the open sea are delayed by 1–2 h as compared to the coastal areas. In mountainous parts of the coast, against the background of the nighttime temperature decrease, its short-term peaks (by 0.5–5 °C during a few tens of minutes) are sometimes observed [4]. The reason for this kind of foehn effect lies in the irregular pulse character of the air downwelling at a nighttime mountain wind and its adiabatic heating. In this process, the moisture content may drop by 5–10%.

Over the entire sea, the daily amplitudes of the temperature variations in the winter period are greater than those in the summertime, except for the eastern region, where it is greatest in the fall. In so doing, the daily amplitudes grow from the southeast to the northwest. The inter-daily temperature variability generally decreases from the north to the south; in the cold season, it is 2–3 times as great as in the warm season. On average, coolings are more intensive than warmings: inter-daily temperature drops may reach 10–15 °C, while temperature rises rarely exceed 10 °C.

Moisture content. The regime of the moisture content over the sea is determined by the processes of interaction between the air and the sea surface. In coastal regions, the diurnal variations in the moisture content are additionally affected by the breeze circulation. The daytime breeze supplies humid air from the sea to land areas. In contrast, the nighttime breeze delivers dry air to the sea surface. The flux of relatively dry air is also provided by foehns and boras.

The intraannual changes in the water vapor partial pressure follow the annual trend of the air temperature over the sea (Fig. 7a). The lowest values are observed in January and February, while the highest are confined to July and August. During the entire year, the spatial distribution of the water vapor contents also corresponds to the air temperature distribution. The lowest values of the partial pressure of water vapor are noted in the northwest of the Black Sea (4.7–20 hPa on the coast and 5.0–21.0 hPa over the sea). The water vapor content grows in the southeastern direction (7.2–23.4 hPa on the coast and 8.0–24.0 hPa over the sea).

The annual trend of the relative moisture content over the greater part of the Black Sea shows its maximum values in the cold season of the year and the lowest values in the warm period. The humid subtropical areas of the eastern coast are characterized by a somewhat distinct regime; here, the highest values are observed in the summer and the intra-annual variations are in-

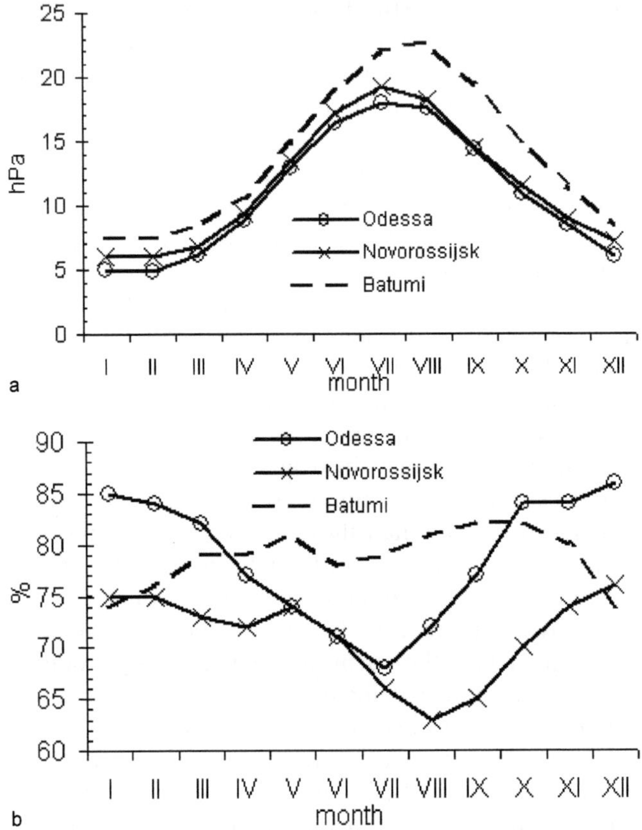

Fig. 7 Annual trends of the **a** partial pressure of water vapor, hPa and **b** relative moisture content, %

significant (Fig. 7b). The regularities of the spatial distribution of the relative moisture content over the Black Sea in the summer season are similar to those of the distribution of the water vapor partial pressure. The lowest values of the relative moisture content are characteristic of the northwestern areas, while the highest values are confined to the southeastern and southwestern parts of the sea. On the contrary, in the wintertime, the relative moisture content grows from the southeast to the northwest.

Atmospheric precipitation. Atmospheric precipitation over the Black Sea is mostly related to the cyclonic activity. The convective process plays a noticeable role only it near-shore band and on the coasts. An additional influence is provided by the topography of the coastal zone. Throughout the year, the precipitation amount grows from the northwest (380–420 mm/year) to the southeast, where the Caucasian ridges approach the coastline and are oriented across the principal moisture-bearing airflows (up to 1500–2500 mm/year) (Fig. 8). The greatest number of days with precipita-

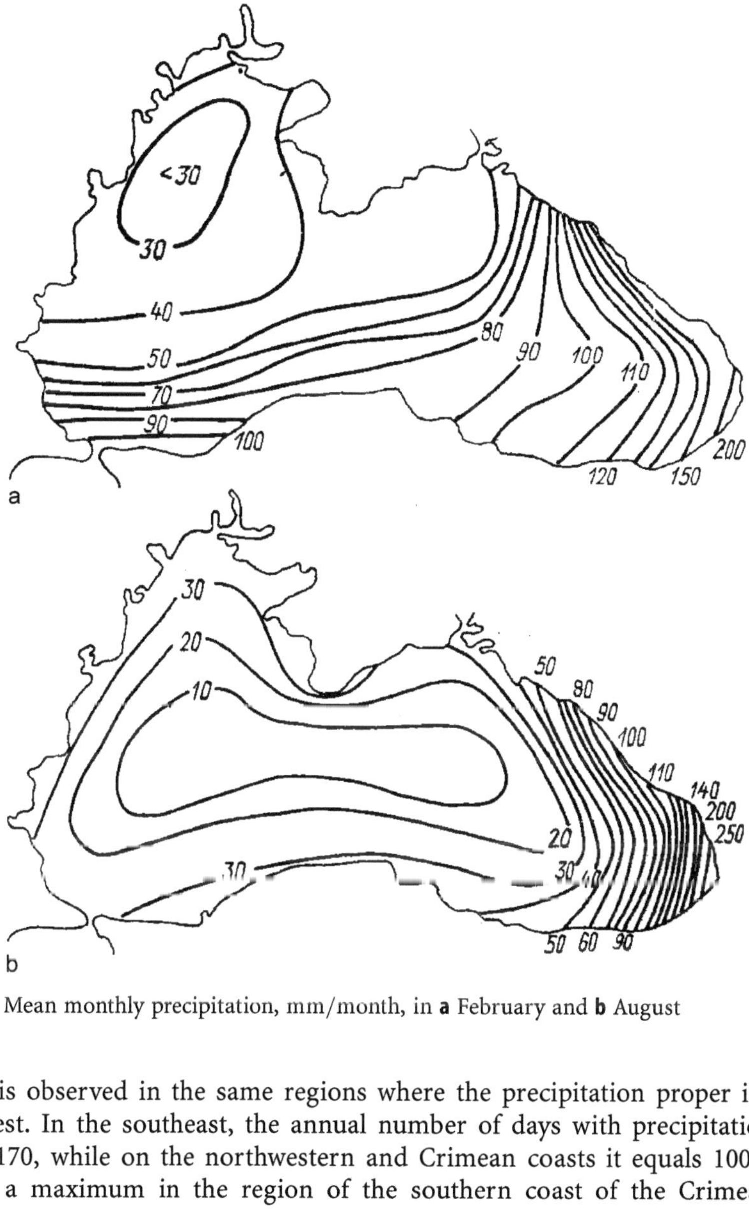

Fig. 8 Mean monthly precipitation, mm/month, in **a** February and **b** August

tion is observed in the same regions where the precipitation proper is the highest. In the southeast, the annual number of days with precipitation is 100–170, while on the northwestern and Crimean coasts it equals 100–125 with a maximum in the region of the southern coast of the Crimea. In the summertime, the precipitation intensity is greater. In the winter season, especially on the northern coast of the sea, precipitation may take the form of a snowfall, though no stable snow cover can be formed. On average, during the winter, 25–40 days with a snow cover on the northwestern coast, 15–25 days in the Crimea (on its southern coast, not greater than 15 days), 14–17 days in the northeast, and less than 15 days in the southeast are observed.

3 Wind Waves

According to the character of the wind activity over the sea, heavy waves develop mostly in the autumn and winter in the northwestern, northeastern, and central parts of the sea. In the sea, depending on the wind speed and wave vector distance, waves with heights of 1–3 m dominate. In open sea regions, the maximum wave heights may reach 7 m; at strong storms, they may be even higher. The southwestern and southeastern parts of the sea are the calmest; here, strong winds are rare and usually wave heights do not exceed 3 m even at storms.

The wind wave regime of the Black Sea is poorly studied, since there were virtually no regular instrumental observations of waves in the open part of the sea. The principal characteristics of the waves were determined using calculations; this also refers to the wave heights cited above. Meanwhile, under extreme conditions, the wave height may be significant. For example, on November 14, 1854, during the Crimean war, in the region of Balaklava, the joint English–French squadron of 34 battleships sunk; the losses reached 1500 seamen. Later, such extreme storms were registered in November of 1969, 1981, and 1992. In these cases, the maximum wave heights in the open part of the sea may reach 14–15 m.

The strong waves that develop during storms create serious obstacles for practical activities in the sea and on the coasts, such as dangers for navigation, destruction of coastal constructions, and, recently, from losses at the prospecting and extraction of hydrocarbon resources. In some cases, the storm activity is enhanced by local winds owing to the orographic effects.

One of the most hazardous aftereffects related to the wind wave action is the appearance of the so-called tyagun phenomenon in selected ports of the Caucasian coast. In these cases, the ships in the ports both moored and anchored start to spontaneously move; they are pressed to the piers or, on the contrary, moved away from the moorings breaking their fastening ropes. This phenomenon may last a day or longer, the reason being supposedly related to the generation of a resonance of natural oscillations of the water mass in the basin of the port caused by the penetration of the long swell waves into it with the free oscillations of the ship moored. The tyagun is most frequently observed in the port of Tuapse (up to 20 cases per year) and in the ports of Poti and Batumi (5–7 cases per year) [1, 5].

4 Water Balance

Calculations of the water balance of the Black Sea were performed by many scientists and their results are naturally slightly different. This depends on the values of the data taken as the basis for the calculations and of the periods of averaging. The component most difficult for its estimation is the water exchange via the Bosporus Strait, due to its strong variability and lack of instrumental data. Here, the information on the water balance mostly corresponds to the most complete dataset for the period 1923–1985 presented in [1]. Meanwhile, recently V.N. Mikhailov obtained refined data on the riverine runoff to the Black Sea (see corresponding chapter), which yields a value 16 km^3/year greater than the commonly accepted one [1]. Due to this, we slightly corrected the loss component of the water exchange via the Bosporus Strait.

The receipt part of the water balance of the Black Sea consists of the riverine runoff, atmospheric precipitation, and marine water supply via the Bosporus and Kerch Straits. A small contribution is also provided by the ground water delivery. The expenditure part of the balance includes evaporation from the water surface and the removal of the Black Sea waters via the Bosporus and Kerch Straits. The mean annual value of these components of the balance (under certain assumptions) comprises about 816 km^3/year, that is, only 0.15% of the total volume of the Black Sea waters. Approximately 354 km^3 of riverine waters is annually supplied to the sea; of them, up to 200 km^3 is contributed by the Danube River. The atmosphere precipitation in the form of rain and snow provides 237 km^3 of water. The lower current via the Bosporus Strait annually delivers about 175 km^3 of saline waters of the Sea of Marmara, while the Kerch Strait supplies approximately 50 km^3 from the Sea of Azov. The mean annual water expenditure for evaporation comprises up to 396 km^3; the upper current in the Bosporus Strait removes about 385 km^3 of the Black Sea water to the Sea of Marmara, and the water removal via the Kerch Strait to the Sea of Azov makes up to 35 km^3. Thus, the receipt part of the balance mostly consists of riverine waters, which comprise about 40% of all the water supplied. This component is characterized by a strong variability. The expenditure part of the balance consists of evaporation and water removal via the Bosporus. Meanwhile, evaporation features a low variability and thus has no significant effect on the variations in the water regime.

The distribution of the waters supplied over the sea area is quite irregular. The riverine runoff is mainly concentrated in the northwestern part of the sea (up to 80%) and, to a smaller extent, in the southeast. The waters of the Sea of Azov with a salinity of 10–14 psu flow via the Kerch Strait to the northeastern part of the Black Sea. They feature a low density and propagate with currents in the upper sea layer. The saline (about 30 psu) waters of the Sea of Marmara are delivered with the lower Bosporus current to the southwestern part of the sea at a level of about 50 m.

The degree of the desalination of the upper water layer of the sea throughout the year mostly depends on the volume of the riverine runoff and its distribution over the sea area. In the spring–summer season, the sea receives about 60% of the annual runoff volume. On average, the maximal and minimal runoff values are noted in May and September, respectively. The seasonal and interannual changes in the components of the water balance finally affect the sea level and the water exchange via the Bosporus Strait.

In the above-described version of the water balance, we assumed equilibrium between its receipt and expenditure parts, which is rather conventional. For example, owing to the sea level rise recently observed in the Black Sea (see this chapter), the water supply should exceed the water loss by approximately 2 km^3/year (or even more). However, the tendencies of the sea level changes at different sites of the coast are different and not everywhere registered. Therefore, in order to generally estimate the water regime, we found it reasonable to present the water balance rather than the water "budget" of the Black Sea.

5
Sea Level

The observations of the Black Sea level started in the middle of the 19th century. The longest observation series are available in Romania for the ports of Constanta and Sulin (since 1858), in the Ukraine (Ochakov, since 1874; Odessa and Sevastopol, since 1875), and Georgia (Poti, since 1874 and Batumi, since 1882). At present, the observation network of the Black Sea level includes 30 stations (of them, 13 in the Ukraine, five in the Russian Federation, four in Bulgaria, three in Romania, three in Turkey, and two in Georgia). During the past decade, the sea level is also studied with the use of satellites.

Multiannual sea level changes. An analysis of the data of observations of the Black Sea level over the last century allowed one to recognize to stages in its multiannual variability [6]. At the first stage (from the beginning of the observations up to the middle 1920s), the level was relatively stable with a slight tendency to fall. Subsequently, it began to rise at an average rate of 0.16 cm/year (Table 1); during selected intervals, this rate was significantly higher. For example, at the Tuapse hydrometeorological station, the general tendency over the entire period of observations is 0.23 cm/year, while during the last decade of the past century, it comprised 1.2 cm/year (Fig. 9).

The present-day rise of the Black Sea level is most often explained by the variability of the components of water balance of this basin [7, 8]. Among the other reasons for this phenomenon one should note the general level rise in the Atlantic Ocean [9].

In the multiannual variability of the Black Sea level, in addition to the tendency to the sea level rise, we can distinguish some reliable interannual periods: 2.5, 3.5–4, and 10–20 years (Fig. 10). These sea level oscillations are most often explained by the changes in the freshwater balance of the Black Sea [9].

Table 1 Tendencies in the sea level changes at different points of the Black Sea

Hydrometeorological station	Observation period	Tendency, cm/year	Standard deviation, cm/year
Odessa*	1923–1995	0.44	
Khorly*	1923–1995	0.14	
Blacksea*	1927–1995	0.17	
Evpatoriya*	1923–1995	0.18	
Yalta*	1927–1995	0.19	
Sevastopol	1910–1994	0.13	0.028
Feodosiya*	1923–1995	0.14	
Anapa*	1923–1995	0.13	
Novorossiisk*	1923–1995	0.13	
Tuapse	1917–2002	0.23	0.027
Sukhumi*	1926–1995	0.12	
Batumi	1882–1996	0.18	0.021
Burgas	1929–1995	0.16	0.043
Varna	1930–1996	0.14	0.044
Constanta	1933–1996	0.13	0.051

(* after [12])

Seasonal sea level oscillations. Interannual level changes in the Black Sea are mainly defined by the variations in the water balance components, seawater density, and atmospheric pressure.

On average, the range of the seasonal sea level oscillations at the coastal hydrometeorological stations reach values of 15–20 cm; in so doing, the highest level is observed in June–July, and the lowest sea level standing is confined to October–November (Fig. 11).

The level changes caused by the atmospheric pressure variations (the inverse barometer effect) comprise approximately 7–8 cm with a maximum in July and a minimum in November–January [1].

The ranges of the steric sea level oscillations related to the changes in the seawater density are different over the Black Sea area [10]. The highest annual ranges of the steric oscillations are observed in the central (up to 20 cm) and southeastern (up to 16 cm) regions; their lowest values are characteristic of the center of the eastern part of the sea. The explanation of this kind of spatial pattern may be found while assessing the phases of the annual harmonics of the total level and its temperature and salinity components.

The maximum of the temperature component of the annual harmonic of the steric sea level is observed approximately simultaneously over the entire sea (in August) because of the location of the Black Sea in the single climatic zone. Meanwhile, the phases of the maximum of the salinity component in different regions of the sea differ by a few months. For example, in the near-

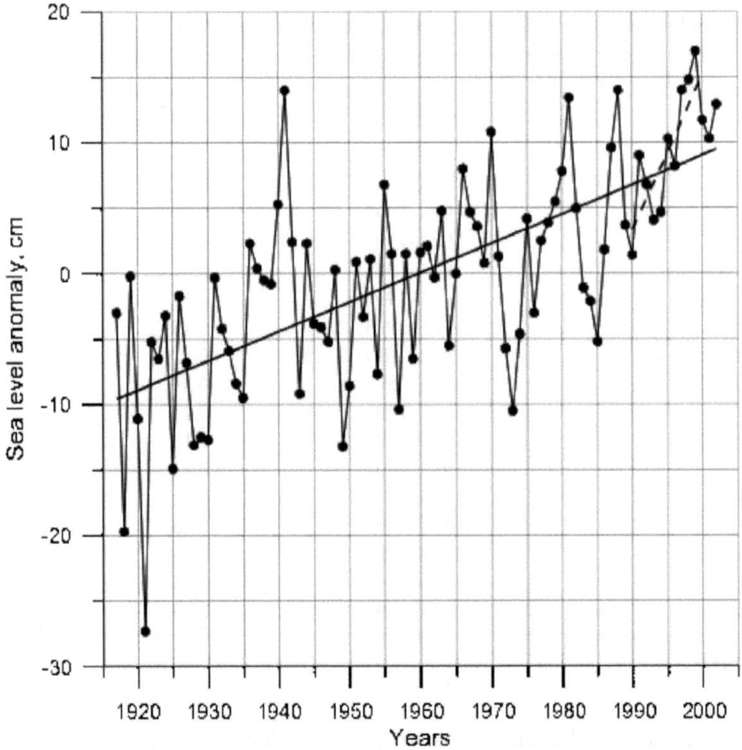

Fig. 9 Mean annual sea level anomalies in Tuapse according to the data of the PSMSL (Permanent Service for Mean Sea Level). The *solid line* shows the linear trend for the entire observation series; the *dashed line* represents the data for 1990–2000. Sea level anomalies were calculated with respect to the mean value over the entire time of observations at the hydrometeorlogical station

shore areas, the range maximum of this component is observed in the spring period, which is related to the riverine runoff, while in the central parts of the sea, the maximum is observed in August, which is the period of the lowest intensity of the cyclonic circulation. In the regions in which these phases coincide, their coupled range is equal to the total range; phase shifts result in discrepancies between these ranges (for example, the southwestern region).

Storm surges. These significant nonperiodical sea level oscillations are caused by coastal winds. In so doing, their range and duration depend on numerous factors such as the time of forcing, the wind direction and speed, the outlines of the coastline, the shelf depth, and the water stratification. The most complete characteristics of the storm surges in the Black Sea are presented in [1, 11].

It was shown that the most significant storm surges are noted off the western and northwestern coasts of the sea over small sea depths near the shore. Here, the storm surges are formed during the cold period of the year. In

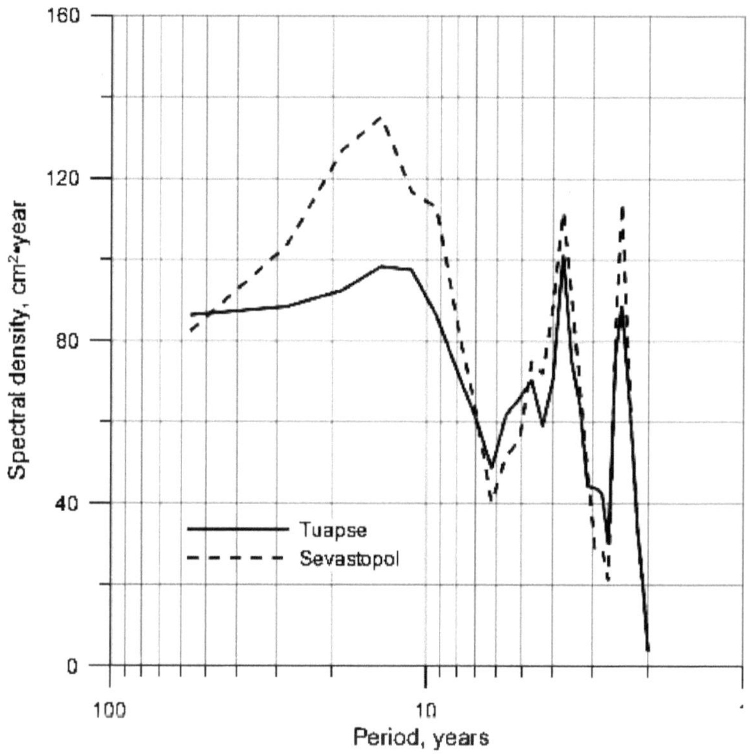

Fig. 10 Spectral density of the mean annual sea level in Tuapse and Sevastopol for 1917–2002

most cases, their values are about 30–40 cm. Storm surges with values greater than 40 cm are rarely observed; they are mostly confined to the autumn-winter period. Information about the maximum and minimum values of storm surges is shown in Table 2. For instance, near Primorskoe, the maximum height (115 cm) of storm surge was observed in the winter season. In the Odessa area it was about 100 cm in autumn. Off the Crimean coasts, the storm surges are small, while off the Caucasian coast their amplitudes may reach 70–90 cm.

The duration of the storm surge events varies over a wide range (2–57 h) and depends on the duration of the wind forcing and on the stability of the wind direction; in the shallow-water northwestern parts of the sea it is lower than in the deeper areas off the Crimean and Caucasian coasts.

Seiche sea level oscillations. The level of any basin, being turned out of its equilibrium state by a certain force, returns to its initial position performing decaying oscillations with respect to one or several horizontal lines (nodal lines) until their energy is expended for bottom and coastal friction. These free oscillations are known as seiches (uninodal or multinodal depending on

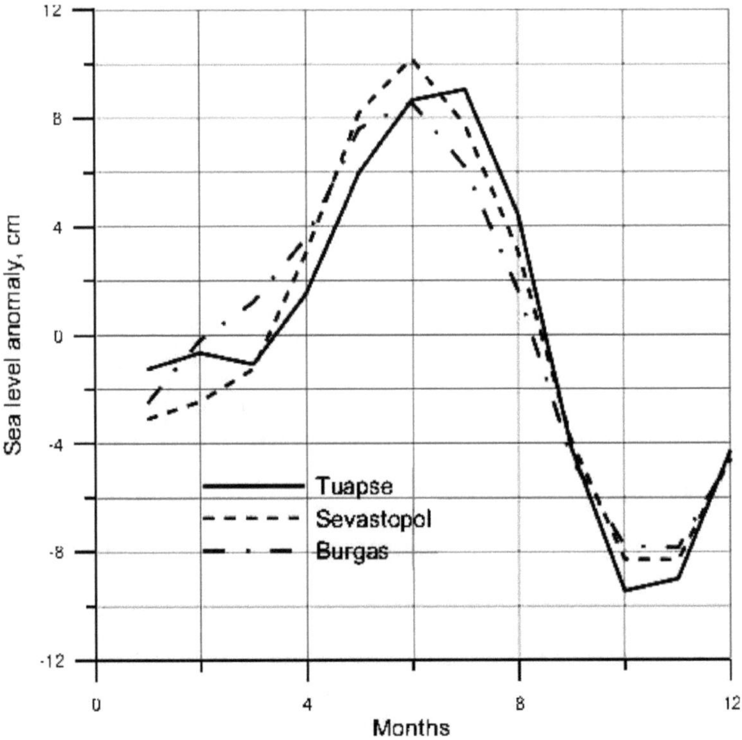

Fig. 11 Mean seasonal sea level variability (cm) in Tuapse, Sevastopol and Burgas

the number of nodal lines). The range of seiche oscillations is defined by the energy of the impact upon the surface of the basin. On nodal lines, no oscillations occur. Their maximal ranges are observed near the coasts of the basin most remote from the nodal lines. The mechanisms of the seiche generation are different for completely closed and semi-enclosed basins. In closed basins, seiches are induced owing to direct impact of an external force (wind, atmospheric pressure, etc.), while in bays and bights seiches are mainly excited directly through the open boundary.

In the Black Sea, the ranges of the seiche oscillations of the basin as a whole are low (up to 7 cm), while in bights and bays they may reach 50 cm [6]. The period of the seiche oscillations depends on the mode of the natural oscillation, the size of the basin, and its depth. For the entire sea, it comprises a few hours (for ten first modes), for bays the periods may be as small as a few minutes up to 1–2 h. The duration of the seiche oscillations, in most of the cases, comprises 6–10 h.

In order to study the spatial structure of seiches and to estimate their periods in the Black Sea, the corresponding sets of equations of fluid motion were numerically simulated. In [13] it was shown that the greatest period

Table 2 Maximal values (cm) of storm surges with respect to the mean monthly level position in 1880–1985 [12]

Site	Month											
	I	II	III	IV	V	VI	VII	VIII	IX	X	XI	XII
	Storm surges											
Vilkovo	111	151	94	62	65	69	69	55	55	71	74	80
	77	77	78	78	53	56	50	69	56	66	54	67
Izmail	189	193	192	157	168	184	186	207	225	233	216	190
	177	178	229	161	165	179	174	126	108	125	148	161
Primorskoe	89	125	107	58	41	31	43	47	50	88	56	91
	64	67	48	41	34	45	36	48	76	89	72	48
Belgorod Dnestrovskii	62	49	46	55	57	50	42	42	54	59	71	64
	56	55	66	59	58	50	43	43	73	60	91	59
Il'ichevsk	72	74	70	57	67	41	38	43	59	92	106	57
	75	71	72	62	39	46	42	59	56	60	101	63
Odessa	106	95	96	87	93	67	66	75	79	120	109	99
	114	113	77	69	88	74	66	70	68	84	174	98
Ochakov	71	68	61	66	63	50	45	40	56	61	78	74
	83	94	83	69	58	48	37	75	52	62	80	79
Nikolaev	91	96	108	79	90	67	99	89	76	77	125	124
	90	94	110	100	72	74	58	60	70	73	105	99
Kherson	96	76	92	148	267	155	83	72	71	89	90	99
	82	87	97	96	113	61	52	38	67	72	88	137
Poti	78	84	65	55	55	53	49	52	53	71	74	85
	67	75	70	69	63	56	60	67	71	77	82	65
Batumi	65	57	91	43	44	39	44	40	55	82	90	102
	35	50	41	39	33	29	40	44	37	36	33	35

of the seiche oscillations in the Black Sea equals 9.7 h. This seiche is represented by a uninodal oscillation with the nodal line running over the seaward edge of the shelf that serves as a natural boundary between the shallow-water northwestern and the deep-water parts of the sea. The greatest ranges of seiche oscillations are observed on the shelf of the northwestern part of the sea, especially in Odessa bay.

Tidal sea level oscillations. In the Black Sea, tides are formed under the action of the tidal forces in the basin proper, which is limited in size; therefore, the tides are small. Over the entire sea area, tides feature a semidiurnal or irregular semidiurnal character. The prevalence of this type of tides is related to the closeness of the semidiurnal period to that of the first mode of free oscillations in the Black Sea (uninodal seishe).

Table 3 Mean and maximal tidal ranges in the Black Sea, cm [3]

Station	Mean values		Maximal tide
	Spring tide	Neap tide	
Poti	9, 6	2, 4	12, 1
Tuapse	3, 5	0, 2	4, 7
Novorossiisk	4, 9	1, 8	6, 3
Il'ichevsk	11, 6	4, 6	14, 5
Batumi	10, 0	2, 0	13, 0
Odessa	14, 0	2, 8	17, 0
Yalta	∼ 0	∼ 0	∼ 0
Sevastopol	∼ 0	∼ 0	∼ 0
Varna	3, 6	0, 2	4, 7
Burgas	8, 2	3, 0	11, 2

The spatiotemporal distribution of the tidal energy over the Black Sea proves the correctness of the inference on the seiche-like character of the Black Sea tides. For example, the tides in the western and eastern parts of the sea are almost precisely in antiphase. In doing so, the highest tides are observed in the northwestern and southeastern parts of the sea. Meanwhile, off the Crimean coast, almost no tides are noted.

The highest tides in the Black Sea are noted in Odessa Bay (up to 17 cm) and in the Poti–Batumi region (up to 13 cm) (Table 3).

Level oscillations caused by tsunamis. Tsunamis are sea waves generated by strong underwater earthquakes or intensive landslides. In the region of the Black Sea, several strong earthquakes accompanied by tsunami wave generation were noted. They include, for example, the Yalta earthquake on September 11–12, 1927, the Turkish earthquake on December 27, 1939, and the Anapa earthquake on July 12, 1966. The height of the wave generated by the Turkish earthquake reached 50 cm in Sevastopol, 53 cm in Novorossiisk, and 40 cm in Tuapse. The tsunamis observed were not hazardous. However, it is supposed that a strong earthquake (such as the Turkish one) with its epicenter located in the sea may generate a tsunami wave a few meters high.

6 Sea Ice

The Black Sea is a partly freezing basin. Ice is formed only in a narrow band in its northwestern part. Even in severe winters, it covers no more than 5% of the sea area, while in moderate winters the coverage comprises 0.5–1.5% of the area. In extremely severe winters, fast ice extends along the western coast to the south up to Constanta and floating ice may driven as far as to the Bosporus Strait. Over the past 150 years, ice flows in the strait were observed five times. In mild winters, only lagoons and selected bights are covered with ice.

In moderate winters, the boundary of the resting ice in the northwestern part of the sea runs at a distance of 10 km from the coast from Dniester Lagoon to the Tendrovskaya Spit. Farther, the ice edge crosses Karkinitskii Bay and reaches the middle part of the Tarkhankut Peninsula. The average thickness of the ice never exceeds 15 cm, but in severe winters it can reach 50 cm. In the Kerch Strait, the ice appears every year. The northern part of the strait up to the Tuzla Spit together with Taman' Bay is its most icy part. Here, the ice is most stable and has a thickness reaching 30 cm. In the southern part of the Kerch Strait, floating ice is observed in the middle–terminal winter period, while local ice is rarely formed. During the winter, the strait may be repeatedly opened and frozen. At strong northerly and northeasterly winds, large masses of compact and hummocky ice are accumulated at the northern entrance to the strait preventing from ship navigation. Meanwhile, under southerly winds, the strait is quickly released from compact ice.

Usually, the ice formation in the sea starts in mid-December while the maximum ice extension is observed in February. The sea is released from ice in March (early release at the beginning of March and late release at the beginning of April). The duration of the ice period ranges from 130 days in extremely severe winters to 40 days in mild winters.

The ice cover of the Black Sea is characterized by instability. In different regions of the northwestern part of the sea, ice can repeatedly appear and disappear. The number of such releases per winter is 2–4 times on average and may reach ten times or more. The ice coverage of the northwestern part of the Black Sea is well correlated with the air temperature in this region [1].

References

1. Simonov AI, Altman EN (1991) (eds) Hydrometeorology and Hydrochemistry of the Seas. vol IV: The Black Sea. Issue 1: Hydrometeorological conditions. Hydrometeoizdat, St. Petersburg (in Russian)
2. Burman EA (1969) Local Winds. Hydrometeoizdat, Leningrad, p 344 (in Russian)
3. Atkinson BW (1981) Mesoscale atmospheric circulation. Academic Press, London
4. Surkova GV, Arkhipkin VS, Mukhametov SS (2006) Coastal mesoscale meteorological processes on the Black Sea shore in summer time. Meteorol Hydrol 3:31 (in Russian)
5. Grinevetsky SF, Zonn IS, Zhiltsov SS (2006) The Black Sea Encyclopedia. Mezhdunarodnye Otnosheniya, Moscow, p 660 (in Russian)
6. Goryachkin YuN, Ivanov VA (2006) The Black Sea Level: Past, Present and Future. MHI NASU, Sevastopol, p 210 (in Russian)
7. Stanev EV, Peneva EL (2002) Regional sea level response to global climatic change: Black Sea examples. Global Planet Change 32:33
8. Tsimplis MN, Josey SA, Rixen M, Stanev EV (2004) On the forcing of sea level in the Black Sea. J Geophys Res p 109

9. Lappo SS, Reva YuA (1997) A comparative analysis of long-term variability of the Black and Caspian Seas level. Meteorol Hydrol p 12 (in Russian)
10. Arkhipkin VS, Berezhnoi VYu (1996) Steric oscillations of the Black Sea level. Oceanology (English Translation) 35:6
11. Fomicheva LA (1975) Storm surges, daily level oscillations, and seishes in the Black Sea. Trudy Gos Okeanogr Inst p 125 (in Russian)
12. Atlas of the Black Sea and Sea of Azov Nature Protection (2006) GUNIO MO RF, St. Petersburg, p 434 (in Russian)
13. Arkhipkin VS, Ivanov VA, Nikolaenko EG (1987) Modelling of the barothropic seishes in the Southern seas. In: Numerical modelling of hydrophysical processes and phenomena in the enclosed basins. Moscow, Nauka p 104 (in Russian)

Hdb Env Chem Vol. 5, Part Q (2008): 159–194
DOI 10.1007/698_5_090

Published online: 17 August 2007

General Circulation

Valentin S. Tuzhilkin

Geographic Department, Lomonosow Moscow State University,
Vorobievy Gory, 119992 Moscow, Russia
tvsmsu@gmail.com

Abstract An analysis and generalization of the published results of the field observations and hydrodynamic modeling (both diagnostic and prognostic) of the Black Sea general circulation are presented. Despite the relatively simple geometry of the basin and its bottom topography, the circulation features a complicated spatial structure and is characterized by strong seasonal, synoptic, and interannual variabilities. The upper 500-m layer is dominated by the jet Main Rim Current (MRC), which runs along the continental slope with mean ("instantaneous") velocities up to 0.5–0.7 (more than 1.0) $m\,s^{-1}$. On the seaward (shoreward) side of the meandering MRC, one finds quasi-stationary cyclonic gyres (anticyclonic eddies) with subbasin (mesoscale) sizes. In the case of their interaction with the MRC, they generate non-stationary mesoscale eddies. From the end of the winter to the fall, the MRC significantly weakens; while the eddy activity is enhanced. In the upper 500-m layer, the general circulation is controlled by the wind forcing, bottom topography, and baroclinicity of the waters. In deeper layers of the Black Sea, the circulation is weaker by an order of magnitude; it is more poorly organized and is still insufficiently studied.

Keywords General circulation · Spatial structure · Temporal variability · Field observations · Hydrodynamic modeling

Abbreviations

ADCP Acoustic Doppler current profiler
BSGC The Black Sea general circulation

CIL	Cold intermediate layer
CTW	Coastal trapped waves
EOF	Empirical orthogonal function
JEBAR	Joint effect of baroclinicity and bottom relief
MRC	Main Rim Current
NSAE	Near-shore anticyclonic eddy
SLE	Sea level elevation
TS	Temperature and salinity.
SBCG	Sub-basin cyclonic gyre
UML	Upper mixed layer

1 Introduction

In this chapter, we present a description of the general circulation of the Black Sea waters (BSGC), its three-dimensional structure, and seasonal and multi-annual variabilities. The studies of the general circulation provide a key to the solution of numerous fundamental and applied problems in different fields of oceanography in any region of the World Ocean, the Black Sea included. This was clearly understood by the pioneers of the Black Sea research; therefore, as early as at the end of the 19th century, the studies of the water circulation have become one of the principal trends in the Black Sea oceanography.

The first reliable schematic of the BSGC was proposed by N.M. Knipovich in his monograph [1]; it resulted from a synthesis of diverse (mostly indirect) information on the large-scale water motions obtained during the expeditions of the 1890s to 1920s headed by I.B. Shpindler, Yu.M. Shokal'skii, and N.M. Knipovich proper. In this schematic, the principal feature of the BSGC is represented by a circular alongshore current flowing in a cyclonic (anticlockwise) direction, whose core is confined to the 500-m depth contour; its width comprises 20–40 km and only off the southwestern coast, it grows up to 50–70 km. On the seaward side of this stream-like flow (which we refer to as the Main Rim Current – MRC) locate two cyclonic gyres: the western (between 29.5°E and 33°E) and the eastern (between 34.5°E and 40.5°E) gyres. West of the southern coast of the Crimea, a branch separates from the MRC, it is directed westward along 45°N; near the mouth of the Danube River, it merges with the current flowing from Odessa Bay and again joints MRC south of Cape Kaliakra. Knipovich termed the areas located inside the central sub-basin cyclonic gyres (SBCGs) as chalistatic areas, owing to the slow water motions in them without a clear general direction. The author of this chapter has to present a such detailed description of the Knipovich's schematic, because, accounting for the lack of information at that moment, it is surprisingly accurate in representing the general features of the BSGC in the upper 500-m layer. During the following 70 years, the scientists managed only to refine the positions and intensities of the elements of this schematic in different seasons and to recover selected additional details.

In the 1930s to 1940s, the well-known dynamical method for calculations of geostrophic currents was first applied to the estimation of the BSGC. The most cited generalization of the shipborne data available at that time with the use of this method was published by G. Neumann, who presented his results in the form of a schematic of the summertime BSGC [2]. As compared to the above-described schematic, it reflects a greater role of SBCGs with respect to the MRC; it explicitly shows local near-shore anticyclonic eddies (NSAEs) in the southeastern corner of the Black Sea, south of the southern coast of the Crimea, and west of the Bosporus. NSAEs located south of the Danube River delta, south of Cape Kaliakra, between Cape Sinop and the mouth of the Kizilirmak River and off Sukhumi and Kerch, are less clearly manifested. The maximal current velocities at the sea surface off the Caucasian coasts (38–40°E), the southern coast of the Crimea, and West Anatolian coast (32–33°E) exceed 0.40 m/s. Below the depth of the calculated zero dynamical surface (100–300 m), Neumann inevitably obtained an anticyclonic general motion of the Black Sea waters, since the dome-shaped structure of the density field in the Black Sea is traced at least down to a depth of 500 m [3].

Meanwhile, it was only in the 1960s when the first climatic fields of the dynamical topography of the Black Sea surface were published [4, 5]. They confirmed the above-listed principal features of the BSGC and complemented the ideas about its seasonal variability. In [4], the most distinct and intensive geostrophic BSGC in the upper layer of the Black Sea was observed in the summer. In [5], mean annual density fields were used to perform calculations for the entire sea area. The amplitude of the values of the dynamical heights of the level surface comprised 0.14 m, which yielded maximal geostrophic velocities of 0.20–0.30 m s^{-1}. More detailed seasonal fields of the dynamical topography in the Caucasian region of the sea, which are most distinct in the winter and the summer, provided velocity values up to 0.40–0.50 m s^{-1}. Taking into account the arguments presented by I.B. Shpindler at the end of the 19th century about the more intensive wind forcing in the winter and the seasonal variability of the density structure of the waters, D.M. Filippov reasonably suggested that the BSGC intensity reaches its maximum at the end of the winter and the beginning of the spring, while its minimum is confined to the autumn [5].

Over the vast northwestern shelf, the above-listed schematics of the BSGC showed a cyclonic water motion with velocities up to 0.20 m s^{-1}. A special research study [6] helped to reveal a strong dependence of the shelf currents on the synoptic wind patterns and a domination of anticyclonic current vorticity in the summertime.

In the 1960s to 1970s, numerous direct current measurements in the Black Sea were performed at 175 mooring stations, of which 86 were 4 to 29 days long. Their results were generalized in [7] in the form of a schematic that differs from the Neumann's one only by the appearance of an isolated southeastern SBCG (between 39°E and 40.5°E). Note that the possible splitting of the eastern SBCG into two gyres was supposed by Knipovich [1]. In the winter and summer, the current velocities off the Crimean and the western coasts of the sea reached 0.8–1.2 m s^{-1}, while off other coasts they were 0.8–1.0 m s^{-1}. In the spring and fall, they decreased down to 0.4–0.6 m s^{-1}. In the SBCGs, the velocities ranged from 0.2 to 0.6 m s^{-1}. The values that high (probably overestimated) have previously been registered only in the profiles normal to the coast with electromagnetic instantaneous current meters (see [5]).

In the middle 1970s, the first results of diagnostic and prognostic numerical modeling of the BSGC were published [8, 9]. Because of the coarse spatial resolution of the model grids and the insufficient reliability of the initial and boundary conditions, only their most general features corresponded to the concepts of the current pattern in the Black Sea that existed at that time.

A subsequent step in the development of the knowledge about the BSGC in the upper 300-m layer was made in the monograph by Blatov et al. [10]. For the first time, it presented the climatic dynamical topography of the Black Sea surface for the principal months of the seasons of the year (February, May, August, and September) with a horizontal resolution of 1° × 1° on the basis of a computer processing of an archived database that contained 25 000 shipborne observation of the vertical profiles of the temperature and salinity. In February, the BSGC is the closest to the above-described schematic by Knipovich. The wintertime range of the dynamical heights of the level surface reaches its maximum (0.28 m) in the western SBCG (in the eastern SBCG it is 0.24 m); at the one-degree resolution, this yields maximum values of the geostrophic velocity in the MRC of 0.20 m s^{-1}. In May, the range of the dynamical heights decreased down to 0.19 and 0.10 m, respectively, and the MRC is traced in fragments mostly in the western part of the sea and off the Caucasian coasts; the Batumi NSAE appears in the eastern part of the sea. In August, the MRC restores its continuity, the ranges of the dynamical heights in the western and eastern SBCGs become almost equal (0.19 and 0.18 m), and the geostrophic velocities reach 0.15 m s^{-1}. In November, the BSGC suffers the most significant weakening and transformation. The range of the dynamical heights decrease down to 0.16–0.17 m, the MRC is traced only along the western and Caucasian coasts, the western SBCG expands eastward up to 36°E, and the eastern SBCG is displaced toward the southeast. The Batumi NSAE disappeared, while an anticyclonic eddy in the northeast of the sea is formed separating the transformed waters of the SBCGs. These results confirmed the suggestion made by D.M. Filippov about the character of the seasonal variability of the BSGC. In addition, a series of new important results were obtained. The Batumi and Sevastopol NSAEs were first recognized as stationary elements of the BSGC. It was established that, in the upper 75-m layer of the Black Sea, the decrease in the geostrophic velocity with depth (the so-called vertical shift) never exceeds 15–30% and the correlation coefficients between the field of the dynamical topography of the 0–300-m layer and the salinity field in the layer 100–200 m are as great as 0.73–0.89. For the first time, it was noted that the seasonal evolution of the BSGC is related not only to the wind speed but also to the intensity of its relative vorticity. A great chapter in [10] was devoted to the observation and modeling of the MRC meandering and to the formation of several types of eddies in the Black Sea (frontal rings, shear eddies, and others). As a result, an important conclusion was made about the great significance of synoptic processes in the Black Sea in the formation of the large-scale dynamics of the BSGC.

By the middle 1980s, a tendency had been formed in world oceanography towards the development of systems for observing temperature, salinity, and current velocity as well as for computer processing and generalizing of their results and numerical modeling of the water circulation. Examples include acoustic Doppler current profilers, satellite-supported drifters, and altimeter observations, computer databases and algorithms for their objective analysis, nonlinear hydrodynamic models for mutual adjustment of the velocity, temperature, and salinity fields, and their later versions that imply data assimilation. It occurred that, starting from the beginning of the 1990s, all the above-listed technologies became applied for the studies of the BSGC, which resulted in a breakthrough in the quantity and quality of our knowledge on this issue. From 1990 alone, more than 200 papers in Russian and more than 100 papers in English have been published concerning the problem of the BSGC.

The subsequent sections of this chapter represent generalizations of the results of the studies of the BSGC performed during the past two decades with the use of various methods for observations and hydrodynamic modeling. In the most general form, they are shown in the schematic of the BSGC of the upper 500-m layer (Fig. 1), where a rather accurate mean annual positions and approximate sizes of its most characteristic elements are plotted. In contrast to the known similar schematics (starting from that by Knipovich), here, the climatic annual mean configuration of the salinity contour lines (from 19.8 to 20.2 psu) is shown at a depth of 100 m, that coincides with the core of the MRC of the Black Sea.

It should be especially noted that, over the 100-year-long history of the studies of the BSGC, the ideas about the circulation of the deep waters of the Black Sea (below a depth of 500 m) have always been kinds of scientific hypotheses or speculations (see [5, 10]).

Unfortunately, it is impossible to present a full list of the publications assessed below. Therefore, we preferred the principal publications that are the latest in series of papers of the same groups of scientists; usually, they contain a complete description of the history of the studies. In addition, publications in English were also considered. Unfortunately, it is impossible to reach any objective point of view; therefore, the author adduces his excuses to those researchers whose scientific efforts didn't find their due assessment in this chapter.

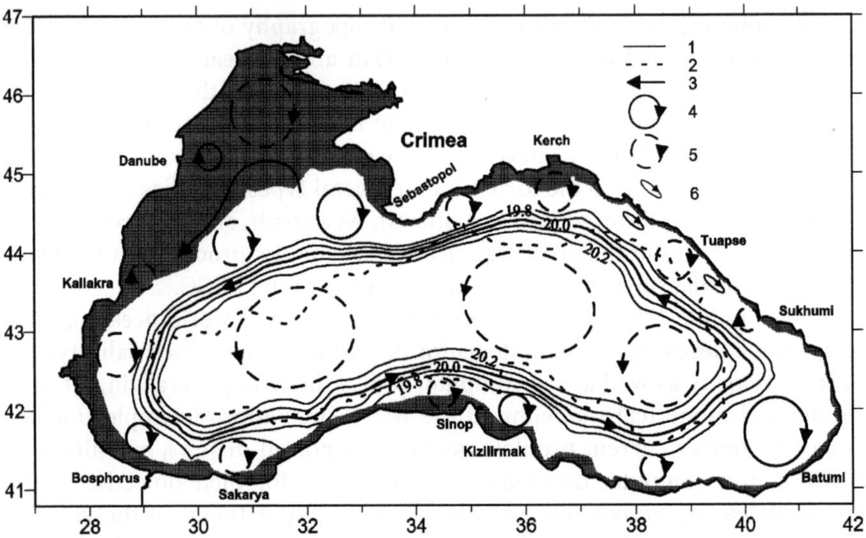

Fig. 1 General circulation schema in the Black Sea upper layer (0–500 m): 1 annual mean climatic isohalines (psu) at a depth of 100 m, 2 1000 m depth contour, 3 bifurcation branch of the Rim Current, 4 quasi-stationary eddies, 5 non-stationary eddies and sub-basin gyres, 6 non-stationary coastal vortices. The areas shallower than 100 m are *shaded*

2 Results of the Field Observations

Field observations of the Black Sea currents were performed using all the techniques available. At the early stages, currents were studied with the help of the bottle post and estimations of ship drifts; electromagnetic induction and propeller current meters lowered from ships were also used (see [5, 10, 11]). From the 1950s through the 1980s, observations at moored autonomous buoy stations with chains of propeller current meters dominated. In the 1990s, they were supplemented by ADCP soundings from ships and satellite observations over the ARGOS drifters and of the topography of the Black Sea level. In this section, we present the results obtained using modern observation technologies such as the mooring, acoustic, drifter, and altimeter measurements. Special attention is paid to the results of the moored buoy observations, since they allow one to most correctly determine characteristic temporal scales of the currents and the current vectors averaged over the period of the observations. In most cases, when the observations are sufficiently long-term (no less than a week long), these vectors provide the most adequate representation of the BSGC.

2.1 Mooring Observations

The total number of moored buoy stations in the Black Sea seems to exceed 1,000. Meanwhile, most of these observations were confidential and unavailable for a reliable quantitative generalization. For example, the numerous autonomous instrumental measurements generalized in the well-known study [7] resulted only in a rather conceptual schematic of the BSGC that, as has been mentioned above, almost does not differ from those compiled in the 1930s to 1940s.

The 1990s were marked by the publication of statistical characteristics of the Black Sea currents by the data of moored stations in various near-shore regions [11–16], including the continuous 5.5-year-long (1976–1981) observations off the northeastern coast in the region of Gelendzhik and those somewhat north of this area in 1997–2001, as well as the observations in the open sea at clusters consisting of 4–5 moored stations [17, 18]. In this section, the results of compilation of the published values of the mean current vectors are presented to illustrate the idea on the spatial structure of the BSGC. In all, about 100 moored stations with observation duration no less than a week were assessed. Observations with a duration of 10–40 days dominate; 12 moored stations operated from 50 to 120 days and observations at 150 stations (mostly in the near-shore zone) lasted from 150 to 256 days.

The long-term buoy observations [12, 16] showed that the most intensive current fluctuations in the Black Sea are characterized by temporal scales not greater than 10–12 days and in the near-shore zone this time is even less (7–8 days). Therefore, one can suggest that weekly averaged values can already represent the condition of the BSGC.

The mean vectors of the winter and summer currents in the surface (0–40 m), subsurface (50–75 m), and intermediate (100–300 m) layers of the Black Sea are presented in Fig. 2. They confirm the cyclonic character of the BSGC in the winter and summer. Even at depths of 1,000 and 1, 500-m (not shown), southwest and southeast of the Crimea, where the currents feature relatively low velocities and diverse directions, in nine and five of the total 19 cases, cyclonically and anticyclonically directed vectors, respectively, were observed (in the remaining five cases, the vectors had quasimeridional directions). In the surface layer near the shore, anticyclonic vector directions were observed only three times: twice in the wintertime off the Danube River mouth (see Fig. 2a) and once in the fall west of the Bosporus (not shown). These regions are known as areas of quasi-stationary NSAEs (see Fig. 1).

In the vertical distribution of the modules of mean current vectors (Fig. 3a), the highest vertical gradient (shear) is observed between 10 and 25 m (total mean values of 0.215 and 0.165 m s^{-1}, respectively). This is probably related to the effect of the wind drift. In the layer 25–50 m, mean velocities are homogeneous and the main shear in their values takes place deeper down

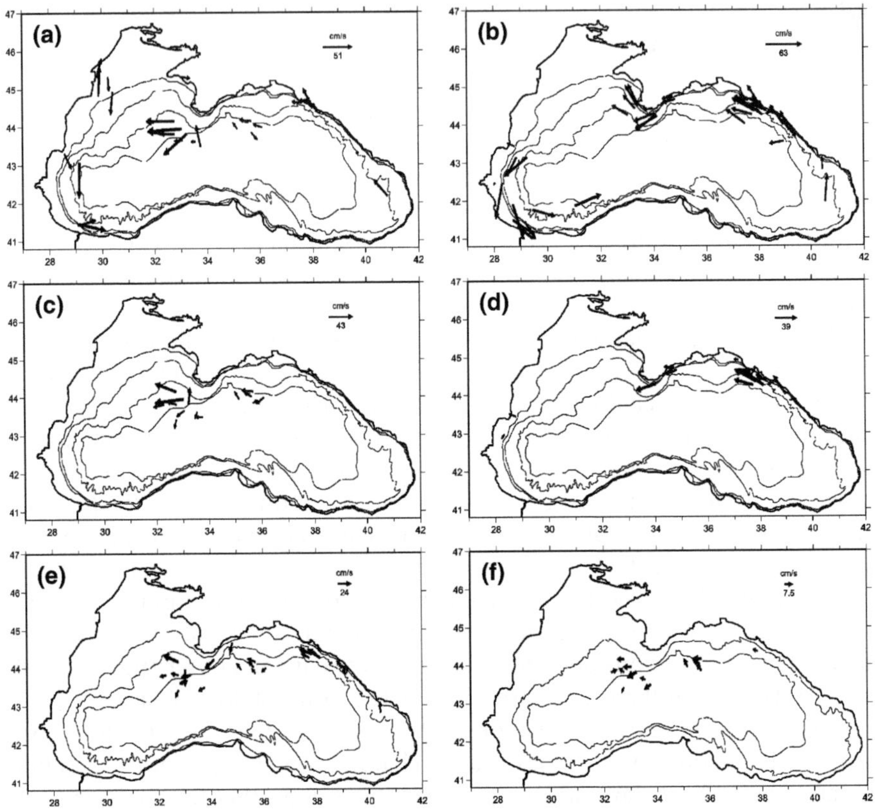

Fig. 2 The Black Sea mean current vectors from mooring observations **a, b** in upper layer (0–40 m): **a** in winter, **b** in summer; **c, d** in subsurface layer (50–75 m): **c** in winter, **d** in summer; **e, f** in intermediate layers: **e** 100–150 m and **f** 200–300 m. *Fine lines* in Fig. 2a–d 50, 100, 500 and 1000 m depth contours, in Fig. 2e 100, 500 and 1000 m depth contours, in Fig. 2f 300 and 1000 m depth contours

to a depth of 300 m. Below, down to a depth of 1500 m, mean velocity values become equally low (not higher than 0.11 m s^{-1}).

The maximal mean velocity (0.63 m s^{-1}, averaged over 13 days) was registered west of the Bosporus in July 1972 at a depth of 25 m. In seven current records at depths of 5–25 m, the "instantaneous" velocities exceed 1 m s^{-1}: in two cases off the Bulgarian coast and the rest of the cases were confined to the coasts of the Caucasus. The absolute maximum (1.41 m s^{-1}) was detected off Cape Kaliakra in September 1976 at a depth of 5 m at a point with a sea depth of 37 m [13].

The ratio of the standard deviation of the mean velocity to its value proper, which is called the variation coefficient, almost does not depend on depth (Fig. 3b). In most cases, the fluctuations of the currents are greater than their mean velocities, which is characteristic of sufficiently long observation at

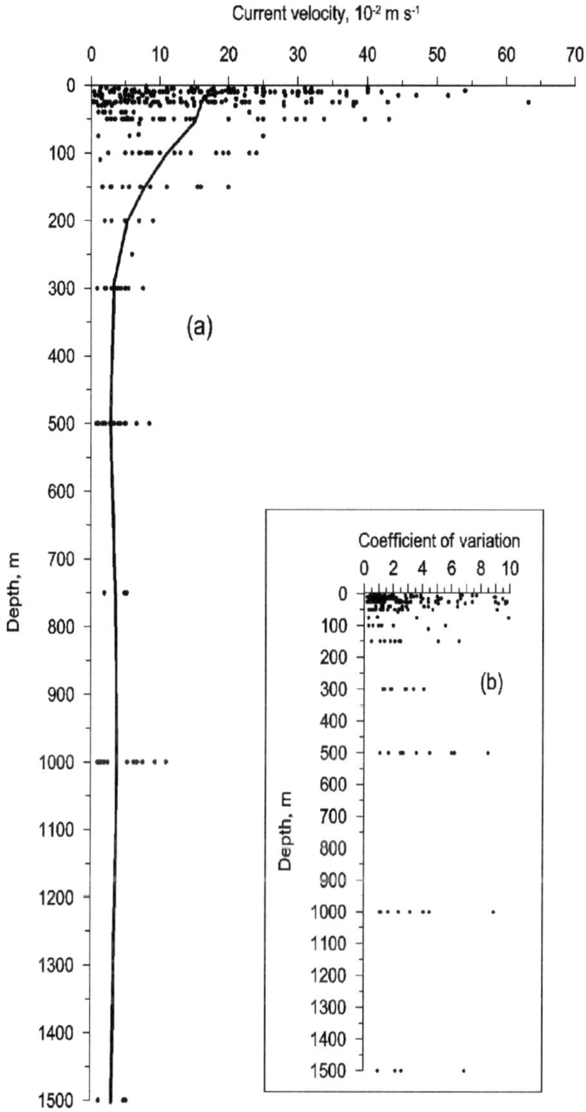

Fig. 3 **a** The Black Sea mean current velocity (10^{-2} m s^{-1}) and **b** its coefficient of variation versus depth (m) from mooring observations. *Thick line* in Fig. 3a total mean current velocity profile

worldwide moored stations. Meanwhile, among 59 mean velocity values that exceeded 0.25 m s^{-1} (most of which were registered in the core of the MRC of the Black Sea), 39 featured variation coefficients significantly lower than unity (0.54 on average) and only nine values were slightly higher than unity (1.34 on average). This clearly points to the very high stability of the MRC.

For example, according to the data of [14], the amplitude of variations of "instantaneous" directions of the MRC was only 30°.

The distribution of the mean velocities over the normal to the coast is presented in Fig. 4. The quadratic trends show that, in the 0–200, 200–750, and 1, 000–1, 500 m layers, the jet of the MRC is located at distances of 50–75, about 75, and 75–100 km from the shore, respectively. On the whole, the

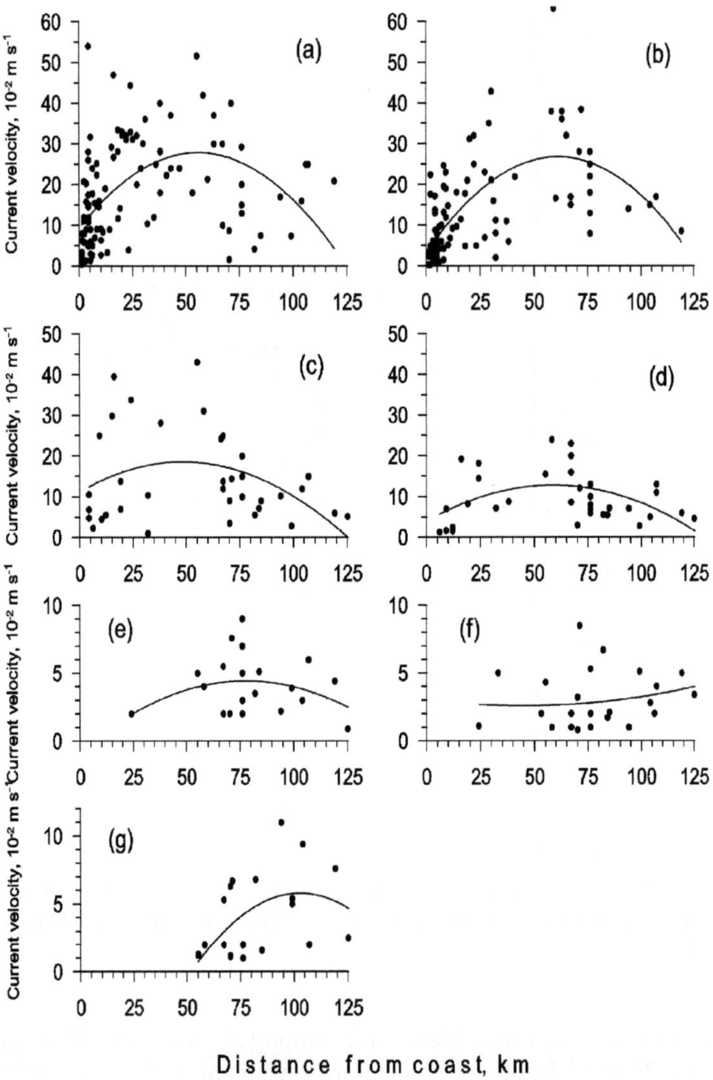

Fig. 4 The Black Sea mean current velocity (10^{-2} m s^{-1}) versus mooring observations distance from coast (km) at a depth ranges: **a** 0–15 m, **b** 20–40 m, **c** 50–75 m, **d** 100–150 m, **e** 200–300 m, **f** 500–750 m, **g** 1000–1500 m. *Solid lines* quadratic fits

trends are not statistically confident, though the curvatures of the parabolas regularly change with depth. In so doing, the distance of the MRC jet from the depth contour that serves as a coastline at the given depth level remains almost constant (about 50 km).

The estimates of the climatic mean annual parameters of the MRC in the sections normal to the northeastern coast of the Black Sea presented in [17] yielded a distance of its core from the shore about 40 km, a full width of the current (with respect to velocity values of 0.02 m s^{-1}) of 75 km, a penetration depth of 275 m, a maximal geostrophic velocity of 0.31 m s^{-1}, and a volume transport of 1.3×10^6 m^3 s^{-1}. These estimates are of the same order of magnitude as shown in Fig. 4a within the velocity interval to 0.20 m s^{-1}. This allows us to suggest a certain geographical universality (self-similarity) of the MRC profile normal to the coast.

The normal to the coast distributions of the variation coefficients of the currents in the upper 150-m and the lower 200–1500-m layers shown in Fig. 5 occurred fundamentally different. In the upper layer, the minimum of the variation coefficient in Figs. 5a–5c coincides with the velocity maximum in Figs. 4a–4d, which confirms the above conclusion on the high stability of the MRC. Meanwhile, in the lower layer, zones of the maximal velocities of mean currents in Figs. 4e–4g coincide with the maximums of the variation coefficient in Fig. 5d. Thus, in the intermediate and deep layers, the MRC is rather instable, which allows some researchers, following Neumann, to seek for an intermediate or deep countercurrent beneath the MRC [17].

One should also note the sharp increase in the variation coefficients in the near-shore zone approximately 15–20 km wide (see Figs. 5a–5c). In numerous publications concerning analyses of the observations off the Caucasian coasts [12, 13, 15, 16, 19, 20], this fact is interpreted as the effect of the chain of NSAEs that travel along the shore in a cyclonic direction. Meanwhile, the distinct alongshore polarization of the velocity fluctuations, their high coherence at distances greater than the horizontal sizes of the near-shore quasiperiodic structures, and their high spatial and temporal periodicities allow us to regard them as manifestations of coastal trapped waves (CTW). The theoretical grounds of their existence and their probable parameters in the Black Sea are presented in [21], and an example of interpretation of the observations of near-shore currents off the Crimean coasts based on the CTW theory is described in [14].

The seasonal variability of the mean current velocities according to the data of moored buoy observations shown in Fig. 6 demonstrates no clearly manifested structure.

This should be regarded as an unexpected fact, the more so as the numerous diagnostic calculations and rare long-term current measurements in the Black Sea suggest the existence of a winter–spring maximum and a summer–fall minimum in the BSGC velocities caused by the features of the wind regime over the Black Sea. For example, in [19], the calculations

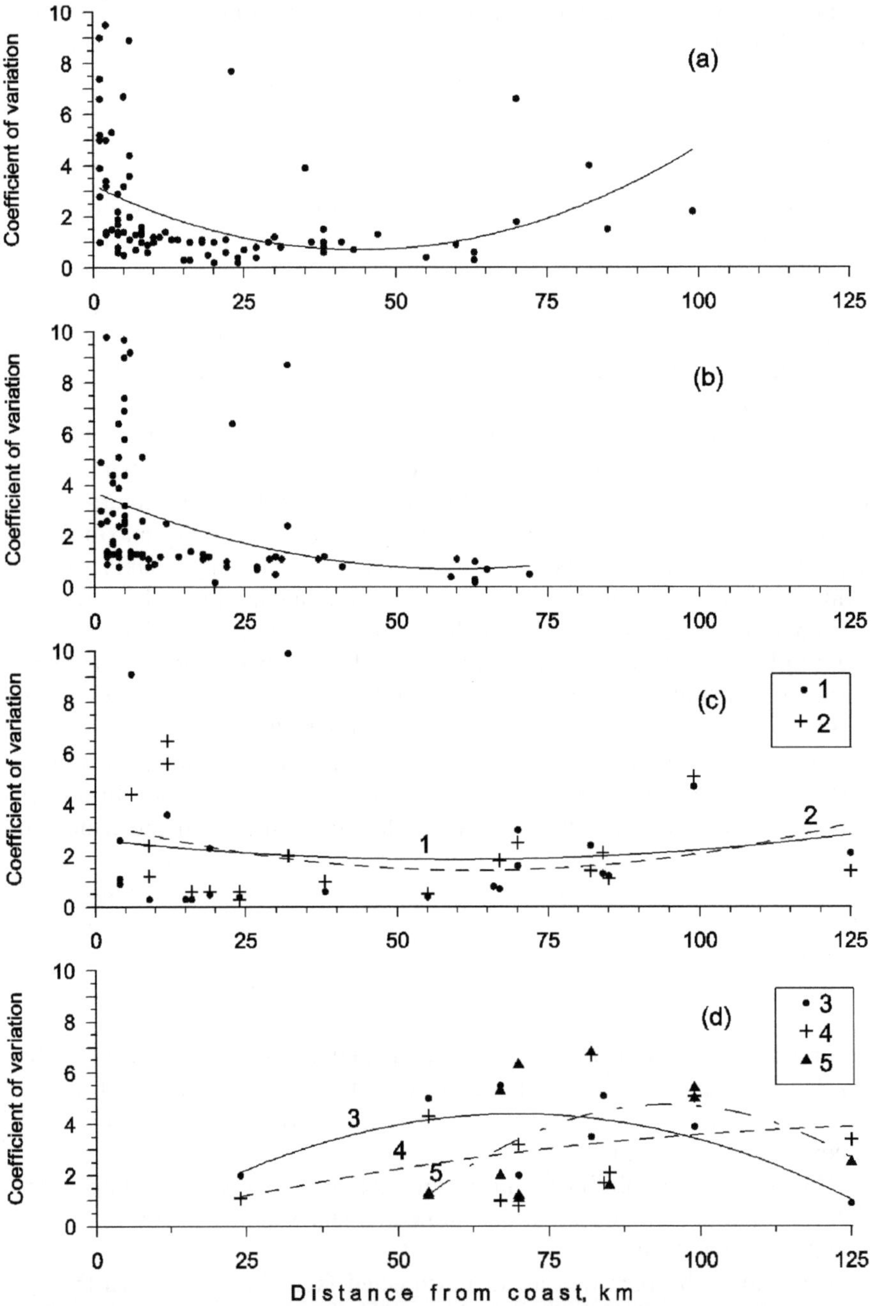

Fig. 5 Coefficient of variation of the Black Sea mean current velocity versus mooring observations distance from coast at a depth ranges: **a** 0–15 m, **b** 20–40 m, **c** 50–75 m (1) and 100–150 m (2), **d** 200–300 m (3), 500–750 m (4), and 1, 000–1, 500 m (5). *Solid*, *dashed* and *dotted-dashed lines* quadratic fits

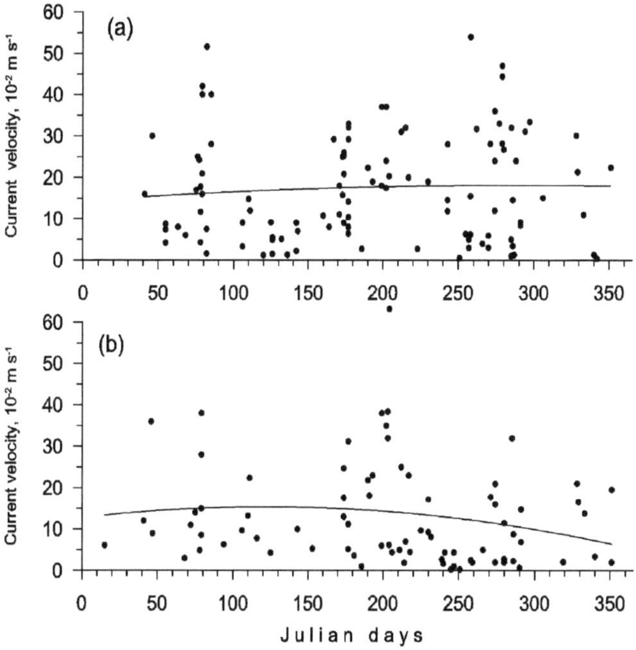

Fig. 6 The Black Sea mean current velocity (10^{-2} m s^{-1}) versus mean Julian days of mooring observations at a depth ranges: **a** 0–15 m, **b** 20–40 m. *Solid lines* quadratic fits

of the geostrophic velocities in 95 profiles normal to the coast performed in the region between Kerch and Sochi in 1954–2000 provided winter and summer velocity values in the core of the MRC of 0.35–0.40 and 0.20–0.25 m s^{-1}, respectively. Multiannual buoy observations in the near-shore zone off Gelendzhik [12] also showed a distinct seasonal variability of the mean current velocities from 0.09 m s^{-1} in December to 0.007 m s^{-1} in August. The most probable reason for the absence of distinct velocity variability in Fig. 6 lies in the combination of the observations performed at different distances from the coast. However, the data available are insufficient to compile reliable individual current patterns at various distances from the coast.

The interannual variability of the currents was estimated from the two above-mentioned cases of multiannual observations off the northeastern coast of the Black Sea. In the period 1976–1981, at a site located 5.5 km away from the coast with a sea depth of 70 m, at a depth of 25 m, the modulus of mean annual velocity increased from 0.027 to 0.041 m s^{-1} [12]. In the period 1997–2001, at a site located 3.5 km away from the coast with a sea depth of 60 m, at a depth of 5 m, the velocity values averaged over 150–250 days also increased from 0.11 to 0.17 m s^{-1} [16].

An estimation of the distribution of the kinetic energy over the variability ranges in the former of these cases [12] showed that the contribution of interannual variations is very low (lower than 1%); the mean annual condition provides about 4%, while seasonal, synoptic, and short-period variations contribute 5, 65 and 26%, respectively. With the distance from the coast and depth growth, the proportion between the energies of synoptic and short-period variations changes to an opposite one; according to the data of [16], it comprises from 1 : 3 to 1 : 4. The short-period range is dominated by inertial motions. Owing to them, in the Black Sea, even at significant depths far away from the coasts, the "instantaneous" velocities reach 0.30–0.40 m s^{-1}, as shown by the results of buoy observations [17, 18].

2.2 ADCP Observations

There are two shipborne surveys of the vertical profiles of the current vectors in the Black Sea using ADCPs in the layer from 8 to 350 m. One was performed in July 1992 in the Turkish economic zone (south of 43°N), where observations were made over a few dozens of latitudinal and longitudinal tracks with an interval between the stations about 3–4 km [22]. The other survey was performed in April 1993 on the northwestern continental slope and off the Turkish coast between 28° and 35.5°E over 18 profiles normal to the shore and 9 alongshore profiles with intervals between the stations from 3–4 to 18 km [23].

In July 1992, off the Turkish coast, a slightly meandering MRC stream was observed; its core was approximately 30 km wide and had maximal velocities in the upper layer 50–75 m thick up to 0.50 m s^{-1} [22]. In the layer from 75 to 125 m, the most rapid velocity drop was observed (by 0.25 m s^{-1}). At a depth of 200 m, the values decreased down to 0.05–0.10 m s^{-1}. Toward the coast, the velocities decreased by 0.20 m s^{-1} per 10 km, the rate of their decrease in the seaward direction was fourfold lower. The maximal geostrophic velocities with respect to the 500-dbar level were 0.20 m s^{-1} lower than the ADCP velocities, which points to significant ageostrophic effects in the MRC dynamics.

In April 1993, the velocity of the MRC and its meandering were twice as great as in July 1992 [23]. The velocities in the upper 100-m layer within the MRC core comprised 0.50–0.60 m s^{-1} on the northwestern shelf and 0.80 m s^{-1} off the Turkish coast (more than 1.00 m s^{-1} at 33°E). Even in the layer between 200 and 350 m, the velocities comprised 0.20–0.25 m s^{-1} (up to 50 m s^{-1} at 33°E). The highest vertical velocity shear (decrease by 0.30–0.50 m s^{-1}) was registered between depths of 125 and 200 m. The lateral shear between the core of the MRC and the coast reached 0.30–0.50 m s^{-1} per 10 km. The dynamical balance in the MRC core was strongly nonlinear. The average amplitudes of the meanders were 75 km (up to 125 km) and the average wavelengths comprised 100–125 km.

At that time, on the northwestern shelf of the Black Sea, weak currents up (to 0.10 m s^{-1}) along the outer front of the desalinated area off the Danube River mouth were observed. As was noted in [23], the MRC features almost no interaction with shelf currents, since its meanders are able to penetrate only to the edge of the shelf.

Thus, the measurements with ADCP helped to estimate the actual parameters of the MRC in the Black Sea and confirmed its strong seasonal variability and the existence of the Sevastopol, Danube, Bosporus, Sakarya (about 31°E), and Sinop (about 35°E) NSAEs (see Fig. 1).

2.3
Drifter Observations

In 1999, observations of the BSGC with the help of drifters began. The drifters had parachutes at a depth of 15 m and were traced from the satellites of the ARGOS system [24]. In 1999–2002, 21 paths up to 277 days long (with an average duration of 135 days) were mapped [25]. Of them, nine drifters fulfilled a complete revolution along the Black Sea coasts for 3.3–9.0 months. Mean vectors averaged over trapezoids 2/3° with respect to the latitude and 1° with respect to the longitude formed a single common cyclonic gyre with a maximum MRC velocity of 0.30–0.35 m s^{-1} over the continental slope. The highest mean velocities (up to 0.50 m s^{-1}) were observed approximately at the same places as those according to the data of the ADCP observations (off the western part of the Turkish coast); elevated values were noted off the coasts of the Caucasus, Bulgaria, and southwest of the Crimea.

In the zone of the MRC, the proportions of the contributions of the mean motion and of the synoptic and inertial variabilities to the total kinetic energy were 50 : 40 : 10, which is close to the estimates based on the data of moored buoy observations (see above). Thus, in the Black Sea, the relative contribution of the kinetic energy of synoptic fluctuations is sixfold lower than in the World Ocean. In the opinion of the authors of [25], this may be related to the small sizes of the sea and to the correspondingly high ratio R_d/L, where R_d is the baroclinic Rossby radius and L is the half-width of the basin. In the Black Sea, $R_d/L = 0.1$, while in the World Ocean $R_d/L = 0.01$. At $R_d/L > 1$, no baroclinic mechanism for eddy formation cross section be implemented.

In the near-shore zone, selected drifters were repeatedly captured by NSAE and performed anticyclonic rotation with velocities up to 0.60–0.80 m s^{-1} (in the Batumi NSAE); nevertheless, no one of these eddies was noted in the averaged schematic.

2.4
Altimeter Observations

From the methodological point of view, the studies of the BSGC using the data of satellite altimeter measurements of sea level elevation (SLE) anomalies occupy the position between direct measurements and diagnostic calculations of the currents. In order to obtain vectors of the surface currents from altimeter SLE values with excluded effects of tides, water budget, atmospheric pressure, and geoid shape, at the first approximation, geostrophic relations are applied in which latitudinal (longitudinal) velocity components are proportional to the longitudinal (latitudinal) gradients of SLE. Since this operation is relatively simple, we can include satellite altimetry into the group of field observations of the currents.

Important results of an analysis of the large-scale altimeter SLE fields in the Black Sea were presented in [26, 27] using the observations of 1992–1997. Both studies confirm that the BSGC in the form of a superposition of a large-scale MRC and a few SBCGs is significantly enhanced by the end of the winter and weakened at the beginning of the fall (1.5–2.5-fold with respect to kinetic energy) [27]. Different methods applied in [26, 27] equally unambiguously showed that this phenomenon is caused by the enhanced supply of cyclonic wind vorticity to the Black Sea in the wintertime. This is observed two months in advance of the maximum of the BSGC intensity. The spring inflow of riverine waters has no effect on the BSGC, as it was supposed earlier in [28]. In the winter, the MRC is 40–60 km wide and the geostrophic velocities are up to 0.40 m s^{-1} [27]. These inferences confirm the point of view on the origin and evolution of the BSGC that was first formulated in [10]. In [26], the interannual variability of the wintertime enhancement of the MRC was followed: from 1993 to 1997, it decreased almost threefold.

Meanwhile, the annual cycle of the intensity of the wind cyclonic pumping in the Black Sea generates, in addition to the large-scale quasi-standing mode, a non-stationary mode in the form of baroclinic Rossby waves with an annual period, a wavelength about 350 km; and an amplitude of the dynamical SLE of 0.10–0.12 m. They originate near the eastern coast and propagate westward at a phase velocity about 0.025 m s^{-1} [27]. For the first time, the possibility of the planetary-wave response of the Black Sea to the wind forcing of an annual period was posed at the beginning of the 1990s [29].

An analysis of the recent observation data [30, 31] shows that baroclinic Rossby waves that are generated off the eastern coasts in the northern parts of the Pacific and Atlantic oceans in a period of about a year represent their dominant non-stationary dynamical response to the annual cycle of the atmospheric forcing in the latitudinal range from 10–15° to 45–50°N. In so doing, their mean phase velocities (0.02–0.03 m s^{-1} at 40–45°N) are higher than the theoretical values (about 0.01 ms^{-1}). A similar situation is observed in the Black Sea as well [27]. In [32], several reasons of this phenomenon were listed such as the interaction with more large-scale non-stationary processes, topographic and nonlinear effects, and insufficient duration and spatiotemporal resolution of the observation data.

The first EOFs mode of the annual cycle of the altimeter SLE fields [27] contain almost all the elements of the BSGC that are schematically shown in Fig. 1. Besides them, local maximums of the sea level spectra at periods about 280 and 125 days are best expressed in the areas of the MRC bifurcations and are minimal at the centers of NSAEs. In [27], it was supposed that the 125-day period is related to the baroclinic instability and meandering of the MRC.

Thus, field observations of the currents provide certain ideas about the general features of the BSGC and its seasonal, synoptic, and interannual variabilities mostly in the upper 300-m layer. In the nearest future, the knowledge about a more detailed spatial pattern is possible only for the surface currents with the use of drifter and altimeter data. Reliable data about the BSGC in deeper layers from observational data represent a matter of the relatively remote future.

3 Results of Diagnostic Modeling

Hydrodynamic modeling represents the most significant alternative to the insufficient field observations when studying water circulation in seas and oceans.

The general property of diagnostic modeling of sea currents consists of the strong dependence of the current field calculated on the water density field used. The differences in the results of different calculation methods (hydrodynamic models) with the same initial conditions (three-dimensional density fields) are usually observed only in details of the current patterns obtained.

The first diagnostic calculations of the BSGC were performed using the dynamical method based on simple geostrophic relations that were briefly discussed above (for more details, see, for example, [33]). When applying this method, the heights of the sea surface (equal SLE) and isobaric surface in the water column are calculated from the three-dimensional water density field with respect to a reference isobaric surface, at which the absence of horizontal water motions is assumed. There are no absolutely objective ways to determine this kind of "zero" surface. This is the principal disadvantage of the dynamical method. Nevertheless, owing to its simplicity, it is still used for express estimates of the BSGC.

In the 1960s, the start of application of computers to the practice of marine research gave a pulse to the development of numerical diagnostic hydrodynamic models [33]. In them, the SLE (or the integral stream function) field is calculated from the three-dimensional density field in the equation of potential vorticity balance over the entire water column from the surface to the bottom. The iterative computational procedure is repeated until a stationary condition of the SLE (or the integral stream function) is reached at the specified fixed density field. Then, from equations of momentum balance, horizontal components of the current vector are obtained, while the continuity equation provides the calculations of the vertical component. The advantage of this approach is related to the absence of the problem of the choice of the "zero" surface and to the account for the coupled effect of the baroclinicity of the seawater and the bottom relief (JEBAR) [33] of the sea; its shortage lies in the insufficient smoothness of the velocity field because of its dynamical misadjustment with the density field.

At the beginning of the 1980s, the idea of mutual hydrodynamic adjustment (adaptation) of these fields was implemented [34]. Actually, the adaptation hydrodynamic models deal with a full set of nonlinear equations of the hydrothermodynamics of the sea. The computation process is stopped when the "fast" adjustment, that is the significant decrease in the energy of inertial and more high-frequency oscillations of the currents and water density ("dynamical noises"), is completed. Usually, this takes a few days of the model time; therefore, there is no need in specifying actual thermodynamic boundary conditions.

Meanwhile, the idea was formulated about resolving the full set of primitive hydro- and thermodynamic equations with all the boundary conditions specified successively correcting the current model fields of the temperature, salinity, and SLE by their observed values with the use of this or that kind of assimilation algorithm [35, 36]. This approach is sometimes referred to as a "four-dimensional analysis." Strictly speaking, it has little in common with the initial diagnostic methods. They are joined only by the common goal – the hydrodynamic calculations of the fields of currents from the data of observations of the temperature, salinity, and sea level. Therefore, in this section, we consider the results of application of all the above-mentioned approaches.

3.1 Diagnostic and Adaptation Modeling

The earlier results of a diagnosis of the climatic BSGC with the help of dynamical method are assessed in the introduction to this chapter. The most detailed schematics of the BSGC were obtained in [22, 37] using the data of multivessel surveys of the temperature and salinity profiles performed over the entire Black Sea in September 1991 and July 1992. The observation points were located at the nodes of a regular grid at an interval of 36 km (in selected near-shore areas, the network was twice as dense). The calculations of the dynamical topography of isobaric surfaces were performed with respect to the 900-dbar surface in 1991 and 500-dbar surface 1992 (depths about 900 and 500 m, respectively). In both cases, the BSGC patterns obtained were close to that presented in Fig. 1 consisting of a single MRC jet with a width of a 1–2 grid steps and a system of SBCGs and NSAEs. In both cases, the range of the SLE of the Black Sea (about 0.26 m) and the maximal (surface) geostrophic velocities of the MRC (0.33–0.35 m s^{-1}) were similar as well as the vertical structure of the currents. At depths of 100, 250, and 500 m, the velocities decreased down to 0.20, 0.05, and 0.02 m s^{-1}, respectively. The absence of noticeable geostrophic velocities below a depth of 500 m in 1991 was the reason for the equal ranges of the SLE in 1991 and 1992.

The difference lied in the meandering of the seaward boundary of the MRC. In 1991, it was twice as high as in 1992. In September 1991, 17 meanders of the MRC were observed with wavelengths of 125 and 250 km off the southern and northern coasts of the Black Sea, respectively. The amplitudes of the meanders were of the same order of magnitude. They split the SBCG into a series of smaller cyclonic eddies. The largest and most distinct NSAEs were the Batumi eddy (about 200 km in size) in the southeastern part of the Black Sea and the Bulgarian eddy (about 100 km) in its western part. In July 1992, SBCGs retained their integrity and were deformed only by weaker MRC meanders; no Bulgarian NSAE was observed. In 1991, the parameters of the meanders corresponded to the theory of the baroclinic instability of the MRC developed in [10].

A characteristic example of early diagnostic modeling of the BSGC is presented in [38]. Winter and summer climatic density fields by the data of 10 000 shipborne measurements over a $1^\circ \times 1^\circ$ grid (about 110 km) in the upper 300-m layer of the Black were used in a linear model. Below 300 m, the density was extrapolated down to the bottom. The horizontal resolution of the model grid was 50×50 km; it included 11 levels over the vertical. In the winter (summer), the range of the SLE in the Black Sea reached 0.55 m (0.43 m), which is twofold greater than those calculated with the dynamical method. In the wintertime, the width of the MRC was of the order of the grid step and the surface velocities were about 0.50 m s^{-1}. In the summertime, it was 2–3 times wider and weaker. In the winter, the distinct western and eastern SBCGs located in the central part of the Black Sea at 31°E and 37°E are separated by an anticyclonic eddy south of the Crimea. In the summer, this eddy is not observed since the eastern SBCG is displaced westward (to 35°E), while its southeastern part is occupied by the Batumi NSAE, which is the only one in the schematics [38]. Over the entire water column, the BSGC featured the same character with a monotonic velocity decrease with depth; this was predetermined by the density extrapolation to the deeper layers. The main inference that can be made based on these and earlier diagnostic model studies consists of the recognition of the leading role of JEBAR in the formation of the dynamical SLE of the Black Sea.

In the middle 1980s, a series of diagnostic calculations of the BSGC were performed (see review [39]) using monthly and seasonal climatic density fields with a discreteness about 70 km, obtained from the archived data from 51 000 stations [11]. The same fields were used in the first adaptation models of the BSGC at the beginning of the 1990s The results of these calculations are best presented in [40]. They are characterized by a higher strength and stationary of the western SBCG throughout the year. Its eastern counterpart is significantly smaller, especially in the summer–fall, and displaced westward to 35°–36°E. A system consisting of the Batumi and Tuapse NSAEs about 100 km in diameter each exists off the Caucasian coast over the greater part of the year. In the summer, this system is complemented by the NSAEs located southeast and southwest of the Crimea. In selected months, NSAEs are also registered off the southern coast of the Black Sea west and east of the Sinop area. The results confirm the suggestions [10] about the homogeneity of the BSGC in the upper 100-m layer and about its summer–fall weakening. An analysis of the adaptation process [40] showed that in the upper layers of the sea it proceeds significantly faster; this also refers to the salinity. This leads to unreasonable results, in particular, to a dissipation of the cold intermediate layer in the summer. An idea was suggested of introducing relaxation terms with respect to the observations as a Newton's type of sources into the thermodynamic equations.

In [41], calculations of the BSGC were performed with the model [34] using monthly climatic density fields with a discreteness about 22 km [11] obtained from the data from about 65 000 stations. For the first time, a clear seasonal variability in the intensity and structure of the BSGC was obtained with a physically reasonable succession of the current fields from one month to another. In February–May, the range of the SLE reached 0.24–0.26 m, while in June and October it decreased down to 0.20 and 0.12 m, respectively. Figures 7–9 represent the fields of current vectors in addition to those published in [41]. The level 0 m characterizes the BSGC in the upper 100-m layer, while the level 300 m best represents the currents at the lower boundary of the layer the maximal velocity decrease with depth; below it, their vertical changes are multifold lower (see Fig. 3a). In order to illustrate this, the current field at a depth of 1000 m in May is additionally shown in Fig. 8.

In March (Fig. 7) the MRC is most intensive along the northern and southern coasts of the Black Sea (up to 0.36 m s^{-1} at 0 m). With approach to the western and eastern coasts, it undergoes bifurcations and a slight weakening. The best developed is the eastern SBCG, which is especially clearly manifested at a depth of 300 m (Fig. 7b), where the MRC is almost indistinguishable in the bifurcation areas. With depth, the NSAEs off the Sakarya River canyon (displaced toward the central area), Cape Sinop, Sukhumi, Tuapse, and southeast of the Crimea become more distinctly expressed.

In May (Fig. 8), the western SBCG, which displaced to the Bulgarian coast, becomes more intensive. South of the Crimea and off the middle part of the Caucasian coast, anticyclonic gyres develop; they cannot be referred to as NSAEs, especially at a depth of 300 m (Fig. 8b). One of them is well manifested even at a depth of 1000 m (Fig. 8c), where the BSGC retains only its most general features intrinsic of it in the upper 300-m layer.

In September (Fig. 9), a disintegration of the BSGC occurs because of the large meandering and weakening of the MRC. On the whole, the structure of the September climatic water circulation is very close to the synoptic situation in September 1991 [37]. This should be especially noted, because the results of this survey were not included into the archived database that was used for the calculations of the climatic thermohaline fields in [41]. This suggests a high recurrence of the seasonal features of the BSGC. This conclusion

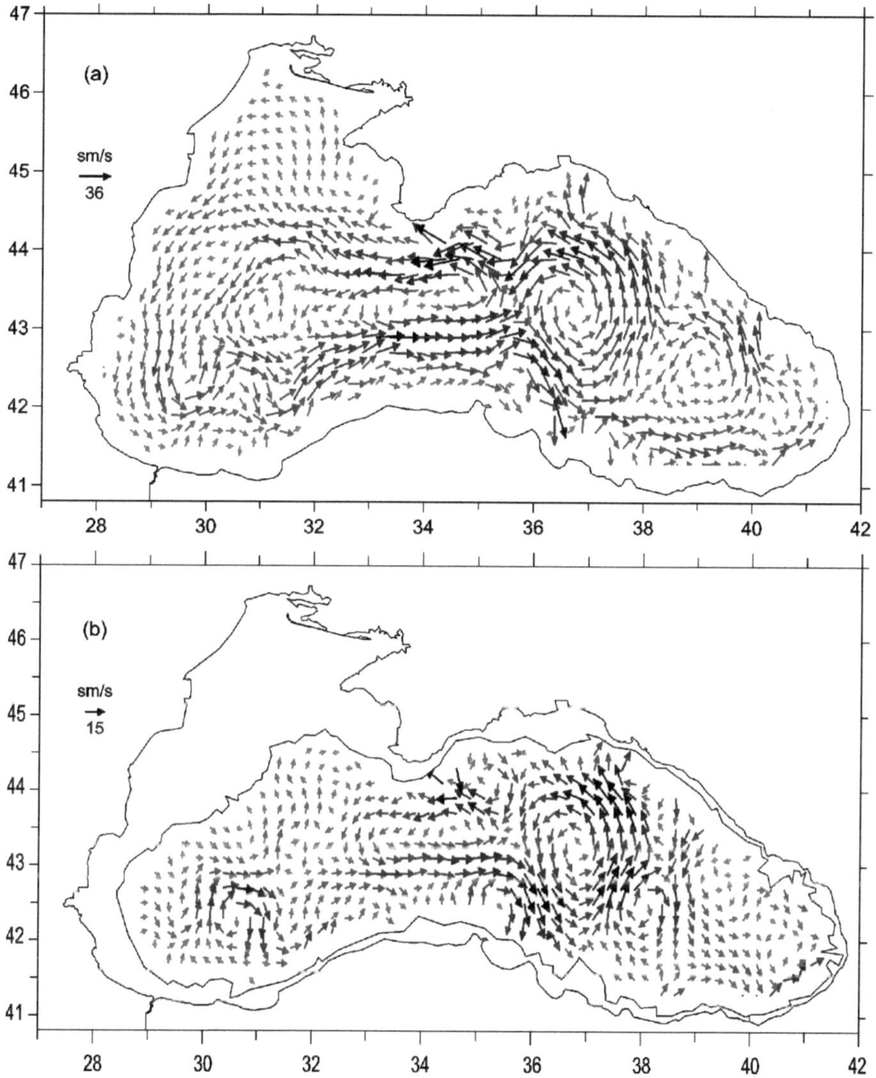

Fig. 7 Climatic current vector fields in the Black Sea at a depth of **a** 0 m, **b** 300 m in March from adaptation modeling

is supported break-up the September configuration of the sub-basin eigenmode of the climatic salinity field at a depth of 100 m shown in the inset in Fig. 9a. The technique of its derivation is described in [3]. The positive (negative, shaded) salinity anomalies correspond to areas with cyclonic (anticyclonic) vorticity of the currents in Fig. 9a. In September, the BSGC rapidly rearranges over depth. At a depth of 300 m (Fig. 9b), only the western half of the Black Sea retains the general cyclonic vorticity of the currents. East of

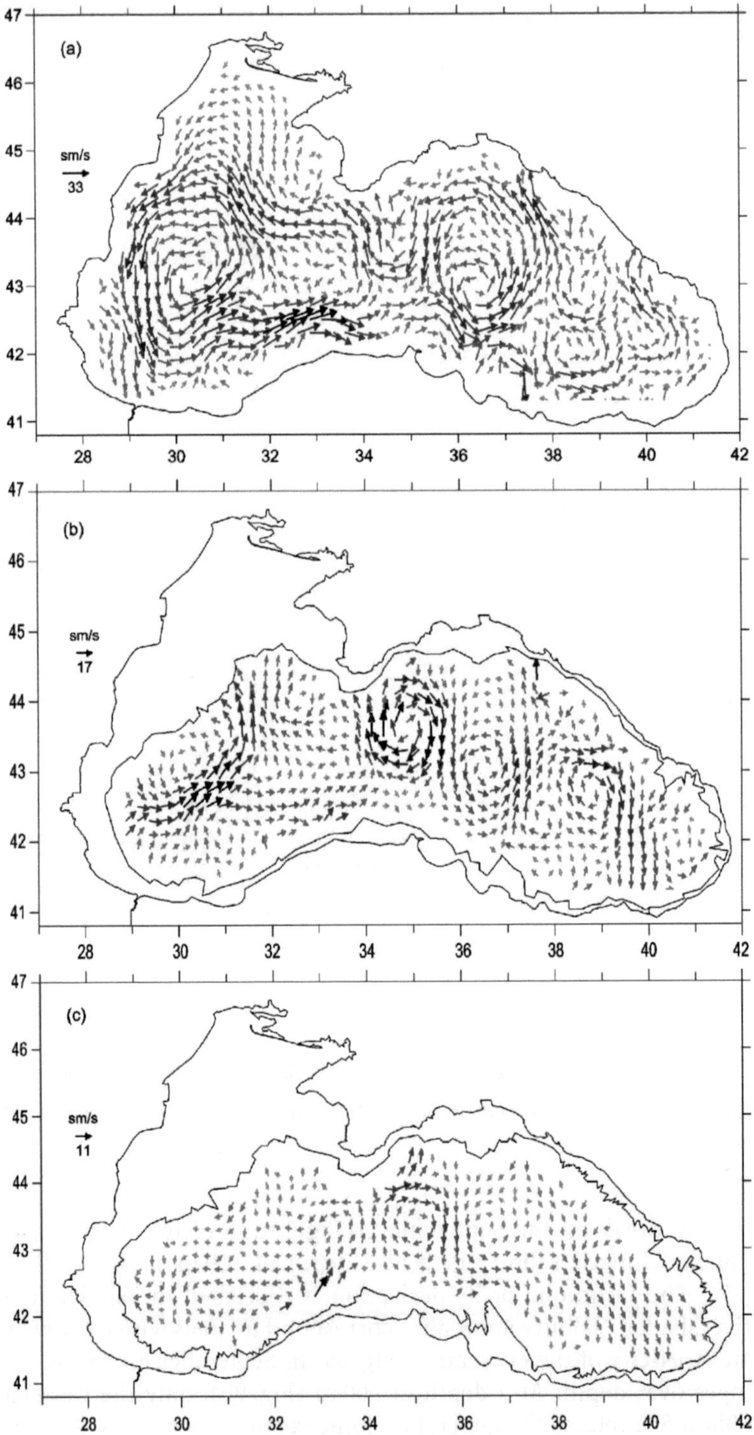

(a)
sm/s
33
(b)
sm/s
17
(c)
sm/s
11
47
46
45
44
43
42
41
28
30
32
34
36
38
40
42

◀ **Fig. 8** Climatic current vector fields in the Black Sea at a depth of **a** 0 m, **b** 300 m, and **c** 1000 m in May from adaptation modeling

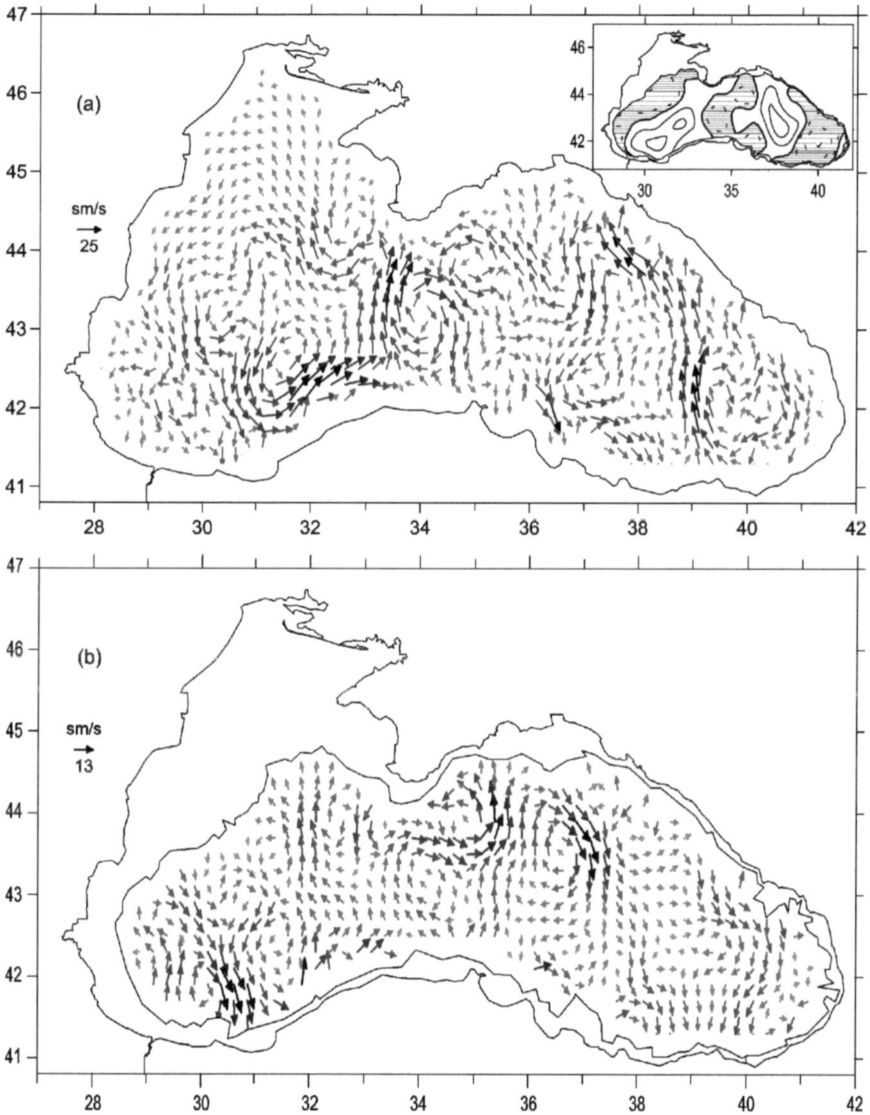

Fig. 9 Climatic current vector fields in the Black Sea at a depth of **a** 0 m, **b** 300 m in September from adaptation modeling. *Inset* in Fig. 9a climatic water salinity anomaly (negative *shaded* and *dotted*) relative sum of annual mean and 1-st EOF at a depth of 100 m in September

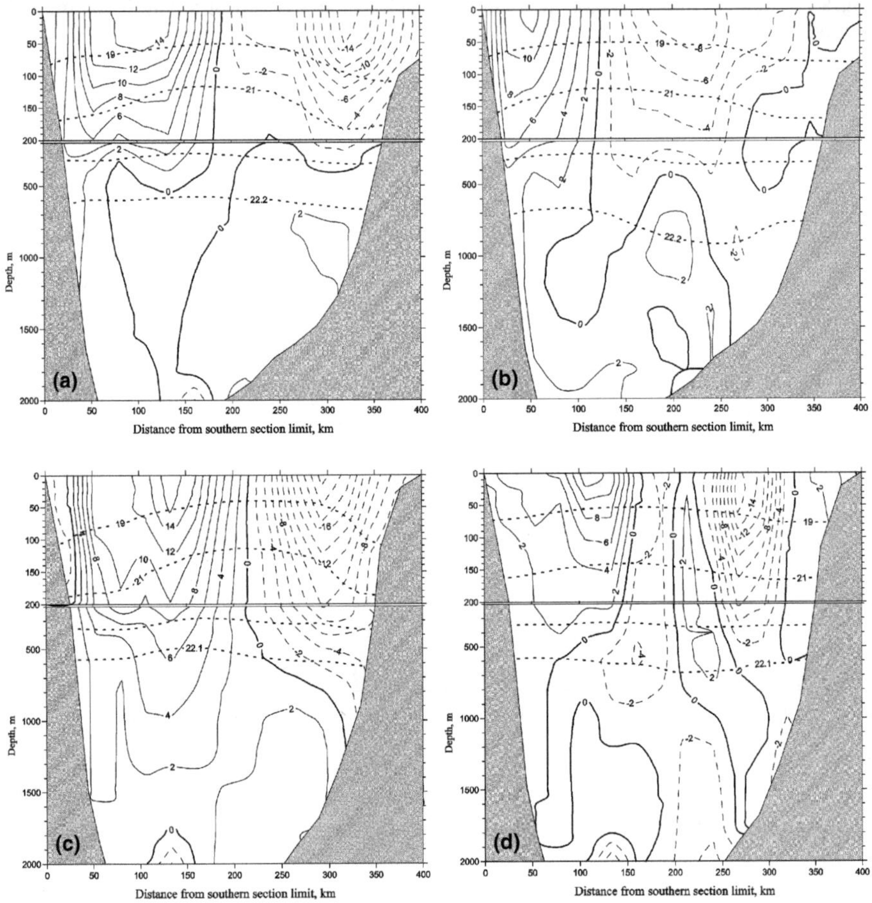

Fig. 10 Vertical sections of the eastern component of the Black Sea climatic currents velocity (10^{-2} m s^{-1}) from adaptation modeling **a** along 32°E in March, **b** along 32°E in September, **c** along 36.5°E in March, **d** along 36.5°E in September. *Dashed lines* westward velocity component, *dotted lines* isohalines (psu)

34°E, the former Crimean, Kerch, and Batumi NSAEs almost merged to form a common anticyclonic water motion, although over a complicated meandering trajectory.

The vertical structure of the zonal velocity components in the meridional sections across the eastern and western SBCGs is shown in Fig. 10. In March, the current structure is simpler than in September. In the eastern part of the sea, surface currents penetrate deeper than in the west. Locally, one can observe weak deep countercurrents beneath the surface streams; however, there are no reasons to suggest a full change in the BSGC in the deep and, the more so, in the intermediate layers. Note that the climatic anticyclonic vorticity of the currents at the center of the Eastern Black Sea in September (Fig. 10d) is confirmed by recent observations [24]. The slope of the climatic salinity contour lines shown with the dashed lines in Fig. 10 agree well (even in its details) with the vertical velocity shears from the point of view of the so-called "thermal wind relations" (see, for example, [33]). A more thorough comparison shows that the geostrophic vertical shears are usually by 25–30% lower than the nonlinear model and observed ones, which explains the above-mentioned underestimation of the geostrophic velocities.

3.2
Modeling with Data Assimilation

The studies of the BSGC with the help of assimilation of shipborne and satellite altimeter data in the full hydrothermodynamic model [42] began simultaneously with the above-described adaptation calculations. In the course of the studies it was realized that the optimal variant of assimilation with respect to its efficiency and simplicity is the above-mentioned introduction of terms with relaxation to observations into the thermodynamic equations [43]. The main goals of the studies became the simulation of the detailed climatic annual cycle of the BSGC with the help of assimilation of climatic monthly temperature and salinity fields and the simulation of the synoptic variability of the BSGC with the use of the SLE acquired by the altimeter observations from the *TOPEX/Poseidon* and *ERS–1,2* satellites.

During the assimilation of climatic shipborne data, the relaxation coefficient featured an inverse dependence on the relative dispersion of the observation errors. Owing to the growth of the latter with depth, this coefficient decreased, which allowed one to smooth the vertical differences in the rates of adaptation recognized in [40]. The climatic temperature and salinity fields were interpolated over the nodes of the calculation grid and were assimilated at each model time step. This way, the degree of the disagreement between the calculated and observed fields at the moment of assimilation was reduced to its minimum.

By the middle 2000s, the model used [42] had been physically and numerically enhanced by the introduction of biharmonic horizontal mixing of the momentum, free sea surface, and actual thermodynamic fluxes at all the open boundaries implemented with a 15-km horizontal resolution, 44 levels over the vertical and a 5-min time step [44, 45]. In the latter papers, instead of the density fields [9], climatic temperature and salinity fields with a twice coarser horizontal resolution (about 37 km) were used based on a twofold greater database (about 100 000 stations).

In the resulting new BSGC fields, the ranges of the SLE in the Black Sea in all the seasons became quite close to the values reported in [41] with a maximum at the end of the winter about 0.26 m and a minimum at beginning of the fall of 0.16–0.18 m. The corresponding maximal velocities of the surface currents decrease from 0.29 to 0.17 m s^{-1}. In the wintertime, the eastern

SBCG is most developed, while the Batumi and Tuapse NSAEs off the Caucasian coast are not observed. They appear in the spring, when the eastern SBCG weakens together with the enhancement of the western SBCG. In this aspect, the results of [44, 45] are close to the evolution of the BSGC presented in Figs. 7–8.

In [45], special attention is paid to the least studied issue of the deep BSGC. A more thorough preparation of the new deep temperature and salinity fields allowed one to learn that the surface structure of the BSGC can be traced down to 250–350 m. With the further depth growth, in all the seasons, the eastern SBCG and the Batumi NSAE are enhanced. Between them, an eastward flow is observed; it is the most prominent element of the deep BSGC and originates off the southwestern coasts of the sea. Starting from a depth of 1000 m, one finds that this kind of structure dominates over the deep layers of the Black Sea throughout the year. Note that its above-listed elements are also seen in Fig. 9c, where, at a depth of 1000 m, general eastward water transport prevails. In contrast to [45], this is a strongly meandering flow and features higher velocities (up to $0.11\ \mathrm{m\,s^{-1}}$), while in [45], the velocities never exceed $0.02\ \mathrm{m\,s^{-1}}$.

Satellite altimeter observations were assimilated in two BSGC models: 1.5-layer model with a reduced gravity acceleration [46] and in the slightly modified model [44, 45] considered above. In the former model, the altimeter sea level was assimilated directly into the equations of continuity at each time step. In the latter model [47], the assimilation was similar to that in [44, 45], where the differences between the model and observed temperature and salinity fields were retrieved from the level increments with the use of corresponding coefficients of proportionality. These coefficients depended on the depth and were determined from the relations between the SLE and the thermohaline fields obtained in [43] from the modeling results. Selected simplifications in the model physics helped to decrease the horizontal step of the grid in both of the calculations down to 7 km.

In [46], the assimilation was performed for the period from May 1992 to May 1999. The most important result of it concerning the BSGC is the stable, from year to year, reproduction of most of its characteristic features considered above, such as the jet character of the MRC, the winter–spring maximum of the intensity (especially in the eastern SBCG), the summer–fall weakening and disintegration with an enhancement of the meandering and eddy formation, and the rapid strengthening of the large-scale circulation in November–December. In the years with severe winters and windless warm period of the year (1993, 1998), seasonal changes in the BSGC are especially strong. The principal objects of the analysis presented in [46] are the NSAEs, for which estimates of the lifetime during the year and particular features of the evolution were obtained. During the year, the most stable elements were the Bosporus, Batumi, Tuapse, Sevastopol, and Danube NSAEs with recurrence frequencies from 150 to 265 days per year.

In [47], the assimilation was performed for the period from 1992 to 2003. As a result, BSGC fields averaged over 12 years were obtained; they were compared with the above-described ones from old and new shipborne climatic TS data. The principal distinctions of the altimeter BSGC are related to its higher horizontal resolution (see above) and are manifested in the narrower and faster MRC (by 10–20%) and in the increase of the number of NSAEs, among which one finds all the above-listed eddies except for the Bosporus NSAE.

The authors of [45, 47] relate the enhancement of the BSGC obtained from the new observation data to its multiannual variability under the increase in the wind cyclonic vorticity in the 1980s–1990s. For the 1960s–1980s, this is indirectly confirmed by the results of the study presented in [3] as well as by direct measurements of the currents [12, 16] off the Caucasian coasts in 1976–1981 and 1997–2001 (see Sect. 2.1). However, in [26], an opposite trend in the wintertime BSGC in 1993–1997 is reported, although also based on indirect considerations (see Sect. 2.4). Probably, a significant part of the BSGC enhancement recognized is related to the decrease in the discreteness of the observation data assimilated and the model grids.

To conclude, it should be noted that the diagnostic (in a broad sense of this term) modeling represents the most fruitful and actively developing line in the BSGC studies. In the future, its most developed branch, i.e. assimilation of the real-time data of satellite and field observations in full models with primitive equations (see Sect. 3.2), may become an important tool for the BSGC monitoring, at least in the upper layers of the Black Sea.

4 Results of Prognostic Modeling

In contrast to diagnostic modeling, which is aimed at the construction of reliable current fields from the specified density fields, the principal goal of the so-called prognostic modeling lies in the understanding of the mechanisms of formation of the circulation in seas and oceans and their possible reproduction in numerical models. Only if thickness problem is resolved, one can speak about the hydrodynamic current forecasting.

Prognostic models reproduce the process of evolution of the initial condition of the current, temperature, and salinity (density) fields under the action of the boundary conditions (momentum, heat, moisture, and mass fluxes) without any correction for the observational data. Usually, the climatic annual cycle of the variabilities in the circulation and thermohaline water structure is modeled. The calculations are performed until the parameters of this cycle stabilize, i.e. the differences between two successive become lower than a certain specified value. Then, the results obtained (model current, temperature, and salinity fields; energy, dynamic, and thermodynamic budgets, etc.) undergo a physical analysis and a comparison against the data of measurements. The proportions between the significance of the processes considered in the model are defined based on an analysis of the above-mentioned budgets or through comparison between the versions of the calculations with different combinations of the processes.

The results of prognostic modeling depend on numerous factors such as the completeness of the processes considered in the model, the parameterization of subgrid processes that are not reproduced explicitly, the quality of the numerical approximation of the model equations, the spatial and temporal resolutions in the calculation domain, and the reliability of the initial and boundary conditions and model constants.

The modern prognostic models have reached their maximum in the fullness of the equations of the momentum, heat, and salt balance considered and in their numerical approximation. They may be joined into multilevel model and quasi-isopycnic model groups. The former models deal with a grid domain fixed in space and time, while the latter models involve layers with fixed water density values varying with respect to depth and time. Below, we present the results of the applications of these two groups of models to the studies of the BSGC separately.

4.1 Multilevel Modeling

Multilevel prognostic models may be divided into two groups with respect to the mode of consideration of the vertical coordinate. One group uses the traditional Cartesian vertical coordinate with horizontal levels; the other considers a vertical coordinate normalized with respect to the sea depth at the site (the so-called σ-coordinate). Let us start the discussion of the results with the first group of models, because they prevail in the studies of the BSGC.

The first prognostic models of the BSGC [8,9] were mentioned in the introduction to this chapter. They reproduced only the most general features of the BSGC.

The first detailed study of the mechanisms of the BSGC formation was performed by E. Stanev [48]. The full nonlinear model suggested by K. Bryan with 12 levels over the vertical were used in two options of the horizontal discretization, one with steps 0.5° over the latitude and 1° over the longitude and the other with steps three times smaller. The coarse discretization was applied to initialization the BSGC from the initial condition of zero motions and horizontally homogeneous temperature and salinity, while the finer resolution was used to reproduce its climatic seasonal variability.

The main result of study [48] is the inference that the principal features of the BSGC and its seasonal variability are generated under the action of the relative wind vorticity. The particular baroclinicity of the Black Sea related to the existence of a thin and intensive permanent pycnocline in the upper layer (from 50 to 200 m) prevents the thick abyssal layer from the penetration of this influence. The bulk of the energy of wind forcing is expended on the doming of the Black Sea pycnocline. Therefore, the intensity of the BSGC, especially in abyssal layers, is significantly lower than it might be at an absence of density stratification. This proved by a special numerical experiment, in which the calculations were continued after reaching the stationary condition with a zero riverine runoff. In 17 years, the permanent pycnocline descended by 500 m and in 34 years, the stationary state became homogeneous over the vertical and the BSGC significantly enhanced.

The actual variability in the heat and moisture fluxes almost does not influence the BSGC [48]. In addition, the former flux suppresses the effect of wind forcing. The weakness of the abyssal currents in the Black Sea results in a decrease in the efficiency of the their interaction with the bottom relief (JEBAR) that generates additional relative vorticity and BSGC energy. One more result of the particular baroclinicity of the Black Sea is the weakness of the vertical water circulation, which, with respect to the integral mass transport, is 25 times lower than that of the horizontal BSGC.

In [48], it was shown that, in order to simulate the cold intermediate layer of the Black Sea (see [3]), one should take into account the dependence of the vertical turbulent mixing coefficient on the density stratification of the waters. In this case, the optimal coefficients in the well-known formula by Munk–Anderson for the Black Sea occurred to be an order of magnitude lower than those for the World Ocean.

These conclusions imply a special significance of the spatiotemporal configuration of the wind forcing for the simulation of an adequate BSGC. In [48], the well-known wind fields after Hellermann–Rosenstein were used; they feature maximums of the velocity and relative vorticity over the western part of the Black Sea. Therefore, the model BSGC in [48] consisted of a single cyclonic gyre centered in the western part of the sea (at 30–32°E). The current velocities were not high (lower than 0.20 m s^{-1}), especially in the eastern part of the sea; the MRC jet was noticeable only in its western part. The horizontal resolution used (equal to 20 km) was not sufficient to generate the known SBCGs and NSAEs (see Sect. 3). Later, E. Stanev and coauthors paid much attention to the search for optimal boundary conditions [49].

In [50], the mean annual wind field compiled according to the data of the Russian Climatic Reference Book was used. The mean wind speeds became two to threefold higher. The maximums of the velocity and cyclonic vorticity of the wind were confined to the eastern part of the Black Sea. The almost twofold decrease in the horizontal grid step (11 km) as compared to [48] allowed one to reproduce in [50] a system of subbasin cyclonic and anticyclonic eddies quasiperiodic over the longitude; it clearly dominated over the large-scale BSGC. The latter is represented in [50] only in the weaker mean annual current fields.

The eddies were formed off the eastern coast of the Black Sea and moved westward showing a decrease in the phase velocity in the narrowest area of the sea south of the Crimea. In the model version with a flat abyssal floor at a depth of 1540 m, the wavelength comprised 250 km in the east and 190 km in the west with a period of 160 days and a phase velocity of - 2.0–2.5 km day^{-1}. The orbital velocities in the eddies in the surface layer reached 0.45 m s^{-1} and deeper decreased down to 0.25 m s^{-1} a depth of 70 m and to 0.05–0.10 m s^{-1} at a depth of 1100 m. The wave regime was more intensive in the eastern part of the Black Sea; in its western part, eddies dissipated above the continental slope and partially reflected from the western coast. In the study [50], the introduction of the abyssal bottom topography increased (reduced) the sizes and intensities of cyclonic (anticyclonic) eddies by a factor of 1.5–2. The cyclonic (anticyclonic) eddies became more alike the SBCGs (NSAEs) in Fig. 1. The period of the eddies grew up to almost two years, while their phase velocity decreased down to 0.4–0.5 km day^{-1}.

The authors of [50] regarded the eddies as manifestations of Rossby waves modified by the bottom topography. The parameters of similar waves obtained from the data of altimeter observations (see Sect. 2.4), except for the period, are close to the model values. The annual wave period, which prevails in the observations, is absent in the model; this is related to the forcing of the model BSGC by a constant mean annual wind field.

The further improvement of the horizontal (to 9 km) and vertical (to 24 levels) resolutions together with the use of long-term series of actual meteorological conditions and current model sea surface temperature for the calculations of the external effects (momentum, heat, and moisture fluxes) allowed one to obtain in [51, 52] synoptic fields of the BSGC that were significantly closer to the actual pattern and to reproduce the processes of formation of the CIL. All the principal elements of the BSGC such as the MRC, SBCGs, and NSAEs actively interact with each other. This is manifested in the complicated and rapidly changing current pattern. In the winter, the formation of the CIL proceeds both in the continental slope areas and at the centers of SBCGs. A special role in this formation belongs to synoptic processes in NSAE areas, where, every few days, active pumping of the CIL with new cold waters occurs. The contribution of cold but desalinated and lighter waters from the northwestern shelf of the Black Sea is significantly smaller than it has been considered before. Along with this, in [51], one more mechanism of the CIL renewal in the Black Sea was recognized; it consists of slope convection (sliding of cold and dense near-bottom shelf waters down the continental slope).

From the beginning of the 1990s, active studies of the BSGC have been performed using the basic Ukranian model [42]. Its modern versions [53, 54] are in many respects close to those assessed above [51, 52]. The main distinctions are related to the use of the free sea surface instead of the stream function as the integral function (which allows one to more accurately calculate currents in the upper layers), to the explicit account for the moisture fluxes and penetrating solar irradiance, and to the numerical realization of the model (see [42, 54]). The specifying of the climatic temperature and salinity fields [11] as initial conditions allows one to rapidly initialize the model; meanwhile, this, to a certain extent, predetermines the results. Along with the specified wind fields [49] this resulted in a more intensive BSGC in the eastern and central parts of the Black Sea. The seasonal changes in the BSGC intensity and the principal mechanisms of the formation and transport of the CIL waters in this model correspond to the present-day concepts. In addition, it was established that the seasonal variability of the BSGC takes place in the abyssal layer of the Black Sea as well. However, the insufficient horizontal resolution of the grid domain (about 15 km) and, especially, of the initial and boundary conditions did not allow the authors of [53, 54] to reproduce the known subbasin inhomogeneities of the BSGC (see Fig. 1). This was reached in [55] based on the same model as a result of a twofold reduction of the horizontal model grid step and a use of actual multiannual series of wind fields from reanalysis data.

In addition to the results presented above, we should also note the studies of the climatic BSGC [56] based on the basic Russian prognostic model [57]. The distinctive features of [56] were related to the dependence of the coefficients of horizontal turbulence on lateral velocity shears and to the specifying of the monthly climatic temperature and salinity field at the surface [29] instead of the heat and moisture fluxes. Despite the relatively coarse horizontal calculation grid (about 22 km), this allowed the authors to reproduce [56] a relatively distinct MRC jet and the known NSAEs off the Turkish and Caucasian coasts and off the Danube River mouth. The results of the tuning in [56] of the Munk–Anderson's formula for the coefficient of the vertical turbulent exchange from the point of view of reproduction of the actual CIL were used in [53, 54].

The well-known Princeton model with a vertical σ-coordinate, a curvilinear horizontal grid adapted to the coastline, a turbulent closure of the order of 2.5 was used for the studies of the BSGC in [58]. Eighteen levels were specified over the vertical and the horizontal spacing was about 10 km. Similarly to [48], various combinations of the surface boundary conditions were specified. The model started with the wintertime climatic temperature analysis salinity fields [11] and three years later reached a quasi-stationary regime in the upper 200-m layer.

Under the wind forcing alone, the surface BSGC, even being averaged over the fourth to eighth calculation years, demonstrated a complicated horizontal structure, especially in the spring and summer. Its most stable elements were the MRC, the SBCGs in the southwestern part of the sea and the NSAEs in the northwestern shallow-water area. In addition to them, in all the seasons, about ten eddies of different signs with horizontal sizes of a few tens of kilometers were also reproduced. Meanwhile, under the action of heat and moisture fluxes across the surface of the Black Sea alone, an anticyclonic BSGC is formed (which corresponds to the results of [48]); on the northwestern shelf, it is enhanced by the riverine runoff. Under the actual forcing, as in [48], the wind effect dominates; meanwhile, the cyclonic BSGC becomes weaker though more large-scale. On the whole, the current patterns presented in [58] have the greatest differences from all the other considered above.

4.2 Quasi-isopycnic Modeling

It is assumed that quasi-isopycnic models have advantages in the description of the evolution of the upper mixed layer and intermediate water masses (temperature and salinity extremes). Instead of constant geometrical levels and layers enclosed between them, these kinds of models consider physical layers with fixed ranges of density changes, whose thicknesses, temperatures, and salinities evolve with time under the action of external forcing up to an absolute disappearance of any of them. The density in the UML is not fixed. Depending on the budget of the turbulent energy, the UML may entrain the water from the underlying layers or vice versa. The latter regime is referred to as subduction.

The study of the BSGC with the help of a quasi-isopycnic model [59] started at the beginning of the 1990s. In the course of its development, the model obtained seven layers and a 20-km horizontal resolution. The initial conditions corresponded to climatic thermohaline fields. Throughout the year, the BSGC was dominated by the MRC. The eastern and western SBCGs were poorly manifested, while the NSAEs were noted only east of Cape Sinop. Along with this, the main objects of the study—the fields of the UML thicknesses and the distribution of the subduction zones at the end of the winter were very close to their counterparts known from observations [3].

The development of quasi-isopycnal models logically resulted in a model with a hybrid grid domain (HYCOM), in which the stratified deep-water area is described by quasi-isopycnal layers, the UML is specified by horizontal levels, and the near-shore area is represented by σ-levels with a horizontal grid adjusted to the coastline. The first results of the application of this kind of model to the studies of the BSGC with a horizontal resolution about 3.2 km are presented in [60]. The initial and boundary conditions were specified from the results of reanalyses of meteorological and thermohaline fields. The initialization of the BSGC took five years. In the BSGC fields averaged over the fifth to eighth model years, one can clearly see the MRC jet with a mean annual position close to that shown in Fig. 1. The western SBCG is two to three times as great as the eastern one; around the latter, one finds the Kizil-Irmak, Batumi, and Tuapse NSAEs of comparable sizes. The details of the BSGC significantly depend on the variations in the boundary conditions and even on the depth of penetration of solar irradiance into the water column.

Among the common physical results of the prognostic models considered we should recognize the leading role of the wind forcing and bottom topography in the formation of the cyclonic BSGC, the damping of abyssal circulation by the baroclinicity of the upper 300-m layer, the lower rate of its stabilization as compared to the surface currents, and the small role of the cold winter shelf waters in the renewal of the CIL owing to their low density and dynamical capturing on the shelf.

From the methodological point of view, the BSGC modeling performed poses three most important problems. First, this is the significance of the quality and spatiotemporal resolution of the initial and boundary conditions for the reproduction of a BSGN close to the actual pattern. In all of the prognostic models with no exceptions, even in the surface layer, current velocities were noticeably lower than the values obtained from the observations and diagnostic calculations. This is related to the insufficient intensity of the BSGC forcing. Second, in order to reproduce subbasin inhomogeneities of the BSGC (see Fig. 1) a horizontal resolution about 5 km (that is, four to five times smaller than the baroclinic Rossby radius in the Black Sea) is required. Third, there is a need in a thorough parameterization of the influence of the water stratification on the turbulent fluxes as well as of the dependence of the penetration of solar irradiance into the water column on the water transparency.

5 Conclusions

The generalization of the results of field and model studies of the Black Sea general circulation presented in this chapter allows one to reach the following conclusions:

- In the upper 500-m layer, the BSGC consists (see Fig. 1) of the MRC along the entire continental slope; several SBCGs in the central area, whose number and position change during the year, and a few NSAEs existing in fixed areas between the MRC and the shore over 5–9 months per year.
- In the winter and early spring, the surface BSGC is most strong with a domination of large-scale features such as the stable jet MRC; in the summer and early fall, it weakens by a factor of 1.5 to two and undergoes disintegration with a domination of subbasin inhomogeneities (SBCG and NSAE) and a strongly meandering MRC, which actively generates mesoscale eddies.
- According to the data of direct observations and diagnostic calculations, in the core of the MRC 30–50 km wide, the winter–spring velocities in the upper 100-m layer comprise 0.30–0.80 m s^{-1} (with maximal values up to 1.20 m s^{-1}); at depths of 200–250 (500) m, they decrease two to four (ten

and more) times; in the summer–fall MRC is wider and slower by a factor of 1.5 to two.

- Below 500 m, the BSGC is poorly studied; it significantly differs from the surface pattern by low mean velocities (not higher than 0.01–0.03 m s^{-1}) and a prevalence of mesoscale eddies with anticyclonic vorticity; "instantaneous" velocities here may reach 0.30–0.40 m s^{-1} only owing to short-period (inertial, etc.) motions.
- The main source for the generation of the BSGC and its seasonal and interannual variabilities is related to the relative vorticity of the tangential wind stress; the influence of the momentum, moisture, and heat fluxes across the sea surface and via the river mouths and straits is significantly lower.
- Direct and indirect observations as well as diagnostic calculations suggest a certain enhancement of the BSGC from 1960 to the beginning of the 1990s related to the increase in the cyclonic relative vorticity of the wind over the sea in this period.

The understanding of more detail and fine features of BSGC requires much more data of direct current observations and higher quality and spatiotemporal resolution of the initial and boundary conditions for numerical models.

References

1. Knipovich NM (1932) Transactions of the Sea of Azov and the Black Sea Expedition, vol 10. VNIRO, Moscow (in Russian)
2. Neumann G (1942) Ann D Hydr Mar Met 70:265
3. Tuzhilkin VS (2007) Thermohaline structure of the sea (this volume)
4. Cheredilov BF (1967) Seasonal dynamics maps of the Black Sea surface. In: Oceanographic investigations of the Black Sea. Naukova Dumka, Kiev, p 119 (in Russian)
5. Filippov DM (1968) Circulation and water structure in the Black Sea. Nauka, Moscow (in Russian)
6. Tolmazin DM, Shnaidman VA, Atsikhovskaya JM (1969) Problems of water dynamics of the Northwestern Black Sea. Naukova Dumka, Kiev (in Russian)
7. Bogatko ON, Boguslavsky SG, Belyakov YM, Ivanov RI (1979) Surface currents in the Black Sea. In: Complex studies of the Black Sea. Naukova Dumka, Kiev, p 25 (in Russian)
8. Marchuk GI, Kordzadze AA, Skiba JN (1975) Izvestya Acad Sci USSR, Phys Atmos Ocean 11:379 (in Russian)
9. Dzhioev TK, Sarkisyan AS (1976) Izvestya Acad Sci USSR, Phys Atmos Ocean Engl Transl 6:217
10. Blatov AS, Bulgakov NP, Ivanov VA, Kosarev AN, Tuzhilkin VS (1984) Variability of hydrophysical fields of the Black Sea. Gidrometeoizdat, Leningrad (in Russian)
11. Simonov AI, Altman EN (eds) (1991) Hydrometeorology and hydrochemistry of the seas, Vol IY. The Black Sea, Issue 1. Hydrometeorological conditions. Gidrometeoizdat, St. Petersburg (in Russian)
12. Titov VB (1989) Characterization of coastal current regime off northern Caucasus coast. SB IO RAS, Gelendjic (in Russian)
13. Titov VB (1991) Marine Hydrophys J 2:41 (in Russian)

14. Boguslavsky SG, Ivanov VA, Yankowsky AE (1995) Marine Hydrophys J 3:36 (in Russian)
15. Titov VB, Savin MT (1997) Oceanology 37:50 (in Russian)
16. Krivosheya VG, Moscalenko LV, Titov VB (2004) Oceanology 44:358 (in Russian)
17. Bulgakov NP, Golubev YN, Nazin AV (1989) Some preliminary results of Main Rim Current in the central Black Sea in winter 1988. In: Complex oceanographic studies of the Black Sea. MHI NASU, Sebastopol, p 26 (in Russian)
18. Titov VB, Udodov AI, Laptev SY (1991) Statistical characteristics of current field in central parts of the Black Sea quasi-stationary gyres. In: Winter state of the central Black Sea ecosystem: Results of 21-th survey of R/V "Vityaz" February 9–March 8 1991. IO RAS, Moscow, p 22 (in Russian)
19. Titov VB (2003) Meteorol Gidrol 8:80 (in Russian)
20. Titov VB (2002) Oceanology (English Translation) 42:637
21. Ivanov VA, Yankovsky AE (1992) Long-wave motions in the Black Sea. Naukova Dumka, Kiev (in Russian)
22. Oguz T, Ivanov LI, Besiktepe S (1998) Circulation and hydrographic characteristics of the Black Sea during 1992. In: Ivanov LI and Oguz T (eds) Ecosystem modeling as a management tool for the Black Sea. Kluwer Acad Publ, Dordrecht, The Netherlands, p 69
23. Oguz T, Besiktepe S (1999) Deep-Sea Res 10(Pt1):1733
24. Zatsepin AG, Ginzburg AI, Kostianoy AG, Kremenetskiy VV, Krivosheya VG, Stanichny SV, Poulain P-M (2003) J Geophys Res 108:2–1 (doi:10.1029/2002JC001390)
25. Zhurbas VM, Zatsepin AG, Grigoryeva YV, Eremeev VV, Kremenetskiy VV, Motyzhev SV, Poyarkov SG, Poulain P-M, Stanichny SV, Soloviev DM (2004) Oceanology 44:34 (In Russian)
26. Stanev EV, Le Traon P-Y, Peneva EL (2000) J Geophys Res 105:17203
27. Korotaev GK, Saenko OA, Koblinsky CJ (2001) J Geophys Res 106:917
28. Bulgakov SN, Korotaev GK, Whitehead JA (1996) Izvestya RAS Phys Atmos Ocean 32:548 (in Russian)
29. Eremeev VN, Ivanov VA, Kosarev AN, Tuzhilkin VS (1994) Annual and semiannual harmonics in the climatic salinity field of the Black Sea. In: Diagnosis of the state of marine environment of the Azov-Black Sea basin. MHI UAS, Sebastopol, p 89
30. Chelton DB, Schlax MG (1996) Science 272:234
31. Osychny V, Cornillon P (2004) J Phys Oceanogr 34:61
32. Zang X, Wunsch C (1999) J Phys Oceanogr 29:2183
33. Sarkisyan AS (1977) The diagnostic calculation of large scale oceanic circulation. In: The Sea, Marine Modelling, Vol 6. Wiley and Sons, New York London Sydney Toronto, p 363
34. Sarkisyan AS, Demin YL (1983) A semidiagnostic method of sea current calculation. In: Large-scale oceanographic experiments in the WCRP. WCRP Publ Series Vol 2, Tokyo, p 201
35. Knysh VV, Demyshev SG, Korotaev GK, Sarkisyan AS (2001) Rus J Numer Anal Math Model 16:409
36. Fukumori I (2001) Data assimilation by models. In: Satellite Altimetry and Earth Sciences. Academic Press, San Diego, p 237
37. Oguz T, Aubrey DG, Latun VS, Demirov E, Koveshnikov L, Diaconu V, Sur HI, Besiktepe S, Duman M, Limeburner R, Eremeev V (1994) Deep-Sea Res 41(PtI):603
38. Gamsakhurdiya GR, Sarkisyan AS (1976) Oceanology (Engl Transl) 15:164
39. Bulgakov SN, Demyshev SG, Korotaev GK (1992) Modelling of the Black Sea circulation and water stratification (review). In: Problems of the Black Sea. MHI UAS, Sebastopol, p 34

40. Eremeev VN, Kochergin SV (1991) Numerical modelling of intra-seasonal variability of the Black Sea water circulation. MHI UAS, Sebastopol (in Russian)
41. Trukhchev D, Tuzhilkin V, Kosarev A (1995) CR Acad Sci Bulgare 48:35
42. Demyshev SG, Korotaev GK (1992) Numerical energy-balanced model of baroclinic currents in no-flat-bottom ocean on C-grid. In: Numerical models and results of calibration calculation of the Atlantic Ocean currents. INM RAS, Moscow, p 163
43. Korotaev GK, Demyshev SG, Knysh VV (2000) Three-dimensional climate of the Black Sea. In: Black Sea Ecosystem Processes and Forecasting. METU IMS, Erdemli, p 1
44. Knysh VV, Korotaev GK, Demyshev SG, Belokopytov VN (2005) Marine Hydrophys J 3:11 (in Russian)
45. Demyshev SG, Knysh VV, Inyushina NV (2005) Marine Hydrophys J 6:28 (in Russian)
46. Korotaev GK, Oguz T, Nikiforov A, Koblinsky C (2003) J Geophys Res 108:3122 (doi: 10.1029/2003JC001508).
47. Dorofeev VL, Knysh VV, Korotaev GK (2006) Marine Hydrophys J 4:3 (in Russian)
48. Stanev EV (1990) Earth-Sci Rev 28:285
49. Rachev N, Stanev EV (1997) Eddy dynamics controlled by basin scale, coastline and topography. In: Ozsoy E, Mikaelyan A (eds) Sensitivity to change: Black Sea, Baltic Sea and Northern Sea. Kluwer, Dordrecht, p 341
50. Staneva JV, Stanev EV (1997) Cold intermediate water formation in the Black Sea. Analysis of numerical model simulations. In: Ozsoy E, Mikaelyan A (eds) Sensitivity to change: Black Sea, Baltic Sea and Northern Sea. Kluwer, Dordrecht, p 375
51. Staneva JV, Stanev EV (1999) Oceanol Acta 21:393
52. Staneva JV, Dietrich DE, Stanev EV Bowman MJ (2001) J Mar Syst 31:137
53. Oguz T, Malanotte-Rizzoli P (1996) J Geophys Res 101:16551
54. Ibrayev RA, Trukhchev DI (1998) Model study of seasonal variability of the Black Sea circulation. In: Ivanov LI, Oguz T (eds) Ecosystem modeling as a management tool for the Black Sea. Kluwer, Dordrecht, p 179
55. Demin YL, Ibrayev RA (1989) Soviet J Numer Anal Math Model 4:179
56. Demyshev SG, Korotaev GK, Knysh VV (2004) Izvestya RAS, Phys Atmos Ocean (Engl Transl) 40:259
57. Demyshev SG, Knysh VV, Sarkisyan AS (2004) Izvestya RAS, Phys Atmos Ocean 40:636 (in Russian)
58. Dorofeev VL, Korotaev GK (2004) Marine Hydrophys J 1:52 (in Russian)
59. Ryabtsev YN Shapiro NB (1997) Marine Hydrophys J 1:12 (in Russian)
60. Kara AB, Wallcraft AJ, Hurlburt HE (2005) J Phys Oceanogr 35:13

Hdb Env Chem Vol. 5, Part Q (2008): 195–215
DOI 10.1007/698_5_062

Published online: 16 December 2006

Mesoscale Water Dynamics

Anna I. Ginzburg · Andrey G. Zatsepin · Andrey G. Kostianoy ·
Nickolay A. Sheremet (✉)

P.P. Shirshov Institute of Oceanology, Russian Academy of Sciences, 36 Nakhimovsky Pr., 117997 Moscow, Russia
sheremet@ocean.ru

Abstract An analysis of the results of hydrographic surveys and satellite measurements performed during in roughly the last two decades is presented in order to demonstrate the diversity of manifestations of the mesoscale water dynamics in the coastal zone and in the deep-water part of the Black Sea and the role of mesoscale structures (eddies and jets) in the horizontal (including that between the coastal zone and the open sea) and vertical water exchange. The characteristic routes of movement of coastal and deep-sea anticyclonic eddies are considered, together with the mechanisms of their formation, spatio-temporal and kinematic parameters, the features of manifestation in the surface temperature field, and their influence on the formation of large meanders and branching of the Rim Current. The factors providing the transformation of coastal anticyclonic eddies into deep-sea ones and the long-term (to about eight months) existence of the latter in the eastern deep-water basin are discussed. An asymmetry in the intensity of vortical activity in the eastern and western Black Sea is noted. Based on the results of recent laboratory modeling, the differences in the local values of the continental slope width are regarded as the possible reason for this asymmetry.

Keywords Black Sea · Coastal anticyclonic eddies · Coastal upwelling · Deep-sea anticyclones · Jets · Rim Current · Wind forcing

Abbreviations

RC	Rim Current
CZCS	Coastal zone color scanner
SeaWiFS	Sea-viewing wide field-of-view sensor
SLA	Sea Level Anomaly
CIL	Cold Intermediate Layer
IR	infrared

1 Introduction

The limited water exchange with the open seas and the small (100–150 m) thickness of the oxygen-containing (active) layer (see Chaps. "Thermohaline Structure of the Sea" and "Hydrochemical Structure of the Sea") make the Black Sea ecosystem extremely sensitive to climate changes and anthropogenic impact. Since the main sources of the anthropogenic contamination of the sea (riverine runoff, waste from cities and recreation zones, oil terminals, etc.) are concentrated in the coastal zone, the studies of the processes of horizontal water mixing in this zone and its exchange with the deep-water basin represent an urgent task.

Numerous satellite images in the IR and visible spectral ranges (including maps of chlorophyll *a* concentration derived from the measurements with CZCS and SeaWiFS) evidently suggest that the synoptic pattern of currents in the Black Sea surface layer is far more complicated than it follows from the known schemes of general circulation (for example, [1–3]), which include the Rim Current (RC), one or two cyclonic gyres in the deep-sea, and coastal anticyclonic eddies between the RC and the coast, or than it is derived from the results of numerical modeling [4]. Mesoscale ($\sim$ 15–100 km) eddies, vortical dipoles (paired eddies with vorticities of opposite signs), jets, and filaments represent typical circulation elements of the Black Sea. The data of hydrographic surveys collected over about two past decades characterized by a mesoscale spatial resolution (for example, [2, 3, 5–15]), the satellite images with a high spatio-temporal resolution [7, 12, 15–25], the altimetric measurements from the TOPEX/Poseidon and ERS-1, 2 satellites [15, 26–29], as well as the measurements with Argos-tracked drifters [12, 15] provide the information on the spatio-temporal and kinematic characteristics of the mesoscale features, on the sites of their most frequent formation and evolution, on the contributory factors for their formation, on their influence on the structure of the RC, and on their role in the horizontal and vertical water exchange in the active layer of the Black Sea. Below, we present the principal results of the studies of the mesoscale dynamics of the Black Sea waters based on an analysis of remote and field measurements.

2 Meanders of the Rim Current and Coastal Anticyclonic Eddies

The mesoscale variability in the Black Sea is mostly related to the meandering of the RC, to the formation of anticyclonic eddies between the RC and the coast, to their transformation into the deep-sea eddies, and to the interaction of the latter with the neighboring anticyclonic and cyclonic circulation elements and with the RC. The meandering jet of the RC together with the eddies (mostly anticyclonic) contained in its meanders is observed along the entire perimeter of the sea except for its southeastern region, where the main flow leaves the slope and propagates to the northeast toward the Caucasian coast. The most probable reason for the RC meandering lies in the hydrodynamic (mostly baroclinic) instability of this current, which interacts with the coastal geometry or with the continental shelf/slope topography. This kind of instability is best manifested along the Anatolian coast between the Sakarya Canyon, where an abrupt termination of the western continental shelf takes place, and the Arkhangelsky Ridge [7, 19, 20]. The formation of the largest meanders (with an offshore extent up to 100–150 km) is related to the main flow rounding the anticyclonic eddies; these eddies propagate over the wide and gentle northwestern slope (the so-called "Danube Fan") [10] or depart from the southeastern coast of the Crimea and from the Anatolian and Caucasian coasts [3, 5, 6, 8–10, 15, 23].

The characteristic wavelength of the meandering RC comprises about 100 km off the Anatolian coast [7, 10, 18, 19] and over the Danube Fan [10] and about 80–90 km in the northeastern part of the sea [9, 21, 23]. The speed of the meander movement in the general direction of the RC ranges from 3–6 cm/s [9, 21] to 10–15 cm/s [19, 20]; sometimes, however, the translation speed is too small to be detectable [7]. Off the northeastern coast, the meandering and the formation of coastal anticyclonic eddies is most intensive in summer and autumn, when, owing to the predominant anticyclonic vorticity of the wind field, the RC proper is least intensive. For example, in winter, the recurrence of meanders off the northeastern part of the Caucasian coast is lower than that in summer (on average, two and four events monthly, respectively) [1]. However, off the Anatolian coast, the RC instability is best manifested in winter and spring, when the current is stronger [7]. For example, in September 1991, the wave pattern here was poorly expressed, while in April 1993, large meanders related to outbursts of filaments and eddies were observed [10]. An intensive meandering (or eddy formation) off the Anatolian coast in the autumn of 2000 caused by the early seasonal enhancement of the RC was also mentioned in [29].

Based on the results of hydrographic surveys performed by the end of the 1980s, the authors of [2] distinguished nine quasi-permanent/recurrent coastal eddies around the basin. They are the so-called Sevastopol, Kaliakra, Bosphorus, Sakarya, Sinop, Kizilirmak, Batumi, Caucasian, and Crimean eddies. Actually, the mesoscale variability in the coastal zone related to the formation and motion of anticyclonic eddies is much more diverse. Two (Fig. 1) or three large anticyclones from 50 to 90 km in diameter may simultaneously exist over the northwestern continental slope [10, 24]. Three anticyclonic eddies with diameters about 40–50 km were simultaneously observed along the northeastern coast between Tuapse and Anapa [9]. A strong variability (interannual, seasonal, and synoptic) of the mesoscale circulation is observed in the Batumi Eddy region. For example, in June 1998, a large an-

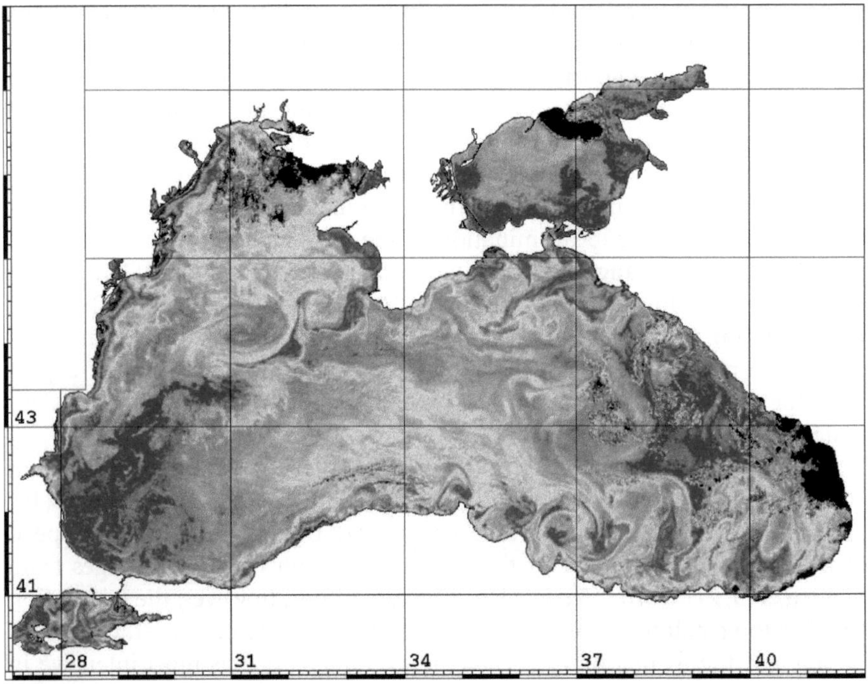

Fig. 1 Mesoscale water dynamics in the summer of 1998 (NOAA IR image for June, 29 1998)

ticyclone about 100 km in diameter terminated a chain of six anticyclonic eddies formed off capes of the Anatolian coast; their diameters increased in the eastern direction (Fig. 1). In this region, a large (more than 100 km in diameter) and intensive anticyclone was registered by a hydrographic survey in June–July 1996 [3, 14]; however, four months later, in November 1996, four anticyclones with diameters of about 45–50 km were located at its place [22].

It should be noted that the geographic names of the eddy features do not unambiguously denote the sites of their formation. These names only indicate regions with enhanced mesoscale activity, which is related not only to the local formation, but also to the passing of the eddies generated at other sites in the course of their propagation in the RC general direction. For example, anticyclones from the region west of Sevastopol move to the southwest over the Danube Fan toward the area located southeast of Cape Kaliakra; therefore, the "Kaliakra Eddy" and the "Sevastopol Eddy" can be the same feature registered at different instants [24]. Similarly, the "Caucasian Eddy" may find itself at the place of the "Crimean Eddy" and the latter feature may occur at the place of the "Sevastopol" one [15, 29] (Fig. 2). The movements of coastal anticyclones in the cyclonic direction in the area located between the Sochi–

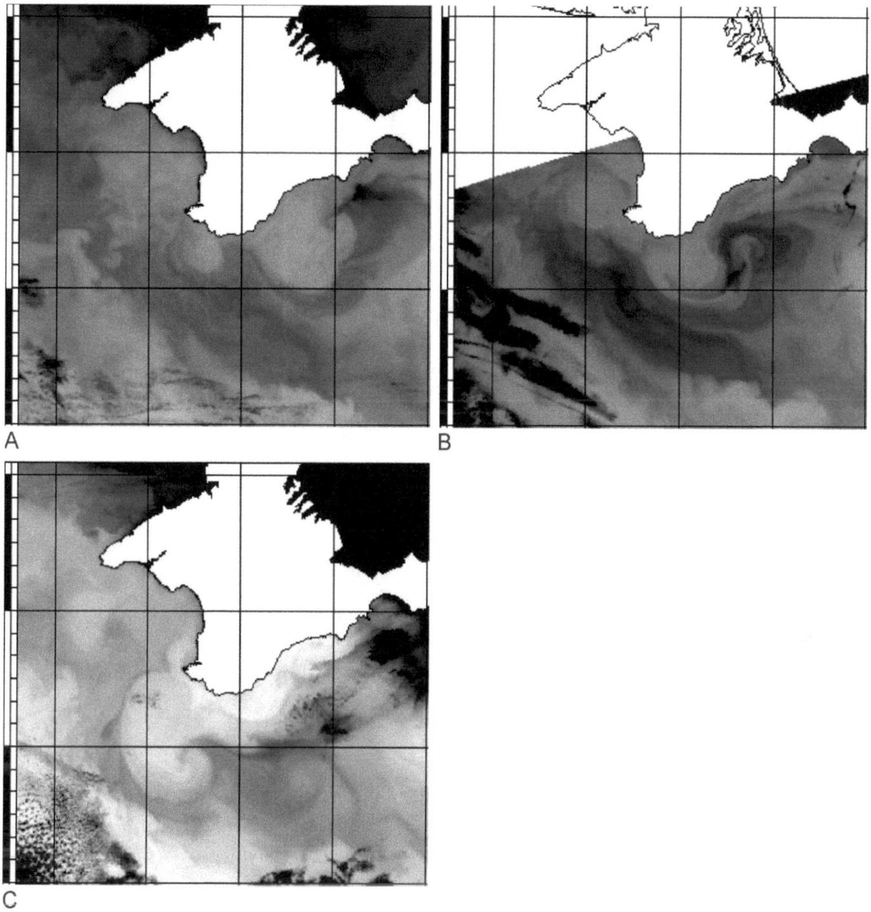

Fig. 2 Transport of anticyclonic vorticity around the southern extremity of the Crimea in October 2000. Fragments of NOAA IR images for **A** October 3, **B** October 5, and **C** October 15

Sukhumi zone in the east and approximately 42.5°N in the west derived from the observations are shown by the arrows in the scheme (Fig. 3).

The mechanisms of formation of coastal eddies have not yet been sufficiently studied. For the short-living (sometimes a few days) anticyclones off the Anatolian coast, this mechanism seems to consist in the interaction between the flow and local coastline/topographic features [10, 19]. In the region off Sevastopol, the most probable reason for the formation of anticyclones is the RC instability near Cape Khersones [4, 28], although, as noted above, they might be a result of the transfer of the anticyclonic vorticity around the Crimean Peninsula from the area off the southeastern coast of the Crimea. In the region of Cape Kaliakra, anticyclones may be observed owing to their supply from the Sevastopol Eddy region, but they may be formed in this

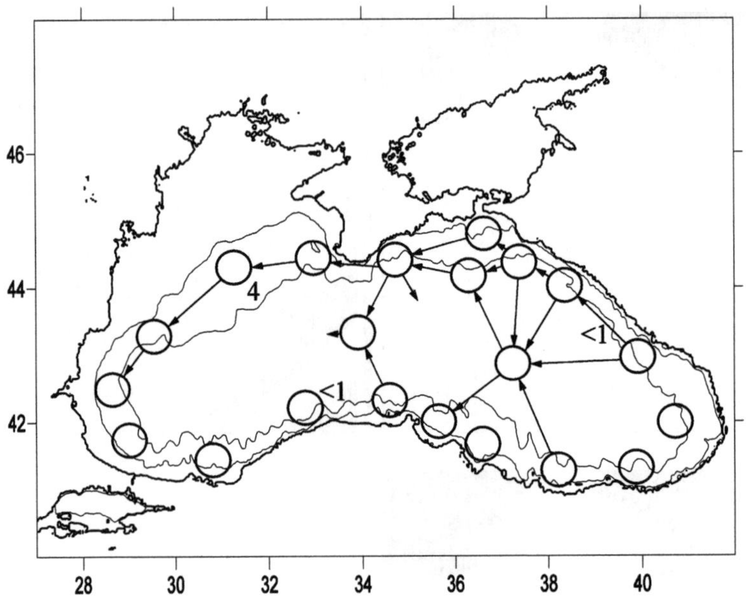

Fig. 3 Scheme of the areas of most frequent observations of anticyclonic eddies in the Black Sea (*circles*) and trajectories of their motion (*lines with arrows*). The numerals show local values of nondimensional width of the continental slope according to [32]. Isobates 100 m and 1500 m are shown

area proper characterized by a sharp decrease in the width of the continental slope, where the RC merges with a coastal current, which propagates in the southward direction along the western coast. (The possible influence of the changing the coastline geometry and the bottom topography on the boundary current instability and the eddy formation was considered in detail in [19].) The generation of the "Bosphorus Eddy" is caused by the splitting of the boundary flow in the region of the Bosphorus Canyon and by the turn of one of its branches northwest of the Bosphorus Strait [7]. The anticyclonic vorticity of the wind field (see [7,28]), the buoyancy effect (excess of precipitation over evaporation [30]), and the slowing down topographic waves propagating cyclonically around the perimeter of the basin in the region of the widened shelfbreak and merging of sequential eddies [4] are regarded as the factors controlling the formation of anticyclonic eddies in the southeastern part of the sea (in the Batumi Eddy region). Off the relatively smooth Caucasian coast north of Sukhumi with its narrow shelf and slope, the anticyclone formation is provided by the anticyclonic vorticity between the RC jet and the coast [1,8,9,14] and by the interaction between the RC and the continental slope topography [3,23]. In addition, some scientists suggest [19,28] that the eddies along the Caucasian coast may be generated owing to their separation from the Batumi Eddy and subsequent propagation toward the northwest. Although this mechanism is possible, it should be noted that the eddy formation off Novorossiisk–Gelendzhik, Tuapse, and Sochi–Sukhumi, i.e., in the regions with the coastline or bottom topography irregularities, is proved by observations [9, 12, 13, 22, 23]. As is shown by an analysis of satellite images, off the southeastern coast of the Crimea (in the Crimean Eddy region), anticyclones may be caused by the transfer of the anticyclonic vorticity from the Novorossiisk–Gelendzhik region along the coast or over the wide continental slope west of Novorossiisk. It is also possible that the factors favoring their local formation include the coastline geometry, the wind forcing, and the delivery of waters with lesser salinity via the Kerch Strait. In September 1999, a merging of eddies supplied from the region of the Kerch Strait resulted in the formation of a large anticyclone about 100 km in diameter off the southern extremity of the Crimea; later, it moved to the Sevastopol Eddy region (Fig. 4B).

The assimilation of satellite altimetric measurements in the model of the Black Sea circulation [28] and the results of the modeling with an eddy-resolving model [4] give an idea of seasonal variability of the vortical activity in the regions considered and the mean residence time of the anticyclones there. The "Batumi Eddy" is mostly formed in March and exists until the end of October–beginning of November (sometimes, it appears in December–January), that is, on average, about 210 days per year. In the summer season, when the RC is the weakest, the probability of its observation is the greatest. On the contrary, for the "Sevastopol Eddy," whose residence time is about 140 days per year, this probability is maximum in the winter season at an intensive RC and is minimum in fall. The appearance of the "Crimean Eddy" (115 days per year) is most probable in August–November. The "Kaliakra" and "Bosphorus" eddies (190 days per year) may be most probably observed

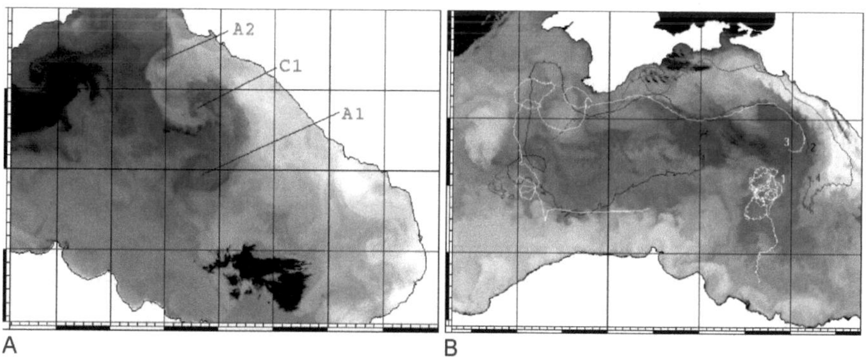

Fig. 4 **A,B** Mesoscale water dynamics in the autumn of 1999: **A** NOAA IR image for September 29, 1999 (*A* and *C* indicate anticyclone and cyclone, respectively); **B** NOAA IR image for December 2, 1999 with drifter trajectories (1–4) for the period from September 29 to December 31, 1999

in fall and winter, while in May this probability is minimum. In the regions of Sukhumi and Anapa, the highest vortical activity is confined to the winter–spring and fall periods (with a minimum in July), while off Sochi, it is minimum from August to December. Korotaev et al. [28] found one more area of anticyclonic vorticity, which is most often manifested as a narrow band along the seacoast between Odessa and Constantsa, though sometimes it can cover the entire northwestern shelf. It is observed virtually throughout the year with a minimum in August–September. Its existence is related to the Danube River runoff and to the anticyclonic vorticity of the surface wind stress.

Hydrographic surveys (for example, [1–3, 6–11, 14, 15, 22]) and satellite observations [7, 11, 12, 15–21, 23, 24, 31] showed the wide spectrum of the spatio-temporal and kinematic parameters of coastal anticyclones. Their diameters may change from ~ 40 to ~ 100 km (which makes $(2\text{–}6)R_d$, where $R_d = 15\text{–}20$ km is the baroclinic Rossby deformation radius [13]), the vertical extension varies from ~ 150 to ~ 400 m, and the orbital velocity ranges from ~ 10 to 60 cm/s. The most typical translation speed of anticyclones along the coast is 2–6.5 cm/s, although at selected sites of their route it may either reach ~ 15 cm/s [23] or fall almost down to zero (when the eddy resides at the same place over a long period [18, 23]). The time of propagation of this kind of structure along the coast (continental slope) ranges from approximately a week (when the eddy merges with a neighboring structure or separates from the coast) to five months (for example, over the wide northwestern slope [24, 29]). The number of simultaneously existing eddies over the perimeter of the sea seems to depend on the season and particular hydrometeorological situation (intensity of the RC and wind field vorticity). For example, in June 1998 (Fig. 1), ten eddies were observed (six of them were located off the Anatolian coast). The numerical modeling [4] results in 12 coastal eddies per annual cycle. The regional statistics of the number of the coastal eddies formed (or passed across the particulate region) based on the field observations are available only for the northeastern part of the sea [14], where an average of 32 eddies were annually registered (with a maximum and minimum numbers of 46 and 19, respectively).

An interesting feature of the surface manifestation of the coastal and deep-sea Black Sea anticyclones is related to the temperature drop by 1–3.5 °C at their centers with respect to their periphery and adjacent waters, which is frequently observed in the warm time of the year. This negative temperature anomaly is caused by a rise of the isotherms in approximately 20-m upper layer and their wedging out toward the surface accompanied by the lowering of the isotherms in the underlying layer [6, 8, 15, 21–23]. In so doing, the maximum of the orbital velocity in the eddy is located immediately under the seasonal thermocline at a depth of approximately 30–50 m rather than at the surface. The negative temperature anomaly at the centers of anticyclones may also be observed at a surface maximum of the geostrophic velocity; however, in such cases, it is related either to the warm coastal waters rounding the eddy [11, 23] or to the entrainment of the cold waters of coastal upwelling.

3 Deep-Sea Anticyclonic Eddies

3.1 Observations of Deep-Sea Anticyclones

A comparative analysis of the hydrodynamical situations that occurred in different years has made it evident that anticyclonic eddies may represent a typical element of the circulation in the eastern deep basin at least during the warm season (April–December). This fact contradicts the traditionally accepted concept (see [1–3]). The appearance and existence of these kinds of anticyclonic eddies in the deep basin is related to the separation of coastal anticyclones from the coast. Some events of this kind together with the subsequent movement of the deep-sea anticyclones were registered with the use of satellite information of high spatio-temporal resolution and derived from the hydrographic surveys of different years.

A large long-living anticyclonic eddy centered at 43°N and 34°E, in the area between the western and eastern cyclonic gyres (approximately abeam the southern extreme of the Crimea), was detected by the hydrographic survey of 1984 [6]. It was formed in September 1984 as a result of coalescence of two other anticyclones formed owing to baroclinic instability of the RC and to detachment of its meanders in the north (from the Crimean coast) and in the south (from the Turkish coast near Sinop). Its diameter exceeded 100 km, the maximum of the orbital geostrophic velocity was 25–45 cm/s, and the rate of the westward displacement was about 1 cm/s. Density and salinity anomalies related to this eddy were traced down to a depth of 1000 m and temperature anomalies were followed down to 300 m.

An example of a long-living (about half a year) deep-sea anticyclone is presented by the eddy observed in 1993 in the northeastern part of the sea with the use of the data of hydrographic surveys [8] and a series of IR images [23]. This feature was generated as a coastal eddy in the region of Tuapse or came from a more southern region of the Caucasian coast at the beginning of June and started to detach from the coast in the Novorossiisk–Gelendzhik region in mid-August. This eddy was characterized by a large diameter ($\sim$ 95 km), low mean translation speed along the coast and in the deep-water basin (about 1 cm/s), large lifetime (no less than 2.5 months near the coast between Tuapse and Novorossiisk and no less than 3 months after its separation from the coast), an orbital velocity about 26 cm/s (with a maximum at a depth of 50 m), and a clearly manifested negative surface temperature anomaly with respect to the periphery at its center (1.4–3.5 °C) throughout the entire time of observations. The vertical extension of the eddy was about 250 m. At the moment of the last hydrographic survey in November 1993, the parameters of the eddy did not allow one to expect its dissipation in the nearest future; therefore, its lifetime might exceed 5.5 months.

One more deep-sea eddy transformed from a coastal anticyclone was observed over a month in the fall of 1997 [23]. This eddy was formed on September 6–8, 1997 west of Novorossiisk in the zone of the shelf/slope widening. Later on, it gradually moved southwestward away from the coast (Fig. 5) at a mean rate of 4.3 cm/s. During this time, its diameter increased from 40 to 75 km due to the entrainment of the adjacent waters. Then, most probably, it merged with an anticyclone off the southeastern coast of the Crimea.

A large and long-living anticyclone, whose evolution was traced in the altimetric maps of sea level anomalies (SLA) [29], was also observed at the center of the eastern part of the sea in 1998 (Fig. 1). It separated from the Anatolian coast at the end of January 1998 approximately at 38°E [27] and resided at the center of the eastern basin over a long period (the coordinates of the center of the eddy were 43°N, 37.5°E). In April–May, this eddy was fed by anticyclones separated from the Caucasian coast near Sukhumi. In August–September, its existence was confirmed by the hydrographic survey performed from R/V *Akvanavt* on August 28–September 5, 1998 [26]. At the end of September, it merged with the anticyclonic eddy located approximately at 44°N, 36.5°E. Thus, the time of residence of the anticyclone at the center of the eastern basin comprised about eight months. The newly formed structure composed

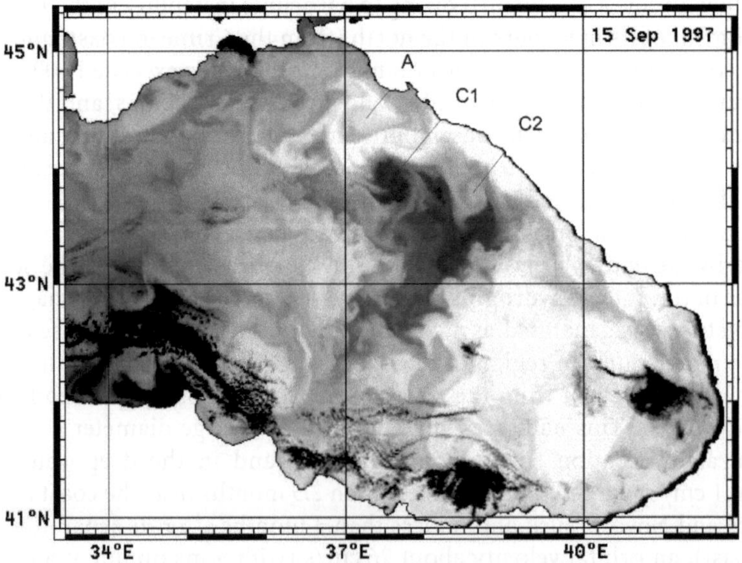

Fig. 5 NOAA IR image for September 15, 1997 (*A* and *C* indicate anticyclone and cyclone, respectively). *Dark* (*light*) *color* corresponds to cold (warm) water

of two merged eddies gradually moved to the west toward the southern extreme of the Crimea and was traced in the SLA maps up to the early middle January 1999.

A long-living anticyclonic eddy was observed at the center of the eastern part of the sea in 1999 as well (Fig. 4A,B). Its life cycle, thermohaline structure, the influence on the structure of the RC, and the distributions of hydrobiological and hydrochemical parameters were studied on the basis of satellite information and multidisciplinary sea-truth surveys including measurements with drifters [12, 15, 29, 31]. This anticyclone was formed as a coastal eddy off Sukhumi at the end of March 1999 and separated from the coast on April 6–9. By the end of May 1999, it had been strengthened owing to its merging with anticyclones from the Sukhumi region and moved to the west (43°N, 38°–38.5°E). Subsequently, the position of the center of this eddy, which repeatedly interacted with anticyclones separated from the Caucasian coast, changed in the zonal direction within 37°–38°E; its maximum intensity was observed in the early June. Approximately from December 23, its interaction with an anticyclone (or an anticyclonic meander) off the Anatolian coast resulted in its gradual southwestward movement, which is confirmed by the trajectory of drifter 1 in Fig. 4B, and, most probably, led to the merging of these mesoscale structures. Thus, the lifetime of this anticyclone comprised about 8.5 months. According to the data of drifter 1 in Fig. 4B, the period of its orbital rotation was 5–15 days.

Throughout the entire time of its existence, this anticyclone alternately interacted with deep-sea cyclonic eddies. For example, during a month (September–October 1999), it formed a large vortical pair A1–C1 with a cyclone that was located northeast of it and interacted with the RC (Fig. 4A). In so doing, the cyclone featured a northwestward displacement with respect to the anticyclone at a mean rate about 3 cm/s. The horizontal size of this vortical pair A1–C1 at the center of the eastern basin equaled about 160 km. According to the results of the hydrographic survey and drifter data, the near-surface orbital velocities at the northern periphery of the cyclone near its interface with the RC, were about 25–30 cm/s; at its southern periphery and in the anticyclone, the corresponding values were 15–20 cm/s. Both of the eddies caused significant deviations in the isopycnal positions down to depths of 350–400 m. At the center of the anticyclone, the negative temperature anomaly with respect to the surrounding waters was 1–2 °C.

The above considerations allow us to conclude that mesoscale areas of anticyclonic vorticity in the open eastern part of the Black Sea are not so rare and the lifetime of individual deep-sea anticyclones may reach approximately eight months. The reason for their long-term existence seems to lie in their "replenishment" owing to the merging of adjacent anticyclones or to the energy transfer from the RC implemented by coherent vortical features (for example, to anticyclone A1 via the neighboring cyclone C1 interacting with the RC, Fig. 4A).

3.2 Factors Influencing the Formation of Deep-Sea Anticyclones

An analysis of the conditions that accompany the formation of the above-considered deep-sea anticyclonic eddies in the eastern part of the sea observed in 1984, 1993, 1997, 1998, and 1999 allows us to suggest that the factors favoring their generation are the wind forcing and the particular features of the coastline/bottom topography. Correspondingly, the seasonal and interannual differences in intensity of the formation of these kinds of mesoscale anticyclones in the Black Sea are related to the variability in the RC intensity, which, in its turn, is controlled by the intensity of the wind forcing.

A comparison of the results of the hydrographic surveys performed in the northeastern part of the sea in the autumn seasons of 1999–2001 [13] and of the SLA maps for 1998–2001 [29] showed that, in 1998, 1999, and 2001, at the absence of intensive cyclonic circulation on the scale of the entire sea and at a baroclinically instable RC in the autumn period, anticyclonic eddies were observed in the eastern deep-water basin. However, no such eddies were registered during the same period of 2000, at an intensive RC and well-manifested western and eastern cyclonic gyres. In 1993, the formation and evolution of a long-living anticyclone also proceeded under a dominating anticyclonic circulation in the atmosphere and weak winds in summer and fall, especially over the northern part of the sea [8]. It should be noted that the greater part of the observations of the coastal eddies separation from the coast and their transformation into deep-sea eddies refers to the warm season (from April to December), when the cyclonic circulation and the wind speed in the atmosphere are usually lower than in winter. However, judging from selected satellite images and altimetric data, separation of anticyclones from the Caucasian and Anatolian coasts may sometimes occur in the wintertime as well. Probably, these kinds of events are caused by local synoptic wind impacts when the wind direction is opposite to the direction of the RC flow. Similar situations are also observed during the warm period. For example, in September 1997 (during strong winds and an intensive RC) and in August 1993 (at a weak RC), events of separations of coastal anticyclones from the coast followed intensifications of northeasterly winds [23]. The severity of the preceding winter seems to have no influence on the RC characteristics in the summer–autumn period and on the existence of long-living deep-sea anticyclones: they were observed both after the very cold winter of 1992–1993 and after the very mild winter of 1998–1999 (the information on the winter severity is presented in Chap. "Sea Surface Temperature Variability").

The movement of anticyclonic eddies over the sea area including their separation from the coast and subsequent evolution in the situations considered is shown in Fig. 3. Frequent events of separation of anticyclones from the coast take place in the eastern part of the sea in the region of Novorossiisk–

Gelendzhik, Tuapse, and Sochi–Sukhumi off the Caucasian coast, off the southeastern coast of the Crimea, and off the Anatolian coast, in the areas with coastline/bottom topography irregularities or close to capes. Deep-sea anticyclones are most often observed in two regions centered approximately at 43°N, 37°–39°E and 44°N, 35°–37°E. In the former case, they are formed by anticyclones that separated from the coast in the Sukhumi region (Gudauta shoal) or from the Anatolian coast, while in the latter case, their parent anticyclones separated from the coast in the Novorossiisk–Gelendzhik region and from the southeastern coast of the Crimea.

It is interesting that in the western basin virtually no observations of long-living deep-sea anticyclones are available. The single case was the registration of an anticyclone centered approximately at 44°N, 33°E (beyond the continental slope) in December 1999 (Fig. 4B); it was related to the separation of an eddy, which came from the region off the southeastern Crimea, from the coast off the southernmost extremity of the peninsula. Nobody managed to follow the further evolution of this anticyclone, which was first recognized with the help of IR images and the trajectory of a single drifter (drifter 3 in Fig. 4B). In a typical situation, the anticyclones that were formed west of Sevastopol propagate along the gentle continental slope and do not enter the deep-water part of the sea; as it has been mentioned before, their lifetime may reach five months.

The asymmetry in the intensity of the mesoscale dynamics between the western and eastern deep-water basins may be explained with the help of the results of a recent laboratory experiment [32]. According to them, the structure and dynamics of the alongshore current strongly depend on the nondimensional width of the continental slope zone L/R_d, where L is the slope width (the least horizontal distance between the 100- and 1500-m depth contours) (Fig. 3). At $L/R_d \leq 1$ (steep slope, eastern coast of the Black Sea), the influence of the inclined bottom on the stability of an alongshore current with a width of $(2–4)R_d$ (as, for example, the RC) is negligible; in this case, at termination or weakening of the wind forcing, the current starts to feature a strong meandering with formation of cyclones and anticyclones. At $L/R_d \geq 2$ (gentle and wide slope, northwestern part of the sea), the flow instability is poorly expressed; the meandering is characterized by waves with small amplitudes; virtually no eddies are formed, and water exchange between the coastal zone and the open sea is suppressed. Note that, in the latter case, we mean the flow instability as a mechanism for eddy formation, while the anticyclones permanently propagating over the wide northwestern slope do not result from the local instability of the RC (Sect. 2).

It is also probable that the gentle character of the northwestern slope favors the long-term coexistence of closely located and interacting anticyclones without their merging [22, 24], while in the eastern basin, anticyclone merging is often observed and seems to be the reason for their great lifetime.

4
Horizontal and Vertical Water Exchange

Mesoscale eddies and jets implement water exchange both within the coastal zone and between this zone and the open sea. In the regions with a narrow continental slope, where coastal anticyclones are enclosed between the RC and the coast (for example, along the Caucasian coast in the northeastern part of the sea), they provide the supply of cleaner waters from the deep sea toward the coast in their frontal parts and the removal of more polluted waters in their rear parts. Anticyclones transport the warmer and lesser saline coastal water together with the admixtures of natural and anthropogenic origins contained in it from the sites of their formation to the sites of the anticyclone dissipation or merging with other vortical structures. The movement of these eddies along the coast provides the change in the direction of the near-shore current from that coinciding with the direction of the RC at the absence of the eddy to an opposite one [1,23].

In the regions where the distance between the RC and the coast exceeds the horizontal scale of the eddies or when anticyclones separate from the coast, the evolution of anticyclones is accompanied by the generation of one to three nonstationary associated cyclonic eddies with diameters changing from ~ 10 km to values comparable with the diameter of the "parent" anticyclone and jets. Because of this and owing to the entrainment of the adjacent waters, the area of the influence of an anticyclone is significantly greater than its diameter. For example, the dense "packings" of vortical dipoles on the basis of a large anticyclone (~ 80 km) in June 1981 [16] or three closely located anticyclones of smaller sizes (~ 45 km) with associated cyclones in November 1996 [22] provided horizontal mixing over the vast southeastern part of the sea (in the Batumi Eddy region). Owing to the water entrainment at their northern and southern peripheries, the anticyclonic eddies over the northwestern slope provide water transport from the western coast to the deep-water region (in particular, to the coast of the Crimea) and southwestward transport of the waters from the coastal zone off the Crimea (including the cold waters of the upwelling off Cape Khersones) toward the deep sea [18,24]. An example of the transfer of shelf waters with enhanced chlorophyll *a* concentrations to the deep-water region by means of their entrainment by an anticyclone over the Danube Fan is shown in Fig. 6. At a constant riverine runoff, this transfer is favored by westerly winds [24].

The anticyclonic eddies over the Danube Fan supply waters rich in nutrients and chlorophyll to the deep sea and also influence the biological productivity in the western part of the Anatolian coastal zone. When the entrainment of the shelf waters by anticyclones is intensive, the southward alongshore flow is weak and the nutrient supply to the south is restricted, while at the absence of eddies it increases [25].

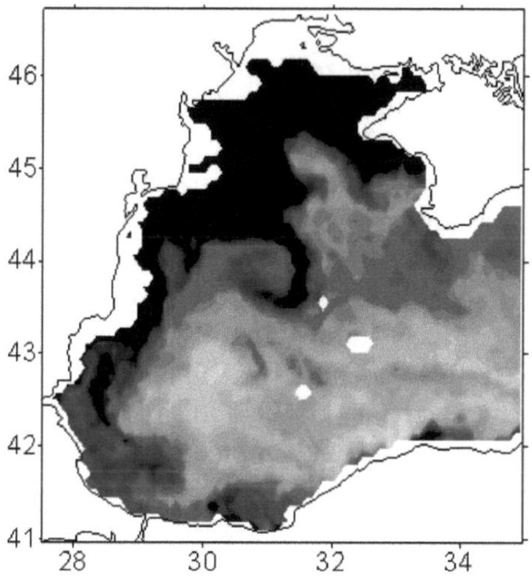

Fig. 6 SeaWiFS image for July 13, 1998. *Dark color* corresponds to the waters with a high chlorophyll concentration

Within the wide northwestern shelf, an efficient mechanism of mixing is also represented by the transversal filaments of the wind-driven coastal upwelling. These cold filaments formed along the northwestern coast, Tendrovskaya Spit, and the western coast of the Crimea are about 40 km long and 10 km wide; the distance between subsequent filaments ranges from ~ 10 to ~ 40 km and their temperature contrast with respect to the surrounding waters is about 1.5–2 °C. Often, they terminate by an eddy or a vortical dipole [22]. In contrast to these ordered structures of the upwelling, which propagate only within the shelf zone, the upwelling filaments up to 150 km long, often observed in summer off the Anatolian coast, carry the cold and lesser saline waters away toward the interior basin [7, 19, 21]. Their formation may be related both to the wind forcing and to the RC deviation from the coast owing to its interaction with the bottom topography features. The entrainment of the transformed waters of the upwelling into anticyclonic eddies over the northwestern slope [18, 24], off the southeastern coast of the Crimea, and off the Anatolian coast (often in the area between Cape Baba and Sinop) also provides their transfer to the deep-water basin.

Another type of ordered structures related to the upwelling in the northwestern part of the sea is represented by cyclonic eddies with a diameter of 10–20 km that leave the coast off Cape Khersones and propagate across the depth contours beyond the shelf zone [22]. An additional contribution to the intra-shelf water exchange in this part of the sea is provided by the eddies (anticyclonic and cyclonic) with diameters about 20–50 km that are formed at the front of the freshened waters related to the Danube River runoff [16, 21].

The contribution of the anticyclones separated from the coast to the water exchange between the coastal zone and the deep-water basin is caused by the entrainment of the coastal waters and their transfer to the open sea as well as by the formation of associated eddies and jets at the peripheries of the anticyclones. For example, waters with relatively low salinity (18.15 psu, at a salinity of the surrounding waters of 18.25 psu) of an evidently coastal origin were observed at the center of an anticyclonic eddy in November 1993, three months after its separation from the coast [23]. The reduced salinity of the waters of an evidently coastal origin observed in separating anticyclonic meanders and eddies was also noted in [5, 11]. The advection of the surrounding waters at the periphery of the eddies at distances of about 150 km also represents an important mechanism for the water exchange between the shelf and the open sea.

The mesoscale structures cause the patchiness in the distributions of chlorophyll and plankton in the coastal zone and in the open sea. For example, on October 8, 1997, a maximum of the chlorophyll concentration was observed in coastal anticyclones that were formed not long before off the Caucasian coast. This maximum was provided by the entrainment of the coastal waters rich in nutrients after heavy rains on October 2–4 and the related enhanced riverine runoff [23]. In this case, in the anticyclone that separated from the coast a month before and was not fed by the coastal waters, the chlorophyll concentration was low. In September 1999, at the center of an anticyclone, an increased concentration of the ctenophores *Mnemiopsis leidyi* and *Beroe ovata* was observed; it was related to the capturing macroplankton-rich waters by the eddy in the course of its formation and separation from the coast [11]. Vortical pair A1–C1 at the center of the eastern basin observed in September 1999 (Fig. 4A) affected the contents of dissolved oxygen and nutrients as well as the total biomass and species composition of plankton (see [12]). For example, in the jet of warm coastal waters entrained by cyclone C1 at its northern periphery (Fig. 4A), elevated chlorophyll *a* contents were noted, while its low concentrations were observed in the cold waters of the vortical pair and in the divergence area between them [33]. The interannual variability of the spatial distribution of the zooplankton biomass related to the interannual variability of the mesoscale water dynamics in the sections perpendicular to the coast was demonstrated in [34]: in 1999 and 2001, at a well manifested eddy dynamics and a weak RC, the biomass of zooplankton in the sections was distributed rather evenly, while in 2000, at a strong RC and the absence of large deep-sea eddies, its maximum was located near the coast and in the midstream of the RC and decreased toward the center of the eastern basin by an order of magnitude.

Short-living (from a few days to about two weeks) cyclones and jets often accompany the evolution of the anticyclonic eddies separating from the coast. For example, a jet $\sim$ 10 km wide terminated with a vortical pair, which moved southwestward from the northwestern periphery of an anticyclone separated from the coast, had a lifetime of 12 days and transported the warm coastal waters over a distance of about 120 km away from the coast (Fig. 5). Jets of another type represent an element of a mushroom-shaped structure and seem to be related to the self-advection of the vortical pair formed as a result of the meandering of the RC (this kind of formation mechanism of a dipole structure with an intensive jet between the eddies of opposite signs and the fast propagation of this structure in the direction perpendicular to the baroclinically instable flow was demonstrated in a laboratory experiment [13]). Such structures with an axis perpendicular to the RC and extended for distances up to $\sim$ 160 km from the coast are often observed off the Anatolian coast (Fig. 1, see also satellite images in [7, 18–21, 23]). Usually, their lifetime does not exceed a few days to two weeks.

Besides, in the IR images of the northeastern part of the sea, one can often recognize warm jets up to $\sim$ 250 km in length that propagate from the coast (often from the Anapa region) southwestward to the deep-water basin (Fig. 1, see also [21, 22]). Sometimes, the bases of the jets are located close to the northwestern peripheries of coastal anticyclones (Fig. 1). The origin of these jets is unknown; however, their orientation and the fact that the subsequent jet propagates sometimes in the wake of the preceding one, which is well seen in the IR image of September 14, 1991 (Fig. 4E in [21]), allow one to suggest that their formation is related to a pulse wind impact (probably, to the Novorossiisk "bora").

The separation of anticyclones from the coast results in the seaward deviation of the RC, formation of large meanders of the current, and its branching. For example, the cyclone that was attached to the RC in September 1999 (C1 in Fig. 4A) made the current between it and the coast narrower and faster [12, 13]. Meanwhile, in the rear of the cyclone (downstream), the flow widened again and, under the influence of this cyclone and a small anticyclone formed northwest of it and separated from the coast (A2 in Fig. 4A), a part of the RC turned toward the open sea. In the process of the movement of A2, at a distance of 60–70 km from the coast, the RC branched forming a few jets and its waters intensely interacted and mixed with the surrounding waters of the open sea [31].

The influence of the anticyclones separating from the coast on the structure of the RC was also well traced in October–November 1993, when the offshore location of a deep-sea anticyclone centered approximately at 44°N, 36.8°E resulted in a departure of the RC from the coast and formation of a large meander [8, 23], and in September 1997 (Fig. 5). Due to the movement of the anticyclonic eddies separated from the coast on the one hand and to the position of the RC changing with respect to the current location of the eddies and the wind direction on the other hand, situations are possible when the same deep-sea anticyclone finds itself now on the right-hand side of the RC mainstream or its branches and now on their left-hand side. In addition, the movement of the anticyclones from the eastern basin to the western one around the Crimea causes significant local displacement of the RC to the south. In this case, the existence of a large anticyclonic meander of the RC or an eddy off the Anatolian coast in the narrowest part of the sea (near 33°E) may result in a short-term isolation of the western and eastern basins of the Black Sea from one another (this kind of situation was observed, for example, in October 2000 [29]).

In addition to deep-sea anticyclones and large cyclonic eddies interacting with them (such as vortical pair A1–C1 in Fig. 4A), an important contribution to the water exchange between the coastal zone and the open sea is made by small cyclones with a diameter of 30–40 km and jets. Two cyclones of this type probably formed as a result of a velocity shear at the seaward boundary of the RC may be seen in Fig. 5. Moving to the northwest at a rate of approximately 10 cm/s and interacting with the RC and the anticyclone separating from the coast, they provided the transport of warm coastal waters over a distance more than 100 km away from the coast [23]. Small cyclonic eddies are also formed in the cyclonic meanders of the RC [8].

Vertical water motion in the coastal and deep-sea eddies affect the topography of the hydrophysical and hydrochemical isosurfaces and the structure of the active layer. For example, in the anticyclones propagating along the northeastern part of the Caucasian coast, the depth of the upper boundary of the Cold Intermediate Layer (CIL) changes, depending on the eddy intensity, from 20 to 70 m (at a mean value of 43 m) and the depth of its lower boundary ranges from 115 to 220 m (at a mean value of 144 m) [14]. In the anticyclone of the vortical pair observed in September 1999 (Fig. 4A), the depth of the temperature minimum in the CIL lowered down to 82 m, while in the cyclone it raised up to 33 m. The characteristic depth of the upper boundary of the hydrogen sulfide zone in coastal anticyclones (isopycnal $\sigma_t = 16.2$) is about 160 m (being 100 and 130 m in the central part of the sea and near the shore, respectively), while in the Batumi Eddy, it may be located as deep as at 200–220 m being 150–180 m deep beyond the eddy [3]. In the core of the anticyclonic part of the vortical pair observed in September 1999 (A1–C1 in Fig. 4A), the upper boundary of the hydrogen sulfide layer lowered down to a depth of 150 m, while in the core of the cyclone it was located at 110 m [15]. This way, the vertical motions in the eddies of the pair led to a total change in the active layer thickness almost by one-third of its value. According to [6], in the summer of 1984, the boundary of the oxygen-containing layer in anticyclones (both coastal and deep-sea) was located at a depth of 190–220 m, while in the surrounding waters, its depth was 100–140 m. Based on the estimates of the oxygen concentration in anticyclones, the author of [6] inferred that anticyclonic eddies represent efficient accumulators and carriers of oxygen. The thickness of the CIL in anticyclonic eddies is also greater than in cyclones and in the adjacent waters; therefore, they may be regarded as efficient carriers of the cold intermediate waters as well.

5 Conclusions

The satellite images and hydrographic surveys suggest a diversity of the hydrodynamical settings (with their interannual, seasonal, and synoptic variabilities) in the Black Sea. Mesoscale eddies up to 100 km in diameter (both coastal and deep-sea, transporting the coastal waters to the open sea over distances up to ~ 200 km) and jets of different origins propagating over ~ 200 km away from the coast significantly affect the intra-basin water exchange in the sea, since the width of the deep-water part of the sea is only a few times greater than the size of the major ordered structures. In particular, water entrainment by the large anticyclonic eddies located over the wide northwestern slope provides the propagation of the desalinated shelf waters rich in nutrients to the deep-water basin of the western part of the sea and controls the biological productivity in the western part of the Anatolian coastal zone. Notice that anticyclonic eddies (sinking water at their centers) do not results in the formation of "biohydrochemical barrier" between the coastal zone and the open sea, as it is argued in some publications (for example, in [35]). Mesoscale features (eddies, jets, and filaments) formed over the entire perimeter of the sea equalize the chemical and biological parameters over its area. Mesoscale eddies (anticyclones, cyclones, and vortical pairs) and related jets also affect the structure of the RC and lead to the formation of large meanders of the current, moving its axis away from the coast over great distances, and branching. The lowering and rising of the upper boundary of the hydrogen sulfide zone in anticyclones and cyclones, respectively, may stimulate the ventilation of the anoxic waters of the Black Sea.

An important result of the comparative analysis of the water circulation in different years and the relevant information on the wind field consists in the establishment of the possibility of appearance and, sometimes, long-term (up to 8 months) existence of anticyclonic eddies in the open sea (beyond the continental slope) and in revealing the factors that favor this phenomenon. Deep-sea anticyclones are characteristic only of the eastern basin (Fig. 3), where their appearance results from the separation of the anticyclones formed due to the RC instability at the sites with a narrow continental slope (Caucasian and Anatolian coasts and the southeastern coast of the Crimea) from the coast. The wide and gentle northwestern slope imposes a stabilizing effect on the RC. The anticyclonic eddies that found themselves over this slope owing to their formation off the southwestern coast of the Crimea or transfer from the eastern basin propagate to the southwest never entering the deep-water part of the sea.

In the eastern deep-water basin, anticyclonic eddies are mostly observed in the warm season (April–December), when the cyclonic vorticity of the atmospheric circulation and the wind speed are lower than in winter and the RC is less intensive. Under these conditions, the separation of coastal eddies from the coast is generally more probable, although significant interannual deviations in the seasonal vortical dynamics are observed.

Local synoptic wind impacts also favor the intensification of the horizontal water exchange in the Black Sea. For example, they cause wind effected phenomena and formation of filaments of coastal upwellings and provide the separation of coastal anticyclones from the eastern coast (under northerly winds) or removal of the shelf waters to the deep-water basin by anticyclones over the northwestern continental slope (under westerly winds).

Acknowledgements The authors are grateful to D.M. Soloviev and S.V. Stanichny (Marine Hydrophysical Institute, National Academy of Sciences, Sevastopol, Ukraine) for providing satellite information and long-term fruitful cooperation. This study was supported by the EC SESAME and ASCABOS International Projects and by the Russian Foundation for Basic Research, project no. 05-05-64927).

References

1. Ovchinnikov IM, Titov VB (1990) Doklady AN SSSR 314:1236 (in Russian)
2. Oguz T, Latun VS, Latif MA, Vladimirov VV, Sur HI, Markov AA, Ozsoy E, Kotovshchikov VV, Eremeev VV, Unluata U (1993) Deep-Sea Res 40:1597
3. Krivosheya VG, Titov VB, Ovchinnikov IM, Kosyan RD, Skirta AY (2000) Oceanology 40:816 (English Translation)
4. Staneva J, Dietrich D, Stanev E, Bowman M (2001) J Mar Syst 31:137
5. Blatov AS, Bulgakov NP, Ivanov VA, Kosarev AN, Tuzhilkin VS (1984) Variability of Hydrophysical Fields of the Black Sea. Gidrometeoizdat, Leningrad (in Russian)
6. Latun VS (1989) Morskoy Gidrofizicheskiy Zhurnal 3:40 (in Russian)
7. Oguz T, La Violette PE, Unluata U (1992) J Geophys Res 97:12569
8. Krivosheya VG, Moskalenko LV, Ovchinnikov IM, Yakubenko VG (1997) Oceanology 37:321 (English Translation)
9. Krivosheya VG, Ovchinnikov IM, Titov VB, Yakubenko VG, Skirta AY (1998) Oceanology 38:492 (English Translation)
10. Oguz T, Besiktepe S (1999) Deep-Sea Res 10:1733
11. Ginzburg AI, Zatsepin AG, Kostianoy AG, Krivosheya VG, Skirta AY, Soloviev DM, Stanichny SV, Sheremet NA, Shiganova TA, Yakubenko VG, Gregoire M (2001) Issledovanie Zemli iz kosmosa 5:3 (in Russian)
12. Zatsepin AG, Flint MV (eds) (2002) Multidisciplinary Investigations of the Northeast Part of the Black Sea. Nauka, Moscow (in Russian)
13. Zatsepin AG, Ginzburg AI, Kostianoy AG, Kremenetskiy VV, Krivosheya VG, Poyarkov SG, Ratner YB, Skirta AY, Soloviev DM, Stanichny SV, Stroganov OY, Sheremet NA, Yakubenko VG (2002) Oceanology 42(Suppl 1):S1
14. Titov VB (2002) Oceanology 42:637 (English Translation)
15. Zatsepin AG, Ginzburg AI, Kostianoy AG, Kremenetskiy VV, Krivosheya VG, Stanichny SV, Poulain P-M (2003) J Geophys Res 108:C8. DOI 10.1029/2002JC001390:2–1
16. Kaz'min AS, Sklyarov VE (1982) Issledovanie Zemli iz kosmosa 5:56 (in Russian)
17. Grishin GA, Makeev IG, Motyzhev SV (1990) Morskoy Gidrofizicheskiy Zhurnal 2:54 (in Russian)
18. Ginzburg AI (1994) Issledovanie Zemli iz kosmosa 2:75 (in Russian)

19. Sur HI, Ozsoy E, Unluata U (1994) Prog Oceanogr 33:249
20. Sur HI, Ozsoy E, Ilyin YP, Unluata U (1996) J Mar Syst 7:293
21. Sur HI, Ilyin YP (1997) Prog Oceanogr 39:109
22. Ginzburg AI, Kostianoy AG, Soloviev DM, Stanichny SV (2000) Remotely sensed coastal/deep-basin water exchange processes in the Black Sea surface layer. In: Halpern D (ed) Satellites, oceanography and society. Elsevier, Amsterdam, p 273
23. Ginzburg AI, Kostianoy AG, Krivosheya VG, Nezlin NP, Soloviev DM, Stanichny SV, Yakubenko VG (2002) J Mar Syst 32:71
24. Ginzburg AI, Kostianoy AG, Nezlin NP, Soloviev DM, Stanichny SV (2002) J Mar Syst 32:91
25. Oguz T, Deshpande AG, Malanotte-Rizzoli P (2002) Cont Shelf Res 22:1477
26. Sokolova E, Stanev EV, Yakubenko V, Ovchinnikov I, Kos'yan R (2001) J Mar Syst 31:45
27. Korotaev GK, Nikiforov AA (2001) Crossfrontal transport by mesoscale eddies in the Black Sea. In: Ecological Safety of Coastal and Shelf Zones and Complex Use of the Shelf Resources. Marine Hydrophysical Institute, Sevastopol, Ukraine, p 40 (in Russian)
28. Korotaev GK, Oguz T, Nikiforov AA, Bekli BD, Koblinski CJ (2002) Issledovanie Zemli iz kosmosa 6:60 (in Russian)
29. Ginzburg AI, Kostianoy AG, Sheremet NA (2003) Issledovanie Zemli iz kosmosa 3:34 (in Russian)
30. Oguz T, Malanotte-Rizzoli P (1997) Wind and thermohaline circulation of the Black Sea driven by yearly mean and monthly climatological forcing. In: Science for stability. NATO TU-Black Sea project: Symposium on scientific results. Crimea, Ukraine, p 156
31. Ginzburg AI, Zatsepin AG, Kostianoy AG, Krivosheya VG, Skirta AY, Soloviev DM, Stanichny SV, Yakubenko VG (2002) Separation of near-shore anticyclonic eddies from the Caucasian shore and their transformation into deep-sea eddies. In: Zatsepin AG, Flint MV (eds) Multidisciplinary Investigations of the Northeast Part of the Black Sea. Nauka, Moscow, p 82 (in Russian)
32. Zatsepin AG, Denisov ES, Emelyanov SV, Kremenetskiy VV, Poyarkov SG, Stroganov OY, Stanichnaya RR, Stanichny SV (2005) Oceanology 45(Suppl 1):S13
33. Vostokov SV, Lisitsyn BE, Konovalov BV, Soloviev DM, Gagarin VI (2002) Mesoscale variability of chlorophyll *a* concentration, particulate organic matter content and spectral index of light absorption by phytoplankton in the upper layer of northeastern part of the Black Sea. In: Zatsepin AG, Flint MV (eds) Multi-disciplinary investigations of the northeastern part of the Black Sea. Nauka, Moscow, p 235 (in Russian)
34. Arashkevich EG, Drits AV, Timonin AG, Kremenetskiy VV (2002) Oceanology 42 (Suppl 1):S79
35. Sapozhnikov VV (1991) Oceanology 31:417 (English Translation)

Hdb Env Chem Vol. 5, Part Q (2008): 217–253
DOI 10.1007/698_5_077

Published online: 20 June 2007

Thermohaline Structure of the Sea

Valentin S. Tuzhilkin

Geographic Department, Lomonosow Moscow State University, Vorobievy Gory, 119992 Moscow, Russia
valver@orc.ru

Abstract Results of statistical and physical analyses of the historical hydrographic data set (more than 90 000 pairs of water temperature and salinity vertical profiles during 1957–1996) are presented. The thermohaline structure of the Black Sea waters consists of a few layers with different thicknesses, origins, and seasonal and interannual variabilities. It is caused by the extremely restricted external water budget, the significant differences in the salinities (and densities) of the waters that are supplied to the surface and deep layers of the sea, and the weak vertical turbulent mixing in the main water layer. In the upper 50-m layer, the seasonal and interannual variabilities of the thermohaline structure are generated by the fluxes of heat and fresh water across the sea surface; in the main pycnocline between depths of 50 and 200 m, they are caused by the flux of the wind relative vorticity, and in the underlying deep layer they are related to the Sea of Marmara waters inflow through the Bosporus Strait and the geothermal heat flow from the bottom. The thermohaline effects of these processes in all the main layers of the Black Sea are described.

Keywords Temperature · Salinity · Structure · Seasonal variability · Interannual variability

Abbreviations

BSRC	Black Sea Rim Current (synonym Main Black Sea Current)
CIL	Cold intermediate layer
CIWM	Cold intermediate water mass

CTD	Conductivity, temperature, depth
DWM	Deep water mass
EOF	Empirical orthogonal function
MFZ	Main frontal zone
NBML	Near-bottom mixed layer
NCAR/NCEP	The National Center of Atmospheric Research/The National Center of Environmental Predictions
psu	Practical salinity unit ≈ part per thousand = ‰
QSNSAG	Quasistationary near-shore anticyclonic gyres
SWM	Surface water mass
T,S	Temperature and salinity
UWM	Upper water mass
UML	Upper mixed layer

1 Introduction

In this chapter, we present the principal large-scale features of the thermohaline structure of the Black Sea waters and their seasonal and interannual variabilities. To a great degree, they define the condition and functioning of other components of the Black Sea ecosystem, in particular the general circulation and chemical properties of the waters and marine flora and fauna.

The description is mainly based on the results of statistical and physical analyses of historical data sets of ship measurements of the water temperature and salinity vertical profiles. Their spatial and temporal distributions are shown in Fig. 1. We processed and analyzed the data for a 40-year-long interval (1957–1996) with the highest measurement density. The data were processed (edited) following the procedures and recommendations given in [1]. The total number of pairs of vertical temperature and salinity profiles in this period exceeded 90 000. Geographically, they are mostly concentrated in the near-shore areas and along standard sections; meanwhile, it should be noted that the coverage of the entire area of the Black Sea is relatively good. The numerals in Fig. 1 denote the areas of standard sections that are assessed in detail: 1—Sevastopol–Bosporus; 2—Yalta–Batumi; and 3—from Gelendzhik to the southwest; where, during the 1950s to 1980s, most regular observations were performed.

The first large generalization of the earlier studies of the Black Sea physical oceanography performed in the 1890s to 1920s under the guidance of S.O. Makarov, I.B. Shpindler, Y.M. Shokal'skii, and N.M. Knipovich was made in monograph [2][1], which describes the major part of the physical and oceanographic features of the Black Sea that are known at present. As early as at that time, many of them found their adequate qualitative interpretation. Later,

[1] For a history of the studies in the Black Sea, see the chapter "Brief history of exploration and oceanographic investigation".

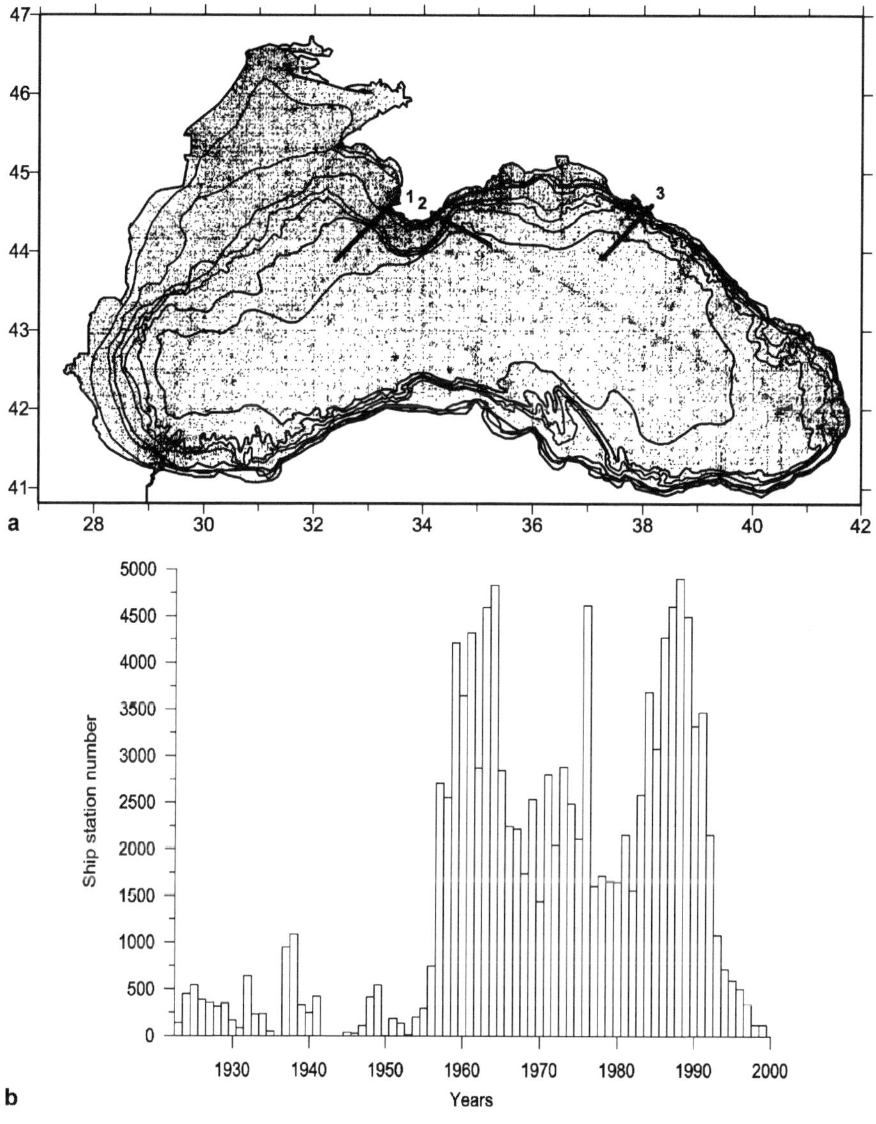

Fig. 1 **a** Location of all shipboard hydrographic observations in the Black Sea. *Thick lines*: the standard sections segments (1–3) discussed in the text. *Thin lines*: depth contours of 20, 50, 100, 200, 300, 500, and 1000 m. **b** Distribution of the shipboard hydrographic stations over the years

the concepts that were formed at the beginning of the twentieth century were quantitatively confirmed and refined, and finer features of the thermohaline structure and related physical processes were discovered. In the middle of the twentieth century we should note the monographs by Leonov [3] and

Filippov [4] devoted to the geographical and physical descriptions of the thermohaline regime of the Black Sea waters. The monograph by Blatov et al. [5] plays a special role in the development of the concepts about the variability of the thermohaline structure of the Black Sea waters and its reasons. For the first time, it presented a systematic quantitative description of the processes of the climatic, seasonal, interannual, synoptic, and short-period variabilities of the temperature and salinity of the Black Sea waters in all the principal layers and regions of the sea. Most of them found their physically justified interpretations. Subsequent refinements of the parameters of the thermohaline regime of the Black Sea waters were generalized in [6–8].

The present-day stage of the studies of the Black Sea differs from the preceding stages by the significant extension of the geography of their participants, the technological possibilities, and the number of publications; during the 1990s to 2000, the latter increased four- to fivefold as compared to the 1970s to 1980s.

This study differs from the latter monographs [6, 8] by a significantly (1.5- to twofold) greater amount of the measurement data used and updated technology for their processing and analysis [1]. This allows one to refine and supplement the existing concepts about the Black Sea thermohaline structure.

2 General Features

The features of the thermohaline structure of the Black Sea waters represent the clear manifestations of the uniqueness of its nature on the whole. Most of them are related to the very restricted water exchange of the Black Sea with the adjacent parts of the World Ocean (the Sea of Marmara and the Sea of Azov), because of which its external water budget is generally small [33].

The fresh waters supplied to the Black Sea with the riverine runoff and precipitation are distributed by currents and turbulence over the upper layer of the sea with a thickness of 5–10 m in the spring and summer and up to 40–60 m at the end of the winter. Usually, the water salinity in this layer is within the range 17.5–18.5 psu. The saline (35–36 psu) waters of the Sea of Marmara flow in the southwestern part of the Black Sea through the Bosporus Strait at a level of 60 m and sink to the deeper layers. Thus, in the multiannual mean (climatic) regime, the depth of 60 m represents the boundary of the direct influence of the surface fresh waters and the saline waters of the Sea of Marmara. They may be referred to as primary water masses, supplied to the Black Sea from outside, which have no direct contact in the Black Sea.

In the most general form, the isolated characters of the primary water masses from one another are expressed in the thermohaline (T,S) diagram shown in Fig. 2a, where each point represents a spatially fixed water volume (within a spherical trapezium with sides 12 min over the latitude and 16 min over the longitude and a thickness of 25 m over the vertical) characterized by certain T,S properties during the selected month of the year.

The relatively rare points in the left-hand part of Fig. 2a with a salinity less than 17 psu represent small in volume freshened surface waters in the northwestern shelf and other near-mouth areas of the Black Sea. The almost vertical segment of the dense cluster of points in the diagram rep-

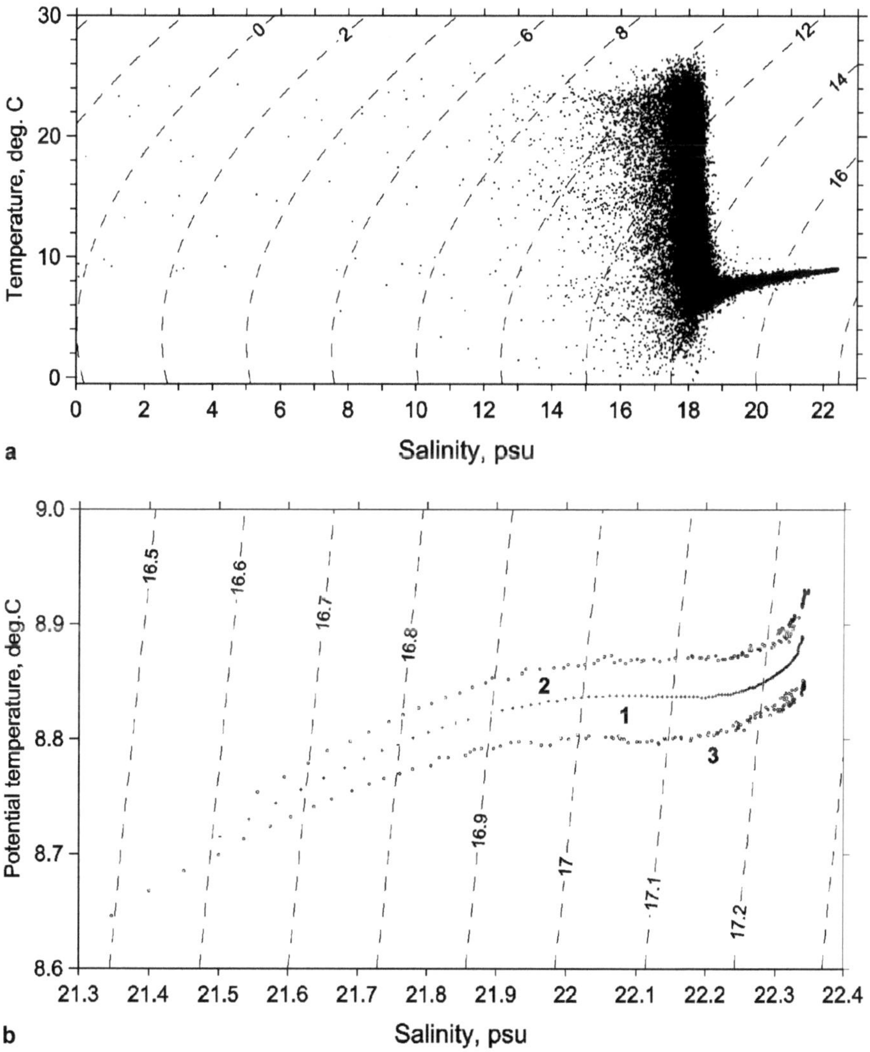

Fig. 2 **a** Temperature–salinity diagram of the Black Sea waters. *Dashed lines*: water specific density (sigma-*t*) contours. **b** Climatic potential temperature–salinity curves in the Black Sea abyssal layer (deeper than 200 m). *1* Mean values, *2* and *3* standard deviations. *Dashed lines*: water specific potential density (sigma-Θ) contours

resents the main part of the surface water mass (SWM) beyond near-mouth areas. The temperature and salinity of both the surface water modes are extremely variable throughout the year owing to the direct influence of the heat and freshwater fluxes across the sea surface, from river mouths, and through straits. The right-hand termination of the almost horizontal segment of the diagram with a salinity higher than 22 psu represents the deep water mass (DWM) below 500 m, which makes up to 70% of the entire volume of the Black Sea waters [10]; it is formed due to the weak undirected influence of the waters of the Sea of Marmara and of the geothermal heat flow from the sea bottom (0.03–0.04 W m^{-2} [11–14]).

The distinct angle between these segments of the T,S diagram in Fig. 2 proves the absence of any active mixing between the SWM and DWM [10]. According to the theory of T,S curves, in the corner region of the Black Sea T,S diagram, an intermediate water mass should be distinguished. It features a temperature lower than those of the SWM and DWM and is referred to as the cold intermediate water mass (CIWM). It is commonly accepted that its boundaries with the surface and deep water masses coincide with the temperature contour line of 8 °C [3–6, 10]. At the end of the winter (February–March), the water temperature in the surface layer of the Black Sea does not exceed 9 °C (see Sect. 3.1) and the SWM together with the CIWM may be regarded as a common water mass with a volume of about 30×10^3 km^3 [10]. By the end of the summer (August–September), about one-third of this volume with temperatures higher than 8 °C is distinguished as the SWM.

The CIWM is located at the left-hand end of the so-called mixing line with the DWM. A more detailed analysis of this line, presented in Fig. 2b, shows that it is nonlinear and, in the depth range from 500 to 700 m, has a near isothermal interval (with temperature about 8.85 °C) [11]. In [12], based on a one-dimensional thermodynamic model, a hypothesis about the mechanism of its formation is posed. According to this hypothesis, in the Black Sea, the temperature growth with depth from the subsurface minimum to a depth of 500 m is caused by the heat supply from lateral intrusions of the waters of the Sea of Marmara in the course of their sinking to greater depths. Below the level of 500 m, the transformed waters of the Sea of Marmara become colder than the adjacent waters that are heated by the geothermal flow from beneath. In the depth range from 500 to 650–700 m, these factors compensate each other, which is expressed by the thermal homogeneity of the water over the vertical. Farther downward, the geothermal factor dominates and the water temperature again increases with depth.

The principal features of the vertical T,S structure of the Black Sea waters are shown in Fig. 3. The upper mixed layer (UML) of the Black Sea in the warm period of the year has a thickness less than 10 m (see Fig. 3a). At this time, it is underlain by the layer of the seasonal pycnocline (thermocline); this layer is also thin (10–20 m) but features high vertical gradients of temperature (0.2–0.3 °C m^{-1}) and, correspondingly, of water density

(0.10–0.15 kg m^{-4}). By the end of the winter, owing to the thermal convection, the thickness of the UML over the greater part of the area increases up to 30–60 m (see Fig. 7a). At this time, the UML is limited from beneath by the layer of the main (constant) pycnocline (halocline), the depth range from

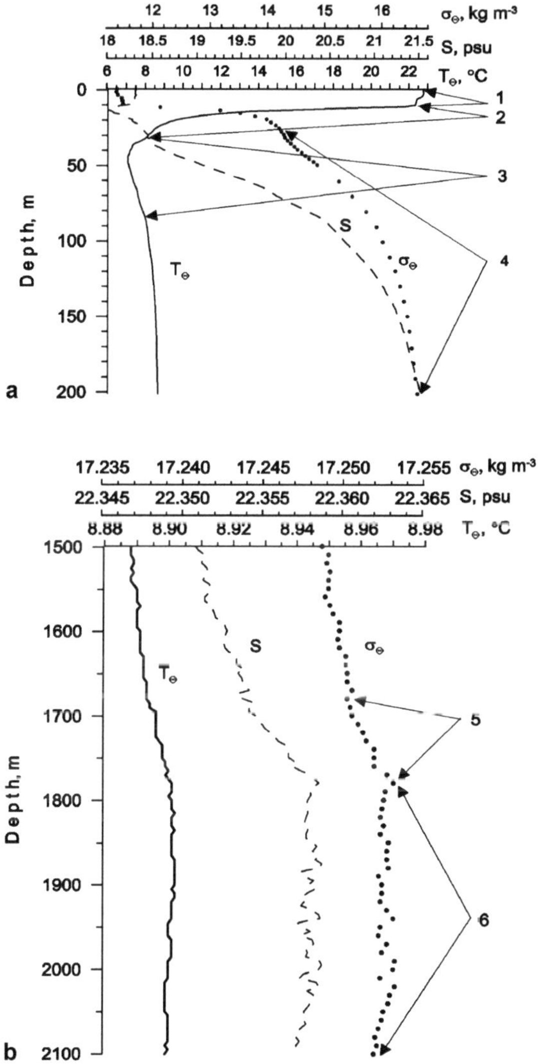

Fig. 3 Vertical profiles of the water potential temperature (T_Θ, degrees Celsius), water salinity (S, practical salinity unit), and water specific potential density (σ_Θ, kg m^{-3}) **a** in upper layer of the Black Sea central area in August 1995 and **b** in deep layer (mean values based on high vertical resolution CTD measurements). *1* Upper mixed layer, *2* seasonal pycnocline (thermocline), *3* cool intermediate layer, *4* main pycnocline (halocline), *5* deep pycnocline, *6* bottom mixed layer

30–60 to 150–200 m of the Black Sea with vertical density gradients up to 0.03–0.04 kg m^{-4}. In the near-mouth areas of the Black Sea, due to the high vertical gradients of the water salinity, the winter thickness of the UML comprises about 10 m.

The absolute minimum of the water temperature in the Black Sea is usually encountered in the upper part of the main pycnocline and has values of 6.5–7.5 °C (see Fig. 3a). Only in severe winters is it located in the UML. The layer with a temperature lower than 8 °C is referred to as the cold intermediate layer (CIL) [2–4]. In the warm period of the year, it is "sandwiched" between the seasonal and main pycnoclines of the Black Sea with a slight local decrease in the vertical density gradients (see Fig. 3a). Over the greater part of the area, at the end of the winter, the upper boundary of the CIL (the upper 8 °C isotherm) is exposed at the sea surface. At this time, the major part of the CIL is located inside the UML and only its lower part is related to the main pycnocline. Thus, in this period, the CIL provisionally becomes the cold surface layer.

Below the main pycnocline, one finds the layer that is named sometimes generally as the deep layer. The present-day concepts about the vertical structure of its upper part have already been considered (see Fig. 2b). Below the intermediate isothermal layer, in the depth range from 700 to 1700 m, one observes a layer with a slow increase in the temperature and salinity with depth sometimes broken by T,S inversions with vertical scales about 10 m, which is typical of the fine T,S structure of the waters [11] (Fig. 3b). Theoretical estimates [13] show that they may result from the thermal type of double diffusion (layered convection), which is the principal mechanism of the vertical heat and salt exchange in this layer.

The deep-water observations with conductivity, temperature, depth (CTD) profilers performed in the Black Sea during the past two decades allowed one to distinguish the near-bottom mixed layer (NBML). In Fig. 3b, we present profiles of the potential temperature (T_Θ), salinity (S), and potential density (σ_Θ) of the Black Sea waters in the layer from 1500 to 2100 m obtained by averaging of 46 CTD profiles observed in 1985–1992 in different regions of the deep-sea area. In all three profiles shown in Fig. 3b, a distinct upper boundary of the NBML is traced at depths from 1750 to 1800 m. Above it, up to a depth of 1700 m, one finds a layer with increased vertical gradients of T_Θ, S, and σ_Θ with a thickness about 100 m; it separates NBML from the deep stratified layer.

The absolute values of T_Θ, S, and σ_Θ in the layer 1500–2100 m significantly vary from one survey to another, which may be explained by the biased calibrations of CTD profilers [14]. In our case, it is manifested in the vertical homogeneity of T_Θ and S standard deviations (respectively, 0.031–0.033 °C and 0.021–0.027 psu) over the depth range from 1500 to 2100 m. According to the measurements carried out at the beginning of the 1990s, the mean values of T_Θ and S in the near-bottom layer of the Black Sea were 8.895 °C and

22.333 psu, respectively [14]. In our case, they are higher by 0.005–0.007 °C and 0.035–0.045 psu because of the greater part of the measurements made at the end of the 1980s. However, the vertical structure of the profiles shown in Fig. 3b is identical to those known from publications [7, 11, 14].

In [13], based on a one-dimensional model, it was shown that the observed parameters of the NBML in the Black Sea are defined by the buoyancy fluxes balance between the destabilizing geothermal heat flux and stabilizing salt flux supplied with the waters of the Sea of Marmara penetrating to great depths.

The T,S structure of the Black Sea waters presented in Fig. 3 is caused by the weak turbulent diffusion below the UML, which is characterized by a diffusivity of about $10^{-5}\ m^2\ s^{-1}$ [6, 7, 11–13], which is one to two orders of magnitude lower than the values usual in the open ocean. The reason for the weak vertical turbulent exchange in the Black Sea is the great differences in the densities of the primary water masses (the freshwater mass and that of the Sea of Marmara).

Over the entire water column of the Black Sea except for the UML and the seasonal pycnocline, 75–95% of the spatial density variations of the waters are defined by their salinity and closely correspond to the latter (see Fig. 3). Therefore, below, we do not consider the horizontal and vertical distributions of the density of the Black Sea waters.

3 Climatic Seasonal Variability

The general condition of the horizontal and vertical T,S structures of the Black Sea waters and their seasonal variability may be derived from the climatic sections of the temperature and salinity along 36.5° E for March and September shown in Fig. 4. In the upper layer of the Black Sea approximately 30–60 m thick the principal thermal processes are represented by the winter renewal of the cold intermediate waters and by the spring formation and autumn destruction of the seasonal thermocline. In the salinity of the upper layer of the Black Sea, the seasonal signal is manifested by its relatively small decrease in the summer in the depth interval from 0 to 30 m (Figs. 4c and Fig. 4d; Table 1). Only in near-mouth areas is this decrease significantly stronger owing to the riverine runoff. The amplitudes of the annual heating/cooling and freshening/salination rapidly decrease with depth down to a level of 30 m and their phases are homogeneous over the Black Sea area (see Tables 1 and 2).

Below 30–50 m, the principal seasonal process is the winter increase and the summer decrease of the dome height of the Black Sea main pycnocline (see Fig. 4). The most reasonable interpretation of this feature of the T,S structure of the Black Sea was suggested in [5] from the point of view of the

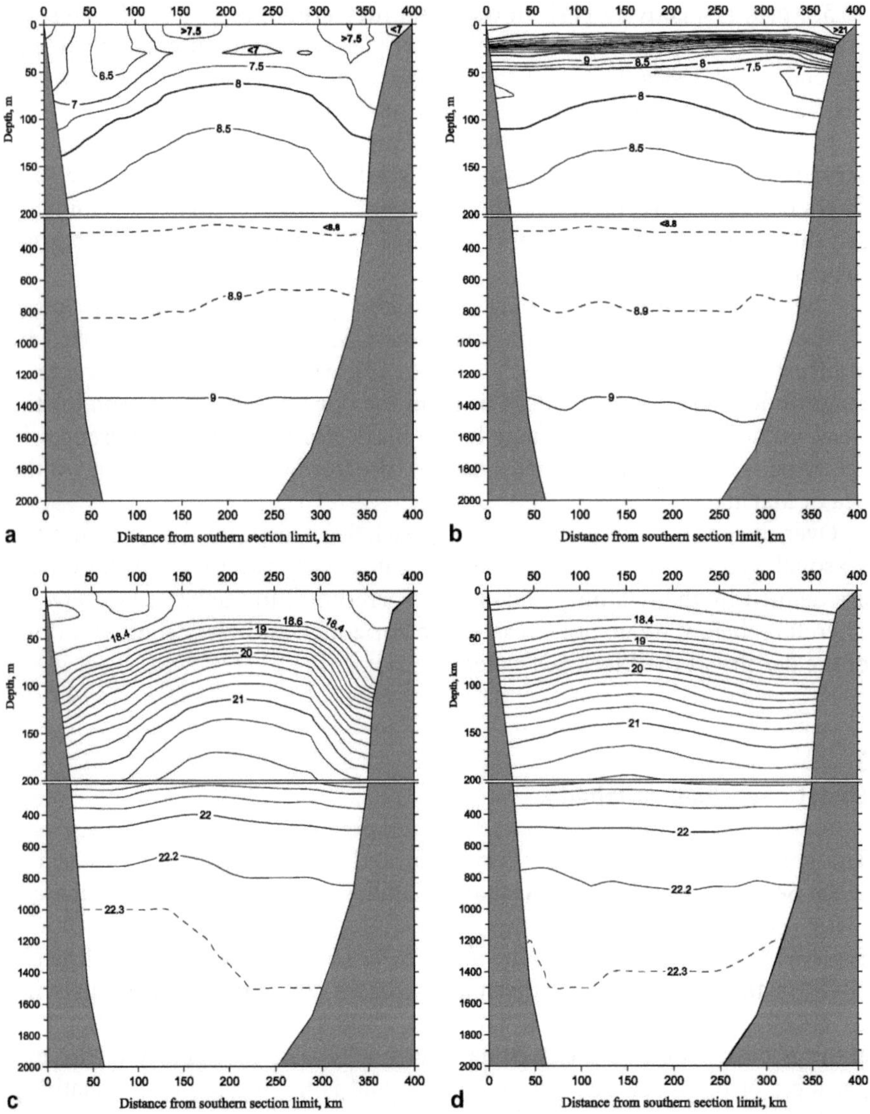

Fig. 4 Vertical sections of (**a,b**) climatic water temperature (degrees Celsius) and (**c,d**) climatic water salinity (practical salinity units) in the Black Sea along 36° E in March (**a,c**) textbfa,c) and September (**b,d**)

response of the large-scale potential vorticity of the Black Sea waters to the seasonal variations in the influx of the relative vorticity from the wind field over the sea surface, which has a cyclonic character in the winter and anticyclonic character in the summer [6, 7].

Table 1 Parameters of the climatic annual cycle of variability of water temperature (degrees Celsius) and water salinity (practical salinity units) at standard levels of main baroclinic layer in the Black Sea central area (offshore end of standard section 3, see Fig. 1). T_{min}, T_{max}, S_{min}, and S_{max}: annual extremes of climatic monthly mean temperature and salinity; mon: corresponding months of extremes; *DT* and *DS*: ranges of temperature and salinity annual cycles

Depth	T_{min} (mon)	T_{max} (mon)	*DT*	S_{min} (mon)	S_{max} (mon)	*DS*
0	6.776 (3)	23.492 (8)	16.716	17.779 (7)	18.555 (3)	0.776
30	6.824 (3)	11.450 (10)	4.626	18.192 (9)	18.618 (2)	0.426
50	6.669 (3)	8.095 (11)	1.426	18.498 (9)	19.243 (2)	0.745
75	7.545 (9)	8.257 (2)	0.712	19.299 (9)	19.925 (2)	0.626
100	8.100 (9)	8.450 (2)	0.350	19.911 (10)	20.404 (2)	0.493
125	8.324 (8)	8.459 (3)	0.135	20.320 (9)	20.723 (3)	0.403
150	8.482 (7)	8.650 (3)	0.168	20.735 (9)	20.957 (3)	0.222
200	8.560 (9)	8.695 (3)	0.135	21.080 (9)	21.314 (3)	0.234
250	8.670 (9)	8.768 (3)	0.098	21.314 (9)	21.523 (3)	0.209
300	8.766 (9)	8.848 (3)	0.082	21.455 (10)	21.697 (3)	0.242

Table 2 Parameters of the climatic annual cycle of variability of water temperature (degrees Celsius) and water salinity (practical salinity units) at standard levels of main baroclinic layer in the Black Sea coastal area (inshore end of standard section 3, see Fig. 1). T_{min}, T_{max}, S_{min}, and S_{max}: annual extremes of climatic monthly mean temperature and salinity; mon: corresponding months of extremes; *DT* and *DS*: ranges of temperature and salinity annual cycles

Depth	T_{min} (mon)	T_{max} (mon)	*DT*	S_{min} (mon)	S_{max} (mon)	*DS*
0	7.660 (3)	23.593 (8)	15.933	17.723 (7)	18.100 (2)	0.377
30	7.529 (3)	13.266 (10)	5.737	18.136 (7)	18.262 (2)	0.126
50	7.296 (7)	9.607 (11)	2.311	18.283 (2)	18.471 (9)	0.188
75	7.286 (7)	8.359 (1)	1.073	18.546 (2)	18.896 (9)	0.350
100	7.742 (8)	7.970 (1)	0.228	18.998 (2)	19.696 (10)	0.698
125	8.020 (2)	8.208 (7)	0.188	19.805 (2)	20.299 (7)	0.494
150	8.245 (2)	8.435 (7)	0.190	20.357 (2)	20.772 (7)	0.415
200	8.517 (2)	8.655 (7)	0.138	21.005 (2)	21.261 (7)	0.256
250	8.650 (2)	8.745 (7)	0.095	21.297 (2)	21.546 (7)	0.249
300	8.737 (2)	8.811 (7)	0.074	21.513 (1)	21.723 (7)	0.210

In the course of this process, the seasonal changes in the temperature and salinity in the main pycnocline layer in the central and near-shore areas of the Black Sea proceed in opposite phases, which is proved by the data from Tables 1 and 2. The amplitudes of the corresponding oscillations reach their maximum at a depth of 50 m in the central area and of 100 m in the near-shore zone. Below 300–400 m no seasonal signals can be traced in the temperature and salinity of the Black Sea waters (see Fig. 4).

In subsections 3.1–3.4, we assess the seasonal variability of the horizontal climatic T,S structure of the Black Sea waters in more detail.

3.1 Upper Layer

The horizontal structure of the temperature and salinity fields in the surface layer of the Black Sea in the extreme months of the annual cycle (February and August) is presented in Fig. 5. In February (similar to March), the surface waters feature the lowest temperature values (Fig. 5a); the temperature decreases from the eastern coast of the Black Sea to its northwestern coast. This decrease is especially sharp over the wide northwestern shelf, where the winter heat losses to the atmosphere are the greatest ($> 6.0 \times 10^8\,\mathrm{J\,m^{-2}}$ per month [6]). The effects of the cyclonic transport of heat by the Black Sea general water circulation are especially noticeable in the winter in the near-shore zone of the deep-water area of the sea. The water temperatures off the northeastern (Caucasian) and northern (Crimean) coasts are higher than those off its southern (Anatolian) coast, where the winter heat losses across the sea surface are minimal ($< 2.0 \times 10^8\,\mathrm{J\,m^{-2}}$ per month [6]). It should be noted that the background decrease in the surface water temperature from the southeast

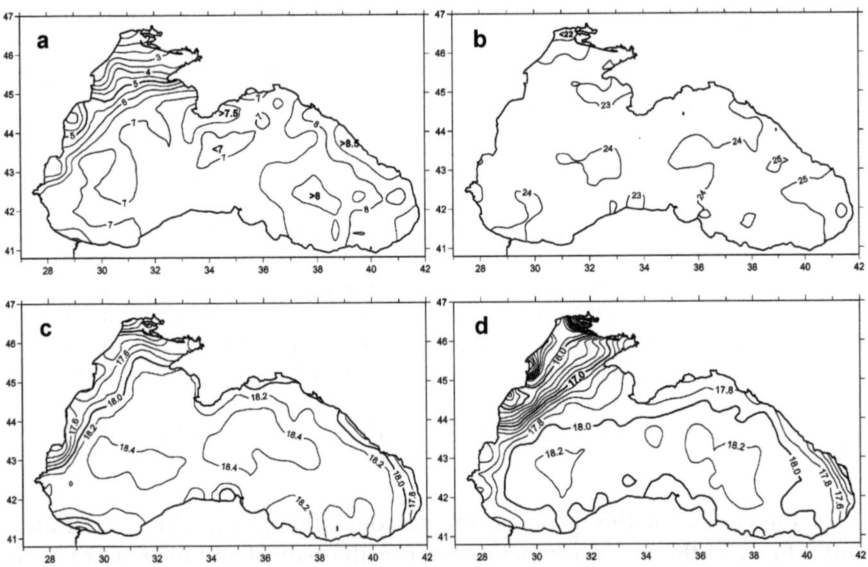

Fig. 5 Climatic fields of (**a,b**) the water temperature (degrees Celsius) and (**c,d**) the water salinity (practical salinity units) in the surface layer of the Black Sea in February (**a,c**) and August (**b,d**)

to the northwest of the Black Sea is retained throughout the entire year [15]. In August, the water temperature is maximal and more homogeneous with respect to the horizontal (Fig. 5b).

In February, the surface salinity (Fig. 5c) features a background level close to the maximal values, despite the fact that the resulting winter freshwater runoff to the Black Sea (20–40 km^3 per month [6]) is only 25% smaller than the maximal spring runoff. The reason for this inconsistency lies in the intensive convective entrainment in the winter UML of the more saline underlying layers. Local salinity minimums exist in near-mouth areas throughout the year; from them, tongues of freshened waters extend along the shores in the general cyclonic (anticlockwise) direction. The local February salinity minimum in the near-Bosporus region is related to the advection of freshened waters from the northwestern part of the sea. The break in the contour lines off the southern part of the Bulgarian coast is caused by the lack of observation data.

In August (Fig. 5d), the influence of the spring riverine waters on the freshening of the surface waters of the Black Sea reaches its maximum, especially in the northwestern part of the sea, where the salinity drop exceeds 1.0 psu. In other regions of the Black Sea, the background salinity level is 0.2–0.4 psu lower in the summer than in the winter, regardless of the maximal precipitation in this season. This is related to the fact that, in the spring and summer, the intensive seasonal pycnocline prevents the surface waters of the Black Sea from mixing with the underlying waters and, owing to the horizontal eddy mixing [16], they efficiently spread within the thin UML (see Figs. 3, 4).

3.2
The Cold Intermediate Layer

The horizontal structure of the field of the minimal water temperatures (the core of the CIL) of the Black Sea in the extreme months of the annual cycle (February and August) is presented in Figs. 6a and 6b. The February field of the minimal water temperatures (Fig. 6a) is similar to the surface temperature field (Fig. 5a). Meanwhile, a detailed analysis shows that they are not fully identical: from the northwest to the southeast the excess of the surface temperature over the minimal value grows up to 1.0 °C. The depth of the temperature minimum location increases in the same direction down to 70–80 m. This points to the absence of CIL water renewal owing to the winter convective mixing over some part of the Black Sea area.

In order to quantitatively estimate this statement, we carried out a statistical analysis of all the vertical temperature and salinity profiles available from January to March in 15 regions of the Black Sea with bottom depths greater than 50 m. In each of them, we determined the recurrences of two types of profiles with different depths of temperature minimums: (1) in the UML and

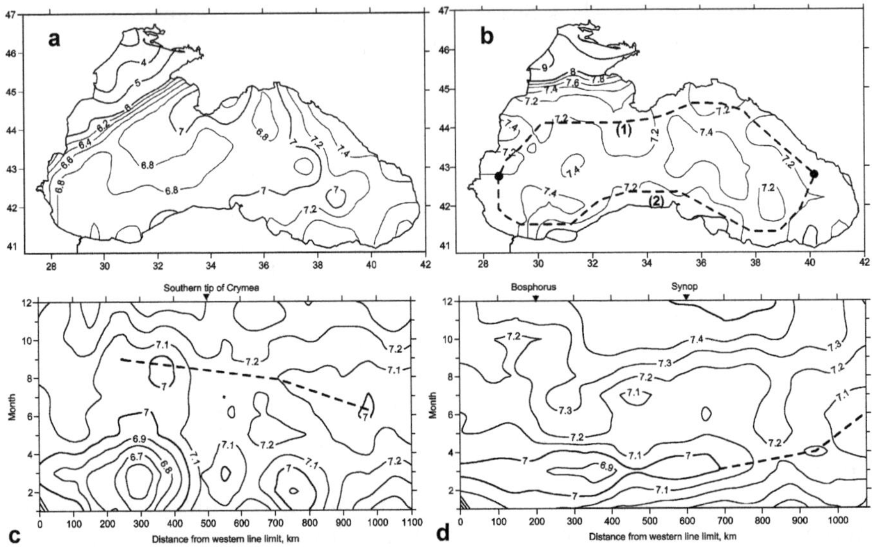

Fig. 6 Climatic fields of the minimal water temperature (degrees Celsius) in the Black Sea **a** in February and **b** in August and the space–time diagrams of its climatic annual cycle along **c** line 1 and **d** line 2. *Dashed lines* in **b**: locations of the space–time diagrams. *Dashed lines* in **c,d**: phase shift of the temperature minimum

(2) in the main pycnocline. The UML was defined as the upper layer with vertical density gradients lower than 0.002 kg m^{-4}. The former of these types of profiles corresponds to the conditions of a geographically local winter renewal of the CIL waters.

From January to March, only on the Black Sea northwestern shelf does the first type of profile dominate over the second type with a proportion of 5 : 2. In February, in the western (Bulgarian) and southwestern (near-Bosporus) near-shore regions, the recurrences of these two types of profiles are almost equal. Over the rest of the Black Sea area, even in February, the mean proportion of the first and second types of profiles is 1 : 3. On the northwestern continental slope and in the southeastern region of the Black Sea, it increases up to 1 : 2, while in the southeast, it decreases down to 1 : 30, which is statistically indistinguishable from zero.

The major part of the first type of profile is observed during cold winters with a sum of negative daily air temperatures lower than – 300 °C. Over the 40-year-long period from 1957 to 1996, more than 70% of the first type of profile in the deep-water part of the Black Sea were observed precisely during ten such winters [17].

Thus, only in cold winters (once every 3–5 years) does the local water renewal in the CIL occur over the major part of the Black Sea, except for its southeastern region. In such winters, the UML values of the temperature, salinity, and σ_t averaged over the deep-water area of the Black Sea comprise 6.96 °C, 18.39 psu, and 14.38 kg m^{-3}, respectively. During other winters, the UML mean values were 7.80 °C, 18.24 psu, and 14.16 kg m^{-3}, respectively. Corresponding mean values at the level of the February temperature minimum in the upper part of the main pycnocline (at an average depth of 58 m) were 7.21 °C, 18.71 psu, and 14.59 kg m^{-3}.

During moderately cold winters, the waters of the CIL are renewed only in some areas of the Black Sea; they may be referred to as focuses of ventilation. It cannot be excluded that, in very warm winters, there may be no CIL water renewal over the entire area of the Black Sea. However, this suggestion is not yet confirmed.

In August, the minimal water temperature averaged over the deep-water area of the Black Sea grows up to 7.31 °C (see Fig. 6b), the depth of its location—to 62 m (that is, slightly greater than in the winter). At the central deep-water area, the mean minimal temperature is observed at a depth of 56 m and comprises approximately 7.4 °C with a corresponding salinity value of 18.9 psu. In the near-shore zone, these values are 70 m, 7.2 °C, and 18.7 psu, respectively. Thus, in the CIL core, the waters are relatively stable with respect to their T,S properties in the winter and summer, which is caused by their efficient thermodynamical isolation provided by the seasonal thermocline from above and the main pycnocline from below. This is especially characteristic of the areas of the quasistationary near-shore anticyclonic gyres (QSNSAGs) of the Black Sea, where temperature minimums are especially deep [5]. Of them, the most intensive and vast eddy—the southeastern (Batumi) QSNSAG—is characterized by the lowest minimal temperature (< 7.1 °C) and its deepest position (78 m) in the summer.

The intra-annual evolution of the minimal CIL temperature along the line shown in Fig. 6b, which corresponds to the February position of the axis of the alongshore main frontal zone (MFZ) related to the Black Sea Rim Current (BSRC), is presented in Figs. 6c and 6d. Over the major part of this line, the absolute temperature minimum is observed almost simultaneously (in February–March). The focuses of the coldest waters at this time are located on the northwestern continental slope (see Fig. 6c, the region 300 km east of the western edge of line 1), in the Kerch area (750 km east of this line) and east of the Bosporus Strait (see Fig. 6d, the region 300–400 km east of the western edge of line 2). During the first half of the year, the phase shift of the absolute minimum explained by the water advection is noticeable only in the eastern part of line 2 (the dashed line in Fig. 6d). At its easternmost point, the absolute minimum is observed in June. In the second half of the year, one can note a phase shift of the secondary temperature minimum along the coasts of the Caucasus and Crimea (the dashed line in Fig. 6c). In both cases, the phase shifts of the temperature minimums proceed in the cyclonic direction (from the left to the right in Fig. 6d and from the right to the left in Fig. 6c), which indirectly proves their origin owing to the transport of the cold waters by the BSRC.

The greatest seasonal transformations are observed in the thickness of the CIL (the vertical distance between the 8 °C isotherms). From the winter to the autumn, they decrease by a factor of 1.5–2.5, mostly owing to the deepening of the upper 8 °C isotherm. As early as in May, the upper CIL boundary is observed at depths of 30–40 m; this depth reaches 35–50 and 45–60 m in August and November, respectively. From the winter to the summer, the lower 8 °C isotherm sinks from 75 to 80 m at the Black Sea central area and rises from 110 to 95 m in the near-shore zone according to the seasonal evolution of the main pycnocline dome (see Fig. 4), where the lower part of the CIL is located. In the summer and autumn, the CIL thickness in the near-shore zone of the Black Sea, especially in the anticyclonic eddies, is 1.5–2 times as great as in its central area [5].

The location of the focuses of the winter renewal of the CIL waters and the mechanisms of their propagation over the Black Sea area are the most disputable issues in the CIL studies. In the 1920s to 1950s, two points of view on the mechanism of the ventilation of the Black Sea CIL waters were formulated: the convective and the advective concepts [4]. According to the first concept, the CIL is renewed over the major part of the Black Sea area due to the winter convective mixing; the second concept implies the advection of cold waters from the main focus of their formation located over the northwestern continental slope. In the 1980s, a hypothesis of the isopycnal spreading of cold waters down the domes of the Black Sea central cyclonic gyres was posed [18]. An analysis of the CTD surveys of the Black Sea performed in 1991–1995 with detailed horizontal and vertical resolutions [19] showed that, during cold winters, the renewal of the CIL waters proceeds both in the central area and in the near-shore zone of the Black Sea. Meanwhile, in the central area, cold waters feature a significantly higher salinity (by 0.6 psu) and density than the near-shore waters and do not mix with them throughout the year. They are separated by the Black Sea MFZ with high gradients of the potential vorticity. In each of the areas, the renewed CIL waters circulate in the cyclonic direction. In so doing, in the central area, the core of the CIL is located closer to the surface and is better heated during the warm period than the waters of the near-shore area (see Fig. 6b), where the greater part of the CIL waters is trapped by QSNSAG [5]. In warm winters, the significance of the central zone of the Black Sea in the CIL water renewal becomes smaller as compared to the near-shore area [19].

In Figs. 7b and 7c, we present the results of the calculations [20] of the rate of the climatic subduction of the cold Black Sea UML waters in March together with the corresponding surface temperature and salinity fields. The subduction means the transfer of cold waters from the UML to the subsurface layer owing to the action of one or several processes: the decrease in the UML thickness in the course of the formation of a new seasonal pycnocline, the water downwelling, and the water transport by currents over descending isopycnal surfaces. Correspondingly, the subduction rate was estimated as the

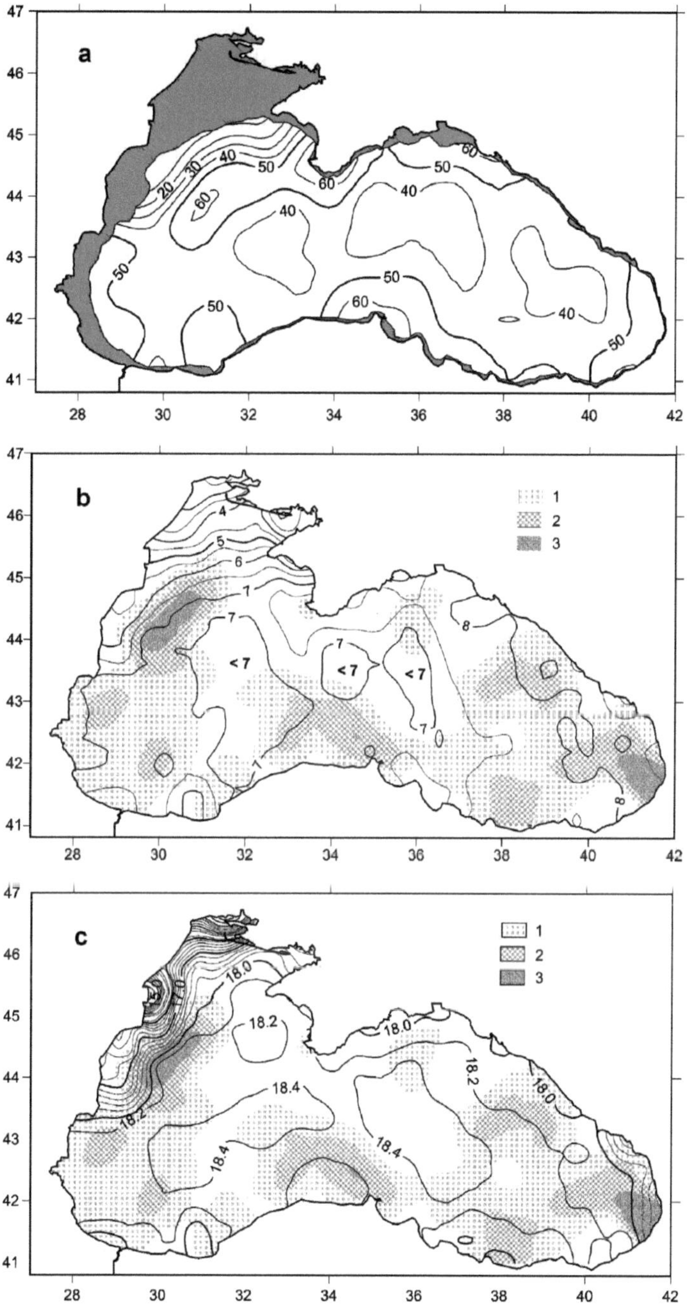

Fig. 7 **a** Climatic fields of the Black Sea upper mixed layer thickness (m) in area with sea depth >50 m. **b** Sea surface temperature (degrees Celsius) and **c** sea surface salinity (practical salinity units), both superimposed on the areas of water subduction with velocities: *1* < 40 m $month^{-1}$, *2* 40–80 m $month^{-1}$, *3* > 80 m $month^{-1}$ in March

sum of the rate of the decrease in the UML thickness (H_{UML}), of the vertical velocity of large-scale currents, and of the scalar product of their horizontal velocity by the horizontal gradient of H_{UML}. Most of the areas with enhanced subduction in Fig. 7 coincide with the maximums of the winter H_{UML} values (Fig. 7a) and, hence, of the volumes of cold surface waters. The exceptions are represented by a few small areas of subduction in the Black Sea central area outlined by salinity contours of 18.4 psu.

In the majority of the areas with enhanced subduction, the surface waters feature similar T,S properties: a temperature of approximately 7 °C and a salinity of 18.2 psu. In the course of their subsequent propagation in the subsurface layer and partial mixing with the warmer and more saline underlying waters, their T,S properties approach the climatic values of the CIL.

In all, in February and March, an average volume of 23.5×10^3 km^3 undergoes subduction in the Black Sea [20]. About 25% of these waters that have an elevated temperature and a lowered salinity in the southeastern part of the sea do not enter the CIL. The remaining part should correspond to the variable part of the CIL volume in the Black Sea, which every year decreases down to zero from the end of the winter to the end of the autumn. According to the estimates presented in [10], the climatic difference between the February (30.2×10^3 km^3) and August (16.9×10^3 km^3) volumes of the CIL waters in the Black Sea comprises 13.3×10^3 km^3. If the autumn continuation of the process of the CIL volume decrease is taken into account, the agreement between the values considered may be regarded as good. More than two-thirds of the renewed waters of the CIL are formed within the sector of the near-shore zone (including the MFZ) from the northwestern continental slope up to the eastern part of the Anatolian coast at approximately 37° E.

Actually, the formation of the major part of the CIL waters takes place during the wintertime synoptic (stormy) enhancement of the winds and heat release to the atmosphere that last from a few hours to a few days. Model studies show that a significant contribution to the subduction process is also provided by synoptic eddies and current meanders [21, 34]. The above-presented results show the location of the areas where the synoptic processes feature greater recurrences and intensities.

To conclude, one can see the following generalized (climatic) schematic of the formation and propagation of the CIL waters in the Black Sea. They are formed from February to April (mainly in March) and result from the loss of a direct contact of the cold waters with the sea surface under the action of the commencement of sea heating, the formation of the upper pycnocline, and the downwelling and diving water motions. The main focuses of these processes are located in the near-shore area including the Black Sea MFZ. In the Black Sea central area, significant CIL volumes are formed during cold winters. In so doing, the T,S properties of the near-shore and central waters of the CIL are different. These two cold-water modes perform a cyclonic circulation within the areas of their formation and almost don't mix with one another. Throughout the year, the waters of the central mode of the CIL are transformed much more strongly than those of the near-shore mode, which are trapped by QSNSAGs and are conserved inside them. After several successive warm winters with insignificant renewal of the CIL, the T,S properties of its waters are equalized over the entire Black Sea area due to the horizontal eddy diffusion.

3.3 Layer of the Main Pycnocline

The seasonal variability of the horizontal structure of the main pycnocline is well manifested in the salinity field at the 100-m level (Fig. 8). In all the seasons of the year, three types of structural elements are recognized: the alongshore MFZ, the central areas of the salinity maximums, and the near-shore areas of the salinity minimums. Each of them is closely related to the corresponding elements of the general water circulation in the upper 500-m layer of the Black Sea [34], such as the BSRC, the central cyclonic gyres, and the QSNSAG [21].

The climatic seasonal variability of the parameters of the MFZ in sections 1–3 shown in Fig. 1 is presented in Table 3. At the end of the winter and the beginning of the spring (from February to May), the MFZ is most intensive. The maximal transverse salinity gradients are located closer to

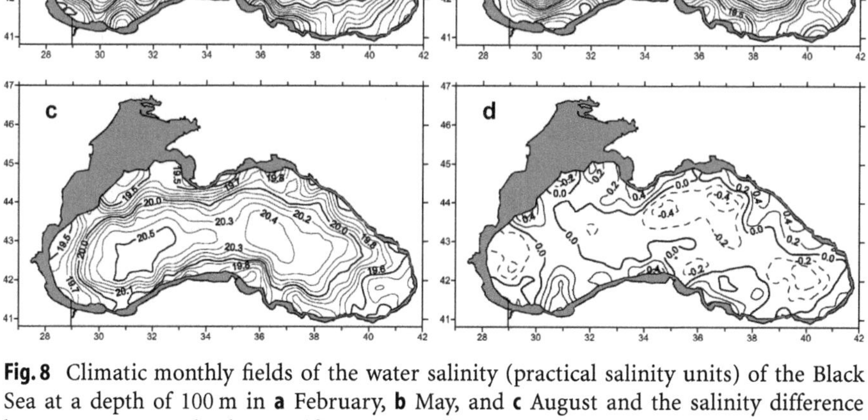

Fig. 8 Climatic monthly fields of the water salinity (practical salinity units) of the Black Sea at a depth of 100 m in **a** February, **b** May, and **c** August and the salinity difference between August and February (**d**)

Table 3 Climatic parameters (mean values ± standard deviations) of the Black Sea main frontal zone in salinity field at a depth of 100 m on standard sections in February and August. For the section locations, see Fig. 1. X_{max}: distance from coast (km); dS/dx_{max}: size (psu km^{-1}) of maximum along section salinity gradient; $X_{offshore}$: distance from coast (km) of the offshore MFZ edge; $X_{inshore}$: the same for the inshore MFZ edge

Section no.	X_{max}	$(dS/dx)_{max}$	$X_{offshore}$	$X_{inshore}$
		February		
1	65 ± 35	0.0165 ± 0.0039	128 ± 36	11 ± 39
2	32 ± 18	0.0290 ± 0.0110	91 ± 11	7 ± 13
3	35 ± 16	0.0169 ± 0.0064	83 ± 21	10 ± 5
		August		
1	72 ± 53	0.0079 ± 0.0059	94 ± 44	31 ± 37
2	43 ± 23	0.0232 ± 0.0067	70 ± 28	11 ± 17
3	41 ± 23	0.0085 ± 0.0045	73 ± 24	18 ± 16

the coastward edge of the MFZ; off the southern coast of the Crimea, they reach a value of 0.03 psu km^{-1}, which is 3–5 times as high as in the frontal zone of the Gulf Stream. At the end of the summer and the beginning of the autumn (from August to October), these values decrease approximately twofold over the entire Black Sea area (see Fig. 8; Table 3), while the maximum gradient zone shifts toward the seaward edge of the MFZ, whose width slightly decreases. This is caused by the winter–spring strengthening and the summer–autumn weakening in the Black Sea general circulation owing to the enhanced cyclonic activity in the autumn and winter and to the anticyclonic weather conditions in the spring and summer [5–7]. The MFZ strengthens with a delay of approximately 3 months with respect to the maximum of the wind forcing, which is about one-fourth of the annual cycle; this kind of delay is characteristic of the processes in "forcing–response" systems.

One can find one more manifestation of the intra-annual evolution of the fields shown in Fig. 8 in the displacement and changes in the intensities of the local salinity extremes—the central maximums and the near-shore minimums. The most distinct change is the westward displacement of the central salinity maximums occurring from February to May (see Fig. 8a,b) with the formation of a common maximum centered at 32° E in August (see Fig. 8c). The August salinity field at the 100-m level is characterized by an alternation of minimums and maximums from the east to the west with a wavelength of 350–400 km, well known as "Knipovich's spectacles" [2–4]. With respect to their sizes, directions, and phase shift rates, they correspond to mid-latitude baroclinic Rossby waves [22, 23].

A more complete presentation of the salinity field evolution from the winter to the summer gives the difference between the August and February fields shown in Fig. 8d. The prevalence of the negative (positive) salinity anomalies in August with respect to those in February in the central (near-shore) areas of the Black Sea reflects the summer decrease of the main pycnocline dome height (see Fig. 4). Meanwhile, in some regions of the central and near-shore areas of the Black Sea this regularity is broken.

For a more detailed study of the intra-annual variability of the horizontal structure of the Black Sea main pycnocline, we decomposed climatic monthly salinity fields at a depth of 100 m over empirical orthogonal functions (EOFs). The results showed that 80% of the total dispersion of the intra-annual variability in the salinity fields are described by five EOFs; three of them are presented in Fig. 9.

The first EOF, which is responsible for 46.1% of the total dispersion, represents the most large-scale mode of the Black Sea main pycnocline response to external forcing (Fig. 9a). The intra-annual variability of the corresponding coefficient (curve 1 in Fig. 9d) shows that the maximal positive (negative) salinity anomalies in the central (near-shore) areas of the Black Sea described by this mode are observed in April, when the main pycnocline dome is especially high. An opposite situation is observed a half-year later, in October, when the dome is most low. The near-shore zones of maximal positive values

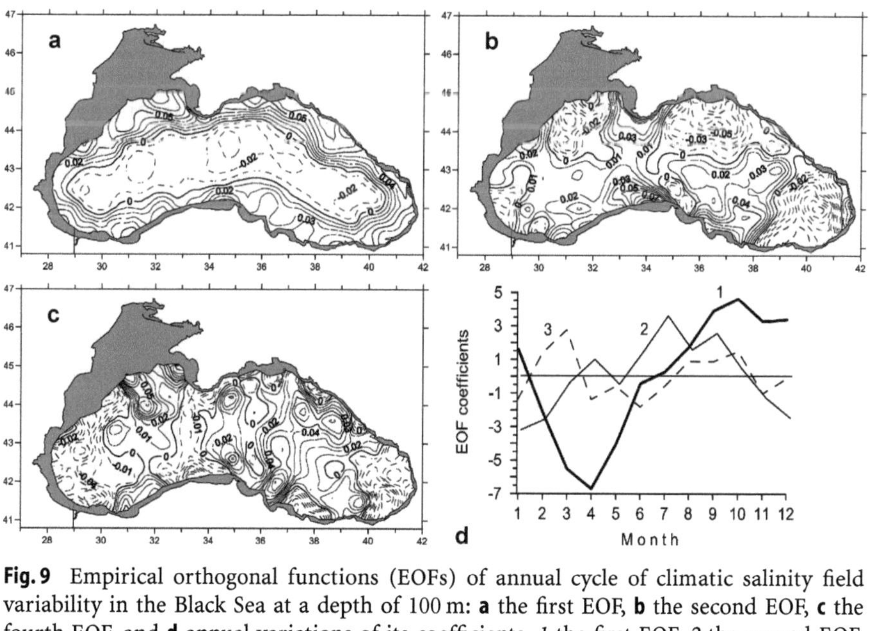

Fig. 9 Empirical orthogonal functions (EOFs) of annual cycle of climatic salinity field variability in the Black Sea at a depth of 100 m: **a** the first EOF, **b** the second EOF, **c** the fourth EOF, and **d** annual variations of its coefficients. *1* the first EOF, *2* the second EOF, *3* the fourth EOF

of the first EOF correspond to the location of the well-known QSNSAGs (the Sevastopol, Batumi, and others).

The second EOF that describes 14.2% of the total dispersion represents an alongshore quasiperiodical structure with a wavelength of 300–400 km and coastal trapping of amplitudes, which decreases with the distance from the coast. The annual cycle of the variability of the second EOF coefficient (curve 2 in Fig. 9d) is shifted by a quarter of the period with respect to the first EOF. This mode of the main pycnocline variability should be most clearly manifested in the summer in the salinity fields at a depth of 100 m.

The following three EOFs, each of which covers from 5.9 to 6.8% of the total dispersion, also feature a wave structure trapped by the coast (Fig. 9c). One may suggest that they represent overtones of the second EOF. The combined effect of the second and higher modes of the intra-annual variability of the main pycnocline was obtained by extracting the contribution of the annual mean salinity field and first EOF from the monthly salinity fields at a depth of 100 m. The results for the first 6 months of the year are shown in Fig. 10.

The areas of the salinity anomalies of different signs shown in Fig. 10 have sub-basin sizes and a complicated spatiotemporal evolution. This is manifested, first, in the cyclonic (anticlockwise) rotation of the pairs of anomalies of opposite signs (dipoles) with respect to their common centers and, second, in the changes of their shapes, sizes, and intensities. While approaching the coast, the anomalies spread about it. For example, the positive anomaly observed in January south of the Crimea (Fig. 10a), after its southward displacement toward the Anatolian coast, was trapped by it in April (Fig. 10d). At the same time (in January), somewhat east of this site, a negative anomaly separated from the Anatolian coast, whose area and intensity significantly increased by April. In the second half of the year, the evolution of the anomalies is identical to that shown in Fig. 10, but with an opposite sign. For example, the negative anomaly observed south of the Crimea in June (Fig. 10f) has the same fate as the positive anomaly observed in January (Fig. 10a).

Hence, one can suggest that, at certain stages of their joint cyclonic rotation, salinity anomalies are trapped by the coast, spread along it, and then are issued to the open sea. The rapid increases and decreases in the sizes and intensities of the anomalies point to their wave origin. The wavelength along the trajectory of the centers of the anomalies that form a dipole pair comprises about 300–350 km, a mean phase speed of the cyclonic motion of - 1.0–1.5 km day^{-1}. Within the alongshore segments of the trajectories, the phase speed of the anomalies is greater, while at the center of the sea it is slower.

Model studies of the medium-scale wave dynamics of the Black Sea waters [23] showed the possibility of the existence of Rossby waves with a period of about half a year, a wavelength of 250 km, and a phase velocity of 2 km day^{-1} in this sea. The model Rossby waves were generated by the wind

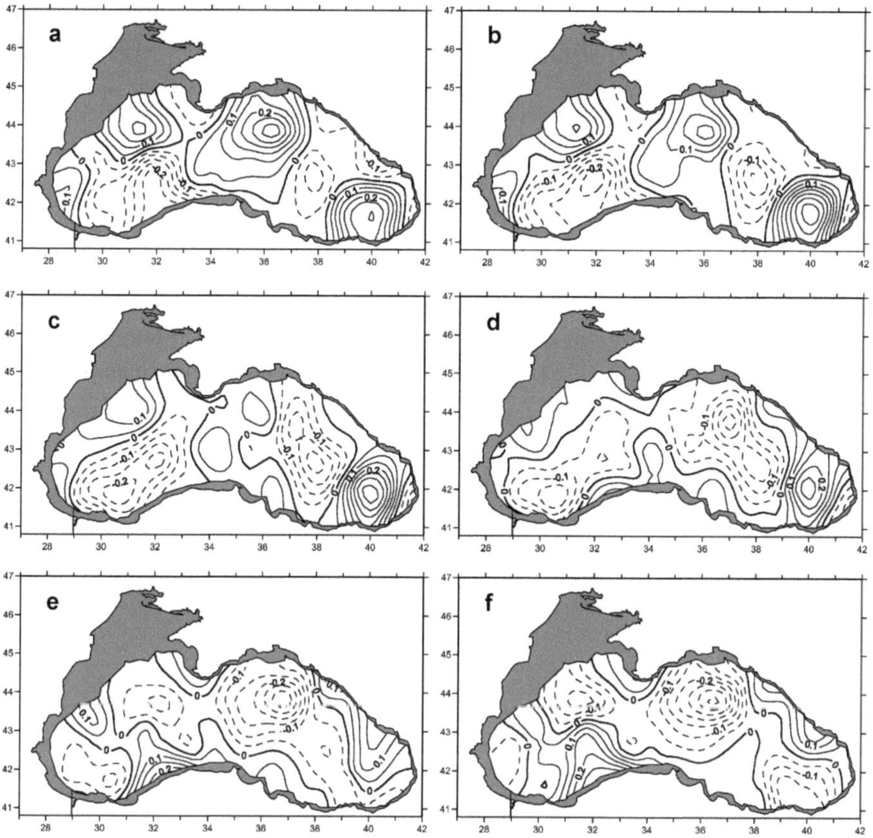

Fig. 10 Climatic monthly fields of the Black Sea water salinity anomaly (practical salinity units) at a depth of 100 m received by subtraction of the annual mean and first EOF fields from the climatic monthly salinity field in **a** January, **b** February, **c** March, **d** April, **e** May, and **f** June. *Dashed lines*: negative anomaly

forcing in the southeastern part of the sea. While crossing the narrowest part of the Black Sea south of the Crimea, their phase speed, sizes, and intensities significantly decreased and their final dissipation occurred off the western continental slope. The topographic effects only slightly modified their evolution. Meanwhile, in [23], the authors reported manifestations of coastal trapped waves and discussed their possible interaction with the Rossby waves. This interaction resulting in the formation of hybrid Rossby–coastal trapped waves was obtained with the model of mesoscale water dynamics in a circular basin [24]. The proportions of the properties of the coastal trapped waves and the Rossby waves [24] changed at different evolution stages of the hybrid waves.

In our case, we may suggest that the response of the Black Sea main pycnocline to the external (wind) forcing of an annual periodicity is manifested in superposition of a basin-scale standing oscillation and sub-basin hybrid Rossby–coastal trapped waves, which form quasigeostrophic cyclonic amphidromic systems.

The next stage in the cascade energy transfer from the external forcing over the horizontal scales is the formation of mesoscale eddies [35].

3.4 Deep and Near-Bottom Layers

As high as a depth of 300 m, the horizontal inhomogeneity and seasonal variability of the climatic fields of the salinity and, especially, of the temperature are very weak (Figs. 11a–11d). Meanwhile, here, one can still trace an increase in the temperature and salinity values from the coasts toward the central part

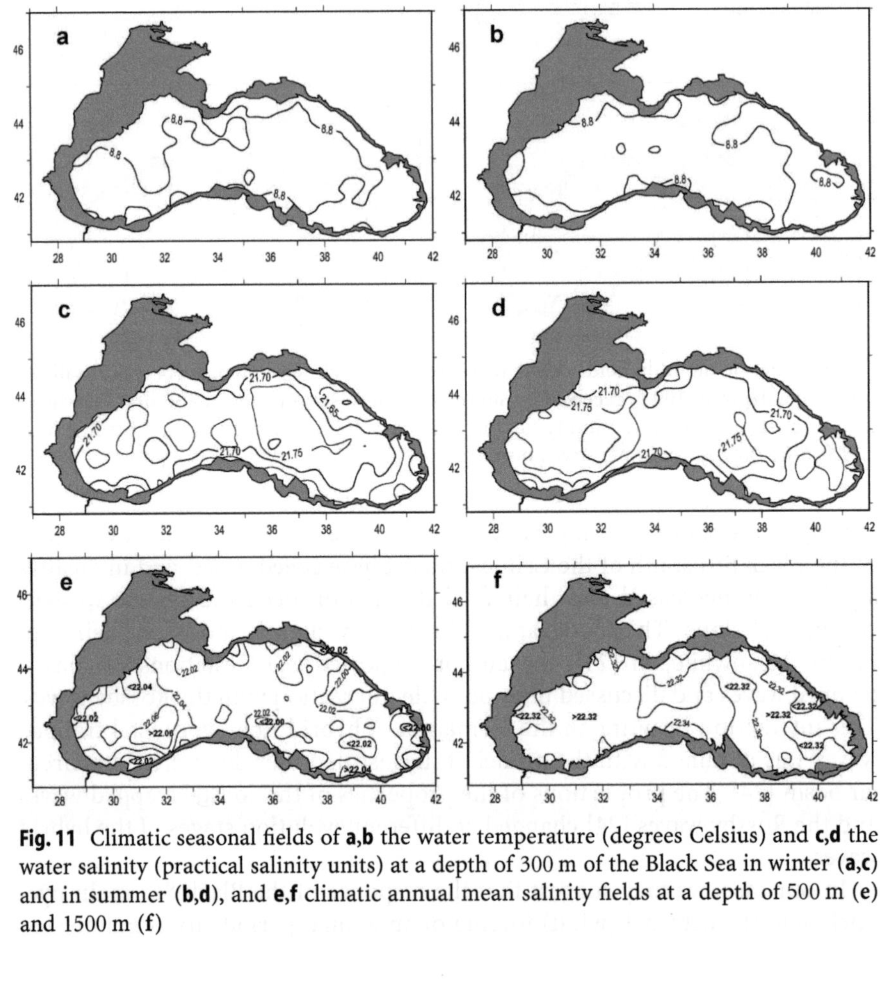

Fig. 11 Climatic seasonal fields of **a,b** the water temperature (degrees Celsius) and **c,d** the water salinity (practical salinity units) at a depth of 300 m of the Black Sea in winter (**a,c**) and in summer (**b,d**), and **e,f** climatic annual mean salinity fields at a depth of 500 m (**e**) and 1500 m (**f**)

of the sea as well as the westward displacement of their absolute maximums from February to August.

At a depth of 500 m (and lower), the seasonal differences in the climatic temperature and salinity fields are statistically indistinguishable; therefore, in Fig. 11e, only the mean annual salinity field is presented. At this level, the temperature field is especially homogeneous (and, therefore, it is not shown) because of the existence of the intermediate isothermal layer in this depth range (see Sect. 2), in which vertical water motions produce no thermal inhomogeneities.

The most large-scale and statistically reliable feature of the salinity field at a depth of 500 m is presented by the vast area of maximal salinity values (> 22.04 psu) in the western deep-water part of the Black Sea. Probably, it is related to the influence of the transformed waters of the Sea of Marmara flow in the Black Sea through the Bosporus Strait.

The range of the horizontal differences in the temperature and salinity of the Black Sea waters decreases with depth. In the deep layer below a depth of 100 m, it becomes close to the observation accuracy of the T,S characteristics. In particular, the amplitude of the salinity changes at a depth of 1500 m comprises 0.035 psu (Fig. 11f). This is manifested in the extremely mosaic character of the fields in Fig. 11 dominated by minor details that can hardly be interpreted. With further accumulation of more accurate measurements with CTD probes, one may expect certain progress in the studies of the regularities of the T,S structure of the deep Black Sea waters.

4 Interannual Variability

The T,S structure of the Black Sea waters is subjected to a strong interannual variability over a wide range of temporal scales from a few years to a few millennia. In the range from a few years to a few decades, it is induced by the atmospheric forcing both directly, by the fluxes of the momentum and relative wind vorticity, and indirectly, through the riverine runoff and water exchange via straits.

The first studies of the interannual variability of the T,S structure of the Black Sea waters and of the external forcing generalized in [5] showed that they feature close oscillation periods (3–5, 7–8, and 10–12 years) characteristic of the entire World Ocean–global atmosphere system [25]. Meanwhile, the multifactor character (a great number of degrees of freedom) and the regional particular features of multiannual global processes prevent one from recognition of statistically confident relations between them from the relatively short time series available. In recent years, more and more evidence has appeared on the nonharmonic character of these processes and on the presence of the so-called regime shifts in them [25], which reduces the efficiency of traditional methods for statistical analysis.

In this section, we consider the most general features of the interannual variability of the T,S structure of the Black Sea waters using published data and hydrographic observations in the regions of the standard sections fragments presented in Fig. 1. Near the coastal and seaward ends of these fragments, in areas 20 km in radius, all the observations available for the winter (from January 16 to March 15) and the summer (from July 16 to September 15) seasons were considered. As a result, we compiled time series of the mean winter and summer values of the temperature and salinity of the water at the sea surface and at a depth of 100 m and of the minimum temperature in the core of the CIL from 1959 to 1989–1994 (the lengths of the series varied depending on the section and season).

For each of the sections performed at a depth of 100 m, we also defined the parameters of the MFZ defined as the domain with a horizontal salinity gradient $dS/dx > 0.005‰\,km^{-1}$, in particular, the distances of its inshore and offshore boundaries from the coast, the corresponding salinity values, and the distance from the coast and the value of the maximum dS/dx. The values of these parameters were averaged over decadal intervals for 1959–1989.

Usually, while analyzing multiannual time series, the greatest interest is focused on the estimates of trends. The local extremes of these series contain the effects of synoptic and more high-frequency processes, which significantly distorts the auto- and intercorrelation and spectral functions of the interannual variability. Trends are the least sensitive to the noise effects. However, the parameters of the trends, such as the order of the approximation, the range, and the angular coefficient, strongly depend on the series length. The present-day oceanographic experience shows that, with the series length growth, the angular coefficient of the linear trend decreases and a second-order trend becomes a better approximation. This is caused by the quasiperiodic character of the interannual variations of the parameters of the ocean and atmosphere on temporal scales about a few decades (50–70 years [25, 26]). Therefore, in this study, we determined quadratic trends of the series assessed. The statistical significance of the trends obtained was defined from Fisher's criterion: to accept a hypothesis for a series 30–35 years long with a 95% probability, its minimum values should be greater than 4.10–4.18 for a linear trend and 3.25–3.35 for a quadratic trend.

4.1 Upper Layer

An analysis of the most long-term observations at coastal hydrometeorological stations of the Black Sea from 1910/1920s to 1985 [6] showed that, in the winter, the water temperature off the entire northern half of the Black Sea coast increased at a rate of 0.02–0.04 °C $year^{-1}$, while its salinity decreased at a rate highest in near-mouth regions (in the northwest, up to – 0.05 psu year^{-1}). According to [26], from 1960 to 1992 the autumn–wintertime salinity in the deep-water part of the sea continued to decrease at the same rate, while the temperature trend was replaced by an opposite one (from – 0.02 to – 0.04 °C year^{-1}). Our results show (Fig. 12a) that the change in the sign of the temperature trend occurred approximately in 1970, both in the western and eastern deep-water parts of the Black Sea. In the eastern part, Fisher's criterion exceeded its threshold values for both linear and quadratic components of the trend. In the western deep-water part, the wintertime salinity values linearly decreased at a rate of – 0.007 psu year^{-1} (Fig. 12b) with a probability of 95%. In the eastern part, a wintertime salinity maximum was observed in the 1970s; however, here, both quadratic and linear trends are not statistically significant.

In contrast, during other seasons, the temperature of the Black Sea waters decreased from the 1910/1920s to 1985 at a rate from – 0.02 to – 0.03 °C year^{-1} [5,6] and subsequently grew in 1992–2000, especially quickly in the summertime (0.07 °C year^{-1}) [27]. Our results show (Fig. 13a) that the increase in the summer temperatures in the western deep-water part and on the northwestern shelf of the Black Sea [28] started at the end of the 1970s.

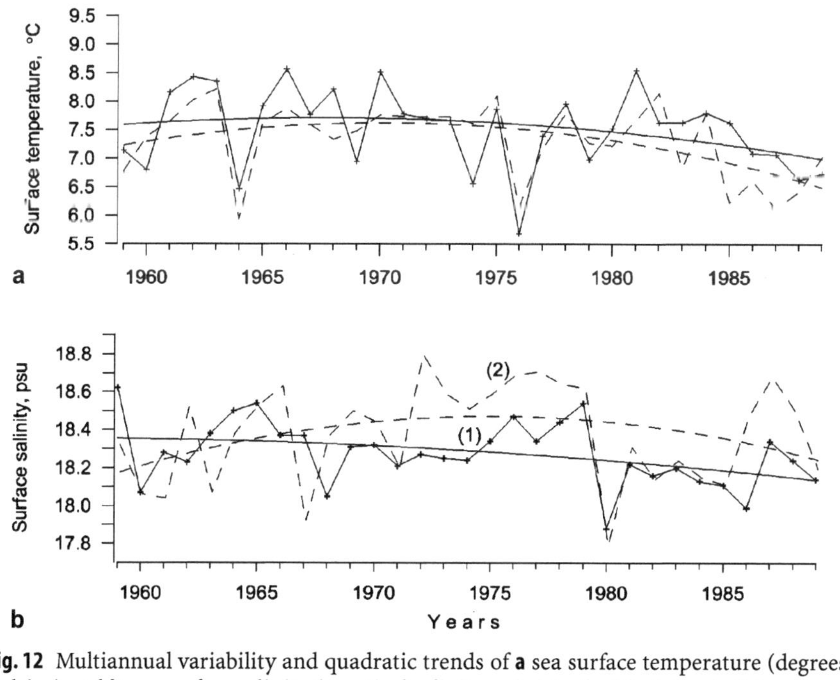

Fig. 12 Multiannual variability and quadratic trends of **a** sea surface temperature (degrees Celsius) and **b** sea surface salinity (practical salinity units) in the Black Sea in February at offshore ends of standard sections *1* from Sebastopol southwestward (see line 1 in Fig. 1) and *2* from Gelendjic southwestward (see line 3 in Fig. 1)

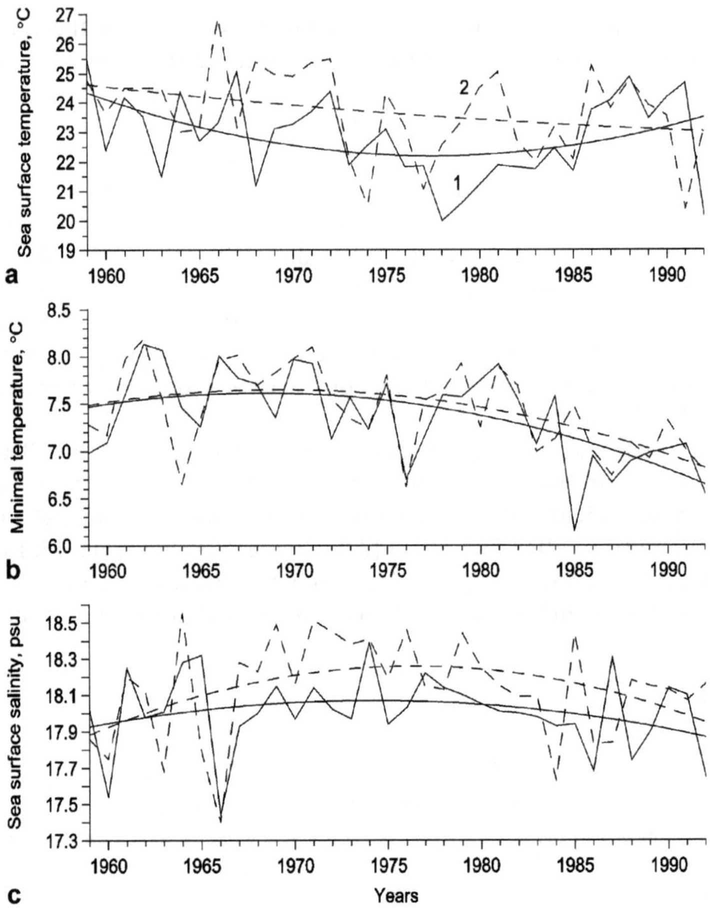

Fig. 13 Multiannual variability and quadratic trends of **a** sea surface temperature (degrees Celsius), **b** minimal temperature, and **c** sea surface salinity (practical salinity units) in the Black Sea in August at offshore ends of standard sections *1* from Sebastopol southwestward (see line 1 in Fig. 1) and *2* from Gelendjic southwestward (see line 3 in Fig. 1)

Here, the quadratic trend is statistically significant, in contrast to the negative linear temperature trend in the eastern deep-water part.

According to the data of [6], in the summer the salinity in the near-shore zone also decreased from the 1910/1920s to 1985 at a rate about -0.03 psu year^{-1} (up to -0.008 psu year^{-1} off the southern Crimean coasts). Meanwhile, in the deep-water part of the sea, positive summer salinity trends (0.015 psu year^{-1}) were observed in 1957–1973 [5]. Our results (Fig. 13c) confirm this observation. In addition, they indicate that, simultaneously with the temperature growth, a salinity decrease started from the end of the 1970s.

Thus, according to our data, in the winter and summer the temperature and salinity trends in the surface layer of the Black Sea matched one another from the point of view of the local mechanism of the cooling/heating of its waters (through the sea surface), in which the salinity plays the role of a trigger. The decrease in the salinity, which enhances the thermodynamical isolation of the surface layer from underlying layers, favors its temperature decrease in the winter and increase in the summer, as is shown by the trends presented in Figs. 12 and 13. Along with this, an analysis of the situations in particular years points to the ambiguity of the relations between individual temperature and salinity anomalies.

According to the data of [17], the interannual variations of the winter sea surface temperature of the Black Sea waters in 1957–1995 were correlated with the winter severity with a coefficient of 0.72 at a zero time shift. The temperature decrease from the beginning of the 1980s to the middle of the 1990s was caused by the coupled influence of the growth in the winter severity and the freshening of the surface layer. In the past few years, the winter severity significantly decreased. In particular, the winter of 1999 was one of the mildest over the entire period of observations [27]. The summer increase in the surface temperature from the beginning of the 1980s corresponds to the general tendency in the World Ocean [36].

During this time, the surface salinity had no statistically significant relations with the riverine runoff or the total freshwater budget. Probably, this is related to the removal of a significant part of riverine waters via the Bosporus Strait as was suggested in [5]. Correspondingly, the role of the precipitation–evaporation increment should grow. To a certain extent, this may be proved by the decrease in the surface salinity of the Black Sea waters in the winter and summer from the beginning of the 1980s (see Figs 12, 13); this may be explained only by a 15% growth in the precipitation and an equal relative drop in the evaporation, since during this period the riverine runoff has decreased by 20% [9].

4.2 Cold Intermediate Layer

The interannual variations of the summer minimal (Fig. 13b) and the winter surface (Fig. 12a) temperatures of the Black Sea waters agree well with one another with a correlation coefficient of 0.81 [17]. In the multiannual time series of the minimal temperatures, this is manifested in the fact that, not rarely, the local minimums (such as, for example, those in 1965, 1969, 1972, and 1985 shown in Fig. 13b) are followed by local maximums. This allows one to suggest that, in the Black Sea, the CIL may be renewed not only by the waters with negative temperature anomalies during cold winters but also in relatively mild winters owing to warmer waters. If it is not so, the core temperature in the CIL of the Black Sea should always monotonically increase between cold winters through mixing with the warmer surrounding waters. This kind of CIL evolution took place either after extremely cold winters (from 1976 to

1981 and from 1987 to 1991 in Fig. 13b) or during series of mild winters, such as that at the end of the 1990s [27].

The growth in the water buoyancy of the summer surface layer of the Black Sea from the end of the 1970s that was caused by the temperature increase and the salinity decrease provided a greater thermal insulation of the CIL expressed in a distinct drop in its minimal temperature (see Fig. 13), which is statistically significant with respect to both linear and quadratic trends in the east and west of the Black Sea. In the surface layer only in the western part of the Black Sea the quadratic trend of the surface temperatures was statistically confident.

4.3 Main Frontal Zone in the Layer of the Main Pycnocline

As early as in the first studies of the interannual variability of the T,S structure of the Black Sea waters, it was found that the most statistically significant linear trends were confined to the main pycnocline [5]. In 1957–1973, the temperature and salinity growth at its lower boundary (200-m level) in the central part of the sea comprised 0.01 °C $year^{-1}$ and 0.02 psu $year^{-1}$, respectively, which corresponded to a water upwelling at a rate of 2–3 m $year^{-1}$. Later, these estimates were confirmed by analyses of the time series starting from 1957 up to the beginning of the 1990s (see [6, 29]). A discussion began about the possibility of the release of the Black Sea hydrogen sulfide at the surface of the sea. However, in the 1990s, an alternation of sinks and rises of the main pycnocline by approximately 10 m at a rate of 3–4 m $year^{-1}$ was observed [27, 29].

In all the studies mentioned, the vertical motions of the main pycnocline were interpreted from the point of view of the external water budget of the Black Sea and of the winter severity. If so, the elevated (reduced) freshwater supply to the Black Sea should cause a decrease (increase) in the inflow of the waters of the Sea of Marmara and related upwelling water motions in the main pycnocline. The increase in the volume of the cold intermediate waters during cold winters should favor a sort of subsidence of the main pycnocline.

Recent studies [30, 31] showed that the interannual vertical migrations of the main pycnocline in the central and near-shore regions of the Black Sea proceed differently, sometimes in opposite phases. In the last decades of the twentieth century, this resulted in a certain enhancement of the general cyclonic circulation of the Black Sea waters [30] and intensification of the MFZ in its western part [31], as a response to the multiannual increase in the relative vorticity of the wind field over the sea.

The results of our consideration of this issue are presented in Fig. 14, where the multiannual salinity characteristics at a depth of 100 m at the inshore and offshore ends of lines 1 and 3 (see Fig. 1) are shown together with the corresponding relative wind vorticity based on the well-known NCAR/NCEP reanalysis data. Figure 14 proves the rise of the main pycnocline in the central

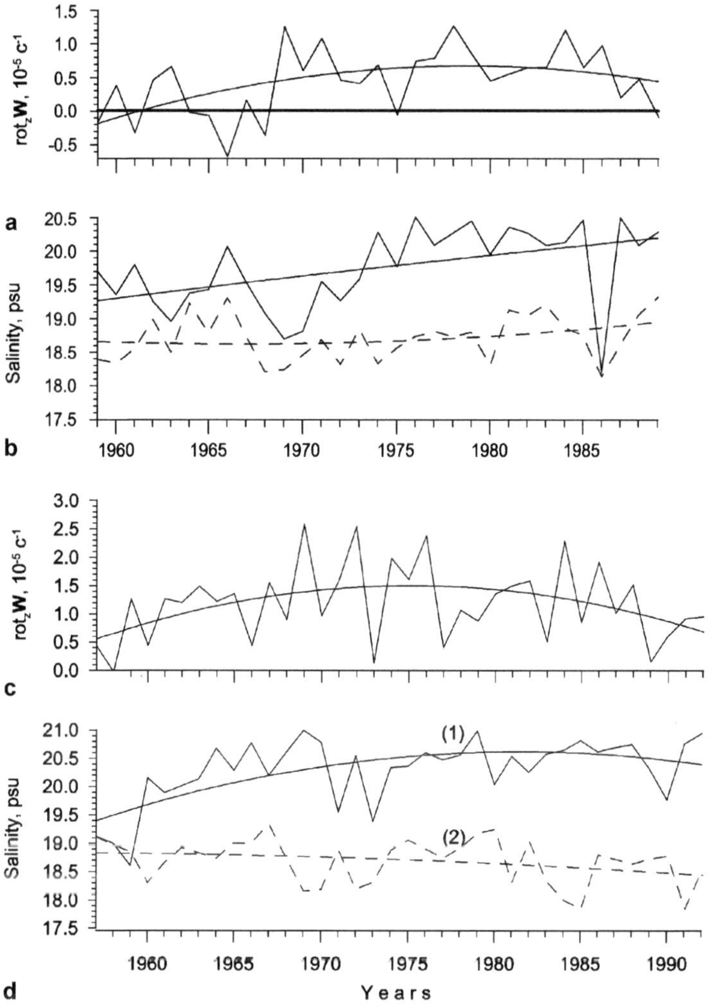

Fig. 14 Multiannual variability and quadratic trends of **a,c** relative vorticity of wind (10^{-5} c^{-1}) on the Black Sea surface and **b,d** water salinity (practical salinity units) at a depth of 100 m in February at offshore (1) and inshore (2) ends of standard sections **a,b** from Sebastopol southwestward (see line 1 in Fig. 1) and **c,d** from Gelendjic southwestward (see line 3 in Fig. 1)

region of the Black Sea by the beginning of the 1980s that has been revealed during previous studies. The multiannual time series presented for its eastern and western parts feature statistically significant linear and quadratic trends. Meanwhile, in the near-shore area, the winter salinity features no noticeable trends, which resulted in the increase in the salinity contrast across the MFZ in both of the sections. The MFZ intensifying proceeded in the period of the significant enhancement of the cyclonic wind relative vorticity over the central part of the Black Sea, which allows one to suggest the existence of a relation between them.

The mean values of these characteristics averaged over decades both for the winter and summer are presented in Table 4, including those for section 2 (see Fig. 1). Table 4 shows that the winter MFZ intensifying in the western section (section 1) is caused by the decrease in the distance of its offshore boundary from the shore with the related decrease in its width, while in the eastern section (section 3) it is caused by the increase in the salinity contrast. In so doing, in section 2, the intensity of the MFZ decreased. In this respect, our results agree with those of [31] where, however, a wrong inference about the weakening of the eastern cyclonic gyre of the Black Sea at the end of the twentieth century was made. Actually, its core displaced eastward to the region of the seaward end of section 3 (Fig. 1).

Table 4 also proves the multiannual intensification of the summer MFZ in the Black Sea related to the two- to threefold decrease in the anticyclonic wind relative vorticity. Meanwhile, statistically, the summer tendencies are less significant than the winter ones.

The results of our studies show that the interannual variability of the wind relative vorticity may be a more important reason for the long-term variations of the main pycnocline of the Black Sea than the similar variations in the components of its external water budget.

4.4 Deep and Near-Bottom Layers

The only study of the interannual variability of the T,S characteristics of the Black Sea waters at depths from 500 to 2000 m published to date [32] suggests the vertical and horizontal homogeneity of the deep temperature variations with standard deviations of 0.01–0.03 °C. The standard deviations of the interannual salinity variations decrease with depth from 0.2–0.3 psu at a depth of 500 m to 0.02–0.03 psu at depths of 1500–2000 m. At all the levels, statistically significant quadratic trends dominate with temperature and salinity maximums confined approximately to 1980. The mean rates of the salinity increase (decrease) before (after) 1980 comprised ± 0.0025 psu year^{-1}. Against the background of the quadratic trends, 6.5-year and 20-year periodicities were recognized.

Thus, in the 1950s to 1990s, the interannual variabilities of the T,S structures in the waters of all the Black Sea layers and of the external atmosphere forcing were dominated by quadratic trends with extremes and changes in the signs of linear trends in the interval 1975–1980. Probably, they represent manifestations of the 50–70-year-long cycles [26] or of a regime shift in the large-scale processes in the World Ocean–global atmosphere system [25].

Table 4 Decadal parameters (mean values ± standard errors) of the Black Sea main frontal zone in salinity field at a depth of 100 m on standard sections in February and August. For the section locations, see Fig. 1. X_{max}: distance from coast (km); dS/dx_{max}: size (psu km^{-1}) of maximum along section salinity gradient; $X_{offshore}$: distance from coast (km) of the offshore MFZ edge; $S_{offshore}$: salinity (psu) in the offshore MFZ edge; $S_{inshore}$ the same in the inshore MFZ edge; $Rot_z W$: sea surface wind relative vorticity (10^{-5} s^{-1})

Depth	X_{max}	$(dS/dx)_{max}$	$X_{offshore}$	$S_{offshore}$	$S_{inshore}$	$Rot_z W$
February						
Section 1: From Sebastopol to southwestward						
1959–1968	84 ± 15	0.0136 ± 0.0006	153 ± 14	20.34 ± 0.06	18.91 ± 0.10	0.15 ± 0.06
1969–1978	60 ± 8	0.0166 ± 0.0011	116 ± 7	20.23 ± 0.09	18.83 ± 0.08	0.37 ± 0.10
1979–1989	53 ± 8	0.0188 ± 0.0013	113 ± 11	20.36 ± 0.12	18.87 ± 0.06	0.62 ± 0.09
Section 2: From Yalta to southeastward						
1959–1968	21 ± 4	0.0360 ± 0.0025	90 ± 4	20.66 ± 0.07	18.90 ± 0.04	0.37 ± 0.05
1969–1978	34 ± 5	0.0276 ± 0.0024	93 ± 4	20.60 ± 0.06	19.07 ± 0.12	0.71 ± 0.11
1979–1989	38 ± 7	0.0239 ± 0.0044	88 ± 4	20.30 ± 0.15	19.13 ± 0.15	0.87 ± 0.10
Section 3: From Gelendjic to southwestward						
1959–1968	28 ± 3	0.0146 ± 0.0017	75 ± 7	20.30 ± 0.08	18.83 ± 0.08	0.79 ± 0.06
1969–1978	41 ± 7	0.0157 ± 0.0019	82 ± 8	20.31 ± 0.14	18.65 ± 0.10	1.24 ± 0.13
1979–1989	34 ± 3	0.0215 ± 0.0017	94 ± 3	20.59 ± 0.04	18.51 ± 0.08	1.09 ± 0.11

Table 4 (continued)

Depth	X_{max}	$(dS/dx)_{max}$	$X_{offshore}$	$S_{offshore}$	$S_{inshore}$	$Rot_z W$
			August			
Section 1: From Sebastopol to southwestward						
1959–1968	77 ± 18	0.0073 ± 0.0011	93 ± 13	19.61 ± 0.10	19.27 ± 0.11	– 1.60 ± 0.12
1969–1978	80 ± 17	0.0068 ± 0.0024	126 ± 15	19.88 ± 0.08	19.58 ± 0.14	– 0.40 ± 0.05
1979–1989	67 ± 16	0.0097 ± 0.0013	68 ± 7	20.04 ± 0.08	19.38 ± 0.07	– 0.50 ± 0.10
Section 2: From Yalta to southeastward						
1959–1968	40 ± 6	0.0215 ± 0.0011	75 ± 7	20.25 ± 0.10	19.35 ± 0.08	– 1.41 ± 0.11
1969–1978	46 ± 7	0.0233 ± 0.0020	90 ± 9	20.51 ± 0.06	19.36 ± 0.11	– 0.46 ± 0.05
1979–1989	52 ± 8	0.0230 ± 0.0027	57 ± 7	20.09 ± 0.10	19.38 ± 0.14	– 0.55 ± 0.12
Section 3: From Gelendjic to southwestward						
1959–1968	38 ± 8	0.0092 ± 0.0004	81 ± 8	19.80 ± 0.06	19.33 ± 0.04	– 0.60 ± 0.09
1969–1978	37 ± 3	0.0085 ± 0.0007	64 ± 7	19.88 ± 0.08	19.52 ± 0.12	– 0.25 ± 0.07
1979–1989	40 ± 7	0.0118 ± 0.0019	71 ± 10	20.01 ± 0.13	19.40 ± 0.09	– 0.38 ± 0.09

5 Conclusions

The generalization of the results of the studies of the T,S structure of the Black Sea waters and of its seasonal and interannual variabilities presented in this chapter allows one to make the following conclusions:

- The T,S structure of the Black Sea waters consists of a few characteristic layers with different thicknesses; top-down: the upper mixed layer, the seasonal pycnocline (thermocline); the cold intermediate layer, the main pycnocline (halocline); the isothermal intermediate layer; the thickest deep layer with a slow temperature and salinity increase with depth; and the near-bottom mixed layer.
- The principal features of this structure are related to the very weak vertical turbulent exchange of the T,S properties between, on the one hand, the freshened surface and the cold intermediate water masses and, on the other hand, the significantly more saline deep water mass.
- The seasonal and interannual variabilities of the UML, the seasonal pycnocline, and the CIL are caused by the corresponding variations in the heat and freshwater fluxes through the sea surface and in the riverine runoff.
- The ventilation of the Black Sea waters is restricted to the CIL; it is renewed over the major part of the area in severe winters and in some regions (focuses of ventilation) over the shelf and continental slope of the western part of the sea in other years.
- The seasonal and interannual variabilities of the main pycnocline are caused by the changes in the flux of the wind relative vorticity.
- The intra-annual response of the Black Sea main pycnocline to the annual forcing by momentum and vorticity fluxes from the wind is manifested in the superposition of two principal modes: a basin-scale standing oscillation and sub-basin hybrid Rossby coastal trapped waves, which form quasigeostrophic cyclonic amphidromic systems.
- In the layers below the main pycnocline, the interannual variability is caused by the variations in the inflow of the waters of the Sea of Marmara related to the rest of the other components of the external water budget of the Black Sea.
- In the 1950s to 1990s, the interannual variability of the T,S structure of the waters of all the layers of the Black Sea was dominated by quadratic trends with extremes and changes in the signs of linear trends in the interval 1975–1980, when a regime shift in the large-scale processes in the World Ocean–global atmosphere system occurred.

Many of the results presented in this chapter are hypothetical. The degree of their validity should be found out from further studies.

References

1. Boyer T, Levitus S (1994) Quality control and processing of historical oceanographic temperature, salinity, and oxygen data. NOAA, Washington DC
2. Knipovich NM (1932) Transactions of the Sea of Azov and the Black Sea Expedition, Vol 10. VNIRO – Russian Marine Fisheries Institute, Moscow (in Russian)
3. Leonov AK (1960) Regional oceanography. Pt 1. The Black Sea. Gidrometeoizdat, Leningrad (in Russian)
4. Filippov DM (1968) Circulation and water structure in the Black Sea. Nauka, Moscow (in Russian)
5. Blatov AS, Bulgakov NP, Ivanov VA, Kosarev AN, Tuzhilkin VS (1984) Variability of hydrophysical fields of the Black Sea. Gidrometeoizdat, Leningrad (in Russian)
6. Simonov AI, Altman EN (eds) (1991) Hydrometeorology and hydrochemistry of the seas. Vol IY. The Black Sea, Issue 1. Hydrometeorological conditions. Gidrometeoizdat, St Petersburg (in Russian)
7. Ozsoy E, Unluata U (1997) Earth Sci Rev 42:231
8. Kosarev AN, Tuzhilkin VS, Arkhipkin VS, Daniyalova Z (2004) Hydrology and ecology of the Black and the Caspian Seas. In: Geography, society and environment. Vol YI. Dynamics and interaction of the atmosphere and the hydrosphere. Lomonosov Moscow State University, Moscow, p 218 (in Russian)
9. Goryachkin YN, Ivanov VA (2006) The Black Sea level: past, present, and future. MHI UAS, Sebastopol (in Russian)
10. Mamayev OI, Arkhipkin VS, Tuzhilkin VS (1994) Oceanology 27:178 (in Russian)
11. Murray JW, Top Z, Ozsoy E (1991) Deep-Sea Res 38 Suppl 2A:S663
12. Samodurov AS, Ivanov LI (2002) Rep Natl Acad Sci Ukraine 6:140
13. Eremeev VN, Ivanov LI, Samodurov AS, Duman M (1997) Bottom boundary layer in the Black Sea: a simple model of formation. In: Ozsoy E, Mikaelyan A (eds) Sensitivity to change: Black Sea, Baltic Sea and Northern Sea. Kluwer, Dordrecht, p 275
14. Ivanov LI, Shkvorez IY (1995) Marine Hydrophys J 6:53 (in Russian)
15. Eremeev VN, Ivanov VA, Kosarev AN, Tuzhilkin VS (1992) Climatic thermohaline fields of the Black Sea waters with mesoscale resolution: three-dimensional structure and annual variability. In: Problems of the Black Sea (international conference). Sebastopol, Ukraine, 10–15 November 1992. MHI UAS, Sebastopol, p 209
16. Ivanov VA, Kubryakov AI, Mikchailova EN, Shapiro NB (1996) Izvestya RAS, Phys Atmos Ocean 32:152 (in Russian)
17. Belokopytov V (1998) Long-term variability of cold intermediate layer renewal conditions in the Black Sea. In: Ivanov LI, Oguz T (eds) Ecosystem modeling as a management tool for the Black Sea. Kluwer, Dordrecht, p 47
18. Ovchinnikov IM, Popov YI (1987) Oceanology 27:739 (in Russian)
19. Ivanov LI, Besiktepe S, Ozsoy E (1997) The Black Sea cold intermediate layer. In: Ozsoy E, Mikaelyan A (eds) Sensitivity to change: Black Sea, Baltic Sea and Northern Sea. Kluwer, Dordrecht, p 251
20. Tuzhilkin VS, Nakolushkin IY (1999) Climatic features of the Black Sea cold intermediate layer ventilation. In: International conference on oceanography of the eastern Mediterranean and Black Sea. Similarities and differences of two interconnected basins. 23–26 February 1999, Athens, Greece. Abstracts, p 175
21. Stanev EV, Staneva JV (2001) Dynam Atmos Oceans 33:163
22. Eremeev VN, Ivanov VA, Kosarev AN, Tuzhilkin VS (1994) Annual and semiannual harmonics in the climatic salinity field of the Black Sea. In: Diagnosis of the state of marine environment of the Azov–Black Sea basin. MHI UAS, Sebastopol, p 89

23. Rachev NH, Stanev EV (1997) J Phys Oceanogr 27:1581
24. Bokhove O, Johnson ER (1999) J Phys Oceanogr 29:93
25. Yasunaka S, Hanawa K (2005) Int J Climatol 25:913
26. Polonsky AB, Lovenkova EA (2004) Izvestya RAS, Phys Atmos Ocean 40:832 (in Russian)
27. Ginzburg AI, Kostianoy AG, Sheremet NA (2004) J Marine Syst 52:33
28. Tuzhilkin VS, Berlinsky NA, Kosarev AN, Nalbandov YR (2004) Ecol Sea 65:75 (in Russian)
29. Ivanov LI, Besiktepe S, Ozsoy E (1997) Physical oceanography variability in the Black Sea pycnocline. In: Ozsoy E, Mikaelyan A (eds) Sensitivity to change: Black Sea, Baltic Sea and Northern Sea. Kluwer, Dordrecht, p 265
30. Knysh VV, Korotaev GK, Demyshev SG, Belokopytov VN (2005) Marine Hydrophys J 3:11 (in Russian)
31. Polonsky AB, Lovenkova EA (2006) Izvestya RAS, Phys Atmos Ocean 42:419 (in Russian)
32. Polonsky AB, Lovenkova EA (2006) Marine Hydrophys J 4:18 (in Russian)
33. Kosarev AN (2007) Hydrometeorological conditions, in this volume
34. Tuzhilkin VS (2007) General circulation, in this volume
35. Ginzburg AI, Zatsepin AG, Kostianoy AG, Sheremet NA (2007) Mesoscale water dynamics, in this volume
36. Ginzburg AI, Kostianoy AG, Sheremet NA (2007) Sea surface temperature variability, in this volume

Hdb Env Chem Vol. 5, Part Q (2008): 255–275
DOI 10.1007/698_5_067

Published online: 13 February 2007

Sea Surface Temperature Variability

Anna I. Ginzburg · Andrey G. Kostianoy · Nickolay A. Sheremet (✉)

P.P. Shirshov Institute of Oceanology,
Russian Academy of Sciences, 36 Nakhimovsky Pr., 117997 Moscow, Russia
sheremet@ocean.ru

Abstract Blended weekly multichannel sea surface temperature (MCSST) (1982–1984) and monthly Pathfinder SST (since 1985) data with spatial resolution of ∼ 18 and 9 km, respectively, were used to investigate seasonal and interannual variability of the Black Sea SST in the period 1982–2002. The 18 yearly (1985–2002) mean SST fields for the central months of four hydrological seasons (February, May, August, and November) based on the Pathfinder data were constructed. The years with the winter and summer SST anomalies were indicated as well as a long-term temperature trend. Minimums of the mean annual basin-averaged SST occurred in the years with the winter SST minimums. A sharp increase in the winter and mean annual SSTs was observed after 1993, the year of the temperature minimum. A linear regression of the mean annual basin-averaged SSTs gave a positive trend of about 0.06 °C year^{-1} in the period 1982–2002. However, a trend of the blended field and satellite winter (February–March) SSTs in a longer period (1957–2002) turned out to be small and negative (∼– 0.008 °C year^{-1}). Most of the marked winter and summer SST anomalies were likely related to the El Nino global events, although a value and a sign of an anomaly were not determined by intensity of El Nino. The winter SST anomalies were better related to the winter indices of the East Atlantic-West Russia (EAWR) atmospheric pattern or to various combinations of the winter EAWR and North Atlantic Oscillation (NAO) indices, which determine predominance of cold or warm air masses over the Black Sea basin, than to the NAO winter indices.

Keywords Atmospheric forcing · Black Sea · Interannual variability · Long-term temperature trend · Sea surface temperature · Seasonal variability

Abbreviations

SST	sea surface temperature
MCSST	multichannel sea surface temperature
AVHRR	advanced very high resolution radiometer
NOAA	National Oceanic and Atmospheric Administration
NASA	National Aeronautics and Space Administration
CIL	Cold Intermediate Layer
ENSO	El Nino-Southern Oscillation
NAO	North Atlantic Oscillation
EAWR	East Atlantic-West Russia atmospheric system
SD	standard deviation

1 Introduction

The Black Sea surface temperature (SST) represents one of the most important hydrophysical parameters that determines the sea–atmosphere heat exchange. It also influences the water circulation and the ecological condition of the basin and may indicate the temperature of the Cold Intermediate Layer (CIL). Because of the small thickness of the upper mixed layer (about 10 m in summer and 80 m in winter) and weak vertical mixing in the sea due to peculiarities of its density stratification (see Tuzhilkin VS "Thermohaline Structure of the Sea", in this volume). The temperature of the surface layer and SST quickly respond the atmospheric impacts; owing to this, SST is subjected to a significant spatial and temporal (interannual, seasonal, and synoptic) variability. Therefore, it is important to acquire reliable information about the horizontal distribution of SST over the sea area during various seasons and months, its interannual variability, and the tendencies of the temporal changes in its mean value (over the entire sea and in its individual regions). The studies of the SST interannual variability in the semi-enclosed Black Sea are especially important owing to the global warming, which has been observed starting from the end of the 1970s [1, 2], to the large-scale atmospheric oscillations such as the El Nino-Southern Oscillation (ENSO) and the North Atlantic Oscillation (NAO), and to the growing anthropogenic impact on the sea.

The long-standing mean (climatic) fields of the surface layer temperature for a month (season) derived from the hydrographic measurements of various years are well known (see Tuzhilkin VS "Thermohaline Structure of the Sea", in this volume, and publications [3–6]). However, the interannual variability of SST (on the entire sea and regional scales) has been insufficiently studied till recently. The reasons for this lie in the irregular distribution of the hydrographic measurements performed previous to the 1990s over years, seasons, and regions, especially in the deep-water part of the sea. Blatov et al. [3] calculated the mean annual temperature anomalies for the period 1957–1973.

In [6], the temperature values averaged over February–March of every year for 1957–1995 are presented. The multiannual history of the anomalies of the mean annual values of the water temperature measured at the hydrometeorological stations of Batumi and Odessa (1950–1985) are reported in [5], while in [7], the interannual variability of the mean annual values of the water temperature at four hydrometeorological stations at the Crimean coast in the period 1972–1992 is considered.

Under the conditions of the reduction of regional hydrographic measurements and surveys on the entire sea scale, the study of the interannual SST variability may be based only on the regular and permanently replenished information with a high spatio-temporal resolution. Such satellite information about the SST is acquired by the measurements with Advanced Very High Resolution Radiometers (AVHRR) onboard the satellites of the National Oceanic and Atmospheric Administration (NOAA). The efficiency of the use of these satellite data for the studies of the spatio-temporal variability of the Black Sea temperature regime was demonstrated in [8–13].

In addition, recent studies were also focused on the response of the Black Sea SST to large-scale atmospheric forcing such as the ENSO (sea level atmospheric pressure difference between Darwin and Tahiti, Pacific Ocean), which determines the fluctuations in the ocean–atmosphere climatic system and the particular features of the global atmospheric circulation, the NAO (atmospheric pressure difference between the Azores High and the Iceland Low, Atlantic Ocean), which is extremely important for the European region [14, 15], and the East Atlantic-West Russia (EAWR) system, which represents the quasi-persistent high and low surface pressure anomaly centers over Western Europe and the Caspian region and modulates the NAO over the Eurasia continent [16]. For example, in [9, 11], it is suggested that the seasonal anomalies of the Black Sea SST are related to the El Nino events, the authors of [6, 13, 16, 17] discuss the influence of the NAO on the variability of the winter SST, while in [16], the role of the EAWR pattern in the long-term SST variability is considered.

In this work, we present the results of the study of the seasonal and interannual variability of the Black Sea SST based on an analysis of satellite information over a 21-year-long period (1982–2002). The tendencies in the winter (February–March) temperature changes are considered for a longer interval (1957–2002) through combining satellite data with those of the field measurements. The results obtained are used to reveal the response of the SST to the El Nino, NAO and EAWR forcing. The relation between the variability of the winter SST and the CIL temperature over about five decades and the response of the Black Sea ecosystem to interannual changes in the SST are also briefly considered.

2
Data

In this study, we used remotely sensed SST data based on the measurements with AVHRR onboard satellites of the NOAA series and produced at the Jet Propulsion Laboratory (JPL, USA). Two types of SST data were used: mean weekly multichannel SST (MCSST) interpolated data with a spatial resolution of about 18 km and mean monthly Pathfinder SST data with quality flags 4–7 (i.e. the “best SST”) and a spatial resolution of about 9 km. (Pathfinder is a joint NOAA/NASA (National Aeronautics and Space Administration) project devoted to the production of a high quality global SST dataset from 1985 to the present.) Both MCSST and Pathfinder SST data have a temperature resolution of about 0.1 °C; they are derived from the same AVHRR measurements but obtained using different algorithms (see [18] and [19], respectively). We used only nighttime SST data (to exclude solar heating effect). Because the subsurface measurements (mainly drifter data) are widely used as sea-truth data to retrieve actual SST values from the AVHRR data, MCSST and Pathfinder SST data correspond, in fact, to the surface layer and can be compared with field measurements with no correction for the temperature drop in the skin layer [20].

The MCSST data from November 1981 to date are available via the Internet; however, they have gaps (for example, data for 11 weeks of 1983 including August and September and for the period from February to September 2001 are missing). The Pathfinder data have no gaps but they are available via the Internet only from 1985 to March 2003 (at this moment). Therefore, in this study, we used blended satellite data on SST: the MCSST data for 1982–1984 and the Pathfinder data for 1985–2002. The validation of the MCSST data for the Black Sea was reported in [11]. In order to validate the Pathfinder data for the Black Sea, in the absence of a special study, we calculated the basin mean satellite-derived SST values averaged for February–March of every year during the period 1985–1995 and compared these values with the appropriate SSTs based on the hydrographic measurements and presented in [6]. The difference between the satellite and in situ February–March SST values mostly did not exceed 0.3 °C, the mean difference being 0.15 °C (the Pathfinder SST values, as distinct from the MCSST ones, were lower than those of in situ measurements).

We also compared the mean seasonal values calculated using the MCSST and Pathfinder datasets for the same period 1985–2000. The average difference between the mean seasonal MCSST and Pathfinder SSTs was determined to be + 0.17 °C for winter (January–March), – 0.76 °C for spring (April–June), + 0.004 °C for summer (July–September), and + 0.51 °C for autumn (October–December). This means that this difference is low for the winter and summer temperature values (MCSST values are somewhat higher) and higher for the seasons with significant temperature variations; in so doing, the spring Pathfinder SST values are, on average, higher than those derived from the MCSST dataset. When calculating mean annual temperature values, these seasonal differences in the SSTs obtained with the use of the MCSST and Pathfinder datasets practically compensate each other.

The data on the monthly indices of the ENSO and NAO atmospheric oscillations were obtained from the Internet (http://www.cpc.NOAA.gov/data/indices/soi, http://www.cpc.NOAA.gov/data/teledoc). The winter (December–March) NAO indices are given at the site http://www.cgd.ucar.edu/~jhurrell/nao.stat.other.html, and the EAWR index data are available at the site http://www.cpc.ncep.noaa.gov/data/teledoc/eawruss.html.

3 Seasonal Variability of the SST Field

The SST fields obtained from averaging the data over a relatively short period (about two decades) depend both on the phase of the global SST changes (warming or cooling, which are characterized by a duration of approximately 30–40 years [2]), and on the data used for averaging. In [11], SST distributions averaged over a 19-year-long period (1982–2000) for the central months of the four hydrological seasons (February, May, August, and November) obtained with the use of the MCSST data are presented. Similar SST distributions for the period 1985–2002 based on the Pathfinder data are shown in Fig. 1.

The general features of the seasonal SST fields in Fig. 1 and in [11] agree well with each other and with the climatic fields based on the hydrographic measurements (see Tuzhilkin VS "Thermohaline Structure of the Sea", in this volume, and publications [3–6]). Nevertheless, owing to the higher spatial resolution of the Pathfinder SST data as compared to the MCSST data, the distributions shown in Fig. 1 have some details missing in the corresponding temperature fields in [11] such as the wintertime temperature minimums in the northeastern part of the sea (SST < 7 °C at ~ 45°N, 36.5°E caused by the propagation of cold waters from the Sea of Azov via the Kerch Strait) and on the northwestern shelf north of 46°N (SST < 4 °C), as well as the manifestation of the upwelling zone off the Anatolian coast in summer. In so doing, the winter SSTs in the deep sea both in Fig. 1a and in [6, 11] are about 0.5 °C lower as compared to the earlier climatic values. On the whole, the SST values in Fig. 1b,c are somewhat higher and those in Fig. 1d (in the northern part of the sea) are somewhat lower than in [11], which is related to the account for two warm years (2001 and 2002) in the averaging presented in Fig. 1 and to certain differences in the seasonal temperatures obtained from the MCSST and Pathfinder datasets (Sect. 2 "Data").

The spatial variability of the SST fields in different seasons can be considered from the corresponding distributions of the SST standard deviation (SD) (not shown here). The maximum values of SD (up to 2.2–2.5 °C) cor-

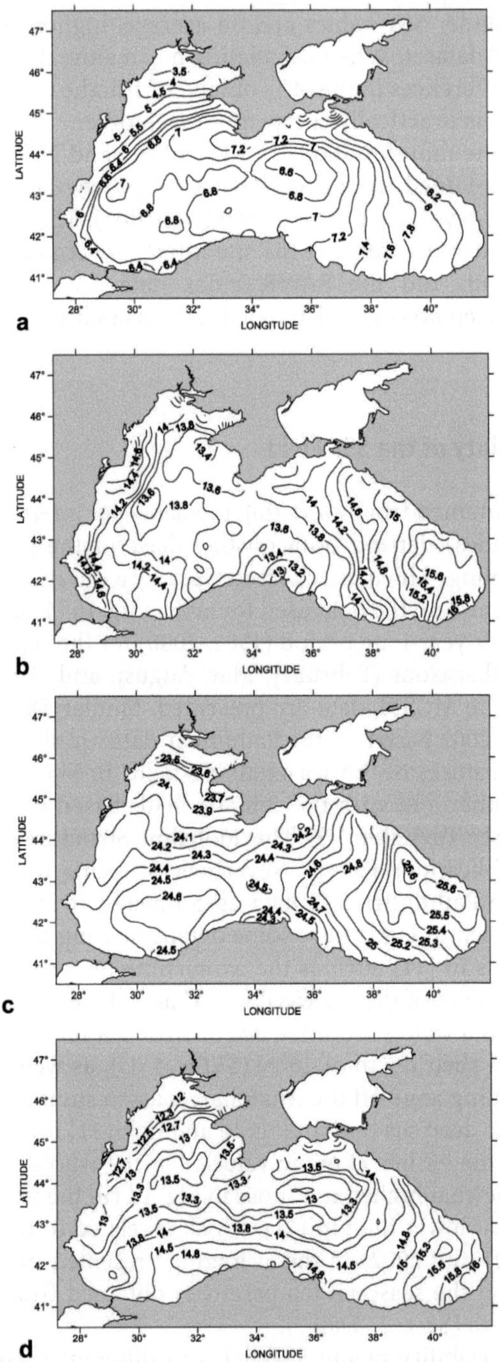

Fig. 1 The satellite-derived (based on the Pathfinder data) SST fields (°C) for the period 1985–2002: **a** in February, **b** in May, **c** in August, **d** in November

respond to the spring and autumn seasons. The spring SD maximum (up to 2.5 °C) on the northwestern shelf is apparently related to the dynamics of riverine waters; the position of their eastern boundary significantly changes on synoptic time scales under wind forcing and entrainment by anticyclonic eddies that propagate over the Danube Fan (see Ginzburg AI, Zatsepin AG, Kostianoy AG, Sheremet NA "Mesoscale Water Dynamics", in this volume, and [21, 22]). The increased variability of SST in the eastern and southeastern parts of the sea (with SD values up to 2.2 °C) is obviously related to the local dynamical features (meanders and branches of the Rim Current, coastal and open sea anticyclones, etc.; see Ginzburg AI, Zatsepin AG, Kostianoy AG, Sheremet NA "Mesoscale Water Dynamics", in this volume).

4 Temporal Variations of SST

4.1 Seasonal and Interannual Variability of the Basin-Averaged SST

Figures 2–5 demonstrate the interannual variability of the mean monthly, mean seasonal, and mean annual values of the basin-averaged SST in the period 1982–2002. Two types of the wintertime SST values are presented averaged either over the winter hydrological season (January–March, Fig. 3a) or over the two coldest months of the year (February and March, Fig. 4a). When constructing Fig. 4a, which covers about five decades, we used the results of the hydrographic measurements of 1957–1983 [6] and satellite-derived SSTs for 1984–2002 (the SST for 1984 was calculated from the MCSST data; starting from 1985, the Pathfinder data were used). Note that the years of the winter averaged SST maximums and minimums (Fig. 3a) generally correspond to those in February–March (Fig. 4a) and in the mean monthly time series (Fig. 2a). The poorer manifested extremum in the winter of 1985 in Fig. 3a as compared with those in Figs. 2a and 4a was caused by a very warm January. The long-term course of the CIL temperature shown in Fig. 4b is also very informative for the consideration of the interannual variability of the wintertime SST. (We compiled this graph according to the data for 1954–1995 in [6] and for 1990–2002 in [23]. A comparison of the temperature data in these two studies for the interval 1990–1995 showed their virtually absolute coincidence.)

From Figs. 2a, 3a, and 4 it follows that, over the approximately 50-year-long period, the coldest winters in the Black Sea were observed in 1954, 1964, 1976, 1985, 1987, 1992, and 1993. Four of them occurred during the last $\sim$ 20 years. The lowest winter SST was registered in 1993. Starting from 1994, mean monthly winter temperatures have never decreased below 6.6 °C (Fig. 2a). High mean monthly and mean seasonal temperatures in winter were

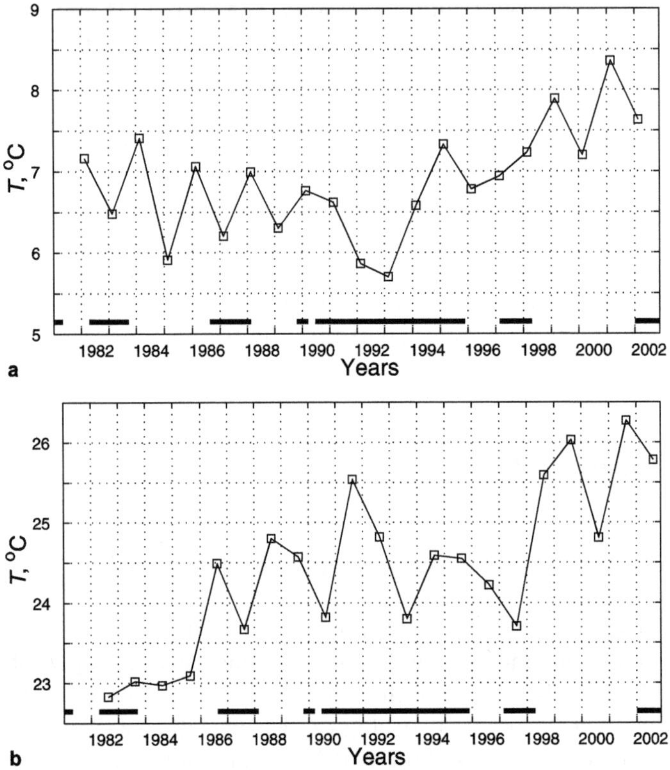

Fig. 2 The peak mean monthly SSTs in the period 1982–2002: **a** winter minimums, **b** summer maximums. Hereafter, the bold horizontal segments mark time periods of El Ninos

observed in 1961, 1962, 1966, 1981, 1984, 1995, 1999, 2001, and 2002, with absolute maximums in 1966 and 2001 (in 2001, the mean monthly temperature in January was 9.6 °C at an average value over the winter season of 8.9 °C).

The least mean monthly and mean seasonal summertime values of SST during the period 1982–2002 were noted in 1982, 1984, and 1985 (rather cold were also the summer seasons of 1987, 1993, and 1997); the greatest values were characteristic of 1999, 2001, and 2002 (Figs. 2b and 3c). Anomalously high summer SST values were also registered in 1966 and 1972 (the latter maximum was observed only in the northwestern part of the sea) [3]. Note that the summer SST anomalies frequently follow the winter anomalies of the same sign in the same year (for example, warm winters and summers in 1966 and 2001 or cold winters and summers in 1985 and 1993). In 1982–2002, the range of the variability of the maximum summer temperatures was 3.4 °C and that of the winter temperatures equaled 2.5 °C (Fig. 2). The maximum annual ranges of the mean monthly SSTs were registered in 1991 and 1992 (18.9 °C) and the minimum ranges were in 1982 and 1984 (15.6 °C).

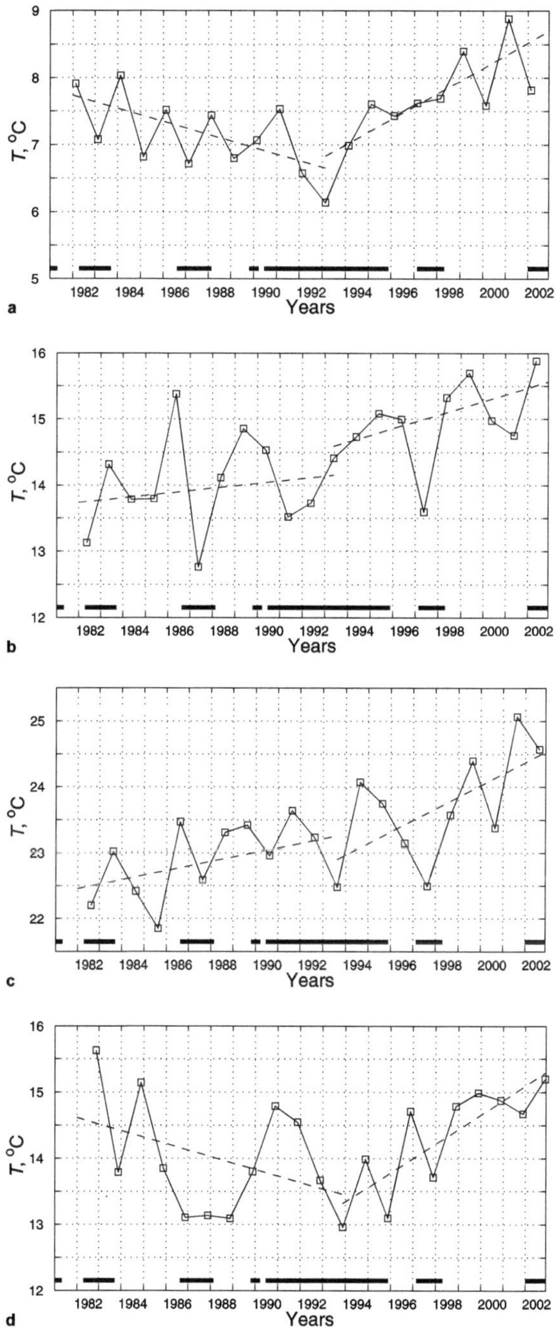

Fig. 3 The mean seasonal SSTs in the period 1982–2002: **a** in winter, **b** in spring, **c** in summer, **d** in autumn. *Dashed lines* show the SST trends in the sub-periods 1982–1993 and 1993–2002

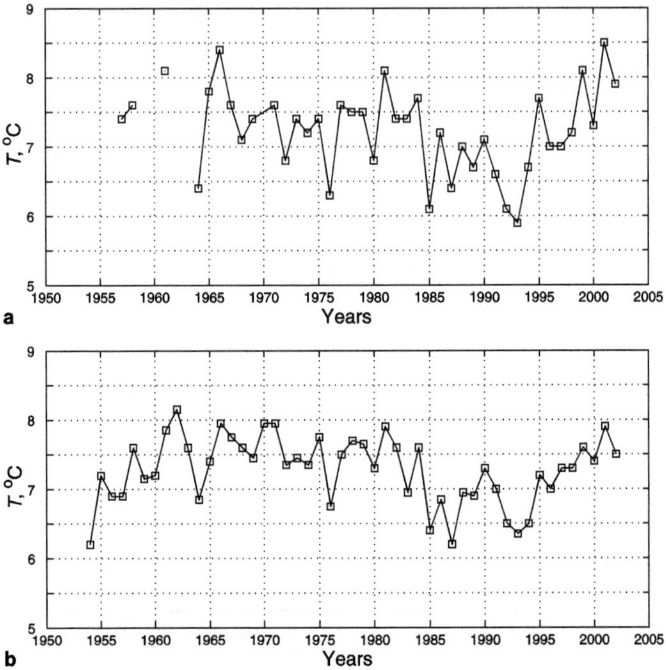

Fig. 4 The winter (February–March) SSTs in the period 1957–2002 (**a**) and the CIL temperature (May–November) in the period 1954–2002 (**b**). Blended hydrographic measurements from [6] (1957–1983) and satellite-derived SST values based on MCSST (1984) and Pathfinder (1985–2002) data were used when constructing Fig. 4a. Figure 4b is based on data from [6] and [23]

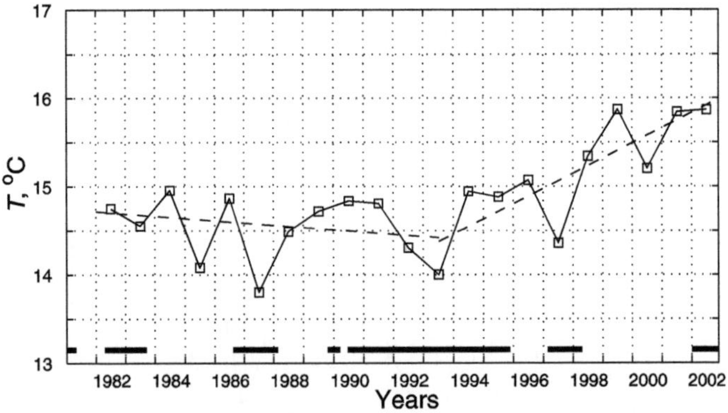

Fig. 5 The mean annual basin-averaged SSTs in the period 1982–2002. *Dashed lines* show the SST trends in the sub-periods 1982–1993 and 1993–2002

The mean annual SST values for the entire sea obtained from averaging mean monthly temperatures in the period 1982–2002 changed within approximately 2 °C (Fig. 5). The lowest mean annual temperatures corresponded to the years with the coldest winters such as 1985, 1987, and 1993 (14.1, 13.8, and 14.0 °C, respectively) and the highest mean annual temperatures corresponded to the years with high both summer and winter temperature values such as 1999, 2001, and 2002 (15.9, 15.8, and 15.9 °C, respectively). High mean annual SSTs were also observed in 1966 (the year with anomalously high temperatures in the winter and summer seasons) and in 1975 (in the northwestern part of the sea) [5], while low temperature values were characteristic of 1956 and 1976 (the latter minimum, at least in the northern part of the sea, was documented at the hydrometeorological stations in Odessa and Yalta) [5, 7].

In order to reveal a trend in the Black Sea SST in 1982–2002, a statistical processing of the satellite-derived data was performed. The linear regression of the basin-averaged mean annual SSTs (Fig. 5) gave a trend of about 0.06 °C $year^{-1}$, that is, during the 21-year-long period, the mean sea surface temperature increased by about 1.3 °C. In addition, seasonal trends of the basin-averaged SST were obtained using the time series of mean seasonal values (Fig. 3). The SST trends for the 21-year-long period appeared to be as follows: ∼ 0.04 °C $year^{-1}$ in winter, ∼ 0.08 °C $year^{-1}$ in spring, ∼ 0.09 °C $year^{-1}$ in summer, and ∼ 0.04 °C $year^{-1}$ in autumn. Thus, in 1982–2002, the warming of the Black Sea surface layer occurred during all the seasons; in so doing, the temperature trends in spring and summer were approximately twice as great as in winter and autumn. However, the character of the temperature changes and the corresponding annual and seasonal trends of the Black Sea SST were significantly different during the periods before and after the coldest year of 1993. For example, for the seasonally averaged time series during the sub-period 1982–1993 (Fig. 3), the winter, spring, summer, and autumn trends were ∼– 0.10, 0.04, 0.07, and – 0.10 °C $year^{-1}$, respectively, whereas for the sub-period 1993–2002, the corresponding values were ∼ 0.19, 0.10, 0.18, and 0.20 °C $year^{-1}$, respectively; the mean annual SST trends for the same sub-periods were ∼– 0.03 and 0.17 °C $year^{-1}$.

The values of the temperature trends obtained refer to relatively short periods 10–20 years long. At greater time intervals, the pattern may be significantly different. At present, we have a relatively long time series containing SST values averaged over February–March (see Fig. 4a), with a minimum also referring to 1993. A linear regression provided trend values of ∼– 0.03 and 0.19 °C $year^{-1}$ for the sub-periods 1957–1993 and 1993–2002, respectively, while for the entire period (1957–2002) the trend was low and negative (∼– 0.008 °C $year^{-1}$).

4.2 Regional and Synoptic SST Variability

Although the tendencies in the changes of the seasonal and annual SSTs over the entire sea and within its individual regions are generally similar (being conditioned by climatic effects), some regional differences related to those in the heat balance and particular features of the local water circulation are also possible [5]. In [11] it was shown that, in the period 1982–2000, the trends in the SST changes in the regions of the western and eastern cyclonic gyres were not identical and the warming of the eastern deep-sea area was going at a higher rate. Here, we present an analysis of the SST variability in two remote regions located in the southwestern and northeastern parts of the sea. The first region (region R1, near-Bosphorus) is limited by the coastline in the west and the south, by the latitude 42°N in the north, and by the longitude 30.5°E in the east. The second region (R2) is bounded by the Caucasian coast in the east, by the latitudes 44.5° and 43.5°N in the north and south, respectively, and by the longitude 36.5°E in the west. Region R1 is, to a significant extent, subjected to the influence of the waters spreading southward along the western coast from the northwestern shelf, while R2 is under the effect of the warm waters of the Rim Current and its branches, as well as of the coastal and deep-sea eddies (see Ginzburg AI, Zatsepin AG, Kostianoy AG, Sheremet NA "Mesoscale Water Dynamics", in this volume).

The analysis performed showed that, during the period 1985–2002, the maximum value of the peak mean monthly SST in winter in region R1 was observed in 2001 (8.4 °C), while the minimum values were observed in 1985, 1987, and 1993 (5.6, 5.8, and 5.7 °C, respectively). In region R2, the highest mean monthly winter SSTs were noted in 1999 and 2001 (9.3 and 9.2 °C, respectively), while the lowest values were observed in 1992 and 1993 (6.2 and 6.3 °C, respectively). The highest mean monthly summer value in R1 was registered in 1999 (25.9 °C), in R2 it was in 2001 (26.8 °C). Thus, in most of the cases, the extreme winter and summer temperature values in both regions were observed in the same years as in the entire sea. Besides, in the southwestern region, these SST values were generally lower than the average values for the sea as a whole (Fig. 2), while in the northeastern region, they were, as a rule, higher than the basin-averaged ones. The ranges of variability of the summer mean monthly maximums in R1 and R2 were 3.0 and 3.1 °C, respectively, and those of the winter minimums were 2.8 and 3.1 °C, respectively. The maximum annual range of the mean monthly SST values in region R1 refer to 1991 (19.6 °C), the minimum range was in 1997 (16.9 °C). In region R2, the maximum values of the annual range were 19.1 and 19.0 °C in 1991 and 1992, respectively, while the minimum value was 16.2 °C in 1990. In the sea as a whole, during the same period 1985–2002, the maximum and minimum SST annual ranges were 18.9 (see above) and 16.8 °C (in 1997), respectively. This means that, in the near-Bosphorus region, the maximum annual range of mean monthly SSTs is higher than its average values over the entire sea and over region R2.

The minimums of mean annual temperatures in regions R1 and R2 were basically observed in the same years as in the entire sea. In R1, the most pronounced minimums were in 1987 and 1997; in R2, they were in 1987 and 1993.

On average, the mean annual SSTs in R1 do not practically differ from those basin-averaged, while in R2 they are 0.4 °C higher. As in the entire sea, the absolute maximums of mean annual temperatures in R1 were observed in 1999 and 2001 (15.9 °C), while in R2 they were observed in 2001 and 2002 (16.3 and 16.7 °C, respectively). The estimated value of the mean annual temperature trend over the period 1985–2002 in region R2 was twice as high as in region R1: 0.12 and 0.06 °C $year^{-1}$, respectively. The similar estimation of the basin-averaged SST trend in the same period gave a value of $\sim$ 0.09 °C $year^{-1}$. This means that, in the northeastern region, the warming was going at a higher rate than in the southwestern region and in the sea as a whole.

In addition to climatic impacts, which determine the seasonal and interannual variability of the sea temperature on the whole, a significant effect on the regional SST variability is produced by local synoptic atmospheric impacts such as atmospheric anticyclones and cyclones, storms, and wind effected phenomena. For example, after the passage of an atmospheric cyclone on September 5, 1997, in the eastern part of the sea seaward of the coastal waters and the Rim Current, a vast area of cold waters appeared, whose location and outlines continuously changed with time (Figure 5 in Ginzburg AI, Zatsepin AG, Kostianoy AG, Sheremet NA "Mesoscale Water Dynamics", in this volume, see also [24]). This area more than 100 km in size was observed in satellite infrared images over at least three months (September–November 1997). In so doing, the maximum SST difference between this area and the warm coastal flow reached 6.5 °C in September and 5.5 °C in November. The patch of cold water $\sim$ 200 km in diameter formed in the western part of the Black Sea after the passage of an atmospheric cyclone on September 25–29, 2005 featured a temperature contrast of 10 °C with respect to the adjacent waters with a temperature of 24 °C [25]. A stormy wind with a speed up to 40 ms^{-1} that occurred on November 10–14, 1993 was accompanied by an air temperature drop down to – 7 °C; it caused a decrease in the surface layer temperature in the northeastern part of the sea by 4–5 °C (from 12–15 °C, a value that is characteristic of the autumn period, down to 8–10 °C [26]).

In the summer months, coastal upwelling results in sharp temperature drops near the coasts (by 10–15 °C off the Crimea [7]). The transformed waters of upwelling driven from the shore by jets and eddies over distances more than 100 km toward the deep-water part of the sea feature a temperature contrast with respect to the adjacent waters of 1–2 °C (see [21, 22], see also Ginzburg AI, Zatsepin AG, Kostianoy AG, Sheremet NA "Mesoscale Water Dynamics", in this volume).

A decrease in SST by 3.5 °C with respect to the adjacent waters may also take place at the centers of anticyclonic eddies (see Ginzburg AI, Zatsepin AG, Kostianoy AG, Sheremet NA "Mesoscale Water Dynamics", in this volume). Local changes in SST are sometimes related to the increased suspended matter content (mainly in near-mouth regions), to the atmospheric precipitation, and to the presence of oil films at the water surface [20].

4.3 Correlation Between Temperature of the Cold Intermediate Layer and SST

A comparison of Fig. 4a and b suggests a generally similar character of interannual variability of the winter SST and the water temperature in the CIL; the latter is formed in the wintertime, is located at a depth of $\sim 50-100$ m and features a temperature lower than 8 °C (see Tuzhilkin VS "Thermohaline Structure of the Sea", in this volume). In the coldest years, the water temperature in the CIL fell down to 6.2 °C (in 1954, 1987, and 1993), while in the warmest years it increased up to 7.9 °C (in 1966 and 2001). In 1962, this temperature exceeded 8 °C, that is, no renewal of the CIL at the entire sea scale occurred.

A linear regression for the period 1954–2002 provided a trend of the CIL temperature of about $-0.005\,°\mathrm{C\,year}^{-1}$. For the sub-periods 1954–1993 and 1993–2002, the trends were ~ -0.01 and $0.12\,°\mathrm{C\,year}^{-1}$, respectively, that is, close to the trend values of the SST averaged over February–March. However, note that the unambiguous correlation between the peaks of the SST and the CIL temperature is not always observed. For example, in 1985, the SST was lower than in 1987 (see Fig. 4a), while in the CIL, lower temperature corresponded to 1987 (Fig. 4b). This equally refers to the relations between the extreme values of the CIL temperature and those of the near-surface atmosphere temperature (winter severity). For example, with respect to the thermal conditions in the atmosphere, the winter of 1987 was moderately cold, while the winters of 1961, 1962, and 2001 were not anomalously warm [27]. It seems that the CIL temperature in a given year is conditioned by this layer state in the preceding year and by the severity of the current winter [27], as well as by the wind field over the Black Sea that determines, in particular, the intensity of cyclonic circulation of the sea waters and their upwelling at the centers of cyclonic gyres, which represent the regions of formation of the CIL waters [28].

5 SST Response to the Large-Scale Atmospheric Forcing

In order to reveal the response of the Black Sea SST to the ENSO and NAO, we compared the years of the extreme seasonal SST values observed (Figs. 3, 4) with the phases of the oscillations cited and the character of their decadal variability. The periods of the El Nino events are indicated in Figs. 2, 3, and 5. The character of the changes in the ENSO and NAO indices in 1950–2002 may be judged from Fig. 6 (the negative values of the ENSO index correspond to the El Nino events).

As follows from Fig. 6, in the period 1954–2002, 12 El Nino events occurred. Of them, the events of 1982–1983 and 1997–1998 were the most intensive and that of 1990–1995 was the most prolonged. Since the time de-

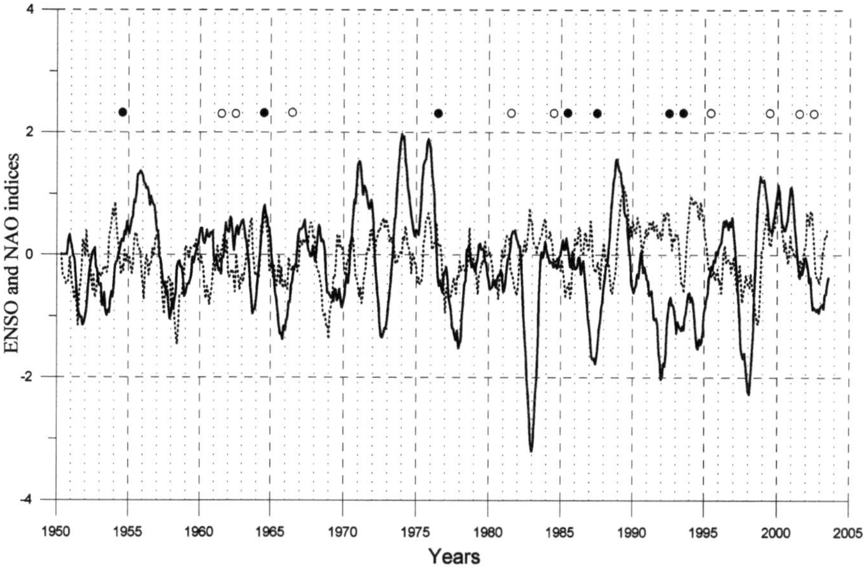

Fig. 6 Six-month running mean of mean monthly ENSO (*solid line*) and NAO (*dashed line*) indices in the period 1950–2003 acquired from the Internet sites (http://www.cpc.NOAA.gov/data/indices/soi, http://www.cpc.NOAA.gov/data/teledoc). *The circles* indicate years with the anomalous winter SST values: the *black circles* correspond to cold winters, the *open circles* to warm winters

lay of the possible response of the Black Sea SST to the ENSO phases (El Nino and La Nina) is unknown, a conventional reference of one or another temperature anomaly to the nearest El Nino/La Nina event is based on the assumption of a significant El Nino influence on the atmospheric circulation in the Atlantic–European region during 15 months after its maximum development (see [14]). In our analysis, as in [29], we do not refer short-term ENSO phases with small negative (in 1961) and positive (in 1985) indices to El Nino and La Nina events, respectively. In Fig. 6, circles mark the years with winter SST anomalies of both signs according to Figs. 2–4.

One can see that, in the period 1954–2002, most of the strongly pronounced winter SST anomalies (11 of 16) occurred either during the El Nino events or in the years immediately after them. Three anomalies (1962, 1976, and 2001) might be related to the La Nina events. The anomalies of 1961 and 1985, in the years with low values of the ENSO index, can hardly be referred to a certain phase of this atmospheric oscillation. Five of the 11 winters related to the El Nino events were cold and six others were warm. During the 1990–1995 El Nino, both cold (in 1992 and 1993), and warm (in 1995) winters were observed. It should be noted that an extreme value of the winter SST is not directly related to the intensity of El Nino. For example, one of the highest values of the winter SST was related to the relatively weak El Nino event of

1965–1966, while the most intensive El Nino of 1982–1983 was accompanied by a weaker anomaly of the same sign (in 1984).

The greater part of the summer SST anomalies (maximums in 1966, 1972, 1994, and 2002 and minimums in 1982, 1984, 1993, and 1997) corresponded to the years of the El Nino events. The warm summer of 1999 was probably also related to El Nino. All the cold SST anomalies in spring (Fig. 3b) and most of them in autumn (Fig. 3d) corresponded to the El Nino periods, and most of the temperature maximums in these transition seasons were likely to be also associated with the events.

It is probable that the cold summer of 1982 in the Black Sea as well as in the Mediterranean Sea [30] and in the northeastern Atlantic [31] was caused by the aerosols from the El Chichon (Mexican volcano) eruption in April 1982 [18]. Similarly, the eruption of Mount Pinatubo (Philippines volcano) in June 1991, whose after-effects were traced in the atmosphere of the Northern Hemisphere up to 1995 (as follows from the modeling results), might also make its contribution to the anomalously cold winter and summer of 1993 [32]. It is interesting that both of these volcanic eruptions coincided with the El Nino events of 1982–1983 and 1990–1995.

The NAO, which strongly influences the hydrometeorological parameters in the winter period, represents more high-frequency oscillation as compared to the ENSO (Fig. 6). (The values of monthly NAO indices, presented on the Internet, are periodically updated with respect to the changing interval of their normalization. When compiling Fig. 6, we used the data on the monthly NAO indices shown in 2003.) Nevertheless, we should note that, during 1950–1980, on average, negative values of NAO indices favorable for warm winters dominated, while the period 1982–1994 was characterized by their predominantly positive values, at which cold and dry air masses enter the Black Sea region [16]. Precisely during this latter period, the coldest winters of 1985, 1987, 1992, and 1993 were observed. On the other hand, at the predominantly positive NAO index values in 1999–2002, the winters in the Black Sea were warm.

A comparison between the anomalous winter SSTs in the period 1954–2002 with corresponding winter values of NAO indices (averaged over December–March) shows that there is no direct relation between them. Only in seven cases of the total 16, the sign of the SST anomaly corresponded to the sign of the NAO index; this was observed in the warm winters of 1962, 1966, and 2001 at negative values of the NAO index and in the cold winters of 1954, 1976, 1992, and 1993 at its positive values. Therefore, one may suggest that, in addition to the NAO, some other factors also exist, which contribute to the appearance of marked winter SST anomalies. The EAWR pattern may represent one of these factors [16]. At positive winter (December–February) EAWR index values, the Black Sea region is subjected to the influence of cold and dry air masses from the north, at its negative values, the sea is affected by warm and moist air masses from the south.

For our analysis, we used winter EAWR indices and different combinations of positive and negative winter NAO and EAWR index values suggested in [33], which are representative of different states of the East Mediterranean–Black Sea atmosphere (see [16]). It is assumed that cold winters correspond to positive EAWR indices regardless of the sign of the NAO index [16]. Indeed, at EAWR > 0 and NAO < 0, the anomalously cold winters of 1964, 1985, and 1987 were observed (Fig. 4). At EAWR > 0 and NAO > 0, the winter minimums of 1954, 1976, 1992, and 1993 were registered; however, the warm winters of 1995 and 2002 corresponded to the same combination of the EAWR and NAO indices. At EAWR < 0 and NAO > 0, both warm and cold winters are probable depending on the relative strengths and spatial sizes of the Iceland low and Caspian high pressure systems [16]. In our case, this combination of indices was characteristic only of the anomalously warm winters of 1961, 1981, 1984, and 1999. Finally, the combination EAWR < 0 and NAO < 0 should result in warm winters, which corresponds to the observation data (the warmest winters of 1962, 1966, and 2001, see Fig. 4). Thus, the major part (14 of 16) of the anomalous winter SSTs that were observed in 1954–2002, fits the combinations of the EAWR and NAO indices suggested in [16, 33].

6 Conclusions

An analysis of interannual variability of the satellite-derived basin-averaged SST values in the period 1982–2002 revealed a mean positive trend of the Black Sea SST of about 0.06 °C year^{-1}. Within this period, the SST trend was slightly negative in 1982–1993 ($\sim$– 0.03 °C year^{-1}) and positive in 1993–2002 (0.17 °C year^{-1}). The warming of the Black Sea water in general occurred in all the seasons. In this case, in the northeastern region of the Black Sea the warming of the surface layer was more intensive than in the southwestern (near-Bosphorus) region.

The warming of the Black Sea revealed is consistent with the warming of the World Ocean during this period [1, 2]. However, in the Black Sea, as in the World Ocean on the whole [2], it was probably preceded by a period of a slight overall change in SST. This may be inferred from the calculated trend of the winter (February–March) basin-averaged SSTs in the period 1957–2002 equal to about – 0.008 °C year^{-1}. This suggestion is also confirmed by the results presented in [5] and [16]. An analysis of the time series of mean annual values of the surface layer temperature observed at 19 hydrometeorological stations located along the western, northern, and eastern coasts of the Black Sea during the period 1923–1985 showed no unidirected changes in them, though, during this period, temperatures off Anapa and Yalta increased by 0.8 °C [5]. In so doing, a general winter warming and summer cooling were observed. According to the data of [16], during the last century, the winter

(December–March) warming trend in the Black Sea was 0.25 °C. Interestingly, similar patterns of changes in the SSTs in 1982–2000 in the closely spaced Black and Caspian inland seas [34] occurred (positive and close in values trends of SST, decrease of the mean annual SST values in 1985–1993 with their subsequent growth, and similar character of distinctly manifested seasonal SST anomalies), which suggests the determining role of climatic rather than anthropogenic factors in the interannual and decadal variability of SSTs in both of the seas.

Large-scale atmospheric oscillations, apparently, influence the temperature regime of the Black Sea, although there is no unambiguous correlation between them and the character of SST anomalies. The greater part of the winter and summer SST anomalies registered in the Black Sea in 1954–2002 coincided with the El Nino events. However, the magnitudes and signs of the temperature anomalies observed were not governed by the intensity of these events (for example, the anomalously warm winter of 1966 and the anomalously cold winter of 1993, although in the latter case an additional contribution to the SST drop might be made by the eruption of Pinatubo volcano).

The phases of the NAO also do not have unambiguous influence on the character of the winter SST anomalies. These anomalies are better correlated with the winter EAWR indices or with a combination of the winter EAWR and NAO indices, which represent the domination of northerly or southerly winds over the Black Sea area. However, in 1995 and 2002, warm winters were observed at the combination of positive EAWR and NAO winter indices with expected domination of northerly winds. Therefore, we can suppose that, in each specific case, various combinations of different global (ENSO, NAO, EAWR, global warming or cooling, volcano eruptions, etc.) and regional factors determine the value and sign of the Black Sea SST anomaly, as was suggested from the analysis of the interannual SST variability in the Caspian Sea [34].

The interannual variability in the Black Sea SST can significantly influence the climatology and ecology of this semi-enclosed basin. Very low May–November temperature values in the CIL (6.2–6.3 °C) followed the cold winters of 1954, 1987, and 1993, whereas after the very warm winters of 1962, 1966, and 2001, the CIL temperature was close to 8 °C. In this latter case, there was practically no renewal of the CIL. In its turn, the decrease in the CIL renewal may cause a decrease in the oxygen content in the core of the layer, which was observed, for example, in 2001, and a lifting of the upper boundary of the anoxic zone by about 5–10 m [35].

The cold period of 1985–1993 and the subsequent steady-state growth in SST were accompanied by important changes in the Black Sea ecosystem. For example, the cold winters of 1985 and 1987 seem to oppose the mass development of the ctenophore *Mnemiopsis leidyi*, which invaded into the Black Sea in 1982–1983 [17]. A sharp decrease in its biomass that followed its mass development at the end of the 1980s, occurred in the cold winters of 1992 and 1993. However, in the warm 1995, its biomass grew again [10, 17, 36]. The mass development of another ctenophore *Beroe ovata*, a specimen of which was noted near the Bulgarian coast and in the northeastern part of the sea as early as in the summer of 1997, was observed in the warm summer (August–September) of 1999 [37]. The warming of the Black Sea since 1995 resulted in a weakening or disappearance of the winter (February–March) peak of annual phytoplankton biomass [17]. An unusually long phytoplankton bloom was observed in the warm 1998–1999 [10, 38] and especially in 2001 [10] that was the year with the highest winter and mean annual SST in the period 1982–2002.

As follows from the above considerations, the monitoring of the Black Sea SST should be continued together with the studies of its response to global atmospheric impacts. On the scale of the entire Black Sea and its individual regions, these studies may be based on regular satellite information with a high spatio-temporal resolution.

Acknowledgements The work was supported by the bilateral Russian-Greece project "Long-term variability of the hydrophysical processes and zooplankton key species in the Black and Aegean seas: interrelations and dependencies upon climate changes", Russian Federal Program "World Ocean" (project 7), the Program N 17 of the Presidium of the Russian Academy of Sciences (Project 5.4), the INTAS funding (Project "ALTICORE", Contract Nr. 05-1000008-7927), and the Russian Foundation for Basic Research (grant N 07-05-00141).

References

1. Levitus S, Antonov JI, Boyer TP, Stephens C (2000) Science 287:2225
2. Rayner NA, Parker DE, Horton EB, Folland CK, Alexander LV, Rowell DP, Kent EC, Kaplan A (2003) J Geophys Res 108(D14):4407
3. Blatov AS, Bulgakov NP, Ivanov VA, Kosarev AN, Tuzhilkin VS (1984) Variability of Hydrophysical Fields of the Black Sea. Gidrometeoizdat, Leningrad (in Russian)
4. Altman EN, Gertman IF, Golubeva ZA (1987) Climatic fields of the Black Sea water salinity and temperature. Gosudarstvennyi Okeanograficheskii Institut, Sevastopolskoye Otdeleniye, Sewastopol (in Russian)
5. Simonov AI, Altman EN (Eds) (1991) Hydrometeorology and Hydrochemistry of the USSR seas. Project: Seas of the USSR. vol IV. The Black Sea, 1, Hydrometeorological conditions. Gidrometeoizdat, St.-Petersburg (in Russian)
6. Belokopytov V (1998) Long-term variability of cold intermediate layer renewal conditions in the Black Sea. In: Ivanov LI, Oguz T (eds) Ecosystem Modeling as a Management Tool for the Black Sea. NATO Science Series. Series 2: Environmental Security 47. Kluwer Academic Publishers, Dordrecht, The Netherlands, p 47
7. Karnaushenko NN, Pogrebnoi AE (2006) Morskoy Gidrofizicheskiy Zhurnal 1:22 (in Russian)
8. Ginzburg AI, Kostianoy AG, Sheremet NA (2001) Issledovanie Zemli iz kosmosa 1:51 (in Russian)

9. Ginzburg AI, Kostianoy AG, Sheremet NA (2002) Seasonal and interannual variability of the Black Sea surface temperature from satellite data (1981–2000). In: Zatsepin AG, Flint MV (eds) Multi-disciplinary investigations of the north-east part of the Black Sea. Nauka, Moscow, p 20 (in Russian)
10. Oguz T, Cokacar T, Malanotte-Rizzoli P, Ducklow HW (2003) Global Biogeochem Cycles 17(3):1088
11. Ginzburg AI, Kostianoy AG, Sheremet NA (2004) J Mar Syst 52:33
12. Babii MV, Bukatov AE, Stanichny SV (2005) Atlas of the Black Sea surface temperature based on satellite data (1986–2002). Morskoy Gidrofizicheskiy Institut, Sewastopol, Ukraine (in Russian)
13. Kazmin AS, Zatsepin AG (2007) J Mar Syst (in press)
14. Nesterov ES (2000) Meteorol Gidrol 8:74 (in Russian)
15. Stanev EV, Peneva EL (2002) Global Planet Changes 32:33
16. Oguz T, Dippner JW, Kaymaz Z (2006) J Mar Syst 60:235
17. Oguz T (2005) Oceanography 18(2):122
18. McClain EP, Pichel WG, Walton CC (1985) J Geophys Res 90(C6):11587
19. Walton CC (1988) J Appl Meteorol 27:115
20. Fedorov KN, Ginsburg AI (1992) The Near-surface Layer of the Ocean. VSP, Utrecht, The Netherlands
21. Ginzburg AI, Kostianoy AG, Soloviev DM, Stanichny SV (2000) Remotely sensed coastal/deep-basin water exchange processes in the Black Sea surface layer. In: Halpern D (ed) Satellites, oceanography and society. Elsevier, Amsterdam, p 273
22. Ginzburg AI, Kostianoy AG, Nezlin NP, Soloviev DM, Stanichny SV (2002) J Mar Syst 32:91
23. Krivosheya VG, Yakubenko VG, Moskalenko LV, Skirta AY (2005) Oceanology 45, Suppl. 1:S3
24. Ginzburg AI, Kostianoy AG, Krivosheya VG, Nezlin NP, Soloviev DM, Stanichny SV, Yakubenko VG (2002) J Mar Syst 32:71
25. Stanichny S, Ratner Y, Soloviev D, Stanichnaya R, Suslin V (2006) Remote sensing of the Black Sea – Mesoscale processes and interannual variability. In: Black Sea Ecosystem 2005 and Beyond. 1st Biannual Scientific Conference. Commission on the Protection of the Black Sea Against Pollution, Istanbul, Turkey, p 178
26. Krivosheya VG, Moskalenko LV, Ovchinnikov IM, Yakubenko VG (1997) Oceanology (English Translation) 37:321
27. Titov VB (2003) Meteorol Gidrol 10:68 (in Russian)
28. Titov VB (2004) Oceanology (English Translation) 44:783
29. Sidorenkov NS (1991) Trudy Gidrometeorol Nauchno-Issled Tsentra SSSR 316:31 (in Russian)
30. Santoleri R, Bohm E, Schiano ME (1994) Coast Estuar Stud 46:155
31. Djenidi S, Kostianoy AG, Sheremet NA, Elmoussaoni A (2000) Seasonal and interannual SST variability of the north-east Atlantic Ocean. In: Oceanic Fronts and Related Phenomena (Konstantin Fedorov International Memorial Symposium), IOC Workshop Report N 159, UNESCO. GEOS, Moscow:99
32. Minnis P, Harrison EF, Stow LL, Gibson GG, Denn FM, Doelling DR, Smith WL Jr (1993) Science 259:1411
33. Krichak SO, Kishcha P, Albert P (2002) Theor Appl Climatol 72:209
34. Ginzburg AI, Kostianoy AG, Sheremet NA (2005) Sea surface temperature variability. In: Kostianoy A, Kosarev A (eds) The Caspian Sea Environment. The Handbook of Environmental Chemistry. V5P. Springer, Berlin Heidelberg New York, p 59

35. Yakushev EV, Podymov OI, Egorov AV, Demidova TP, Pakhomova SV, Chasovnikov VK, Chelysheva MV (2005) Oceanology 45, Suppl 1:S61
36. Shiganova TA, Mirzoyan ZA, Studenikina EA, Volovik SP, Siokou-Frangou I, Zervoudaki S, Christou ED, Skirta AY, Dumont HJ (2001) Mar Biol 139:431
37. Vinogradov ME, Shushkina EA, Anokhina LL, Vostokov SV, Kucheruk NV, Lukashova TA (2000) Oceanology (English Translation) 40:46
38. Nezlin NP (2001) Oceanology (English Translation) 41:375

Hdb Env Chem Vol. 5, Part Q (2008): 277–307
DOI 10.1007/698_5_088

Published online: 11 July 2007

Vertical Hydrochemical Structure of the Black Sea

E. V. Yakushev[1] (✉) · V. K. Chasovnikov[1] · J. W. Murray[3] · S. V. Pakhomova[2] · O. I. Podymov[1] · P. A. Stunzhas[2]

[1]Southern Branch of P.P. Shirshov Institute of Oceanology, Russian Academy of Sciences, Okeanologiya, Gelendzhik-7, 353470 Krasnodarsky Kray, Russia
e_yakushev@yahoo.com

[2]P.P. Shirshov Institute of Oceanology, Russian Academy of Sciences, 36 Nakhimovsky Ave., 117997 Moscow, Russia

[3]School of Oceanography, University of Washington, Box 355351, Washington, WA 98195-5351, USA

Abstract The Black Sea is the largest marine anoxic basin in the world. It has an oxygenated surface layer overlying a sulfide-containing (anoxic) deep layer. This condition has evolved because of the strong density stratification on the water column: water with high salinity enters from the Bosporus, while the upper layer water is of riverine origin. This mixture of Bosporus outflow with overlying cold intermediate layer (CIL) water forms the Bosporus plume which ventilates the deep layers of the Black Sea. The rate of CIL formation is variable in response to changing climate.

The hydrochemical structure of the Sea is determined by these peculiarities of the hydrophysical regime. Here we will describe the main features of the Black Sea biogeochemical structure from the point of view of the changing of redox conditions. We will also describe the main features of the different scale temporal variability of this structure on the basis of recent expedition data received in 1997–2006.

Keywords Biogeochemical structure · Black Sea · Nutrients · Redox interface · Temporal and spatial variability

Abbreviations

CIL Cold intermediate layer
FPL Fine particle layer
NAO North Atlantic oscillation
NBS US National Bureau of Standards
OM Organic matter
RC Rim Current

1 Introduction

The Black Sea is the world's largest semienclosed marginal sea with permanent anoxic zone (about 85% of the total water volume). Its physical and chemical structure is determined by its hydrophysical balance [1]. The narrow (0.76–3.60 km) and shallow (< 93 m) Bosporus Strait provides the only pathway of water exchange between the Black Sea and the Mediterranean. The sill depths of the Bosporus are 32–34 m at the southern end and 60 m at the northern end [2, 3]. The seawater that flows out of the Bosporus Strait is the only source of salty water to the basin. Deep-water salinity values increase to $S = 22.33$ psu. Freshwater inflow from several European rivers (especially the Danube, Dniester, and Dnieper) and brackish water inflow from the Sea of Azov keep the salinity low in the surface layer ($S \approx 18.0$–18.5 psu in the central region). As a result, the water column is strongly stratified with respect to salinity, and thus density.

The Black Sea cold intermediate layer (CIL) has two sources that are highly variable in intensity depending on climate (Fig. 1). The first is the shallow northwest shelf where the water gets very cold (< 5.5 °C in the winter) [4]. The second site is in the central gyres region where surface water can become sufficiently cold to rejuvenate the CIL during some winter storms [5, 6]. Gregg and Yakushev [7] observed in the Western Gyre in March 2003 that the surface water had a uniformly cold temperature ($T = 6.1$ °C) from the surface to the depth of the CIL core (density $\sigma_\theta = 14.5\ \mathrm{kg\,m^{-3}}$). The intensity and relative importance of these two sources is probably variable on a year to year basis depending on climatic conditions.

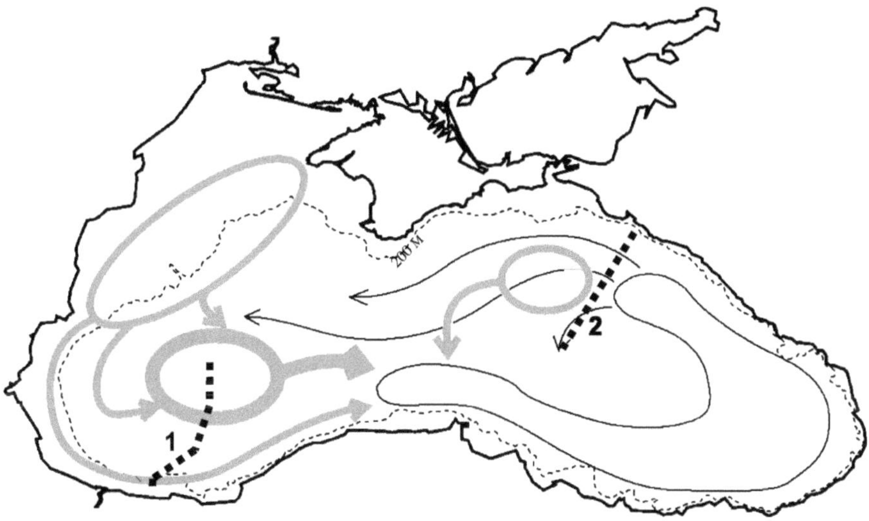

Fig. 1 General scheme of seasonal evolution of the core of CIL (*gray lines*, after Belokopytov, 2004) and position of transects from the SW coast to the Eastern Gyre (1) and from the northeastern coast (near Gelendzhik) to the Eastern Gyre (2)

Most of the mixing between the Bosporus outflow and the CIL occurs on the continental shelf just north of the Bosporus [8]. The bottom layer of high-salinity water from the Marmara Sea comes in from the south and thins as it enters the Black Sea. Salinity gradients are sharp at its upper boundary indicating mixing with overlying water. The overlying water is characterized by the temperature minimum properties of the CIL. Most mixing occurs before the Bosporus outflow reaches the shelf break. Based on the salinity balance for the deep Black Sea (50 to 2200 m) the ventilating water is composed of an average CIL to Bosporus entrainment ratio of $\sim 4:1$ [9]. Buesseler et al. [10] estimated this ratio as 7 : 1 and Eremeev et al. [11] as 10 : 1.

The resulting Bosporus plume ventilates the interior of the Black Sea at the depth represented by its density when it reaches the shelf break [12, 13]. Generally the Bosporus origin waters are pushed eastward by Western Gyre waters [14] following the Rim Current (Fig. 1). The most common entrainment conditions result in ventilation of the upper 500 m, but evidence suggests there must occasionally be rare ventilation events that reach the bottom.

The vertical turbulent flux below the CIL is too low to replace the oxygen consumed by respiration. Thus the Black Sea has an oxygen-containing surface layer and a sulfide-containing deep layer. The presence of hydrogen sulfide in the deep waters of the Black Sea was first described by Andrusov [15] in his report of the scientific expedition to the Black Sea on the Russian gunboat "Chernomorets" in 1890. He proposed that the reason for this occurrence is that the Bosporus restricts the exchange of deep waters between the Black and Mediterranean Seas. He also proposed that the hydrogen sulfide originated due to "reaction of non-mineralized organic matter with sulphurous salts of the sea water".

A consequence of the vertical stratification is that the surface layer (from 0 to 50–200 m) is well oxygenated while the deep layer (50–200 to 2000 m) is anoxic and contains high sulfide concentrations. At the boundary between the oxic surface and anoxic deep layers, there is a suboxic zone (at approximately 50 to 100 m depth) where the concentrations of oxygen are lower than the detection limit.

The suboxic zone is defined as the region between where oxygen decreases to near zero ($O_2 < 10\,\mu M$) and where sulfide first appears ($H_2S > 1\,\mu M$) [16, 17]. Many important redox reactions involving Fe, Mn, N, and other intermediate redox elements occur in the suboxic zone. Similar redox reactions take place in sediments throughout the world's oceans, but they are easier to study in the Black Sea because they are spread out over a depth scale of tens of meters (rather than centimeter or millimeter scales as in sediments). The Black Sea suboxic layer hydrophysical structure is very stable compared with other ocean redox regions such as Cariaco Trench, which is influenced by mesoscale eddies, or the Baltic Sea that is influenced by inflows of the North Sea saline oxygenated waters in cold winters.

The various oxidation–reduction reactions in the Black Sea occur in narrow layers of water of similar density and form features that are characteristic of the hydrochemical structure (e.g., maxima and minima, onset points). The position of these features in the density field is very stable [17–20] and it is possible to name this feature "chemotropic" [21]: the connection between the water density and properties of the chemical structure (by analogy with barotropic—the connection between density and pressure). In Table 1 we summarize the correspondence of the key features of the chemical structure with the density values. These values have served as a benchmark for subsequent cruises to evaluate the stability of the characteristic features.

The feature of chemotropicity of the Black Sea redox layer is well known, and this Sea is successfully used as a natural laboratory making the chemical species and the sequence of microbes easy to study on repeated cruises. The Black Sea is also an ideal site to study the effect of climate on the ocean structure. It is of small enough scale that variability in climate can vary physical forcing and thus chemical fluxes and biological processes.

Over the past few decades the Black Sea has been seriously perturbed by climatic change and intensive anthropogenic contamination. Some nutrients have increased (e.g., NO_3 due to eutrophication) while others have decreased (e.g., Si due to Danube river dams construction) [22]. Organisms imported as part of international shipping (e.g., *Mnemiopsis* and *Beroe*) have contributed to modification of the natural ecosystem. Understanding the natural temporal variability of the hydrochemistry of the Black Sea is important when trying to determine the effects of these anthropogenic perturbations.

Table 1 Characteristic density values ($kg\,m^{-3}$) of features associated with the biogeochemistry of the northeastern Black Sea in comparison to the central and western Black Sea

Feature	Central and western Black Sea (Murray et al. 1995)	Northeastern Black Sea (Chasovnikov 2002, 2005; Yakushev et al. 2002; Pakhomova 2005, and our unpublished data)
NO_3 maximum	15.35–15.45	15.27–15.45
PO_4 shallow maximum	15.45–15.55	15.50–15.60
$O_2 < 10\,\mu M$	15.65–15.75	
$O_2 < 2\,\mu M$		15.85–15.95
$NO_3 < 0.2\,\mu M$	15.90–16.00	
$NO_3 < 0.1\,\mu M$		15.85–15.95
NO_2 maximum	15.80–15.90	15.85–15.95
PO_4 shallow minimum	15.90–16.00	15.90–15.95
CH_4 onset		15.88–15.98
NH_4 onset	15.90–16.00	15.85–15.95
Mn(II) onset	15.80–15.90	15.80–15.90
Mn(IV) maximum	15.80–15.90	15.80–16.20
$Fe_{diss} < 10\,nM$	16.00	16.10
H_2S onset	16.10–16.20	16.08–16.14
PO_4 deep maximum	16.15–16.25	16.18–16.23
PO_4 deep minimum		16.50–16.70

Here we will describe the main features of the Black Sea biogeochemical structure from the point of view of the changing of redox conditions. We will also analyze the main peculiarities of the seasonal and interannual variability of this structure primarily on the base data received in 1997–2006. The main attention we will placed on describing the structure and temporal variability of the chemical structure of the redox layer, the boundary layer between oxic and anoxic waters.

2 Main Features of the Vertical Distributions of the Hydrochemical Parameters

The feature of chemotropicity allows analysis of the vertical distribution of chemical structure versus pressure and versus density and comparison of the results obtained in the different expeditions. In Fig. 2 we used both the scales to present the typical distribution of the chemical parameters. Here we will describe these typical distribution features of the concrete parameters.

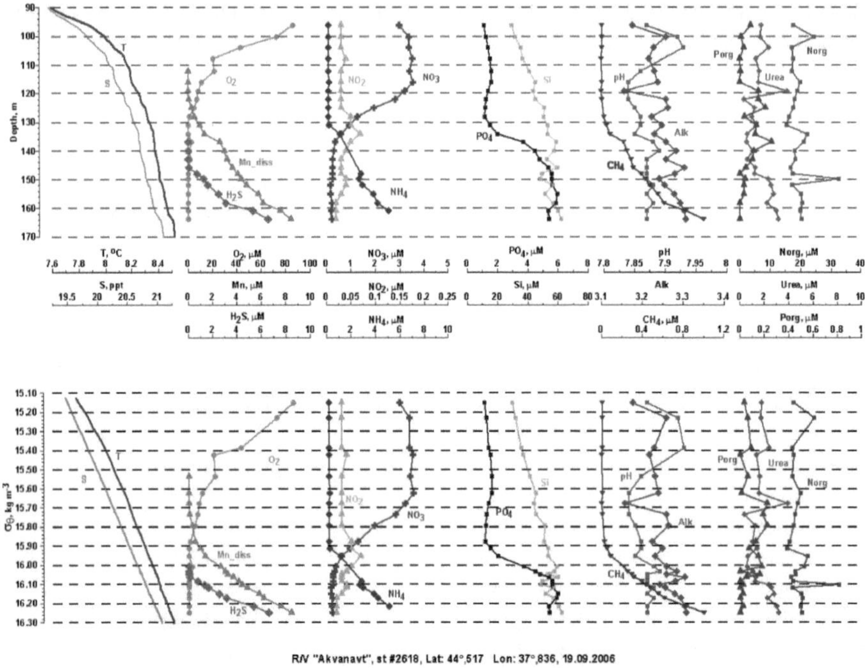

Fig. 2 Vertical distribution of temperature (*T*), salinity (*S*), dissolved oxygen (O_2), hydrogen sulfide (H_2S), dissolved manganese (Mn_diss), nitrate (NO_3), nitrite (NO_2), ammonia (NH_4), phosphate (PO_4), silicate (Si), pH (pH), total alkalinity (Alk), methane (CH_4), organic phosphorus (Porg), organic nitrogen (Norg), and urea (Urea), at a station near Gelendzhik (St. 2618, September, 2006). Concentrations of chemical parameters are in μM. Distributions are plotted versus depth (m) at the *top* and versus density (σ_θ, kg m^{-3}) at the *bottom*

2.1 Dissolved Oxygen

The vertical distribution of dissolved oxygen in the Black Sea reflects its specific features as the density stratified basin, which has a permanent H_2S zone under the pycnocline [23]. The thickness of the oxic zone varies between 70 and 100 m in areas of central cyclonic gyres with elevated isopycnal surfaces, and between 120 and 200 m in peripheral areas.

A layer of coexistence of oxygen and hydrogen sulfide (C-layer) was observed in the earliest hydrochemical studies in the Black Sea, and it was assumed that oxidation of hydrogen sulfide takes place mainly by oxygen reaction within this layer [23–26]. During the RV "Knorr" 1988 expedition, it was found that oxygen concentrations measured with the Winkler technique were significantly lower than found earlier [16, 27]. This fact of absence of detectable oxygen at the hydrogen sulfide boundary was subsequently confirmed [20, 28–30].

It became possible to significantly increase the accuracy of measurements of dissolved oxygen in the 1990s, because oceanographers started to use the 5-L PVC Niskin bottles instead of 1-L bottles (the larger volume of water allowed better protection of the portion of water sampled in the flask from the oxygen inside the bottle).

The accuracy and detection limit of the oxygen technique is an acute problem for studying the processes that occur in the redox zone. The formal accuracy of the Winkler technique is 0.9 μM (0.02 ml L^{-1}) [31, 32] and its detection limit is about 3.0 μM (0.07 ml L^{-1}) [33]. These values are significantly higher than the similar characteristics (in molar concentrations) for such parameters as hydrogen sulfide (correspondingly 0.1 and 0.3 μM), nitrates (0.02 and 0.02 μM) [31], dissolved manganese (0.2 and 0.2 μM) [34], and others. The detection limit of oxygen sensors and voltammetric techniques is also about 3 μM [33]. Therefore, the improvement of oxygen measuring techniques remains an issue.

The error of the Winkler technique increases in conditions of work at sea. Bezborodov and Eremeev [26] showed that the errors connected with contamination of reagents with oxygen and consumption of oxygen during the sampling procedure can reach 0.15 ml L^{-1} (6.6 μM). The main part of these errors is contamination of manganous sulfate and alkaline iodate reagents with oxygen. According to Broenkow and Cline [35], the oxygen concentrations in these reagents are about 30% of the saturation value for distilled water. Based on this estimate, reagent blank values can reach 2–3 μM. In addition, a correction for concentrations of reducing and oxidizing substances present in the suboxic layer (oxidized manganese, thiosulfate, elemental sulfur, etc.) that can react with iodine should be used. The procedure of this correction is described in several papers [31, 32, 35]. We have found the typical value of this correction is negative with values about 0.01–0.02 ml L^{-1} (0.45–0.90 μM) in the layer from the onset of hydrogen sulfide to about 10 m shallower (probably due mostly to oxidized manganese) and positive inside the hydrogen sulfide zone, increasing with depth from about 0.01 to 0.65 ml L^{-1} in a layer about 10 m below the hydrogen sulfide zone.

From 2000 we applied for studies of the vertical distribution of oxygen with a membraneless oxygen sensor [30]. This sensor allowed us to measure oxygen with a vertical resolution of 10 cm down to the detection limit of about 1 μM [36]. The application of this sensor allowed the following results to be obtained (Fig. 3):

- Dissolved oxygen is absent at the hydrogen sulfide onset level and there is a zone of absence of detectable concentrations of both dissolved oxygen and hydrogen sulfide.
- Oxygen distribution below the oxycline is characterized by small vertical gradients (or uniform distribution or local maxima).

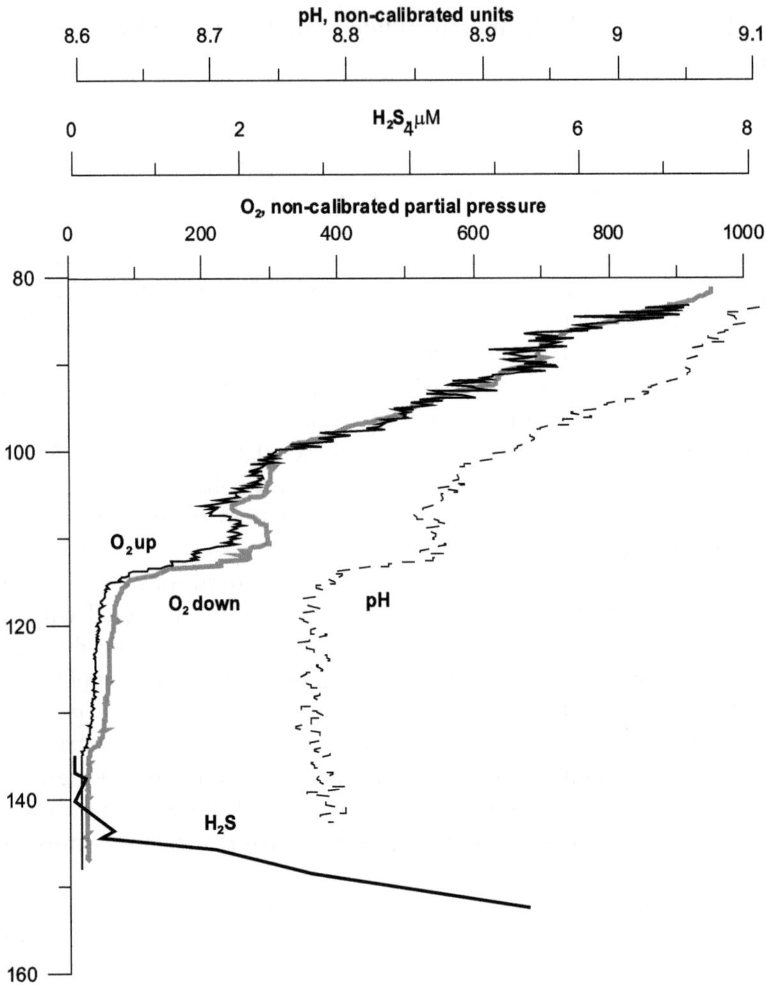

Fig. 3 Vertical distribution of dissolved oxygen (in noncalibrated partial pressure units) measured with open oxygen sensor "Kisa" downcast (O_2 down) and upcast (O_2 up); FSI sensor measured pH (in noncalibrated pH units) and photometrically measured H_2S (in μM) at St. 1770

- In the general case oxygen depletes with a steep gradient in the vicinity of the onset depths of reduced Mn(II) and ammonia about 10 m shallower than the appearance of hydrogen sulfide.

The measurements with this sensor allowed us to demonstrate that the mentioned features could be smoothed or destroyed probably with an intensified turbulence. In this case we observe correlated anomalies of distributions of both oxygen and temperature [36].

Based on the modern data, the vertical distribution of dissolved oxygen can be described as follows. Oxygen concentrations in the surface layer

($\sigma_\theta \sim 14.0\,\mathrm{kg\,m^{-3}}$) are at or above atmospheric solubility due to gas exchange and biological production (Fig. 2). They vary from 300–370 μM in February–April to 200–250 μM in July–August depending on the seasonality of temperature influenced air–water exchange and organic matter (OM) production and decomposition. A layer of decreasing oxygen concentration (oxycline) coincides with the main pycnocline (halocline). Oxygen decreases quasi-linearly with depth from 250–300 μM above the main pycnocline to 10–20 μM at the density level of $\sigma_\theta \sim 15.50\text{–}15.60\,\mathrm{kg\,m^{-3}}$. The vertical gradient of oxygen in this layer is 7–10 $\mu\mathrm{M\,m^{-1}}$. Below this depth the vertical gradient of oxygen decreases significantly to 0.5–1.5 $\mu\mathrm{M\,m^{-1}}$. This can probably be explained by the fact that here the oxygen reaches concentrations that might be too low for the aerobic respiration and too high for the anaerobic respiration of OM [37]. Around the densities of 15.90–16.00 $\mathrm{kg\,m^{-3}}$ dissolved oxygen decreases to below detection.

The absence of a C-layer is observed at the majority of stations. Only in the part of the Black Sea that is influenced by the Bosporus can both oxygen and hydrogen sulfide be found in one sample [38]. Oxygen concentrations of 4–5 μM, slightly greater than the detection limit, can be traced to the hydrogen sulfide boundary in the northeastern part of the Sea in the case of intensive mixing connected with the eddies. In some such cases, slight concentrations of hydrogen sulfide (0.3 μM) can also be found about 10–15 m higher than its onset. These situations are presumably not stable.

At this time, some investigators consider that oxygen practically disappears at the density of nitrate maximum [14] or quite the contrary traces, that oxygen penetrates down into the hydrogen sulfide zone for 10–20 m [23]. In our opinion oxygen disappears at the level of onset of deep ammonia and dissolved Mn(II) and is consumed in reactions with these species. This situation appears stable from a hydrophysical point of view.

As follows from the ratios of molar concentrations and vertical gradients (Fig. 2) oxygen plays the leading role in oxidation of reduced compounds and OM, but its role decreases in the middle part of the redox zone. The role of oxygen in the lower part of the redox zone cannot be estimated because the measurement techniques require improvement.

2.2
Reduced Sulfur Compounds

Because of increased interest in the position of the hydrogen sulfide boundary in recent years, the depth of this feature was studied carefully. It was found that the hydrogen sulfide with formal boundary of 0.3 μM appears in density layer $\sigma_\theta = 16.10\text{–}16.20\,\mathrm{kg\,m^{-3}}$ [17, 19, 26, 33, 39, 40]. According to [40] the concentration of hydrogen sulfide increases from its first appearance defined as 0.3 μM quasi-linearly to 300–400 m with a vertical gradient of 0.6–0.7 $\mu\mathrm{M\,m^{-1}}$. The more frequent sampling performed during the past few years has allowed us to show that in the vicinity of the onset of hydrogen sulfide, the values of the vertical gradient decrease and become equal to about 0.30–0.40 μM m^{-1} [41].

Elemental sulfur and thiosulfate are rarely measured due to difficulties of the analytical technique [42]. Such measurements completed in summer 1989, winter 1991 [43], and summer 2004 testify that these parameters are being formed as intermediate products or from hydrogen sulfide oxidation. Their concentrations in the upper part of the anoxic layer were about 1–7 μM. Directly at the density of hydrogen sulfide onset, elemental sulfur is the dominant form of these three species (about 50–80% of total reduced sulfur) [43]. A step of sharp decrease in concentrations of elemental sulfur and thiosulfate to about 0.5–1 μM is usually observed about 3–5 m higher than the hydrogen sulfide onset. Volkov et al. [43] noted that the distributions of reduced sulfur forms were more pronounced in summer than in winter.

Thus, hydrogen sulfide is the main reducing agent that comes from the anoxic zone. Its vertical gradients (0.5–1.0 μM m^{-1}) below $\sigma_\theta = 16.2$–

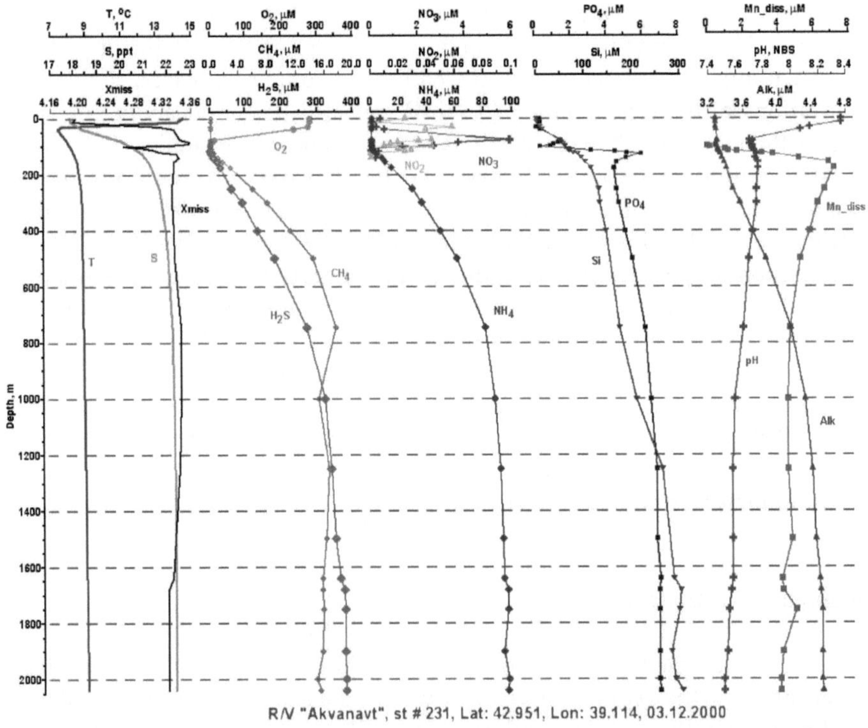

Fig. 4 Vertical distribution versus depth (m) of temperature (T), salinity (S), transmission (Xmiss), dissolved oxygen (O_2), hydrogen sulfide (H_2S), methane (CH_4), nitrate (NO_3), nitrite (NO_2), ammonia (NH_4), phosphate (PO_4), silicate (Si), pH (pH), total alkalinity (Alk), and dissolved manganese (Mn_diss) at a station in the Eastern Gyre (St. 231, December 03, 2000). Concentrations of chemical parameters are in μM

16.5 kg m^{-3} are significantly higher than of other reductants and in the layer $\sigma_\theta = 16.0$–16.1 kg m^{-3} are equal to those of ammonia and Mn(II) (0.5–1.0 μM m^{-1}) (Fig. 2).

The concentrations of hydrogen sulfide rapidly increase down to depths of 800–1000 m where they reach about 300–320 μM, then their growth decreases (Fig. 4). The maximum concentrations of hydrogen sulfide (380–390 μM) are observed in the bottom mixed layer [44].

2.3 Nitrogen Compounds

In the upper layer of the sea subjected to photosynthesis the concentration of nitrate varies from 4 to 6 μM in winter to less than 1 μM in summer. Concentrations of nitrite in this layer are very low (less than 0.1 μM). The whole basin mean content of ammonia in the upper mixed layer is close to 0.2 μM in summer, and to 0.4 μM in winter [23].

Below the photic layer the nitrate concentrations start to increase, from the depth where nitrite and ammonia maxima are usually observed. In the regions affected by the coastal influx concentrations of inorganic nitrogen, compounds are larger and they may suddenly increase during the rains [45].

In oxic conditions nitrate is produced by respiration and reaches a maximum (3–10 μM) at $\sigma_\theta = 15.30$–15.50 kg m^{-3}. Below this depth the concentrations of nitrate decrease sharply with vertical gradients 0.2–0.5 μM m^{-1}. After oxygen, nitrate is the second most abundant oxidizing agent in the oxic–anoxic interface. Nitrate disappears in the vicinity of $\sigma_\theta = 15.90$–16.00 kg m^{-3}.

A sharp maximum of nitrite with concentrations of 0.02–0.30 μM is usually observed at the same level. This maximum is characterized by very large temporal and spatial variability, probably because nitrite is very labile and 2–3-m sampling intervals are comparable to the thickness of the nitrite maximum which, according to our observations, is usually less than 5 m. The increase in ammonium starts at approximately the same depth ($\sigma_\theta = 15.90$–16.00 kg m^{-3}) with a vertical gradient of about 0.15–20 μM m^{-1}.

The decrease in the sum of nitrate, nitrite, and ammonium that is observed in the $\sigma_\theta = 15.90$–16.00 kg m^{-3} density layer is usually explained by denitrification [26, 46], consumption by chemosynthesis [23], or/and anammox, the reaction between nitrite and ammonia [47, 48]. From the comparison of the vertical gradients (Fig. 3) the role of nitrate becomes comparable with that of oxygen only in the lower part of the redox layer. Nitrate can be consumed for denitrification and reduction by thiosulfate, elemental sulfur, and sulfide [49, 50]. The role of nitrate as a potential oxidizer of reduced manganese and iron is actively being discussed now [51], but the presence of these reactions has not been proved or widely accepted. If they exist, these reactions will probably not play a dominant role compared with the oxygen reaction. Bacteria that mediate the anammox reaction have been found in the Black Sea [47, 52]. However, the total share of anammox in the total production of N_2 theoretically should not exceed 35% [53].

In the anoxic zone of the Sea ammonia increases in a similar way as hydrogen sulfide (Fig. 4) and reaches maximum concentrations of about 90–100 μM in the bottom mixed layer. The losses of combined nitrogen due denitrification and the absence of nitrification in deep waters has transformed the Black Sea into an "ammonium" basin, where about 98% of total inorganic nitrogen stock is composed of NH_4, while in the ocean 98–99% of it belongs to NO_3 [23].

2.4 Phosphate

Phosphate is usually considered as a parameter that limits photosynthesis in the Black Sea. According to Sorokin [23], the mean content of phosphate in the upper mixed layer down to the lower boundary of the euphotic zone is close to 0.10–0.20 μM in spring–summer. In autumn its mean content varied between 0.01 and 0.02 μM in cyclonic eddies and between 0.12 and 0.18 μM at their periphery over the slopes. In winter the phosphate content in the upper water layer usually rises due to the vertical mixing, thus attaining 0.15 to 0.40 μM.

Phosphate does not change its own oxidation state but its distributions in the Black Sea are clearly influenced by changes in the redox environment. The profile of phosphate has the most complicated structure of the profiles of the basic chemical properties. The profiles in the central Black Sea are characterized by two maxima and two minima [26, 54], and the positions of these extrema in the density field are very stable. The density of the upper maximum of phosphate is usually found at the depths of the maximum of nitrate (at about $\sigma_\theta = 15.50–15.60\,\mathrm{kg\,m^{-3}}$) or several meters below. There is a clearly defined phosphate minimum that occupies a thin layer at $\sigma_\theta = 15.90–15.95\,\mathrm{kg\,m^{-3}}$ about 10 m higher than the depth of onset of hydrogen sulfide. A deep maximum in phosphate is observed at $\sigma_\theta = 16.18–16.23\,\mathrm{kg\,m^{-3}}$, about 5–10 m below the hydrogen sulfide onset. In July 2002, the upper phosphate minimum layer had a thickness of 3 m and a concentration of 0 μM. It had a concentration of 0.3 μM in September, and was not observed during the winter cruise in January 2004. Chasovnikov [55] described that in December 2001, the upper minimum phosphate concentrations increased and that the magnitude of the lower maximum had a decreasing trend from the open sea to the coast.

On the basis of anomalies in the Si/P ratio Yakushev et al. [39] found that at the phosphate minimum the phosphate decreased by 2 μM and that at the phosphate maximum the phosphate increased by 2 μM. This structure of vertical distribution of phosphate was called the "phosphate dipole" [56]. The formation of such a structure is still uncertain. It was considered that it can be connected with (1) chemosynthesis [23] and/or (2) phosphate coprecipitation with the metal hydroxides [56]. Both these theories are probably not correct, because (1) the chemosynthesis has maximum values below the sulfide onset, where the phosphate maximum is observed, and (2) this decrease of phosphate probably cannot be explained by coprecipitation of phosphate on iron and manganese hydroxides, because the concentrations of these oxides are of the order of 0.01–0.1 μM. The Fe/P ratio for the iron hydroxides scavenging process is about 4 [57]; the effect of manganese scavenging is much less than that for iron [58]. If these ratios hold for the Black Sea, insufficient PO_4 could be scavenged. This point of view was criticized also by Bezborodov and Eremeev [26], because there is no evidence of a significant maximum of particulate phosphorus in this layer.

Production of oxidized Mn in the form of Mn(III) has been observed for Mn(II)-oxidizing bacteria and in incubations with Black Sea suboxic zone water [59]. Dissolved Mn(III) has been directly observed in the suboxic zone in

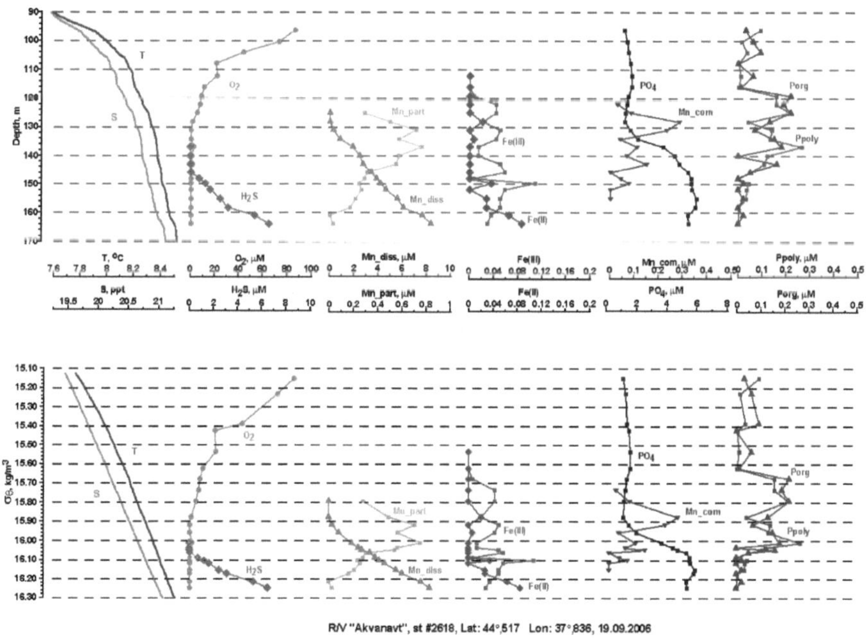

Fig. 5 Vertical distribution of temperature (T), salinity (S), dissolved oxygen (O_2), dissolved manganese (Mn_diss), particulate manganese (Mn_part), bivalent iron (Fe(II)), trivalent iron (Fe(III)), phosphate (PO_4), manganese complex (Mn_com), organic phosphorus (Porg), and polyphosphate (Ppoly) at a station near Gelendzhik (St. 2618, September, 2006). Concentrations of chemical parameters are in μM. Distributions are plotted versus depth (m) at the *top* and versus density (σ_θ, kg m^{-3}) at the *bottom*

the SW Black Sea [14], where Pakhomova [60] measured the increase of the manganese-containing complexes at the depth of the redox interface (Fig. 4). It is known that Mn(III) easily produces complexes with OM and pyrophosphate [61]. Mn(III)–pyrophosphate complexes are characterized by the ratio Mn/P = 0.25 for $Mn(HP_2O_7)_2^{3-}$ or Mn/P = 0.17 for $Mn(H_2P_2O_7)_3^{3-}$ [61].

The possible formation of Mn(III) complexes with pyrophosphate may explain this "phosphate dipole". The phosphorus minimum is located at the same depth, where Mn(II) depletes due to possible oxidation with oxygen, and its maximum is located just below the sulfide interface, where Mn(III) should be reduced by sulfide. In 2006 we observed the maximum of polyphosphate (measured according to Grasshoff [31]) in the same layer (Fig. 5). These depths coincide with the likely limits of the Mn(III) maximum. These processes should be studied.

We also point out that the phosphate maximum/minimum dipole was absent in the region near the Bosporus [56], and according to our data was not found in the winter period (December, January) in eddies near the NE Black Sea coast. The latter appears to be connected to the increased mixing connected with the coastal anticyclones [62].

Thus, there are multiple mechanisms for the origin of the PO_4 minima and maxima. The details of the processes that form the phosphate anomalies are still uncertain. These mechanisms must be connected not only with intensity of mixing, but also with the seasonal decrease of the flux of OM to the redox layer and a decrease in the flux of OM decay. In the deep layers of the Sea the phosphate concentrations increase (Fig. 4) and reach values of about 7 μM in the bottom layer.

2.5 Silicate

In the upper water layer silicate is depleted by phytoplankton down to 0.4–2 μM, e.g., down to a limiting level [23]. Below the euphotic zone it continuously increases up to 40–60 μM at 100 m depth.

Silicate has conservative characteristics in the suboxic zone. Its vertical distribution practically coincides with that of salinity and density because silicate is not involved in the processes connected with changes in redox conditions. Silicate is not consumed in the processes of chemosynthesis, and its distribution reflects the degradation of OM produced only in the euphotic zone. Silica concentrations are low in the surface and increase smoothly from 50 to 100 μM across the suboxic zone.

Its concentrations increase with depth and attain 180–220 μM at 1000 m and about 280–300 μM near the bottom (Fig. 4). This buildup of silicate stock results from a large inflow of terrestrial material from the mountain shores of the Black Sea [23].

2.6 Manganese and Iron

Cycles of manganese and iron in the oxic/anoxic environment are similar. Reduced forms of these compounds are dissolved. They diffuse upward where they oxidize and transfer to oxidized particulate forms that sink down and reduce in the hydrogen sulfide zone.

The oxidation–reduction potentials and reaction constants of oxidation of iron and manganese differ and these reactions can occur in different amounts of oxygen. That is why the level of appearance of particulate manganese is situated higher than that of particulate iron [63]. Bacteria have been shown to oxidize manganese [64], whereas iron oxidation is possible without bacteria but can be carried out with bacteria [50]. Reduced iron can be oxidized by particulate manganese, forming complex compounds [65].

The main features of the observed distribution are as follows (Figs. 2, 4, 5). The onset of reduced Mn(II) is situated about 10 m shallower than the first appearance of hydrogen sulfide ($\sigma_\theta = 15.8–15.9\,\mathrm{kg\,m^{-3}}$, Table 1). The depth of this onset varied in different regions of the Sea. The increase of dissolved Mn(II) concentration started deeper at the continental slope region (140–155 m depth) than in the central part of the Sea (60–70 m), which was connected with the density field distribution. The position of the oxygen deficiency layer in the field of depth is characterized by seasonal variability: in the coastal regions of the sea lowering of this layer happens in winter and shallowing in summer [66]. The vertical gradient of Mn(II) is maximal and varies from 0.35 to $0.5\,\mu\mathrm{M\,m^{-1}}$ above the hydrogen sulfide onset. It decreases in the anoxic zone while the concentration of Mn(II) reaches its maximum of 8–9 μM at $\sigma_\theta = 16.5\,\mathrm{kg\,m^{-3}}$ (~ 200 m). The position of this maximum appears to be controlled by $MnCO_3$ saturation [67]. Mn(II) then decreases slowly with depth to approximately 4.5 μM at 2000 m. The deep water Mn(II) concentrations appear controlled by MnS_2 (haurite) solubility, rather than MnS (alabandite) or $MnCO_3$ (rhodochrosite) solubility [63].

The concentrations of particulate manganese in the layer of its maximum in the central part of the Sea are usually about several tens or hundreds of nM [63, 66, 68]. They increase to 2–5 μM in the regions that are influenced by the Bosporus and to 1.5–2.0 μM in the connected with the Rim Current eddies [38, 60, 64]. According to our and other investigations in the Black Sea [60, 69] and in the Gotland Deep in the Baltic Sea [70], the concentrations of particulate manganese (and iron) increase in the winter–spring period and decrease in summer.

Recent studies have shown that bacterial oxidation of Mn(II) to Mn(IV) can occur through Mn(III) formation [61], which in seawater can be stabilized by strong organic and inorganic complexes. It was found that Mn(III) forms a layer of high concentration (0.5–4.5 μM, $\sigma_t = 15.8–16.2\,\mathrm{kg\,m^{-3}}$) below the maximum of particulate manganese, and its distribution is characterized by the presence of one or two maxima [14]. According to our data [60], organically complexed manganese is characterized by two maxima of concentration (up to 1–3 μM) located shallower or directly at the dissolved Mn(II) onset depth and at the depth of the maximum of dissolved Mn(II).

Distribution of dissolved Fe(II), as well as in the case of dissolved Mn(II), is characterized by the increasing of its content in the redox zone and by formation of an intermediate maximum within the dissolved Mn(II) maximum. Fe(II) is oxidized rapidly in the presence of oxygen to Fe(III) that exists as oxides and hydroxides with low solubility. The dissolved Fe(II) appears at $\sigma_\theta = 16.2\,\mathrm{kg\,m^{-3}}$. Its concentration increases toward the maximum (about 0.3 μM at $\sigma_\theta = 16.5$–$16.6\,\mathrm{kg\,m^{-3}}$), and then decreases to 0.05–0.07 μM at $\sigma_\theta = 16.8\,\mathrm{kg\,m^{-3}}$. The deep concentrations appear to be controlled by solubility with FeS (mackinawite) or Fe_3S_4 (greigite), even though FeS_2 (pyrite) is more insoluble and is present in the water column [71].

The maximum in the vertical distribution of Fe(III) ($\sigma_\theta = 15.8$–$16.3\,\mathrm{kg\,m^{-3}}$) is characterized by smaller values—usually less than tens of nM [68] reaching 0.3 μM as maximum [72]. Our data for the northeastern (January, July 2004) and southwestern (March–April 2003) Black Sea showed the following. The profile of Fe(III) was characterized by two maxima with values reaching about 100–150 nM at 150 m ($\sigma_\theta = 15.8$–$15.9\,\mathrm{kg\,m^{-3}}$) and 170–175 m ($\sigma_\theta = 16.0$–$16.25\,\mathrm{kg\,m^{-3}}$). Sometimes a third maximum is observed shallower at 120 m ($\sigma_\theta = 15.35$–$15.40\,\mathrm{kg\,m^{-3}}$). These maxima are correlated with layers of high contents of particulate iron and could be present as colloidal iron goes through the filters (0.45 μm) and is measured as dissolved iron.

The obtained concentrations of particulate iron are 0.02–0.3 μM, except for stations affected by river input where its contents could be rich at 2 μM [60, 70]. At most stations the maximum of particulate iron is located at the same depth as particulate manganese at $\sigma_t = 16.0$–$16.2\,\mathrm{kg\,m^{-3}}$. At the surface layer at the coastal region the content of particulate iron was on average 0.03 μM (up to 0.07 μM), that is, one order of magnitude higher than particulate manganese [60, 68], reflecting its different contents in the suspended matter coming into seawater with river input.

2.7 Methane

Methane forms microbiologically in strictly anaerobic conditions, but it is oxidized in both oxic and anoxic conditions [73]. The typical values for methane concentrations in the oxic layer are about 0.006–0.009 μM. The increase in methane concentration starts at $\sigma_\theta = 15.92\,\mathrm{kg\,m^{-3}}$. At the onset of hydrogen sulfide, methane concentrations reach 0.350–0.500 μM. The vertical gradient of methane over the density range $\sigma_\theta = 16.20$–$16.40\,\mathrm{kg\,m^{-3}}$ is equal to 0.036–0.054 $\mathrm{\mu M\,m^{-1}}$. In the layer of $\sigma_\theta = 15.90$–$15.93\,\mathrm{kg\,m^{-3}}$ the methane gradient decreases to 0.005–0.007 μM m^{-1}. Therefore, in contrast to dissolved Mn(II), the vertical gradient of methane decreased near its onset.

The role of methane in the formation of the hydrochemical structure of marine interfaces is very small because its molar concentrations and vertical gradients are much lower than those of other species (Figs. 2, 4). Nevertheless, it can be used as an indicator of changes of redox conditions.

In the anoxic water methane reaches about 16 μM. The vertical distribution of methane differs from that of hydrogen sulfide, ammonia, and phosphate: its profile curve bends at 500–600 m and keeps similar concentrations deeper toward the bottom [73].

2.8 Carbonate System

The Black Sea is characterized by an increased content of inorganic carbon. The carbonate alkalinity in the Black Sea surface waters (3.0 mM) is greater than in the oceanic waters (2.15 mM) by 1.4 times. In the deep waters it increases at more than 1 mmol attaining 4.25–4.30 mM [23]. The rise of alkalinity results from OM decomposition in the water column. An additional amount comes from dissolution of carbonates, contained in sedimenting suspended matter, and in particular in feces of zooplankton [23].

The composition of the carbonate system in the water column changes on the vertical profile depending on the decrease of pH with depth resulting from the CO_2 produced during sulfate reduction and during other processes of anoxic OM decomposition. In the upper layer pH values vary from 7.8 to 8.6 (in situ, NBS units), with maximal values in winter. The vertical distribution of pH in the suboxic layer is remarkable for its minimal values. The pH values decrease together with the oxygen content from 8.00–8.20 above the oxycline to 7.80–7.90 at $\sigma_\theta = 15.60$ kg m^{-3}. This decrease of pH values is connected with the respiration of OM to CO_2 and the oxidation of reduced forms of sulfur, nitrogen, carbon, manganese, and iron [26]. In the hydrogen sulfide zone the pH values increase slightly. At some stations (particularly during the RV "Akvanavt" survey in May 2002), we observed a small maximum of pH and minimum of alkalinity at $\sigma_\theta = 15.88$ kg m^{-3}.

We used these data for calculation of carbonate system parameters. The calculations were made with a standard set of equations [32]. For these calculations we corrected the alkalinity changes for phosphate, ammonia, and sulfide as is recommended for the Black Sea [74]. The results of these calculations revealed the formation of two minima of CO_2. A well-pronounced shallower minimum with relative decrease of concentrations by 0.015 mM and a total inorganic carbon (TIC) decrease by 0.040 mM was observed at the density layer $\sigma_\theta = 15.85–15.95$ kg m^{-3}, and a smoothed deeper minimum was observed below the sulfide boundary. In the anoxic water the pH slightly decreases down to 7.5–7.6, and alkalinity increases to 4.50–4.55 mM (Fig. 4).

2.9 Organic Matter

According to the present estimates the Black Sea is considered to be a mesotrophic basin [23]. In the upper layer the content of dissolved organic carbon is 250–420 μM, organic phosphorus varies from 0 to 1 μM, organic nitrogen is in the limits of 8–20 μM, and urea varies from 0 to 4 μM. In the coastal waters we observed a sudden increase of organic phosphorus, organic nitrogen, and urea concentrations after rains, to correspondingly 5–6, 10–25, and 6–8 μM [45]. It is difficult to reveal the seasonal variability of these characteristics, because of large variations in daily or weekly scales.

In the redox zone the role of OM in the balance of oxidizers and reductants is very important. In fact, Rozanov [75] argued that OM is the main reductant in the suboxic layer. Sorokin [23] explained that OM distributions in the redox zone were due to dense bacterial populations in detrital particles and marine snow.

Practically all the variety of bacteria from the point of view of their relation to substrate and energy can be found in the redox interfaces [49, 50]. The presence of autotrophic and heterotrophic organisms in the same sample makes it very difficult to estimate the production or consumption of inorganic and organic nutrient compounds. The activity of these bacteria results in intensive hydrolysis and absorption of organic molecules resulting in their disintegration and consumption, and transformation of dissolved OM into particulate forms. The detritus of the redox layer includes the remainders and shells of phytoplankton, remainders and pellets of zooplankton, and microaggregates of marine snow that are formed from the remainders of gelatinous organisms. Depending on the density of these particles they distribute themselves at the layers of their neutral buoyancy [23].

The vertical distribution of dissolved OM (as carbon) is characterized by a minimum (170–400 μM) in the vicinity of the H_2S onset layer that is superimposed on the overall increasing of values from the surface waters (250–420 μM) to the deep waters (500–800 μM) [23]. On the other hand, the vertical distribution of particulate OM is characterized by values of 160–250 μM directly above the hydrogen sulfide zone. These values are approximately twice as high as in the CIL [23]. Burlakova et al. [76] observed a maximum of particulate organic nitrogen (to about 0.3–0.4 μM) above the hydrogen sulfide boundary. Coban-Yildiz et al. [77] also observed the increase of particulate organic nitrogen and hypothesized that the largest part of this increase is of autochthonous origin with a small part from descending particles. In our studies the concentrations of dissolved organic nitrogen decreased from 14–16 μM in the upper layers to 10–12 μM above the H_2S boundary in summer, and decreased from 12–14 μM to 8–10 μM in the winter. The concentrations of urea also had a maximum above the H_2S boundary (about 4–6 μM compared with 2–3 μM above and below) in the warm period of July and August, 2003. Organic phosphorus (dissolved plus particulate) differs from total carbon and nitrogen and is characterized by increasing concentrations in the vicinity of the onset of hydrogen sulfide (Fig. 2). A layer with high concentrations of organic phosphorus and urea was observed within the depth range of the phosphate maximum and minimum. This layer was also characterized by low light transmission. The urea and organic phosphorus concentrations at this depth were correspondingly about 2–3 and 0.3–0.6 μM compared with concentrations less than 1 and 0.2–0.3 μM in the upper and lower layers. In July 2001, the maximum concentrations of urea and organic phosphorus reached 4–5 and 2 μM.

Probably these anomalies of the OM parameters might be connected with the process of chemosynthesis, and the layer of bacterial chemosynthesis should play an important role in the formation of the vertical distribution of nutrient species there. The results of measurements of the dark CO_2 fixation [78, 79] usually reveal the primary maximum of chemosynthesis (about 0.4–2.0 μM d^{-1}) in a 20–30-m layer below the hydrogen sulfide boundary. The less pronounced secondary maximum is observed about 5–10 m shallower than the hydrogen sulfide boundary and is likely to be connected with nitrification [78].

In the anoxic zone the content of organic phosphorus decreases down to negligible amounts less than 0.1 μM. Organic nitrogen is characterized by concentrations of 4–8 μM and the urea content decreases down to analytical zero [23].

3
Spatial Variability of the Redox Layer Chemical Structure in the Different Regions of the Sea

In the previous section we described the typical vertical hydrochemical structure of the Sea. The observations data [17, 39, 80] show that the density values of the typical chemical structure features vary slightly in the central, and the largest part of the peripheral regions of the Sea (Table 1). In Fig. 6a,b we present a transect from the NE coast to the Eastern Gyre. These waters are chemotropic: we observe significant differences in the depth of the hydrogen sulfide onset and the thickness of the suboxic zone between the coastal and central parts of the Sea (Fig. 6a), while the isolines of oxygen and hydrogen sulfide are parallel to the isopycns (Fig. 6b).

An exception of the chemotropic structure of the Sea is the southwestern region. This part is influenced by the Bosporus and is an area of intensive redox processes in a multilayered oxic/anoxic transition zone (Fig. 7). The chemical structure here is very unstable because of variations in the Bosporus water influx. The structure shown in Fig. 7 with the Bosporus water was observed in the RV “Knorr” cruise on April 4, 2003, at 21^h00^m. This water was

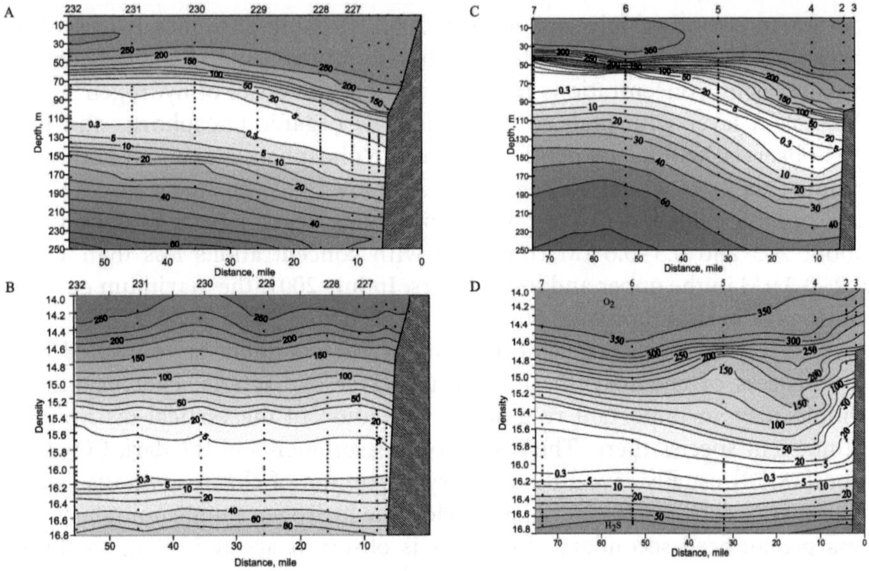

Fig. 6 Distribution of dissolved oxygen and hydrogen sulfide (μM) versus depth (**a, c**) and density (**b, d**) in transects from the Eastern Gyre (*left*) to the NE coast near Gelendzhik (*right*) (**a, b**) and from the SW coast (*left*) to the Eastern Gyre (*right*) (**c, d**)

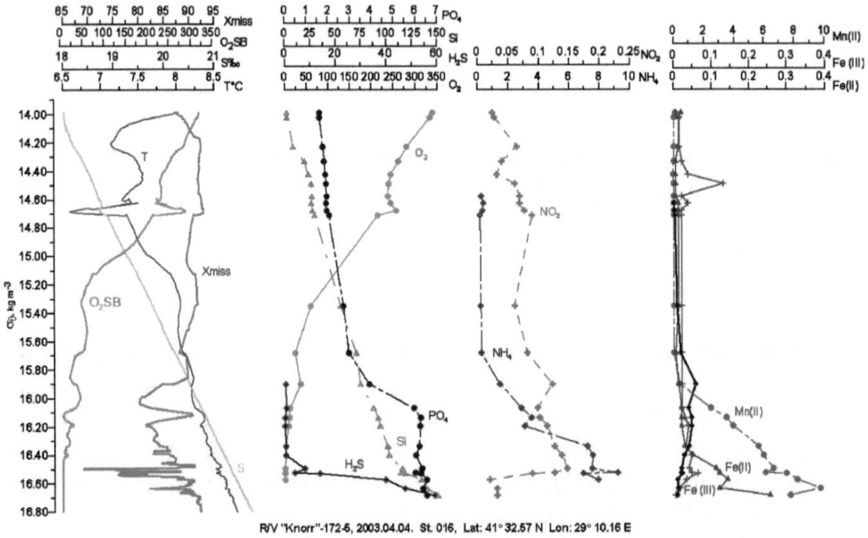

Fig. 7 Vertical distribution versus density (σ_θ, kg m^{-3}) of temperature (T), salinity (S), transmission (Xmiss), dissolved oxygen measured with YSI oxygen sensor (O_2SB), dissolved oxygen measured by Winkler titration (O_2), hydrogen sulfide (H_2S), phosphate (PO_4), silicate (Si), nitrite (NO_2), ammonia (NH_4), dissolved manganese (Mn_diss), bivalent iron (Fe(II)), and trivalent iron (Fe(III)) at a station near the Bosporus (Cast 16, RV "Knorr" 172–05 cruise, April 04, 2003). Concentrations of chemical parameters are in μM

initially found in this region during the daytime on April 3. In the morning of April 4 this water was absent in this region only to reappear in the evening of the same day.

Multiple 5–10-m-thick cold water intrusions were notable over a 150-m-thick (15.20–16.70 $kg\,m^{-3}$) layer. This interleaving led to intensification of interactions between oxic and anoxic waters, which is clearly seen in the distributions of all the parameters studied (Fig. 7). This interaction especially resulted in an increase in the concentrations of Fe(III) and particulate manganese. At this station the hydrogen sulfide boundary was displaced down to a deeper position at 16.40–16.50 $kg\,m^{-3}$ (180–200 m). Basturk et al. [38] observed similar displacement to the same density level in the region of the Sakarya, which is east of this location.

The distributions shown in Fig. 6b,d in a transect from the SSW Black Sea to the Western Gyre demonstrate that the coastal waters there are influenced by the Rim Current that brings water here from the Bosporus situated about 200 km west. The chemical structure keeps some consequences of the Bosporus influence. The hydrogen sulfide onset returned back to the typical isopycnal level of 16.10–16.15 $kg\,m^{-3}$, but there are still significant horizontal gradients of oxygen at density surfaces 15.6–15.9 $kg\,m^{-3}$ from the coast to the center of the Sea. In the SW region sulfide could be consumed by direct reaction with the injected O_2 or indirectly by reacting with oxidized Mn(III,IV) formed from Mn(II) by the O_2 injection [81]. As follows from the scheme of evolution of CIL waters (Fig. 1), the region with violations of the vertical hydrochemical structure connected with the Bosporus water occupies coastal waters in the SW and probably southern Black Sea. Only this region could be a zone subjected to the processes initiated by the Bosporus water. The hydrochemical structure of the other regions of the Black Sea is chemotropic, and can be mainly formed and supported by a combination of biogeochemical processes with the processes of vertical transfer (turbulence, advection, sedimentation). The interactions of the suboxic water with shelf and slope can produce in certain situations the anomalies of suboxic layer structure that can be observed in the coastal regions bordered by the Rim Current [36, 39].

4 Temporal Variability

4.1 Seasonal Variability of the Redox Layer

The seasonal variability at the depths of the redox zone was analyzed in [21]. We found that the main differences of these winter distributions from those in the summer can be summarized as follows:

- The vertical distribution of phosphate in the winter is characterized by the absence of the shallower minimum.
- The concentrations of nitrate at the nitrate maximum were lower in winter (2.5 μM) than at other times of the year (> 4.5 μM).
- The maxima of organic phosphorus and urea content typically observed in summer at the onset of hydrogen sulfide vicinity were absent in the winter. The concentrations of organic nitrogen were lower in winter than in summer. These observations may be related to the decrease in the number of bacteria and a reduction of the rates described by Sorokin [23].
- Changes of the vertical gradients of hydrogen sulfide, ammonium, dissolved Mn(II), and methane at their onset depths were smoother in the winter than in the summer.

Two factors seem to affect the seasonal variability of the hydrochemical structure of the oxic/anoxic interface: seasonality of OM production and seasonality of intensity of mixing. The winter decrease of OM flux can result in a decrease of the number of bacteria and slowing down of the reaction of biogeochemical transformation, because OM is a substrate for many heterotrophic bacteria mediated reactions. An enhanced mixing connected with anticyclonic eddies leads to smoothing of vertical gradients and extrema and changing of conditions of certain biogeochemical reactions, such as formation of the phosphate dipole.

4.2 Interannual Variability

As a part of the World Ocean the Black Sea is suppressed by the climate-induced fluctuations. These fluctuations are superimposed on the 1960–1980 intensive anthropogenic forcing, connected with eutrophication. The reaction of the Black Sea biogeochemical system to these factors is being actively studied [22, 79, 82].

The nutritional statuses of the upper layer of the Sea have changed significantly during the last few decades. The detailed analysis of the influence of these changes on the biological processes is presented in [79]. In the regions subjected to the influence of the Danube River the NO_3/PO_4 ratio was 11.7 in the 1970s and 22–23 in 1988–1992 [83]. Because of a decrease of the phosphate input this ratio increased in 1995 to as high as 100. After 1996 it decreased significantly (e.g., down to 20 in 2000) due to a decrease of the annual input of nitrate from the Danube from 770 000 tons in 1991 to 108 900 tons in 2000 [83]. The NW shelf region was characterized by phosphorus limitation of primary production in the 1980s to 1990s, but the tests completed in 2001 did not show clear P limitation on the shelf [79].

In the open waters the low NO_3/PO_4 ratio in the nutricline below the euphotic zone resulted in nitrogen limited phototrophic production. This can be connected with intensive loss of inorganic nitrogen in suboxic conditions. According to the modern studies [79] the NO_3/PO_4 ratio in the euphotic zone and the upper nutricline is very low (2–6.5), which can make favorable the process of nitrogen fixation.

The depletion of silicate in surface waters during the 1970s and 1980s has also been found to have an important impact on shifts of phytoplankton species composition from siliceous (mainly diatoms) to non-siliceous (coccolithophorids and flagellates). Recent changes in nutrient concentrations of the river inputs have resulted in changes in the coastal waters: concentrations of nitrate decreased considerably and concentrations of silicate increased [83].

We studied the changes of the position of the boundary of the anoxic zone on the basis of data of regular observations received in the northeastern part of the Black Sea near Gelendzhik (more than 1400 stations with the results of field observations from 1989 to the present). This part of the Sea is far from the influence of the Bosporus input and Danube River inflow. Therefore the vertical structure in this region is more stable and reflects "integrated", rather than local, changes of the Sea.

The results of these calculations for monthly averaged intervals are shown in Fig. 8. The values obtained differed slightly from the estimates that were reported by other authors who used visual or linear regression criteria for estimating of the onset of hydrogen sulfide [17, 19, 39, 40]. The Akima spline-based method we used [84] should be better because it is nonlinear and based on an objective approach for every station, which is necessary in analysis of the large data arrays.

The results calculated here show that the depth of disappearance of hydrogen sulfide was characterized by values of $\sigma_\theta = 16.15–16.25\,\mathrm{kg\,m^{-3}}$ in 1991–1998 (Fig. 8). In 1999–2000 the shoaling of this boundary appeared. The value of this shoaling was about $\sigma_\theta = 0.05–0.15\,\mathrm{kg\,m^{-3}}$ (corresponding to about 5–15 m at these depths). After 2000 the position of hydrogen sulfide stabilized. The same tendency can be marked in the other studied reductants: ammonia, total manganese, and methane (Fig. 8). The calculated vertical gradients of hydrogen sulfide, ammonium, total manganese, and methane were stable in both periods [41].

These changes may be related to the two warm winters that occurred in 1998 and 1999, which could affect the balance between input of freshwater from the rivers and saline water from the Bosporus and the winter formation of the oxygen-rich CIL. These years are remarkable for the increase of the Sea surface temperature (Fig. 8), increase of temperature in the core of the CIL [82, 85–87], and shoaling of the CIL in the density field [48]. All these events can be connected with the weather condition oscillations, as follows from North Atlantic oscillation (NAO) index behavior (Fig. 8).

The decrease of intensity of CIL formation should lead to an increase of temperature in its core and decrease of oxygen content there. To check it we

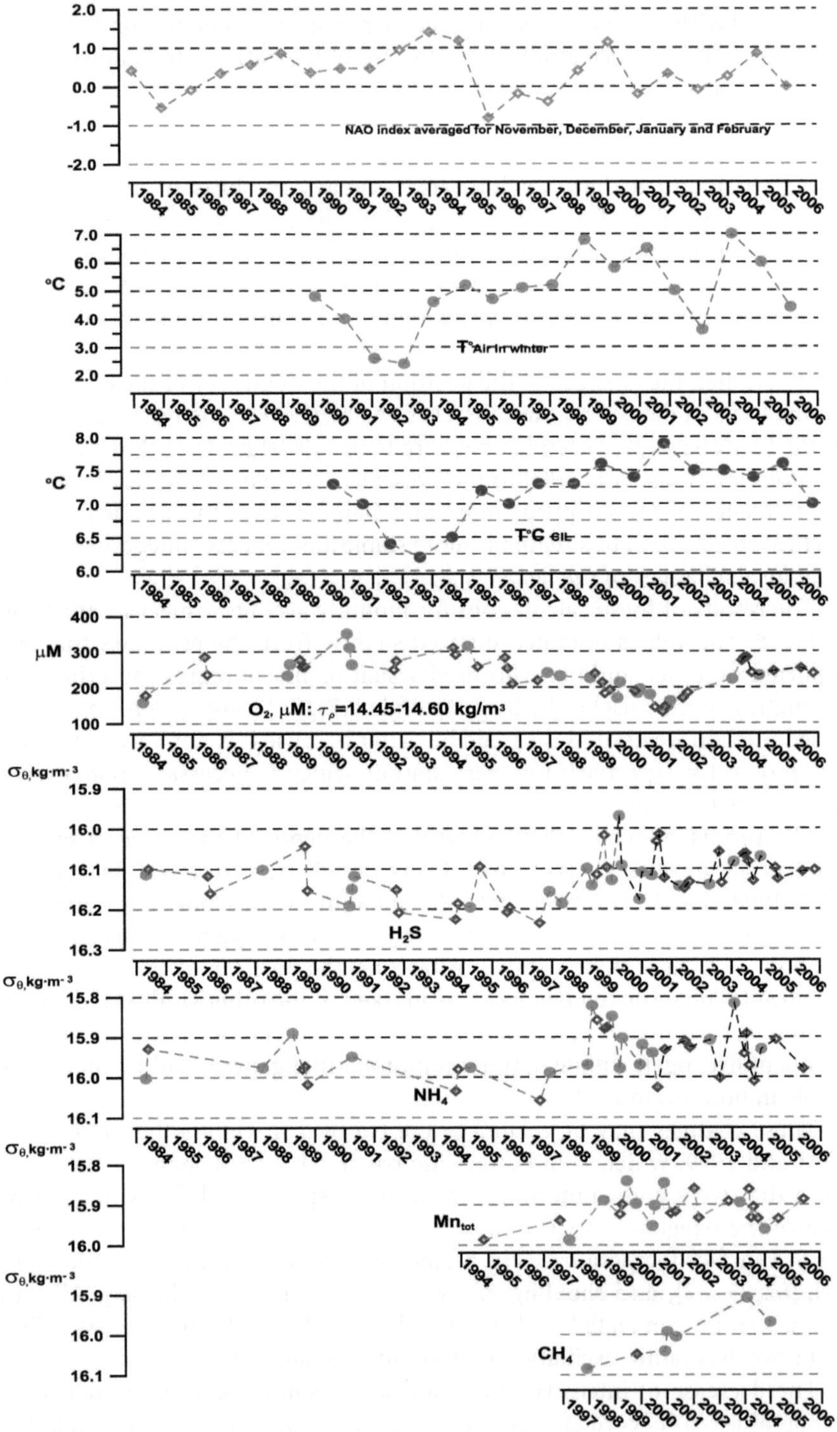
NAO index averaged for November, December, January and February
T°Air in winter
°C
T°C CIL
μM
O2, μM: σθ=14.45-14.60 kg/m3
σθ,kg·m-3
H2S
NH4
Mntot
CH4

◀ **Fig. 8** Interannual variability of the winter NAO index (averaged for November–February), winter air temperature in Gelendzhik, temperature in the CIL core in the northeastern Black Sea (data of V.G. Krivosheya), the averaged content of oxygen in the CIL (in the layer $\sigma_\theta = 14.45–14.60\,\text{kg}\,\text{m}^{-3}$), and onsets in the density field of hydrogen sulfide, total manganese, ammonia, and methane (from *top* to *bottom*)

calculated the average concentrations of dissolved oxygen in the CIL (for the layer $\sigma_\theta = 16.15–16.25\,\text{kg}\,\text{m}^{-3}$) (Fig. 8). These results reflect both changes of concentrations in the CIL and the vertical shifts of the CIL core in the density field [48]. In 1999–2000 when the shoaling of reductants occurred a decrease of oxygen content was marked. The minimal concentrations were found in 2001–2002. In 2003–2004 we observed the increase of oxygen content in this layer to values typical for the beginning of the 1990s.

The obtained results illustrate the mechanism of reaction of the natural system of the Black Sea on the global climate changes. As follows from the analyzed estimates, the changes of the Sea surface temperature lead to changes of winter CIL formation process intensity and to oxygen renovation there. The oxygen inventory in the CIL acts as a specific accumulator that supports the consumption of oxygen for OM decay and downward diffusive flux during all the year. The interannual variations of this oxygen renovation in the CIL lead to changes of suboxic layer hydrochemical structure [22] and, in particular, of the position of the anoxic boundary in the density field. Therefore, the distribution of the chemical parameters in the density field in the Black Sea might be a good indicator of global climate variations.

5 Conclusions

The Black Sea is the largest permanently stratified basin in the world. Its vertical hydrochemical structure is stable and is remarkable for the presence of an upper oxic layer and a lower anoxic layer.

The property of chemotropicity testifies to the balance of the redox layer system with respect to the vertical fluxes of the oxidants and reductants supplied. This should be the well-defined sequence of changes with depth of the favorability of the potential redox reactions [17, 75] that can be realized by the bacterial community. The development of bacteria in this case should affect the distributions of nutrients. By modern estimation [79] the chemosynthetic production is comparable with photosynthetic production, and that should in the same manner affect the consumption of inorganic nutrients and production of their organic forms. Besides this the possible abiotic chemical reactions and the sedimentation of particulate matter of different densities should also play their roles in this mechanism.

It is possible to assume the following hypothesis for the functioning of the different layers of the transition zone between oxic and anoxic layers, the redox layer.

Density layer $\sigma_\theta = 15.50\text{–}15.70\,\text{kg m}^{-3}$. In the upper part of the redox zone, concentrations of dissolved oxygen decrease to 15–20 μM, and its vertical gradient abruptly decreases and becomes equal to that of nitrate. In this layer, nitrate, instead of oxygen, becomes the main oxidizer. Nitrate is rapidly consumed and its concentrations decrease rapidly. The reason for the decrease in the vertical gradient of oxygen is because there is a decrease in the rate of reactions that consume oxygen, probably the mineralization of OM [37].

Density layer $\sigma_\theta = 15.85\text{–}15.95\,\text{kg m}^{-3}$. In the middle of the redox zone, oxidizers diffusing from the upper layer (oxygen and nitrate) decrease to zero. This occurs simultaneously with the disappearance of reductants (ammonia, Mn(II), methane) diffusing up from the anoxic zone. A minimum of phosphate is also found here. This layer may be very thin, probably only 3–5 m, and its position may vary over the density range specified.

The co-occurrence of the phosphate minimum and the depletion depths of ammonium, Mn(II), and methane with the upper boundary of the fine particle layer (FPL) at the same density level suggests the possible existence of some unifying controlling mechanism. This common mechanism may be the redox reactions completed with oxygen. The presence of even 1.5 μM of oxygen (50% smaller than the detection limit of the voltammetric and Winkler techniques [31, 33]), can explain the oxidation of dissolved Mn(II), ammonia, and methane at this depth. Nitrate (which can be measured with accuracy down to 0.05 μM) is evidently also actively consumed in this layer to erratically zero. Denitrification and anammox reactions result in a decrease of total fixed nitrogen and a corresponding increase of N_2. The formation of the phosphate minimum may result from its removal for formation of complexes with Mn(III), which can appear in the water in the case of disappearance of dissolved oxygen. The scavenging by Fe, Mn hydroxides [56] or consumption for chemosynthesis [23] probably do not play a significant role in the formation of the phosphate "dipole" structure.

The main result of the geochemical reactions in this layer is the formation of new oxidizers—dissolved oxidized Mn(III) and particulate oxidized Mn(IV) and Fe(III). In addition to diffusive transport they have a sinking rate that can accelerate the downward transport of these electron acceptors.

Density layer $\sigma_\theta = 16.10\text{–}16.15\,\text{kg m}^{-3}$. This layer constitutes the lower part of the redox zone. The onset of hydrogen sulfide occurs just below the depths of maximum particulate manganese and iron. The reduction of Mn(III) and Mn(IV) by sulfide is very intensive [63, 75] and model estimates [88] suggest these reactions can balance the hydrogen sulfide flux from below. A deeper phosphate maximum occurs about 5–10 m below the appearance of hydrogen sulfide. The vertical gradient of hydrogen sulfide increases at this depth (Fig. 2).

According to modern observations chemosynthesis has a maximum rate in the 15–20-m layer below the sulfide onset [78, 79] and its value is comparable with the rate of photosynthesis [79]. That should lead to the significant consumption of inorganic carbon, phosphate, and ammonia.

This described vertical structure is typical for the regions of the Black Sea distant from the Bosporus influence and not affected by intensive vertical mixing connected, for instance, with the eddies.

Our studies showed that the biogeochemical system of the redox layer is subjected to temporal variability on a seasonal scale (connected with the seasonality of OM production) and interannual changes. Surface ventilation of dissolved oxygen down to the depth of the CIL ($\sigma_\theta = 14.5\,\mathrm{kg\,m^{-3}}$) occurs in the winter from a combination of the NW shelf and the centers of the gyres. The intensity of ventilation is determined by climate forcing which may be determined by large-scale climate patterns like the NAO. This ventilation sets the upper boundary conditions for the downward transport of O_2. Therefore, the position of the hydrogen sulfide boundary in the density field is connected with the climate variability, related to the NAO index.

Another factor that might affect the interannual dynamic of hydrogen sulfide position is the eutrophication. The main sink for oxygen is respiration of sinking particulate organic carbon (POC). Dissolved organic carbon (DOC) may also be important but much less is known about its distributions. Variability in the flux of POC (export production) is influenced by nutrient concentrations and food web structure (which are not unrelated). The late 1970s was a period of increased nutrient levels in the Black Sea (eutrophication) and this appeared to result in smaller oxygen concentrations in the CIL [80].

It is necessary to stress that the direct result of the observed anoxic boundary oscillations for 5–10 m is the change of the volume of the oxic waters of about 5–10%, where the Black Sea oxic ecosystem is situated. Such oscillations are vitally significant and should be studied.

It is important to maintain time series of biogeochemical distributions in the Black Sea. Much is learned about the oceanography of a system when you can watch its response to a perturbation. Two important perturbations we want to continue to watch are climatic forcing and eutrophication.

Acknowledgements This research was supported by the Shirshov Institute of Oceanology of the Russian Academy of Sciences, Russian Foundation for Basic Research grants 05-05-65092, 06-05-96676-Yug, 07-05-01024, CRDF grant RUG1-2828-KS06.

References

1. Sorokin Yu (1983) The Black Sea. In: Ketchum BH (ed) Ecosystems of the world 26: estuaries and enclosed seas. Elsevier, Amsterdam, p 253
2. Gunnerson CG, Ozturgut E (1974) The Bosporus. In: Degens ET, Ross DA (eds) The Black Sea—geology, chemistry and biology. Am Assoc Petrol Geol Mem, Tusla, p 99

3. Latif MA, Özsoy E, Oguz T, Unluata U (1991) Deep Sea Res 38:711
4. Tolmazin D (1985) Prog Oceanogr 15:17
5. Ovchinnikov IM, Popov YI (1987) Oceanology 27:739
6. Belokopytov VN (2004) PhD thesis. Marine Hydrophysical Institute, Sevastopol (in Russian)
7. Gregg MC, Yakushev EV (2005) Geophys Res Lett 32:1
8. Tolmazin D (1985) Prog Oceanogr 15:277
9. Murray JW, Top Z, Ozsoy E (1991) Deep Sea Res 38:663
10. Buesseler KO, Livingston HD, Casso SA (1991) Deep Sea Res 38:725
11. Eremeev VN, Ivanov LI, Konovalov SK, Samodurov AG (2001) Morskoi Gidrophysicheskiy Zhurnal 1:64 (in Russian)
12. Ozsoy E, Unluata U, Top Z (1993) Oceanography 31:275
13. Stanev EV, Staneva J, Bullister JL, Murray JW (2004) Deep Sea Res I 51:2137
14. Trouwborst RE, Brian GC, Tebo BM, Glazer BT, Luther GW III (2006) Science 313(5795):1955
15. Andrusov NI (1890) Izvestiya Russkogo Geograficheskogo Obschestva (Proceedings of the Russian Geographical Society) 26:398 (in Russian)
16. Murray JW, Jannasch HW, Honjo S, Anderson RF, Reeburgh WS, Top Z, Friederich GE, Codispoti LA, Izdar E (1989) Nature 338:411
17. Murray JW, Codispoti LA, Friederich GE (1995) Oxidation–reduction environments: the suboxic zone in the Black Sea. In: Huang CP, O'Melia CR, Morgan JJ (eds) Aquatic chemistry: interfacial and interspecies processes. Adv Chem Ser. ACS, Washington, DC, p 157
18. Nalbandov YR, Vintovkin VR (1980) In: Vinogradov ME (ed) Ecosystems of the open part of the Black Sea. Nauka, Moscow, p 50
19. Vinogradov ME, Nalbandov YR (1990) Oceanology 30:769
20. Turgul S, Basturk O, Saydam C, Yilmaz A (1992) Nature 359:137
21. Yakushev EV, Chasovnikov VK, Podymov OI (2005) Oceanography 18:44
22. Konovalov SK, Murray JW (2001) J Mar Syst 31:217
23. Sorokin YI (2002) The Black Sea: ecology and oceanography. Backhuys, Leiden
24. Skopintsev BA (1975) Formirovanie sovremennogo gidrokhimicheskogo sostava Chyornogo morya (Formation of the modern chemical composition of the Black Sea). Hydrometeoizdat, Leningrad, p 336
25. Brewer PG, Murray JW (1973) Deep Sea Res 20:803
26. Bezborodov AA, Eremeev VN (1993) Chernoe more. Zona vzaimodeistviya aerobnikh I anaerobnikh vod (Black Sea. The oxic/anoxic interface). MHI NASU, Sevastopol (in Russian)
27. Codispoti LA, Friederich GE, Murray JW, Sakamoto CM (1991) Deep Sea Res 38:691
28. Ivanov LI, Konovalov SK, Belokopytov V, Ozsoy E (1998) Regional peculiarities of physical and chemical responses to changes in external conditions within the Black Sea pycnocline: cooling phase. In: Ivanov L, Oguz T (eds) NATO ASI Series. NATO TO Black Sea project ecosystem modeling as a management tool for the Black Sea, symposium on scientific results. Kluwer, Dordrecht, p 53
29. Lukashev YF, Yakushev EV (1999) PACON-99 Symposium. Russian Academy of Sciences, Moscow
30. Stunzhas PA (2002) Fine structure of vertical oxygen distribution in the Black Sea. In: Zatsepin AG, Flint MV (eds) Complex investigation of the northeastern Black Sea. Nauka, Moscow, p 133 (in Russian)
31. Grashoff K, Kremling K, Ehrhard M (1999) Methods of seawater analysis. Wiley, Weinheim, p 600

32. Bordovsky OK, Chernyakova AM (1992) Sovremenniye metody gydrokhimicheskikh issledovaniy okeana (Modern techniques of the hydrochemical studies of the Ocean). IO RAN, Moscow, p 200 (in Russian)
33. Glazer BT, Luther GW III, Konovalov SK, Friederich GE, Trouwborst RE, Romanov AS (2006) Deep Sea Res II 53:1756
34. Brewer PG, Spencer DW (1971) Limnol Oceanogr 16:107
35. Broenkow WW, Cline JD (1969) Limnol Oceanogr 14:450
36. Stunzhas PA, Yakushev EV (2006) Oceanology 46:629
37. Naqvi SWA (2006) Gayana 70:53
38. Basturk O, Volkov II, Gokmen S, Gungor H, Romanov AS, Yakushev EV (1998) Oceanology 38(3):429
39. Yakushev EV, Lukashev YF, Chasovnikov VK, Chzhu VP (2002) Modern notion of the vertical hydrochemical structure of the Black Sea redox zone. In: Zatsepin AG, Flint MV (eds) Complex investigation of the northeastern Black Sea. Nauka, Moscow, p 119 (in Russian)
40. Volkov II, Kontar' EA, Lukashev YF, Neretin LN, Nyffeler F, Rozanov AG (1997) Geochem Int 6:618
41. Yakushev EV, Chasovnikov VK, Debolskaya EI, Egorov AV, Makkaveev PN, Pakhomova SV, Podymov OI, Yakubenko VG (2006) Deep Sea Res II 53:1764
42. Volkov II, Zhabina NN (1990) Oceanology 30:778
43. Volkov II, Rozanov AG, Demidova TP (1992) Reduced inorganic sulphur species and dissolved manganese in the water of the Black Sea. In: Vinogradov ME (ed) Winter state of the ecosystem of the open part of the Black Sea, Shirshov Institute of Oceanology RAS, Moscow, p 38 (in Russian)
44. Volkov II, Skirta AY, Makkaveev PN, Demidova TP, Rozanov AG, Yakushev EV (2002) On hydrophysical and hydrochemical uniformity of the deep waters of the Black Sea. In: Zatsepin AG, Flint MV (eds) Complex investigation of the northeastern Black Sea. Nauka, Moscow, p 169 (in Russian)
45. Yakushev EV, Arkhipkin VS, Antipova EA, Kovaleva IN, Chasovnikov VK, Podymov OI (2007) Chem Ecol 23(1):29
46. Rozanov AG, Demidova TP, Egorov AV, Lukashev YF, Stepanov NV, Chasovnikov VK, Yakushev EV (2000) Oceanology 40:30
47. Kuypers MM, Sliekers AO, Lavik G, Schmid M, Jorgensen BB, Kuenen JG, Damste JS, Strous M, Jetten SM (2003) Nature 422:608
48. Murray JW, Fuchsman C, Kirpatrick J, Paul B, Konovalov SK, Callahan A (2003) Oceanography 18(2):36
49. Nealson KN, Stahl DA (1997) Microorganisms and biogeochemical cycles: what can we learn from layered microbial communities? In: Banfield JF, Nealson KN (eds) Reviews in mineralogy, 35. Geomicrobiology: interactions between microbes and minerals. Mineralogical Society of America, Washington DC, p 5
50. Canfield DE, Thamdrup B, Kristensen E (2005) Aquatic geomicrobiology. In: Southward AJ, Tyler PA, Young CM, Fuiman LA (eds) Advances in marine biology, 48. Elsevier, Amsterdam, p 640
51. Luther GW III, Sundby B, Lewis BL, Brendel PJ, Silverberg N (1997) Geochim Cosmochim Acta 61:4043
52. Kirkpatrick J, Oakley B, Fuchsman C, Srinivasan S, Staley JT, Murray JW (2006) Appl Environ Microbiol 72(4):3079
53. Dalsgaard T, Thamdrup B, Canfield DE (2005) Res Microbiol 156:457
54. Fonselius SH (1974) Phosphorus in the Black Sea. In: Degens EJ, Koss DA (eds) The Black Sea—Geology, chemistry and biology. Am Assoc Petrol Geol, Tusla, p 144

55. Chasovnikov VK (2002) PhD thesis. Shirshov Institute of Oceanology, Moscow (in Russian)
56. Shaffer G (1986) Nature 321:515
57. Savenko AV (1995) Geochem Int 9:1383
58. Yao W, Millero FJ (1996) Environ Sci Technol 30:536
59. Kostka JE, Luther GW III, Nealson KH (1995) Geochim Cosmochim Acta 59:885
60. Pakhomova SV (2005) PhD thesis, Shirshov Institute of Oceanology, Moscow (in Russian)
61. Webb SM, Dick GJ, Bargar JR, Tebo BM (2005) Evidence for the presence of Mn(III) intermediates in the bacterial oxidation of Mn(II). In: Fridovich I (ed) Proc Natl Acad Sci USA 102(15):5558
62. Debolskaya EI (2002) Analysis of the Black Sea redox zone turbulent structure in the Black Sea based on the RV Akvanavt 18th cruise data. In: Zatsepin AG, Flint MV (eds) Complex investigation of the northeastern Black Sea. Nauka, Moscow, p 140 (in Russian)
63. Lewis BL, Landing WM (1991) Deep Sea Res 38:773
64. Tebo BM (1991) Deep Sea Res 38:883
65. Dubinin AV (2005) Geokhimiya redkozemelnikh elementov v okeane (Geochemistry of the rare earth elements in the Ocean). Nauka, Moscow
66. Rozanov AG, Volkov II (2002) Manganese in the Black Sea. In: Zatsepin AG, Flint MV (eds) Complex investigation of the northeastern Black Sea. Nauka, Moscow, p 190 (in Russian)
67. Spencer DW, Brewer PG (1971) J Geophys Res 76:5877
68. Erdogan S, Yemenicioglu S, Tugrul S (2003) Distribution of dissolved and particulate forms of iron and manganese in the Black Sea. In: Yilmaz A (ed) Oceanography of the eastern Mediterranean and Black Sea. Tubitak, Ankara, p 447
69. Muramoto J, Honjo S, Fry B, Hay BJ, Howarth RW, Cisne JL (1991) Deep Sea Res 38(2):S1151
70. Pohl C, Loffler A, Hennings U (2004) Mar Chem 84:143
71. Pilskaln CH (1991) Biogenic aggregate sedimentation in the Black Sea basin. In: Izdar E, Murray JW (eds) Black Sea oceanography. Kluwer, Dordrecht, p 293
72. Yemenicioglu S, Erdogan S, Tugrul S (2006) Deep Sea Res II 53:1842
73. Egorov AV (2002) On distribution of methane in the Black Sea water column. In: Zatsepin AG, Flint MV (eds) Complex investigation of the northeastern Black Sea. Nauka, Moscow, p 144 (in Russian)
74. Makkaveev PN (2002) Calculations of the component of the total titrated alkalinity in the Black Sea waters. In: Zatsepin AG, Flint MV (eds) Complex investigation of the northeastern Black Sea. Nauka, Moscow, p 447 (in Russian)
75. Rosanov AG (1995) Oceanology 35:544
76. Burlakova ZP, Eremeeva LI, Konovalov SK (1999) Phys Oceanogr 10:419
77. Coban-Yildiz Y, Chiavari G, Fabbri D, Gaines AF, Galetti G, Turgul S (2000) Mar Chem 69:55
78. Pimenov NG, Neretin LN (2006) Composition and activities of microbial communities involved in carbon, sulfur, nitrogen and manganese cycling in the oxic/anoxic interface of the Black Sea. In: Neretin LN (ed) Past and present water column anoxia. NATO Sciences Series. Springer, Dordrecht, p 501
79. Yilmaz A, Coban-Yildiz Y, Morkoc E, Bologa A (2006) Deep Sea Res II 53:1988
80. Konovalov SK, Luther GW, Friederich GE, Nuzzio DB, Tebo BM, Murray JW, Oguz TB, Glazer RE, Trouwborst B, Clement KJ, Romanov A (2003) Limnol Oceanogr 48:2369

81. Murray JW, Yakushev EV (2006) The suboxic transition zone in the Black Sea. In: Neretin LN (ed) Past and present water column anoxia. NATO Sciences Series. Springer, Dordrecht, p 105
82. Oguz T, Dippner JW, Kaymaz Z (2006) J Mar Syst 60:235
83. Cociasu A, Popa L (2002) Significant changes in Danube nutrient loads and their impact on the Romanian Black Sea Shelf. In: Yilmaz A, Salihoglu I, Multu E (eds) Oceanography of the eastern Mediterranean and Black Sea. METU-IMS, Ankara, p 402
84. Podymov OI (2005) PhD thesis, Shirshov Institute of Oceanology, Moscow (in Russian)
85. Ginzburg AI, Kostianoy AG, Sheremet NA (2004) J Mar Syst 52:33
86. Krivosheya VG, Ovchinnikov IM, Skirta AY (2002) Intraannual variability of the cold intermediate layer of the Black Sea. In: Zatsepin AG, Flint MV (eds) Complex investigation of the northeastern Black Sea. Nauka, Moscow, p 27 (in Russian)
87. Ginzburg AI, Kostianoy AG, Sheremet NA (2007) Sea Surface Temperature Variability. In: Kostianoy AG, Kosarev An (eds) The Black Sea Environment. Springer-Verlag, Berlin Heidelberg New York Tokyo (this volume)
88. Yakushev EV, Pollehne F, Jost G, Kuznetsov I, Schneider B, Umlauf L (2007) Mar Chem doi:10.1016/j.marchem.2007.06.003

Hdb Env Chem Vol. 5, Part Q (2008): 309–331
DOI 10.1007/698_5_083

Published online: 29 June 2007

Hydrogen Sulfide in the Black Sea

Igor I. Volkov[1] · Lev N. Neretin[2] (✉)

[1]Laboratory of Geochemistry,
P.P. Shirshov Institute of Oceanology of Russian Academy of Sciences,
36 Nakhimovsky prosp., 117851 Moscow, Russia

[2]Moscow Office, United Nations Environment Programme (UNEP),
28 Ostozhenka, 119034 Moscow, Russia
neretin.unep@undp.ru

Abstract The Black Sea hydrogen sulfide inventory is about 4600×10^{12} g, which makes this sea the largest anoxic basin on earth. Anoxic conditions in the basin have been established 7500 years ago. This review presents the contemporary inventory of dissolved sulfide and sulfur intermediates and discusses mechanisms of physical mixing in the anoxic interior. Special emphasis is given to concentrations of dissolved sulfide and other chemical species in the bottom convective layer located at depths below 1700–1750 m. The width and concentrations of dissolved sulfide in the bottom layer are directly proportional to the heat flow from the bottom. The mechanism of double diffusion driven by geothermal heat flux is the main mixing process in bottom waters. Hydrogen sulfide production in the water column by sulfate-reducing bacteria is the main source of dissolved sulfide, and sulfate reduction is the dominant process of organic matter mineralization in the Black Sea anoxic zone. Sulfur budget calculations suggest that Bosporus flux cannot be considered to be the major sulfur sink and factor for deep basin ventilation. Mesoscale physical dynamics along the periphery of the basin as well as pycnocline erosion during exceptionally severe winters are probably major ventilation mechanisms for the anoxic zone. The sulfur isotopic studies of the water column also support the importance of ventilation processes below the oxic/anoxic interface. Physical mixing processes and global climate change impact on the thermohaline structure of the Black Sea water column control the magnitude and direction of the processes within the sulfur cycle.

Keywords Black Sea · Bottom convective layer · Budget · Hydrogen sulfide · Sulfur isotopes

Abbreviations

BCL	Bottom convective layer
BP	Before present
CIL	Cold intermediate layer
RNL	Redox-nepheloid layer
SMMW	Shelf modified Mediterranean water
SR	Sulfate reduction
SRR	Sulfate reduction rate

1 History of the Black Sea Anoxic Conditions

Over last three million years the Black Sea experienced at least eight marine flooding events. The event during the last Pleistocene/Holocene transition was of the highest magnitude [1]. The basin was a freshwater lake with minimum water levels estimated between 20 and 110 m below the present sea level during the Neoeuxinian period, which lasted from 17 000 to 9000–11 000 years BP (before present) [2, 3] (Fig. 1). The inflow of glacial meltwaters from rivers and the Caspian Sea contributed to continuous seawater rise over the entire Neoeuxinian. There is a growing consensus among paleontologists, geomorphologists, and geologists that a unidirectional outflow from the Black Sea to the Sea of Marmara and the Mediterranean Sea existed during most of this time [5–9].

Due to melting of glaciers on the northern Eurasian continent and the postglacial rise of the global sea level, two-layered flow across the Bosporus was established. There is no agreement in the literature on the precise timing and the intensity of this event. Its dating varies from 10 000 years BP [10] to about 7200 years BP [11]. Ryan and colleagues [11] suggested that the influx of Bosporus waters into the Black Sea occurred on a catastrophic scale of some 100 m in just a few years. The authors hypothesized later in a popular book that the event was associated with Noah's flood, described in the Bible [12]. In contrast, the sedimentary record for the postglacial sea-level rise along the southwestern Black Sea shelf does not support the catastrophic refilling of the Black Sea [6]. Gorur and co-authors [13] also argued for a gradual rise in the level of the Black Sea from about 8000 years BP until it attained a surface level of – 18 m around 7200 years BP, when the most recent Mediterranean influx happened. Other than the Bosporus, routes such as the

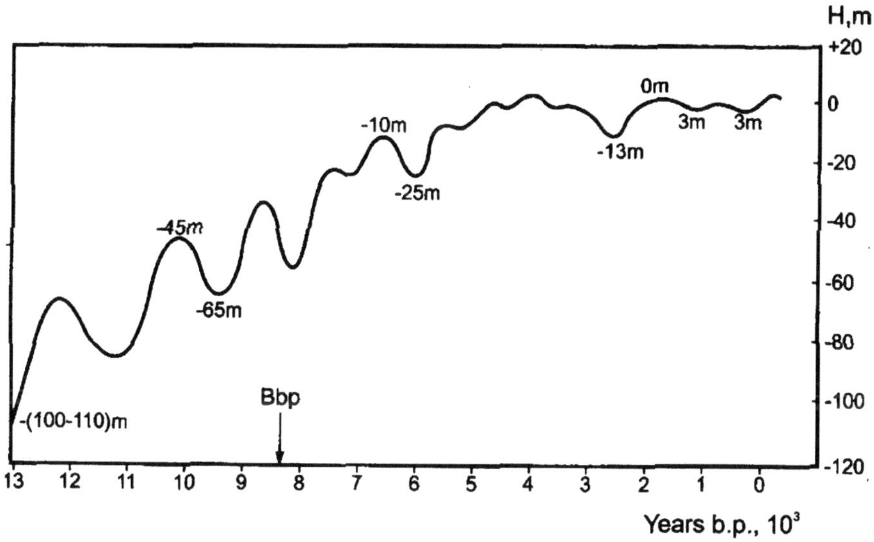

Fig. 1 Changes of the Black Sea level during Late Pleistocene–Holocene. The *arrow* indicates the timing of the Bosporus breakthrough (modified from [4])

Gulf of Izmit-Sapanca and Lake Sakarya Valley may have connected the Black Sea and the Sea of Marmara at that time [8].

The evolution of the Black Sea anoxic zone is closely connected with the evolution of its stratification pattern, presently characterized by the existence of a strong pycnocline separating the upper freshwater-influenced surface layer, with a salinity of 17.5–18.5‰, and the deep water mass below ca. 150–200 m, with a salinity of 22.3‰ at the bottom. Models for evolving Black Sea salinity after the opening of Bosporus agree that salinity in bottom waters reached 90% of present-day values about 3000 years after the Bosporus opening [14–17].

Due to the stable stratification, anoxia developed below the pycnocline, which corresponded to the deposition of an organic-rich sapropel after 7800 years [18], or 7500 years BP [19], through the entire Black Sea area. Since that time, bottom waters have remained anoxic. Development of anoxic conditions over time after the Bosporus inflow was modeled by Dueser [20] and recently by Leonov and Shaporenko [21]. Dueser suggested that present anoxic conditions were achieved within 2000–4000 years after Mediterranean waters reached the Black Sea. More sophisticated modeling approach combining water balance and hydrogen sulfide oxidation kinetics developed by Leonov and Shaporenko demonstrated that complete dissolved oxygen removal in the bottom layer was achieved about 3500 years after the inflow. After this time, model results showed an “abrupt” increase of the upper anoxic boundary up to about 1000 m over a 5-year time period. It took more

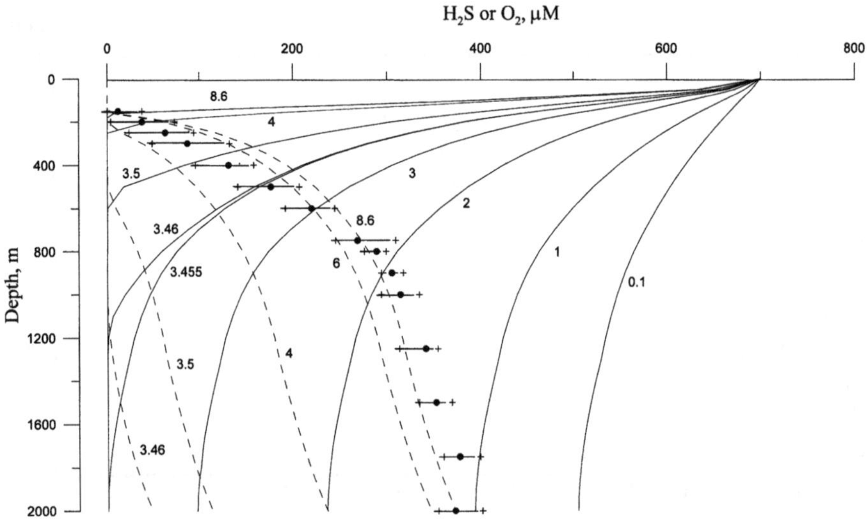

Fig. 2 Development of anoxic conditions in the Black Sea water column over time, based on modeling results. *Solid* and *dashed lines* represent vertical profiles of dissolved oxygen and hydrogen sulfide, respectively. *Numbers* are ×1000 years since the first appearance of the Mediterranean waters in the Black Sea. *Filled circles* and *horizontal bars* represent the average and the range of the observed $\sum H_2S$ concentrations in the present Black Sea, respectively

than 5000 years until present redox stratification became established in the water column (Fig. 2).

Hydrogen sulfide, $\sum H_2S$ ($\sum H_2S = [H_2S] + [HS^-] + [S^{2-}]$, where $[HS^-]$ represents ca. 80% at pH 7.5–7.65 in the Black Sea anoxic interior), is the key chemical compound that defines the direction and origin of many biogeochemical cycles in the anoxic zone of the Black Sea. The main goal of this review is to present the contemporary inventory of hydrogen sulfide and sulfur intermediate species, the results of recent physiochemical studies of the bottom convective layer, and to discuss the sulfur isotopic composition of dissolved sulfide and sulfate. This review concludes by presenting the sulfur budget of the Black Sea.

2 H_2S Inventory

The total sulfide inventory of the contemporary Black Sea is about 4.6×10^3 Tg ($T=10^{12}$), the main part residing between 500 and 2000 m [22]. The average dissolved sulfide concentrations at different depths are given in Table 1. The averaged data for the period before 1996 represent basin-wide averages, whereas more recent data were obtained only in the north-eastern part of the sea (station locations are given in [25]). The lower sulfide concentrations given by Skopintsev [23] are explained by the underestimation of sulfide concentrations in water samples collected with the metal bottles routinely used before the 1980s [26].

The H_2S vertical distribution is quasilinear above 500–600 m. Dissolved sulfide concentration increases gradually with depth and has an average vertical gradient of about 0.5 mmol m^{-4} above 500 m, decreasing with depth (Table 1). The vertical sulfide gradient at the boundary between the entire anoxic water mass and the bottom convective layer (ca. 1700–1750 m) increases sharply and is only two times less than the vertical gradient in the upper 500 m.

Table 1 Average H_2S concentrations in the Black Sea water column

Depth (m)	H_2S concentration 1950–1960s [23] (μM)	1984–1992 [24] (μM)	1989–1995 [22] (μM)	1997–2002 Shirshov Inst. data (μM)	1999–2002 Average conc. (μM)	Vertical gradient[b] (mmol m^{-4})
150	5.6 (135)	10	13 ± 12(91)[a] 0–38(92%)	11.5 ± 10.9(34) 0–39(94%)	12.5 (125)	0.6
175				27.4 ± 11.2(27) 4.1–42.2(41%)	27.4 (27)	0.5
200	24 (156)		40 ± 13(82) 3.8–73(38%)	39.8 ± 9.0(27) 18.9–59(24%)	40 (109)	0.47
250			63 ± 15(60) 24–93(24%)	64.0 ± 9.1(28) 43.0–81.2(14%)	63.3 (88)	0.47
300	69 ± 18(158) (26%)	80	87 ± 16(62) 49–133(18%)	86.4 ± 8.3(35) 68–101(10%)	86.8 (97)	0.43
400			130 ± 14(58) 96–159(11%)	129 ± 12(35) 85–144(9%)	130 (93)	0.46
500	148 ± 29(156) (20%)	166	176 ± 13(58) 142–205(7%)	176 ± 8(33) 160–196(4%)	176 (91)	0.41
600			216 ± 12(14) 192–232(6%)	217 ± 8(19) 206–240(4%)	217 (33)	0.41
700				258 ± 8(7) 249–269(3%)	258 (7)	0.33
800			291 ± 9(5) 276–299(3%)		291 (5)	0.18

Table 1 (continued)

Depth (m)	H_2S concentration 1950–1960s [23] (μM)	1984–1992 [24] (μM)	1989–1995 [22] (μM)	1997–2002 Shirshov Inst. data (μM)	1999–2002 Average conc. (μM)	Vertical gradient [b] ($mmol\,m^{-4}$)
900			309±9(5) / 294–317(3%)		309 (5)	0.05
1000	249±28(155) / (11%)	298	314±13(18) / 294–334(4%)	314±8(31) / 297–331(2.4%)	314 (49)	0.10
1250			344±16(6) / 313–354(5%)	339±7(32) / 326–351(1.9%)	340 (38)	0.05
1500	281±26(115) / (9%)	336	351±11(8) / 334–369(3%)	354±5(27) / 344–364(1.5%)	353 (35)	0.08
1700				369±8(19) / 358–390(2.1%)	369 (19)	0.14
≥ 1750	282±42(65) / (15%)	360	377±15(15) / 354–401(5%)	376±4(72) / 367–400(1%)	376 (87)	

[a] Numerator: average ± standard deviation (number of measurements); denominator: range (coefficient of variability)
[b] Gradients for the depth intervals given in column "Depth"

3 H_2S Vertical Distribution and Mixing Processes in the Anoxic Zone

The sulfide vertical distribution correlates with vertical distributions of temperature, salinity, and density in the Black Sea. As a consequence, the H_2S vertical distribution vs. salinity (Fig. 3a) and temperature (Fig. 3b) is consistent with the θ–S curve (Fig. 3b). It is evidence that the thermohaline structure of the water column controls the vertical distribution of hydrogen sulfide in the basin [27]. Physical mixing processes "dominate" over the in situ sulfide production. Identifiable on the θ–H_2S and S–H_2S diagrams, the boundaries of three water masses in the anoxic water column correspond strictly to the boundaries on the θ–S diagram (Fig. 3b). The temperature–salinity relationship in the Black Sea is a result of large-scale external factors such as water and heat balance of the basin.

In the upper part of the sulfide zone, the correlation between H_2S and σ_θ is smaller due to a larger hydrophysical inhomogeneity of the upper 300 m. The location of the upper sulfide boundary and distribution of other chemical parameters are also density-dependent [28, 29].

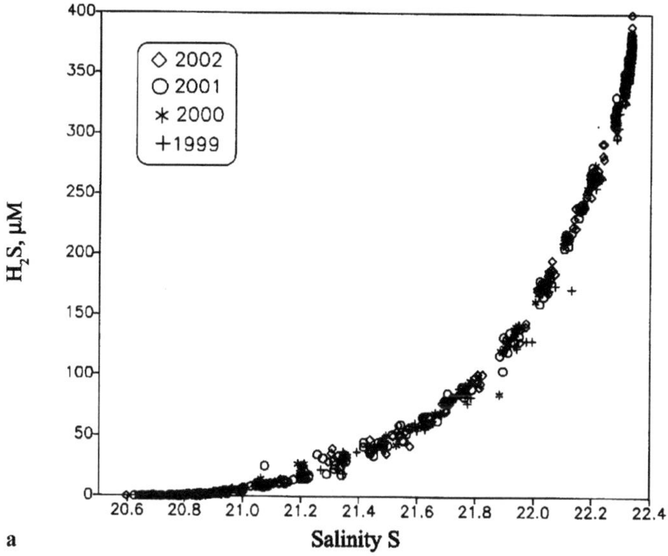

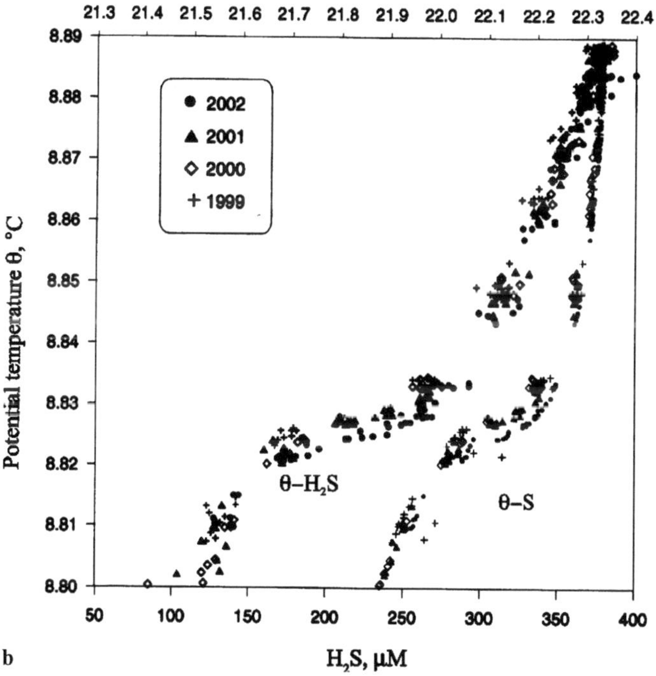

Fig. 3 **a** S–H_2S and **b** θ-S and θ-H_2S diagrams for the Black Sea anoxic zone (station locations are given in [25])

In the centers of cyclonic gyres, the location of the H_2S upper boundary decreases to 90–110 m, whereas at the periphery of the basin and in the centers of anticyclonic gyres it can deepen up to 160–240 m. The spatial differences of the H_2S topography recognizable at the oxic/anoxic interface can be traced to depths below to 1000 m [22]. Spacial and temporal variability of the upper anoxic boundary and global climate change impact on its location are considered elsewhere in this volume (Yakushev et al., in this volume).

The interface between the oxic and anoxic waters, often defined as a suboxic zone [29], is characterized by high bacterial numbers and enhanced microbial production primarily through chemosynthesis [30]. Some of the highest rates of the redox processes within carbon [32], sulfur [33], nitrogen [34], and manganese [35, 36] cycles were also observed in this zone [31]. Recently, the microbial communities involved in anammox (anaerobic ammonium oxidation) [34], denitrification [37], aerobic and anaerobic methane oxidation [38, 39], and anoxygenic photosynthesis [40] were intensively investigated. The suboxic zone is often characterized by decreased transparency and can be easily distinguished during CTD-turbidity profiling in the sea. The authors studying the geochemistry of this zone in the north-eastern Black Sea suggested calling it the "redox nepheloid layer" (RNL) [41]. Bacterial biomass and manganese oxyhydroxides were hypothesized to be the major components responsible for the origin of turbidity in the RNL. Schematic density-dependent distributions of the light attenuation coefficient and some chemical species in this layer are given in Fig. 4.

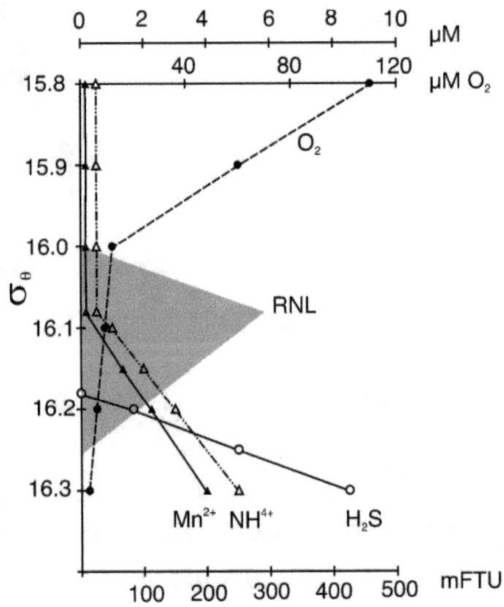

Fig. 4 Schematic density-dependent distributions of the light attenuation coefficient (encompassing *shaded area*), dissolved oxygen O_2, hydrogen sulfide H_2S, ammonium NH_4^+, and dissolved manganese Mn^{2+} in the RNL

4
Black Sea Bottom Convective Layer

The existence of a homogeneous bottom water mass – bottom convective layer (BCL) – at water depths below 1740–1800 m was first reported, based on detailed CTD profiling, by Murray et al. [42] and since then it has been intensively studied [43–52]. Based on the data obtained in 1999–2002 in the north-eastern Black Sea, the bottom water mass was characterized by the following parameters: potential temperature $\theta = 8.883–8.888\ ^\circ C$, salinity $S = 22.330–22.334$ psu, and potential density $\sigma_\theta = 17.233–17.236\ kg\,m^{-3}$ [51]. On average, the water column below 500 m is about 0.01 °C warmer and 0.003–0.005 psu saltier in the western part than in the eastern part of the sea. Recent detailed studies have shown that not only the thermohaline characteristics, but also concentrations of several chemical species in bottom waters

Table 2 Hydrochemical characteristics of the BCL of the Black Sea in 2001 [52]

Parameter	Depth $\geq$ 1750 m[a]
S (psu)	22.332–22.333
Θ (°C)	8.884–8.886
σ_θ ($kg\,m^{-3}$)	17.234–17.236
pH	$\frac{7.51-7.56(32)^b}{7.53 \pm 0.02(0.3\%)}$
Alk, µM	$\frac{4375-4551(32)}{4450 \pm 45(1.0\%)}$
H_2S (µM)	$\frac{369-381(34)}{376 \pm 3(0.9\%)}$
NH_4^+ (µM)	$\frac{85.8-100.6(24)}{94.4 \pm 4.0(4.3\%)}$
PO_4^{3-} (µM)	$\frac{7.42-7.67(34)}{7.52 \pm 0.06(0.8\%)}$
SiO_3^{2-} (µM)	$\frac{321-335(35)}{330 \pm 3(1.0\%)}$
Mn^{2+} (µM)	$\frac{4.0-4.3(27)}{4.1 \pm 0.1(2.0\%)}$
CH_4 (µM)	$\frac{11.8-13.4(29)}{12.5 \pm 0.4(3.4\%)}$

[a] Seven stations
[b] Denominator: average ± standard deviation (number of measurements); numerator: range (coefficient of variability)

below 1670 m are uniform. As an example, the average thermohaline characteristics, alkalinity, and concentrations of some chemical species in BCL measured in 2001 are given in Table 2.

Transport between the BCL and the overlying waters occurs via a single diffusive interface. A destabilizing geothermal heat flux at the bottom acts against stable salinity stratification resulting in double diffusion [42, 43]. Mixing inside the BCL occurs on a scale of about 40 years [53]. The BCL in the Black Sea is the largest known example of bottom convection in the world ocean [43]. Eremeev and Kushnir [44] have found that the width of the BCL is directly proportional to the intensity of heat flux from the bottom. Shoaling of the upper boundary of the BCL and its higher volume were indeed observed in the areas of the increased surface heat flow [51]. H_2S concentrations were higher by about 10–15 μM in these areas [52]. The existence of BCL has important implications for physical and chemical exchange at the sediment/water interface and at the interface between intermediate and bottom water masses. Mixing processes at the interface between deep and bottom waters are particularly important for hydrogen sulfide dynamics and its balance in the sea, because about 30% of the Black Sea sulfide is concentrated in the layer below 1500 m.

5 Intermediate Sulfur Species in the Black Sea Water Column

Inorganic sulfur species are important intermediates in the sulfur cycle. In euxinic environments, the most commonly detected sulfur species are elemental sulfur, S^0, and thiosulfate, $S_2O_3^{2-}$. Sulfite, SO_3^{2-}, is usually of minor importance. Polythionates, $S_nO_6^{2-}$ (n =2–5), have not been reliably determined in the anoxic water due to analytical difficulties. Furthermore, they are not stable in the presence of H_2S, yielding $S_2O_3^{2-}$ and polysulfides [54]. In the presence of hydrogen sulfide, elemental sulfur usually reacts with HS^- and S^{2-} to form polysulfides, S_n^{2-} (n =2–6). In the presence of iron and dissolved sulfur species, a range of non-soluble iron sulfides such as iron monosulfide (FeS), greigite (Fe_3S_4), and pyrite (FeS_2) are formed. Insoluble iron sulfides have been detected in water columns of the Black Sea and Framvaren Fjord [55–57]. Organic sulfur in the suspended phase and thiols have been also measured in several anoxic marine basins including the Black Sea [57, 58].

The formation of sulfur intermediates is both chemically and biologically mediated. Oxidation of hydrogen sulfide with oxygen and Fe(Mn) oxyhydroxides and sulfate reduction are the main processes responsible for sulfur intermediates formation in the euxinic water columns, except for elemental sulfur, which is formed only during hydrogen sulfide oxidation. Chemically mediated reactions of thiosulfate formation are elemental sulfur(polysulfides) hydrolysis, polythionate disproportionation, and the reaction between elemental sulfur and sulfite [59].

Trace concentrations of inorganic sulfur species were measured throughout the oxic water column in the Black Sea [60]. Their occurrence in oxic waters was explained as being a result of redox reactions occurring in microniches of organic detritus. Anoxic microniches can be formed in organic-rich particles through the activity of sulfate-reducing bacteria producing H_2S. Observed irregular distributions of the inorganic sulfur species in the oxic zone indicated the possibility of redox reactions characterized by a non-equilibrium kinetics.

The average concentrations of reduced inorganic sulfur species in the anoxic zone of the Black Sea measured using a new colorimetric method developed by Volkov [61, 62] are summarized in Table 3. Presented elemental sulfur data refer to the sum of elemental sulfur allotropes (zero-valent sulfur) and the zero-valent sulfur derived from some fraction $(n-1)$ of the original polysulfide S_n^{2-}. Thiosulfate data in the table represent the total amount of thiosulfate, sulfite, and polythionates. At some stations in the Black Sea, Volkov [61] observed a concentration maximum of elemental sulfur at the oxic/anoxic interface associated with sulfide oxidation by dissolved oxygen and/or Mn oxyhydroxides. Increasing with depth, elemental sulfur concentrations are probably explained by the ongoing process of polysulfide formation

Table 3 Average concentrations of sulfur intermediates in the Black Sea anoxic zone [61]

Depth (m)	Number of samples	H_2S Range (μM)	H_2S Average (μM)	S^0 Range (μM)	S^0 Average (μM)	$S_2O_3^{2-}$ Range (μM)	$S_2O_3^{2-}$ Average (μM)
115–150	16	0.5–22.1	8.1	0.06–5.4	0.66	0–1.7	0.45
150	7	13.5–28.8	15.3	0–1.2	0.38	0.05–1.5	0.54
160	8	16.5–34.6	20.1	0–2.2	0.53	0–3.4	1.1
170	5	3.7–36.9	23.5	0–1.0	0.41	0–1.6	0.60
180	4	11.4–42.1	24.5	0.13–0.63	0.44	0.04–3.7	1.3
200	5	n.d.		0.38–3.1	1.1	0.5–3.1	1.3
250	7	n.d.		0.34–5.1	1.3	0.26–5.9	1.6
300	5	n.d.		0.75–1.6	1.2	0.42–4.5	1.7
400	1	n.d.		n.d.	1.4	n.d.	1.6
500	4	n.d.		1.0–5.0	2.4	0.11–1.7	1.0
750	3	n.d.		0.59–1.4	0.91	1.3–2.7	2.1
1000	4	n.d.		1.5–2.5	2.1	1.0–4.2	2.1
1500	2	n.d.		2.3–2.9	2.6	1.7–2.7	2.2
2000	3	n.d.		2.1–4.6	3.4	1.5–3.5	2.8

nd: not detectable

during the reaction between zero-valent sulfur and $\sum H_2S$. The vertical distribution of thiosulfate had similar shape as that of the elemental sulfur.

Comparison of the existing data on the distribution of inorganic sulfur intermediates in the Black Sea anoxic zone [33, 61, 63–65] shows a confused picture, which may reflect both the non-equilibrium kinetics of chemical transformations within the sulfur cycle and the differences in methods employed. Future progress in the field will depend on the advance in analytical techniques capable of producing reliable data for individual sulfur intermediates in natural waters at low concentrations, and on sampling and preparation methods without oxidation artifacts.

6 Sulfur Isotopic Composition of Sulfate and Hydrogen Sulfide in the Water Column

Sulfate-reducing bacteria produce sulfide depleted in ^{34}S compared to the initial sulfate [66]. Pure cultures of sulfate-reducing bacteria grown in media with unlimited sulfate produce hydrogen sulfide depleted in ^{34}S by 2–47‰ compared to the initial sulfate [67]. There are different factors that affect the sulfur isotopic fractionations. Among them, specific rate of sulfate reduction per cell is the most important, which in turn is dependent on temperature, cell size, substrate concentration, type of the electron acceptor, and growth phase of bacteria. The observed isotope difference between dissolved sulfate and sedimentary pyrite (usually, the final product of hydrogen sulfide transformations in marine sediments) in modern and ancient sediments is about 51 ± 10‰, and far exceeds fractionation factors observed in bacterial cultures [68]. This discrepancy is usually ascribed to oxidation processes within the sulfur cycle [69]. Several studies have shown that a sulfur isotope difference between sulfate and hydrogen sulfide in the Black Sea water column is 60‰ and therefore comparable to the upper range observed in marine sediments. This difference was also much higher than the fractionation factors produced by enrichment cultures of sulfate-reducing bacteria isolated from the Black Sea water [70].

The average isotope compositions of the sulfate sulfur in the oxic and anoxic zones are + 18.5‰ and + 19.5%, respectively. The isotopic composition of sulfate in the Black Sea forms from two distinct sources. The sea receives annually about 2.82×10^6 tons of sulfate with river discharge with the average isotope composition of + 4.6% [71]. The annual input with Mediterranean waters of 540×10^6 tons of sulfates has an isotopic composition of about + 19.8‰ [18]. The isotopic composition of dissolved sulfide averaged over all depths is -39.6 ± 1.3‰ and varies between – 42.0‰ and – 32.6‰ for all stations [65] (Fig. 5). There is no indication that the sulfur isotopic composition of hydrogen sulfide changes spatially and/or seasonally.

Data analysis suggests that some slight ^{34}S enrichments do exist in the upper and lower parts of the anoxic column [75]. The upper enrichment was explained by several authors as the effect of mixing with ^{34}S-enriched sulfide produced near the oxic/anoxic interface by chemical oxidation with Mn oxyhydroxides and/or dissolved oxygen, or as a result of small fractionation during the biological sulfide oxidation. Decreased sulfur fractionation due to higher sulfate reduction rates in the upper anoxic zone can be of importance too [65, 70]. The lower trend was explained as a result of the mixing between

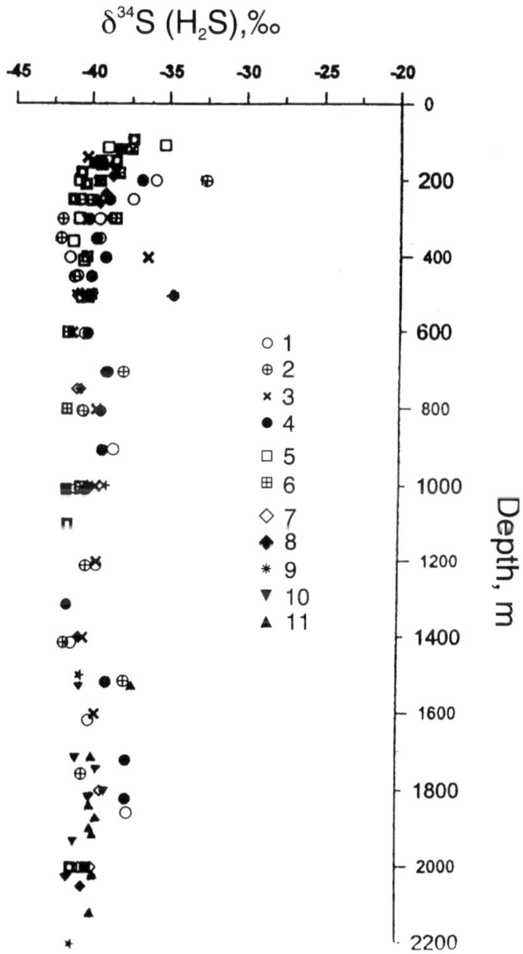

Fig. 5 Depth distribution of the sulfur isotopic composition of the Black Sea dissolved sulfide $\delta^{34}S$ (vs. CDT standard). Data are from different sources: *1* St. 4024 [73]; *2* St. 4037 [73], *3* St. BS2-2 [70], *4* St. 4010 [73], *5* St. BS2-3 [70], *6* St. BS2-1 [70], *7* St. 1135 [72], *8* St. 1136 [72], *9* St. 3 and 4 [74], *10* St. 1915 [52], *11* St. 1916 [52]

the ambient sulfide and ^{34}S-enriched sulfide diffusing up from the bottom sediment [65]. This explanation, however, cannot be supported by sufficient data because there have been only four measurements of the isotope composition of dissolved sulfide in Black Sea sediment pore waters ranging between - 16.2‰ and - 38.7‰ [75].

The most recent studies of the isotopic composition of the sulfate and dissolved sulfide in the BCL conducted in 2004, however, did not confirm the existence of the lower isotopic trend in the isotopic composition of sulfide. The average sulfur isotopic composition of sulfide below 1750 m was - 40.0‰ and varied between - 39.1 and - 41.5% [52], which is very similar to the isotopic composition of most of the anoxic zone (Fig. 6). In contrast, the isotope composition of sulfate in the BCL varied between + 19.9 and + 21.6% (average + 20.8‰) suggesting a slight ^{34}S enrichment of about 1.3‰ compared to the entire anoxic zone. Volkov and Rimskaya-Korsakova [52] hypothesized that the observed enrichment in sulfate sulfur was a result of the relative depletion of sulfate during bacterial sulfate reduction in this zone. These data are supported by the 2% decrease in the sulfate/chlorinity ratio observed in the BCL.

Neretin and co-authors [65] proposed that the high isotope difference between hydrogen sulfide and sulfate of about 60‰ observed in the Black Sea

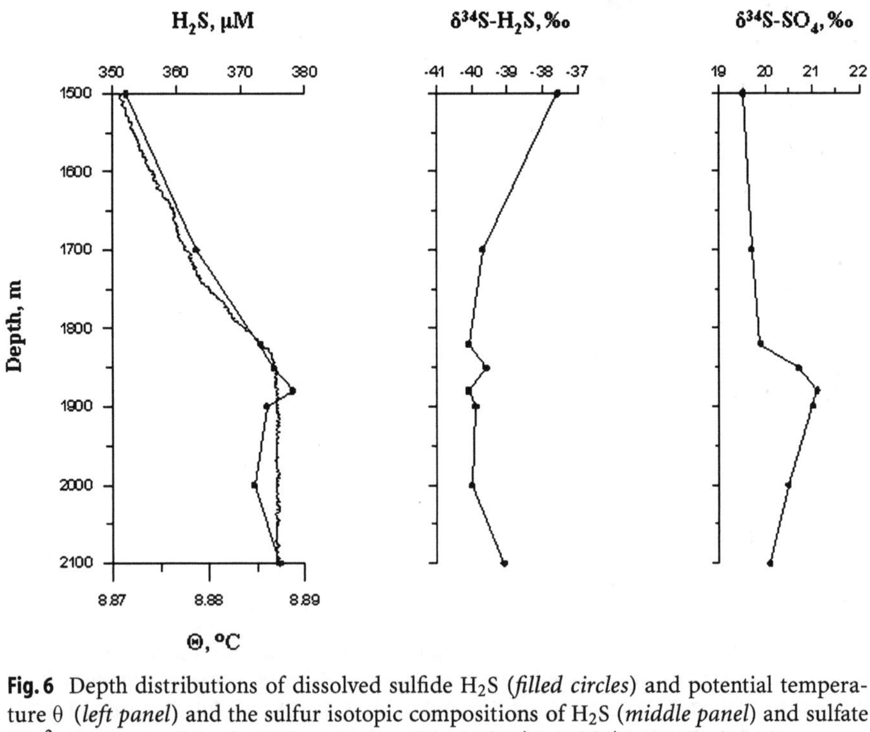

Fig. 6 Depth distributions of dissolved sulfide H_2S (*filled circles*) and potential temperature θ (*left panel*) and the sulfur isotopic compositions of H_2S (*middle panel*) and sulfate SO_4^{2-} (*right panel*) in the BCL at Station 1916 (43°42′N; 37°36′E; NE Black Sea)

water column results from a combination of very low sulfate reduction rates throughout the anoxic water column (less than 1 nM day^{-1} [76]) and efficient mixing mechanisms below the chemocline. The oxidative part of the sulfur cycle in the anoxic interior can be facilitated by the influx of the modified Bosporus waters containing dissolved oxygen, as well as by enhanced mixing and entrainment of oxygen-containing waters along the periphery of the basin, and during intensified winter convection.

7
Sulfide Budget in the Black Sea

The sulfur budget for the Black Sea has been considered in several papers [23, 24, 74–77]. Sulfide sources are sulfide production in sediments, sulfide flux at the sediment/water interface, and sulfide production in the water column. Sulfide sinks are sulfide oxidation at the oxic/anoxic interface and in the basin interior by dissolved oxygen of the modified Mediterranean water and iron sulfide formation in the water column.

7.1
Sulfide Production in Deep-Sea Sediments

Based on measurements by Sorokin [78], Deuser [79] calculated an average annual sulfide production in Black Sea sediments of 3.6 Tg. No sulfate reduction was measured below the uppermost 5 cm of sediment at that time [78]. Recent data on the presence of the anaerobic methane oxidation have shown that sulfate-reducing bacteria can also be active in deep sediments of the Black Sea [80]. However, a contribution of hydrogen sulfide produced in the deeply buried sediment is probably insignificant for its upward flux at the sediment/water interface for most of the Black Sea. Lein and co-authors [77] calculated an average hydrogen sulfide production in the anoxic sediments of the Black Sea of about 560 mmol m^{-2} $year^{-1}$, or 5.9 Tg $year^{-1}$. This estimate is higher than Deuser's, because the whole Holocene sequence was considered. Recent measurements by Albert and co-authors [76] gave an average sulfide production in the upper 20 cm (including 2 cm fluffy layer) of about 5.2 Tg $year^{-1}$. Lein and Ivanov [71] have estimated the total sulfide burial in the Black Sea of 2.4 Tg $year^{-1}$ including about 1 Tg $year^{-1}$ that is buried in the anoxic zone. Using these data and integrated over the upper 20 cm of sediment sulfate reduction rates, Neretin and co-authors [75] concluded that the annual sulfide flux into the water column from sediments of the anoxic zone is between 3 and 5 Tg $year^{-1}$. The value is likely to be overestimated due to spatial differences in pyrite burial rates and possible sulfide diffusion downward into the deeper sediment layers.

7.2
Hydrogen Sulfide Production in the Water Column

A maximum in sulfate reduction rates (SRRs) in the water column was usually observed in the upper (200–300 m down to 600–700 m) part of anoxic column and in the layers adjacent to the bottom. The highest rate measured for the upper anoxic zone was 1569 nmol L^{-1} day^{-1} [81]. The lowest SRRs in the water column were reported by Albert and others [76] and did not exceed 3.5 nmol L^{-1} day^{-1}. With a sensitivity of the method of about 0.2–0.6 nmol L^{-1} day^{-1} [76, 77], reduction of sulfate in the intermediate zone (600(700)–2000 m) comprising the main part of the Black Sea hydrogen sulfide pool was not revealed at all [77, 78], or SRRs in these layers were one to two orders of magnitude lower than in the proximity to the upper anoxic boundary [76, 82]. Existing data sets on SRRs are characterized by significant seasonal variations and yield an average sulfide production in the water column of 41 ± 31 (95% CI) Tg $year^{-1}$. Despite the fact that the gross SR activity is highest close to the interface and concentrated in the upper 500 m, most of the net water column sulfide production occurs in the middle and lower parts of the sulfidic zone, because of the volume and continuous activity of sulfate-reducing bacteria throughout the water column [75].

Sulfate-reducing bacteria are key players in anoxic waters [83]. In the first compiled organic carbon budget of the Black Sea, Deuser [79] suggested that at least half of the total particulate carbon that is transported into the anoxic water column is oxidized there by sulfate-reducing bacteria. Alkalinity, a measure of the buffering capacity of seawater reflecting the net effect of organic matter mineralization processes, is very high in the Black Sea compared to other marine basins. Analyzing the alkalinity of the Black Sea water column and C : S stoichiometry, Volkov and co-authors [84] found that sulfate reduction is almost entirely (95%) responsible for the total inorganic carbon production in the anoxic zone. The Black Sea water column probably hosts not only the most active and diverse microbial communities in the pelagic ocean, as suggested by Jannasch [85], but also the largest reservoir of sulfate-reducing prokaryotes in the world.

7.3
Hydrogen Sulfide Oxidation Processes

Integration of the measured $H_2{}^{35}S$ oxidation rates in the Black Sea chemocline yielded values between 53 and 125 Tg $year^{-1}$ [33, 86, 87]. Rate measurements and modeling data gave median sulfide oxidation rates at the oxic/anoxic interface in the range 20–50 Tg $year^{-1}$ [75].

The annual Bosporus flux into the Black Sea is estimated to be 120–312 km^3 [88]. The role of dissolved oxygen intrusions below the oxic/anoxic interface is a controversial issue since the magnitude and physiochemical properties of the Lower Bosporus current are variable. Some data, however, suggest a substantial role of the modified Bosporus waters in the entrainment of the Black Sea interior, at least in the western Black Sea [31, 36]. Sulfide oxidation in the anoxic interior was estimated using entrainment ratios reported in the literature and simple chemistry of the sulfide oxidation [75]. The authors have found that sulfide consumption with Bosporus intrusions may vary in the range 4.4–9.2 Tg year^{-1}, which roughly represents 10–20% of the sulfide oxidation at the oxic/anoxic interface. In contrast, the modeling approach has shown that the oxygen flux below the anoxic interface may be responsible for as much as 50–70% of the total sulfide consumption in the Black Sea water column [89].

Apart from oxygen intrusions with modified Bosporus waters, other lateral sources of oxygen have been discussed in the literature. Mesoscale dynamics, characterized by the existence of anticyclonic gyres described elsewhere in this volume (Ginzburg et al., in this volume), may provide an efficient mechanism for ventilation of the anoxic zone. In addition, intensified density convection during severe winters superimposed by internal wave forcing may cause erosion of the upper pycnocline and thus ventilate the upper anoxic layers [90, 91]. Lateral intrusions of oxygen below the pycnocline are not specific for the Black Sea and have been reported in other anoxic marine basins such as the Cariaco Basin [92], Framvaren Fjord [93], and the Mariager

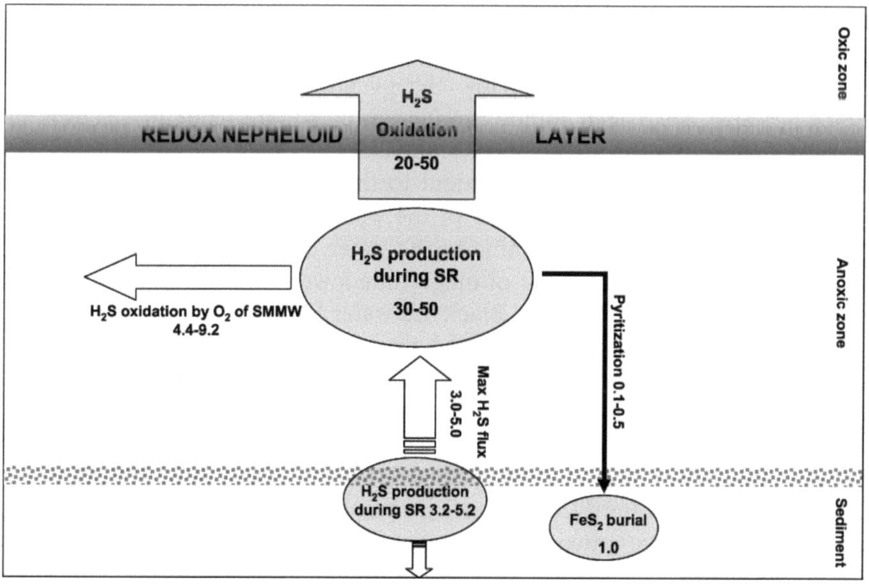

Fig. 7 Sulfur budget for the Black Sea anoxic zone. The *width of arrows* and *dimension of ovals* represent the relative magnitudes of respective processes. *SMMW* stands for shelf modified Mediterranean water. Processes rates are in 10^{12} g S year^{-1} (modified from [75])

Fjord [94]. Future studies of these processes will provide important insight into the scales and spatial distribution of oxidation processes in the anoxic interior.

The magnitude of the particulate sulfur flux in the Black Sea water column is estimated to be 0.1–0.5 Tg year^{-1} [55, 56].

A general scheme of the processes within the sulfur cycle discussed above with their assigned annual fluxes is presented in Fig. 7. If the average sulfide production in the Black Sea is 30–50 Tg year^{-1} and represents an average annual figure and the total sulfide inventory is about 4.6×10^3 Tg, the residence time of hydrogen sulfide in the water column would be about 90–150 years. This value is comparable to the water exchange rate between oxic and anoxic layers [23]. Hydrogen sulfide inventory and the location of the upper sulfide boundary is a delicate balance between complex physical and biological processes. Among them are ventilation of the cold intermediate layer (CIL), the entrainment ratio between CIL and the shelf modified Mediterranean waters (SMMW), and the rates of organic matter respiration processes, specifically oxygen consumption in the oxic zone.

8 Conclusions

Between 10 000 years BP and about 7200 years BP, due to melting of glaciers on the northern Eurasian continent and the post-glacial rise of the global sea level, Bosporus waters reached the Black Sea interior. After ca. 7500 years BP anoxic conditions developed below the pycnocline throughout the entire Black Sea area. Hydrogen sulfide production by sulfate-reducing bacteria in the water column is the main source of the Black Sea dissolved sulfide. Anoxic sediments contribute to a minor extent to the water column H_2S pool. The residence time of hydrogen sulfide in the water column of the Black Sea is 90–150 years, which is comparable to the water exchange rate between oxic and anoxic layers and an order of magnitude lower than the residence times of major salt components of the Black Sea water [23].

Recent studies of the BCL showed homogeneous distributions of physiochemical parameters including hydrogen sulfide. The existence of the BCL has important implications for the physical and chemical exchange at the sediment/water interface and at the interface between intermediate and bottom water masses. Twofold increased vertical gradients of dissolved sulfide at the upper boundary of the BCL suggest the presence of the "anoxic interface" separating the entire anoxic water mass, dominated by turbulent diffusion from underlying waters of the BCL where double diffusion is the main mixing mechanism. Mixing processes at this interface are particularly important for hydrogen sulfide dynamics and its balance in the sea, because about 30% of the Black Sea sulfide is concentrated in the layer below 1500 m.

Temporal variations in the average depth of the chemocline in the Black Sea and the upper sulfide boundary particularly are mainly the result of climatic changes in the density structure of the water column. The upper anoxic boundary location versus density for this basin did not change over the period from 1910 to 1995 [24]. However, recent data have shown a prominent increase in sulfide concentrations, as well as nutrient levels, within the anoxic zone (at 1000–2000 m) supposedly due to anthropogenic impacts [95] or climatic variations. Regular basin-wide monitoring of sulfide and nutrient concentrations in the anoxic interior, with a special emphasis on the bottom zone, are required to reveal how significant these changes are, or how they reflect statistical artifacts and measurement bias.

This review put forth the importance of mixing and ventilation processes in the Black Sea anoxic zone, reflected in sulfide concentrations and its sulfur isotopic composition. The Bosporus flux cannot be considered as a main factor for deep basin ventilation as suggested by the sulfide budget. Nearshore mesoscale dynamics associated with the propagation of anticyclonic gyres along the Rim Current and their influence on chemocline processes and horizontal exchange between shelf and open waters, as well as pycnocline erosion during exceptionally cold winters, are additional or probably major ventilation mechanisms for the anoxic zone [41, 96]. Global climate change has an impact on most of the aforementioned ventilation processes as well as on the replenishment of the CIL and its dissolved oxygen content (Yakushev et al., in this volume) and therefore on the magnitude and direction of the processes within the sulfur cycle in the Black Sea.

Acknowledgements The authors wish to thank T.P. Demidova, M.N. Rimskaya-Korsakova, and N.N. Zhabina (P.P. Shirshov Institute of Oceanology, Moscow) for invaluable analytical help and the crews of RV *Akvanavt*, *Yantar* and *Petr Kotzov* for collaboration. This work was financially supported by the Russian Fund for Basic Research (grant 05-05-65092), Project 17.4.3. "World Ocean" of the Russian Academy of Sciences, Project "Black Sea" of the Russian Ministry of Science, and Science School grant no. 4376.2006.5 to IIV.

References

1. Ryan WBF, Major CO, Lericolais G, Goldstein SL (2003) Ann Rev Earth Planet Sci 31:525
2. Svitoch AA, Selivanov AO, Yanina TA (1998) Paleogeograficheskiye Sobytiya Pleistotsena Ponto-Kaspiya i Sredizemnomoriya (Pleistocene Palaeogeographic events in the Ponto-Caspian and Mediterranean Basins). Moscow State University Press, Moscow (in Russian)
3. Kaplin PA, Selivanov AO (2004) Paleogeogr Paleoclimat Paleoecol 209:19
4. Balabanov IP, Kvirkeliya BD, Ostrovsky AB (1981) Recent history of the formation of the engineering-geological conditions and long-term forecast of the coastal zone evolution of the Pitzunda Peninsula. Metsniereba, Tbilisi, Georgia (in Russian)
5. Aksu AE, Hiscott RN, Yasar D (1999) Mar Geol 153:275

6. Aksu AE, Mudie PJ, Rochon A, Kaminski M, Abrajano T, Yasar D (2002) GSA Today 12(5):4
7. Mudie PJ, Rochon A, Aksu AE (2002) Mar Geol 190(1–2):233
8. Kerey IE, Meric M, Tunoglu C, Kelling G, Brenner RL Dogan AU (2004) Paleogeogr Paleoclimat Paleoecol 204:277
9. Shimkus KM (2005) Sedimentation processes in the Mediterranean and Black Seas during Late Cenozoic. Science World, Moscow (in Russian)
10. Ostrovskii AB, Izmailov YaA, Shcheglov AP, Arslanov KhA (1977) New data on Pleistocene stratigraphy and geochronology from marine terraces of the Caucasus Black Sea coast and Kerch-Taman region. In: Paleogeografiya i otlozheniya pleistotsena yuzhnykh morei SSSR (Paleogeography and sediments of Pleistocene in Southern Seas of USSR). Nauka, Moscow, p 61
11. Ryan WBF, Pitman WCI, Major CO, Shimkus K, Moskalenko V et al. (1997) Mar Geol 138:119
12. Ryan WBF, Pitman W (1998) Noah's Flood: The new scientific discoveries about the event that changed history. Simon and Schuster, New York
13. Görur N, Cagatay MN, Emre Ö, Alpar B, Sakinc M, Islamoglu Y, Algan O, Erkal T, Keser M, Akkok R, Karlik G (2001) Mar Geol 176:65
14. Mamaev OI (1995) Oceanology 34(6):756
15. Boudreau BP, LeBlond PH (1989) Paleoceanogr 4:157
16. Karaca M, Wirth A, Ghil M (1999) Geophys Res Lett 26:497
17. Ayzatullin TA, Leonov AV, Shaporenko SN (2003) Mathematical modelling of the formation and evolution of the Black Sea anoxic zone. In Aktualnie problemy okeanologii (Current problems in oceanography). Nauka, Moscow, p 431 (in Russian)
18. Vinogradov AP, Grinenko VA, Ustinov VI (1962) Geochem Int 10:973
19. Jones G, Gagnon A (1994) Deep Sea Res I 41:531
20. Deuser WG (1974) Evolution of anoxic conditions in the Black Sea during Holocene. In: Degens ET, Ross DA (eds) The Black Sea – geology, chemistry and biology. AAPG Tulsa, Oklahoma, p 133
21. Leonov AV, Shaporenko IS (2005) Water Res 32(3):276 (in Russian)
22. Neretin LN (1996) PhD thesis, Shirshov Institute of Oceanology Moscow (in Russian)
23. Skopintsev BA (1975) Formirovanie sovremennogo khimicheskogo sostava vad Chernogo morya (Formation of contemporary chemical composition of the Black Sea waters). Hidrometeoizdat, Leningrad (in Russian)
24. Bezborodov AA, Eremeev VN (1993) Chernoe more. Zona vzaimodeystviya aerobnykh i anaerobnykh vod (The Black Sea: oxic/anoxic interface zone). Marine Hydrophysical Institute AS of the Ukraine, Sevastopol, Ukraine (in Russian)
25. Neretin LN, Volkov II, Rozanov AG, Demidova TP, Falina AS (2006) Biogeochemistry of the Black Sea anoxic zone with a reference to sulphur cycle. In: Neretin LN (ed) Past and present water column anoxia, NATO science series IV, vol 64. Springer, Dordrecht, p 69
26. Novoselov AA, Romanov AS (1988) Present state of the Black Sea anoxic zone. In: The origin and seasonal variability of hydrophysical and hydrochemical parameters in the Black Sea. MHI, Sevastopol, Ukraine, p 148 (in Russian)
27. Neretin LN, Volkov II (1995) Oceanology 35:60
28. Vinogradov ME, Nalbandov YuR (1990) Oceanology 30:567
29. Murray JW, Codispoti LA, Friederich GE (1995) Oxidation-reduction environments: The suboxic zone in the Black Sea. In: Huang C et al. (eds) Aquatic chemistry. Kluwer Academic, Amsterdam, p 157

30. Pimenov NV, Rusanov II, Yusupov SK, Friedrich J, Lein AYu, Wehrli B, Ivanov MV (2000) Microbiol 69:436
31. Murray JW, Yakushev EV (2006) The suboxic transition zone in the Black Sea. In: Neretin LN (ed) Past and present water column anoxia, NATO science series IV, vol 64. Springer, Dordrecht, p 105
32. Pimenov NV, Neretin LN (2006) Composition and activities of microbial communities involved in carbon, nitrogen, sulfur and mangaenese cycling in the oxic/anoxic interface of the Black Sea. In: Neretin LN (ed) Past and present water column anoxia, NATO science series IV, vol 64. Springer, Dordrecht, p 501
33. Jørgensen BB, Fossing H, Wirsen CO, Jannasch HW (1991) Deep Sea Res 38(2A):S1083
34. Kuypers MMM, Sliekers OA, Lavik G, Schmid M, Jørgensen BB, Kuenen JG, Sinninghe Damsté JS, Strous M, Jetten MSM (2003) Nature 422:608
35. Tebo BM (1991) Deep Sea Res 38(2A):S883
36. Schippers A, Neretin LN, Lavik G, Leipe Th, Pollehne F (2005) Geochim Cosmochim Acta 69:2241
37. Oakley BB, Francis CA, Roberts KJ, Fuchsman CA, Srinivasan S, Staley JT (2007) Environ Microbiol 9:118
38. Durisch-Kaiser E, Klauser L, Wehrli B, Schubert CJ (2005) Appl Environ Microbiol 71:8099
39. Schubert CJ, Coolen MJL, Neretin LN, Schippers A, Abbas B, Durisch-Kaiser E, Wehrli B, Hopmans ES, Sinninghe Damsté JS, Wakeham S, Kuypers MMM (2006) Environ Microbiol 8:1844
40. Manske AK, Glaeser J, Kuypers MMM, Overmann J (2005) Appl Environ Microbiol 71:8049
41. Volkov II, Kontar EA, Lukashev YuF, Neretin LN, Nyffeler F, Rozanov AG (1997) Geochem Int 6:618
42. Murray JW, Top Z, Özsoy E (1991) Deep Sea Res II 38A:S663
43. Özsoy E, Rank D, Salihoğlu I (2002) Coast Shelf Sci 54:621
44. Eremeev VN, Kushnir VM (1998) Mar Hydrophys J 1:50 (in Russian)
45. Ivanov LI, Shkvorets IYu (1995) Mar Hydrophys J 6:53 (in Russian)
46. Ivanov LI, Samodurov AS (2001) J Mar Syst 31:159
47. Kelley DE, Fernando HJ, Yargett AE, Tanny J, Özsoy E (2003) Progr Oceanogr 56:461
48. Volkov II, Skirta Ayu, Makkaveev PN, Demidova TP, Rozanov AG, Yakushev EV (2002) On physical and chemical homogeneity of bottom waters in the Black Sea. In: Multidisciplinary investigations of the northeastern part of the Black Sea. Nauka, Moscow, p 161 (in Russian)
49. Volkov II, Falina AS, Skirta AYu, Yakubenko VG (2002) Hydrophysical and hydrochemical structure of the Black Sea deep waters. In: Current problems in oceanology. Nauka, Moscow, p 414 (in Russian)
50. Falina AS, Volkov II (2003) Oceanology 43:516
51. Falina AS, Volkov II (2005) Oceanology 45:21
52. Volkov II, Rimskaya-Korsakova MN (2007) Doklady RAN (in press)
53. Özsoy E, Top Z, White G, Murray JW (1991) Double diffusive intrusions, mixing and deep sea convection processes in the Black Sea. In: Izdar E, Murray JW (eds) The Black Sea oceanography. NATO/ASI series. Kluwer Academic, Dordrecht, p 17
54. Zopfi J, Ferdelman TG, Fossing H (2004) Distribution and fate of sulfur intermediates – sulfite, tetrathionate, thiosulfate, and elemental sulfur – in marine sediments. In: Amend JP, Edwards KJ, Lyons TW (eds) Sulfur biogeochemistry – past and present. Geological Society of America special paper 379, p 97
55. Tambiev SB, Zhabina NN (1988) Dokl Akad Nauk SSSR 299:1216 (in Russian)

56. Muramoto J, Honjo S, Fry B, Hay BJ, Howarth RW, Cisne JL (1991) Deep Sea Res II 38(A):S1151
57. Cutter GA, Kluckhohn RS (1999) Mar Chem 67:149
58. Dyrssen D (1985) Chem Script 25:199
59. Volkov II (1984) Sulphur geochemistry in ocean sediments. Nauka, Moscow (in Russian)
60. Volkov II, Demidova TP (1991) Dokl Akad Nauk 320:977 (in Russian)
61. Volkov II (1991) Reduced sulphur species in the Black Sea water. In: Vinogradov ME (ed) Variability of the Black Sea ecosystem: natural and anthropogenic factors. Nauka, Moscow, p 53 (in Russian)
62. Volkov II, Rozanov AG, Demidova TP (1992) Inorganic reduced sulphur species and dissolved manganese in the Black Sea water column. In: Vinogradov ME (ed) Black Sea ecosystem in winter. Nauka, Moscow, p 38 (in Russian)
63. Vairamamurphy A, Mopper K (1990) Env Sci Technol 24:333
64. Luther III GW, Church TM, Powell D (1991) Deep Sea Res II 38(A):S1121
65. Neretin LN, Böttcher ME, Grinenko VA (2003) Chem Geol 200:59
66. Kaplan IR, Rittenberg SC (1964) J Gen Microbiol 34:195
67. Detmers J, Brüchert V, Habicht K, Küver J (2001) Appl Environ Microbiol 67:888
68. Canfield DE, Teske A (1996) Nature 382:127
69. Jørgensen BB (1990) Science 249:152
70. Fry B, Jannasch HW, Molyneaux SJ, Wirsen C, Muramoto J, King S (1991) Deep Sea Res II 38(A):S1003
71. Lein AYu, Ivanov MV (1983) Reduced sulphur accumulation in sediments of marine basins with high rates of sulphate reduction. In: The global biogeochemical sulphur cycle, SCOPE 19. Wiley, Chichester, UK, p 413
72. Sweeney RE, Kaplan IR (1980) Mar Chem 9:145
73. Neretin LN, Grinenko VA, Volkov II (1996) Dokl Aked Nauk SSSR 349A:1015 (in Russian)
74. Lein AYu, Ivanov MV (1991) On the sulphur and carbon balances in the Black Sea. In: Izdar E, Murray JW (eds) The Black Sea oceanography. Kluwer Academic, Amsterdam, p 307
75. Neretin LN, Volkov II, Böttcher ME, Grinenko VA (2001) Deep Sea Res I 48:2569
76. Albert DB, Taylor G, Martens C (1995) Deep Sea Res I 42:1239
77. Lein AYu, Ivanov MV, Vaynshtein MB (1990) Microbiol 59:656 (in Russian)
78. Sorokin YuI (1962) Microbiol 31:329 (in Russian)
79. Deuser WG (1971) Deep Sea Res I 18:995
80. Jørgensen BB, Böttcher ME, Lüschen H, Neretin LN, Volkov II (2004) Geochem Cosmochim Acta 68:2095
81. Il'chenko SV, Sorokin YuI (1991) K otsenke intensivnosti obrazovaniya serovodoroda v Chernom more (To the estimate of hydrogen sulphide production in the Black Sea). In: Vinogradov ME (ed) Variability of the Black Sea ecosystem: natural and anthropogenic factors. Nauka, Moscow, p 73 (in Russian)
82. Gulin MB (1991) PhD thesis, Institute of Biology of the Southern Seas, Sevastopol, Ukraine (in Russian)
83. Volkov II (2000) Oceanology 40:535
84. Volkov II, Dyrssen D, Rozanov AG (1998) Geochem Int 1:78
85. Jannasch HW (1991) Microbial processes in the Black Sea water column and top sediment: an overview. In: Izdar E, Murray JW (eds) The Black Sea oceanography. Kluwer Academic, Dordrecht, p 271
86. Sorokin YuI (1972) J Conseil Int Explor Mer 34:423 (in French)

87. Sorokin YuI (1983) The Black Sea. In: Ketchum BH (ed) Ecosystems of the world, vol 26. Estuaries and enclosed seas. Elsevier, Amsterdam, p 253
88. Ünlüata U, Oguz T, Latif MA, Özsöy E (1990) On the physical oceanography of the Turkish straits. In: Pratt LG (ed) The physical oceanography of sea straits. Kluwer Academic, Amsterdam, p 25
89. Konovalov SK, Ivanov LI, Samodurov AS (2001) J Mar Syst 31:203
90. Krivosheya VG, Titov VB, Ovchinnikov IM, Kos'yan RD, Skirta AYu (2000) Oceanology 40:816
91. Titov VB (2000) Oceanology 40:826
92. Scranton MI, McIntyre M, Astor Y, Taylor GT, Müller-Karger F, Fanning K (2006) Temporal variability in the nutrient chemistry of the Cariaco Basin. In: Neretin LN (ed) Past and present water column anoxia, NATO science series IV, vol 64. Springer, Dordrecht, p 139
93. Yao W, Millero FJ (1995) Aquatic Chem 1:53
94. Zopfi J, Ferdelman TG, Jørgensen BB, Teske A, Thamdrup B (2001) Mar Chem 74:29
95. Konovalov SK, Eremeev VN, Suvorov AM, Khaliulin AKh, Godin EA (1999) Aquatic Geochem 5:13
96. Kempe S, Diercks AR, Liebezeit G, Prange A (1991) Geochemical and structural aspects of the pycnocline in the Black Sea (R/V Knorr 134-8 Leg 1, 1988). In: Izdar E, Murray JM (eds) Black Sea oceanography, NATO/ASI series. Kluwer Academic, Dordrecht, p 89

Hdb Env Chem Vol. 5, Part Q (2008): 333–349
DOI 10.1007/698_5_063

Published online: 16 December 2006

Seasonal and Interannual Variability of Remotely Sensed Chlorophyll

Nikolay P. Nezlin

Southern California Coastal Water Research Project, 7171 Fenwick Lane,
Westminster, CA 92683-5218, USA
nikolayn@sccwrp.org

Abstract Seasonal and interannual variability of remotely sensed by Sea-viewing Wide Field-of-View Sensor (SeaWiFS, aboard OrbView-2 satellite) and MODerate Resolution Imaging Spectroradiometer (MODIS-A, aboard Aqua satellite) surface chlorophyll *a* (CHL) in the Black Sea was analyzed since the start of SeaWiFS mission in 1997 till the spring of 2006. The spatio-temporal patterns analyzed by the empirical orthogonal functions (EOF) method revealed four main regions where CHL dynamics was correlated: northwest shelf (NW); southwest coastal region (SW); eastern region (E); and central region (C). Seasonal variability in the NW region had evident maximum in late spring resulting from the maximum of nutrient-rich Danube discharge. In three deep regions (SW, E, and C) the seasonal cycles were characterized by summer minimum and autumn-winter maximum. This pattern is typical to subtropical ocean areas where phytoplankton growth is nutrient-limited as a result of water column stratification. Danube discharge (correlated in 1990s–2000s with ENSO climatic cycle) looks like a most important factor regulating chlorophyll concentration on the northwestern shelf and, after a time lag of 2–2.5 years, in the deep Black Sea regions.

Keywords Black Sea · SeaWiFS · MODIS · Chlorophyll · Seasonal variations · Interannual variations

1 Introduction

In this paper, the principle features of seasonal and interannual variability of the surface chlorophyll *a* (CHL) concentration in the Black Sea are derived from the observations of Sea-viewing Wide Field-of-View Sensor (SeaWiFS, aboard OrbView-2 satellite, September 1997–May 2006) and MODerate Resolution Imaging Spectroradiometer (MODIS-A, aboard Aqua satellite, July 2002–May 2006). The principle features of the Black Sea physical geography and hydrology are described in detail in chapters by Kosarev and Kostianoy, Tuzhilkin and Kosarev.

First, we divide the Black Sea area into several regions and analyze the chlorophyll dynamics in each of them. To classify the Black Sea area into the sub-areas where CHL varies synchronously, we use the empirical orthogonal functions (EOF) statistical method, which provided good results in the analysis of spatio-temporal variability in different ocean regions [1, 2]. The Black Sea is temperate region where seasonal variability dominates; as such, for each sub-region we estimate seasonal climatology, then subtract this climatology from the observed values and operate anomalies rather than absolute values. We also explore the time-lagged correlations between CHL and other measured by satellite sensors factors, which hypothetically can influence CHL dynamics: photosynthetically available radiation (PAR) measured by SeaWiFS (September 1997–May 2006), and sea surface temperature (SST) measured by Advanced Very High Resolution Radiometers (AVHRR) aboard NOAA satellites (1997–2005) and MODIS-A (July 2002–May 2006). PAR is expected to directly influence the rate of phytoplankton photosynthesis. SST anomalies indicate water column stratification, which is closely connected to light and nutrient limitation of phytoplankton growth [3, 4].

Analyzing the interannual variability of CHL in different Black Sea regions, we explore the correlations between CHL and global climatic indices, including North Atlantic oscillation (NAO) and El-Niño-Southern Oscillation (ENSO) indices: the Southern Oscillation Index (SOI) and NINO3. Among other environmental factors influencing the seasonal and interannual dynamics of remotely sensed phytoplankton biomass in the Black Sea, we focus on the Danube river runoff, because the volume of freshwater discharged into the Black Sea ($> 200\ \text{km}^3\ \text{year}^{-1}$) is very high as compared with other basins, and the Danube River contributes about 70% to this discharge ([5], see also chapter by Mikhailov).

2 Methods

The analysis of spatio-temporal variations of phytoplankton biomass in the Black Sea was based on the remotely sensed data collected by the SeaWiFS (September 1997–May 2006) and MODIS-A (July 2002–May 2006) satellite sensors. The data were obtained from National Aeronautics and Space Administration Goddard Space Flight Center Distributed Active Archive Center (NASA GSFC DAAC) [6]. We used monthly averaged Level 3 standard mapped images (SMI) data of SeaWiFS surface chlorophyll calculated at GSFC during reprocessing #5. The format of the Level 3 SMI data is a regular grid of equidistant cylindrical projection of 360°/4096 pixels (about 9.28 km resolution) for SeaWiFS and 360°/8192 pixels (about 4.5 km resolution) for MODIS-A. The basic algorithm used at GSFC for calculating CHL was described in [7]. In this study, we use CHL data (i.e., surface chlorophyll *a* concentration derived from water color) to analyze the dynamics of phytoplankton biomass in the water column. The remotely sensed surface pigment concentration and total pigment concentration in water column are correlated [8] but not identical. However, for this study, the absolute values are not as important as spatial and temporal gradients of phytoplankton biomass. We understand that the absolute values of CHL derived from satellite measurements are subject to significant inaccuracy due to technical difficulty of remotely sensed observations. The standard SeaWiFS and MODIS CHL algorithms were developed for clean open ocean waters (Case 1), where the color of ocean surface results mainly from chlorophyll concentration [9]. The Black Sea is classified as coastal waters (Case 2), where the sea surface color depends also on the dissolved and suspended matter concentrations, uncorrelated with chlorophyll. Standard algorithms developed for open ocean (Case 1) overestimate chlorophyll concentration in Case 2 waters [10, 11]. Regional algorithms based on in situ chlorophyll concentration measurements provide more reliable results in Case 2 waters, but these studies were not yet carried out for the Black Sea [12]. In this study, we do not compare the remotely sensed CHL data directly to in situ absolute values of chlorophyll concentration and phytoplankton biomass. Instead, we consider the variations of satellite-measured CHL as an indicator of dynamics of phytoplankton biomass in the study region.

Before statistical processing, CHL values were log-transformed, because the distribution of remotely sensed chlorophyll in the world ocean has been shown to be asymmetric [13] and log-transformation makes them much closer to normal.

To estimate the SST variations, we used AVHRR and MODIS-A data from NASA Jet Propulsion Laboratory Physical Oceanography Distributed Active Archive Center (NASA JPL PODAAC). The AVHRR data were produced at JPL within the scope of National Oceanic and Atmospheric Administration (NOAA)/NASA AVHRR Oceans Pathfinder Project. SST was calculated on the basis of the AVHRR radiometer observations using an enhanced nonlinear algorithm [14]. We used monthly data for both ascending and descending satellite passes (i.e., daytime and nighttime observations, respectively). The data format of Pathfinder Version 5 is a regular grid of equidistant cylindrical projection of 360°/8192 pixels (about 4.5 km resolution). Only the data with quality flags 4–7 (i.e., the "best SST") were used.

Global monthly maps of PAR (Einstein m^{-2} day^{-1}) estimated from SeaWiFS observations were obtained from GSFC DAAC and processed in the manner similar to CHL and SST.

Each pixel of global monthly maps of CHL (SeaWiFS and MODIS-A), SST (AVHRR and MODIS-A), and PAR (SeaWiFS) represents an average of all measurements obtained during the calendar month. To reveal the principle features of spatial distribution and to avoid the gaps in the data grids (resulting from the absence of observations due to cloudy weather and inappropriate satellite view angles) the 9.28 km and 4.5 km resolution maps were transformed into the maps of 0.5° ($\sim$ 50–55 km) spatial resolution. For this, a standard kriging method [15, 16] was used. To avoid interpolation to the regions where observations were absent (such interpolation inevitably provides erroneous results), missing data code (MD) was assigned to the grid nodes located > 0.5° from the nearest observed data point. Also, MDs were applied to the grid nodes located over land areas. After all these processing steps, we obtained for each parameter (CHL, SST and PAR) 104 or 105 monthly grids (from September 1997 to April–May 2006) of 12×29 size with up to 213 nodes filled with CHL, SST or PAR values. PAR array contained no MDs. In CHL and SST, a small number of grid nodes (total 14 for CHL and 86 for SST) contained MDs, which produced a problem for statistical analysis of these data.

To reveal the areas where CHL and SST vary synchronously, we used the modification of empirical orthogonal functions (EOF) method suggested by Beckers and Rixen [17], which solves the MD problem. In conventional EOF the entire data matrix contains only valid data. All MDs must be removed either with the entire row or with the entire column, which significantly decreases the volume of statistically analyzed data. In the Beckers–Rixen's method, the data gaps (i.e., MDs) are reconstructed on the basis of the correlations between the data measured simultaneously at different locations. This is achieved by an iterative procedure. At the first step, the MDs in each column are replaced by the column mean. Then, at each iteration step, the data matrix is decomposed into factor loadings and a complete set of EOFs, which number is equal to the number of variables. In accordance with the fundamentals of principle component analysis [18], leading EOF modes contain maximum of variance, trailing EOF modes containing mostly noise. The product of factor loadings and several ("significant") leading EOF modes results in a new matrix, which principle features are similar to the initial data matrix, but a substantial part of noise is removed. The MDs from the initial data matrix are substituted by the corresponding values from the new matrix; then the resulting matrix is decomposed again and the process is repeated until converged. The number of "significant" EOF modes used in iterative process was estimated from a series of experiments with different number of "significant" EOF modes (from one to 30). In each experiment, 50 randomly selected points were changed to MD; the mean difference (rms) between these 50 values and the corresponding newly estimated values was used as a measure of accuracy of missing data filling. According to these experiments, 11 EOF modes were selected as significant for MD filling of CHL and 19 for SST.

Correlations between CHL, SST and freshwater discharge were assessed from monthly estimations of Danube discharge ($m^3\,s^{-1}$) for 1960–2005 obtained from the Danube Hydrometeorological Observatory of the Ukrainian State Committee of Hydrometeorology.

To relate the long-scale variations in the Black Sea to the global climatic meteorological cycles, we used NAO, SOI, and NINO3 climatic indices, obtained from the International Research Institute of Climate Prediction (Columbia University, USA). NAO is defined as the monthly averaged difference between the standardized measurements of the sea level atmospheric pressure in Azores and Iceland. SOI is the difference between the standardized measurements of the sea level atmospheric pressure in Tahiti and Darwin, Australia. NINO3 index is determined as the SST anomalies over the eastern tropical Pacific (5°S–5°N; 150°W–90°W).

To analyze the interannual variability of CHL, SST, PAR, and Danube discharge, each parameter was transformed to seasonal anomaly. For this, climatic monthly values were subtracted from each time series.

To compare the changes in correlation between the NINO3 index and the Danube discharge seasonal anomalies, we used cross-wavelet transform software, produced by Aslak Grinsted. The details of cross-wavelet analysis methodology are given in [19–21].

3 Black Sea Regions Based on Chlorophyll Variability

The map of cumulative CHL distribution over the Black Sea area measured by SeaWiFS in 1997–2005 (Fig. 1) illustrates that the lowest CHL ($\sim 0.5\,mg/m^3$) was observed in the central part of the sea, gradually increasing in the coastal regions. The highest CHL was observed over the northwestern shelf.

Three leading EOF modes of log-transformed CHL explained totally 43.8% of variability (19.4%; 14.6%; 9.8%, respectively). The spatial distributions of all three modes exhibited clear patterns (Fig. 2a–c). Each of the remaining EOF modes explained < 5% of variability and had no clear interpretation. The first EOF mode shows the difference between the northwestern shelf and the remaining part of the Black Sea (Fig. 2a). The second EOF mode distinguishes the coastal zone in the western and southwestern parts of the sea (Fig. 2b), indicating the zone of influence of coastal current transporting high nutrient concentration and phytoplankton biomass from the northwestern shelf along the coast in counterclockwise (cyclonic) direction. The third EOF mode reveals the difference between the central deep part of the sea and its eastern and northeastern regions (Fig. 2c).

On the basis of spatial distribution of three leading EOF modes we classify the Black Sea area into four regions (Fig. 2d): northwestern shelf (NW); southwestern coastal region (SW); eastern region (E), including the Batumi

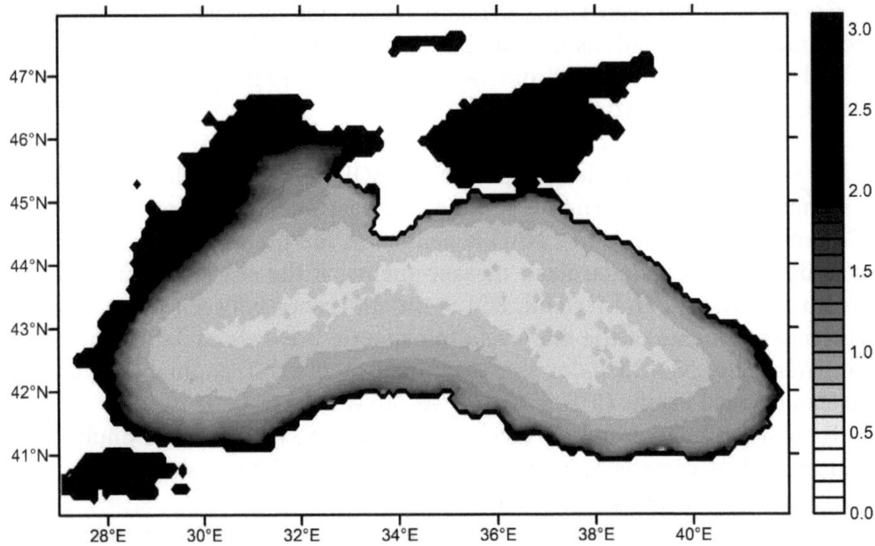

Fig. 1 Cumulative map of surface chlorophyll *a* distribution (mg/m^3) measured by SeaWiFS radiometer during September 1997–December 2005

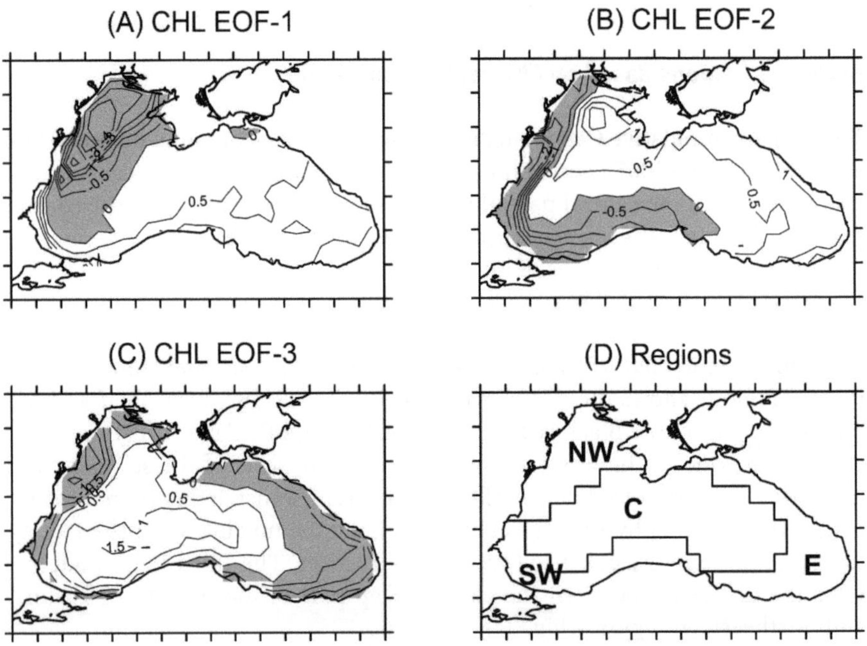

Fig. 2 Spatial distribution of three leading EOF modes of log-transformed CHL over the Black Sea area (**a**, **b**, **c**) and the regions of the Black Sea where seasonal and interannual variability was analyzed (**d**)

eddy and southeastern, eastern, and northeastern coastal areas; and open central region (C). This classification emphasizes the hydrological structure of the Black Sea, characterized by a basin-scale cyclonic boundary Rim Current of < 75 km width ([22], see also Chapter 9). The Rim Current separates the cyclonic circulation in the interior of the basin from anticyclonic eddies in the narrow coastal zone (see chapter by Ginzburg, Zatsepin, Kostianoy, Sheremet).

4 Seasonal Variability

Seasonal cycle of CHL in deep regions (SW, E, and C) is characterized by a minimum in summer (July–August) (Fig. 3a) coinciding with SST maximum (Fig. 3b). This type of seasonality is typical to subtropical ocean regions [3, 4]. Summer minimum is a result of stratification hindering penetra-

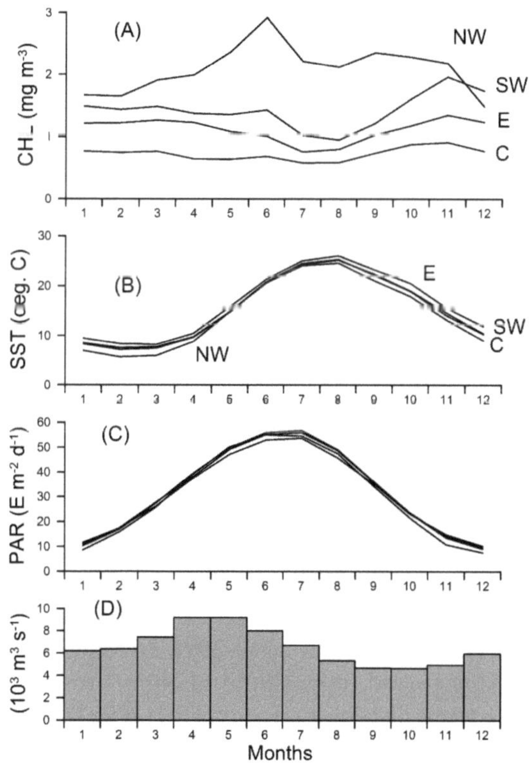

Fig. 3 Seasonal variability of CHL (**a**), SST (**b**), PAR (**c**), and Danube discharge (**d**). Literals indicate (1–4) indicate the regions of the Black Sea: NW–Northwest; SW–Southwest; E–East; C–Central (see Fig. 2)

tion of nutrients into the upper mixed layer. In the wide subtropical ocean regions summer stratification is typically a result of solar heating of the ocean surface. In the Black Sea pycnocline results mostly from low salinity of the upper mixed layer [23], produced by intensive freshwater discharge (see chapter by Tuzhilkin, Kosarev). During winter enhanced wind mixing and surface cooling erode pycnocline and improve phytoplankton nutrition.

Seasonal cycle of SST is similar in all Black Sea regions (Fig. 3b). Phytoplankton growth rate depends on the rate of photosynthesis, which theoretically depends on PAR. However, maximum PAR in June–July does not result in CHL increase (Fig. 3c).

Maximum of Danube discharge is in April–May (Fig. 3d). This maximum can explain CHL maximum over the northwestern shelf one–two months later (Fig. 3a, NW).

A comparison between absolute CHL concentrations in different Black Sea regions illustrates that CHL is highest over the northwestern shelf and lowest in the central region (Fig. 3a). In the two coastal regions with narrow shelf, CHL gradually decreases from NW to SW to E, which coincides with the general pattern of cyclonic circulation in the Black Sea.

5
Interannual Variability

Seasonal anomaliess of CHL were positive (i.e., CHL exceeded seasonal averages) during 1997–2001 in all four regions (Fig. 4). Starting from 2002, CHL seasonal anomalies changed to negative in all four regions. The decrease in CHL was especially pronounced in the shallow NW area in 2002–2003 (Fig. 4a). In three deep regions the period of negative CHL anomalies lasted longer, by the end of 2005 (Fig. 4b–d).

The interannual variations of SST anomalies were less consistent than CHL, but exhibited similar trends (Fig. 5). Negative SST anomalies persisted from mid-2002 by the end of 2004.

No evident trends were revealed in PAR anomalies (Fig. 6).

Danube discharge was higher than normal from autumn 1998 till spring 2000 (Fig. 7a). During summer 2000–summer 2002, it was slightly lower than normal, higher than normal in the second half of 2002, and substantially lower than normal in 2003. NAO time series did not reveal evident trends (Fig. 7b). SOI and NINO3 show the 1997–1998 El-Niño, strongest in the XX century [24], La-Niña period (opposite to El-Niño) from the second half of 1998 to 2000, and a weak El-Niño period in 2002–2005 (Fig. 7c,d).

Danube discharge was a principle factor influencing CHL concentration over the entire Black Sea area. CHL anomalies were significantly correlated with Danube discharge anomalies in all Black Sea regions (Table 1). Positive sign of correlation indicates that nutrients transported with freshwater

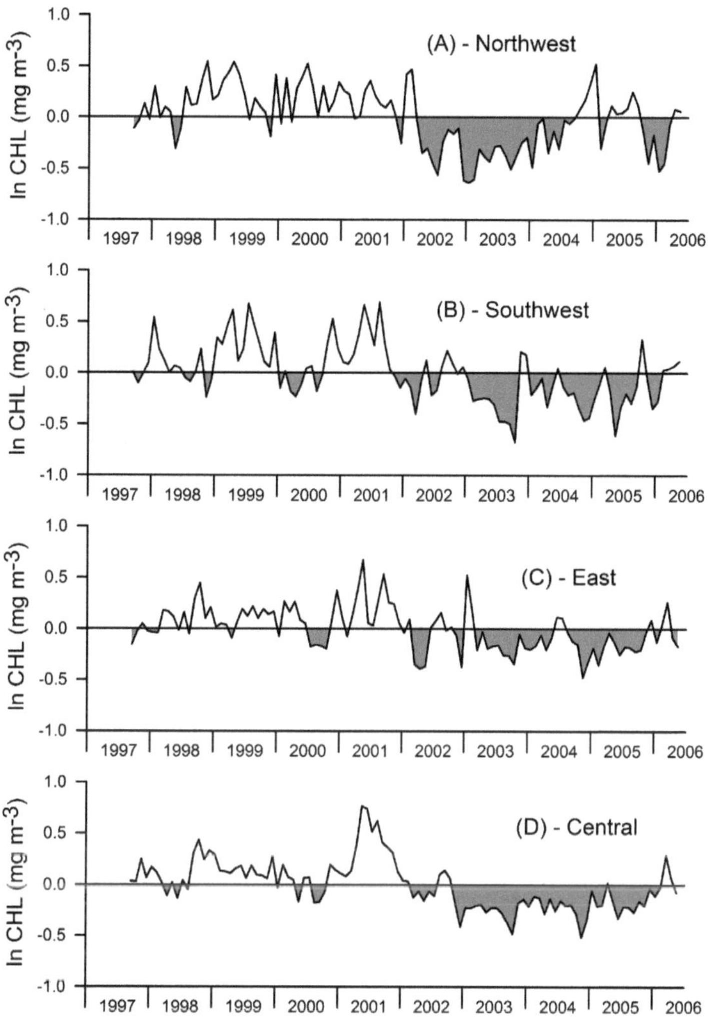

Fig. 4 Interannual variability of log-transformed CHL seasonal anomalies in different Black Sea regions: **a** Northwest; **b** Southwest; **c** East; **d** Central

runoff increase primary production and phytoplankton biomass. However, the time lags of maximum correlation were different in different regions. In the shallow NW region directly influenced by Danube discharge, the time lag of maximum correlation was 0, indicating an immediate (on a monthly time scale) response of phytoplankton biomass to nutrient discharge. In other regions, the CHL response to Danube discharge was significantly longer: 2 years in SW region and 2.5 years in E and C regions.

The correlation between NINO3 and CHL anomalies was highest as compared with other indices of global climate. This correlation was especially

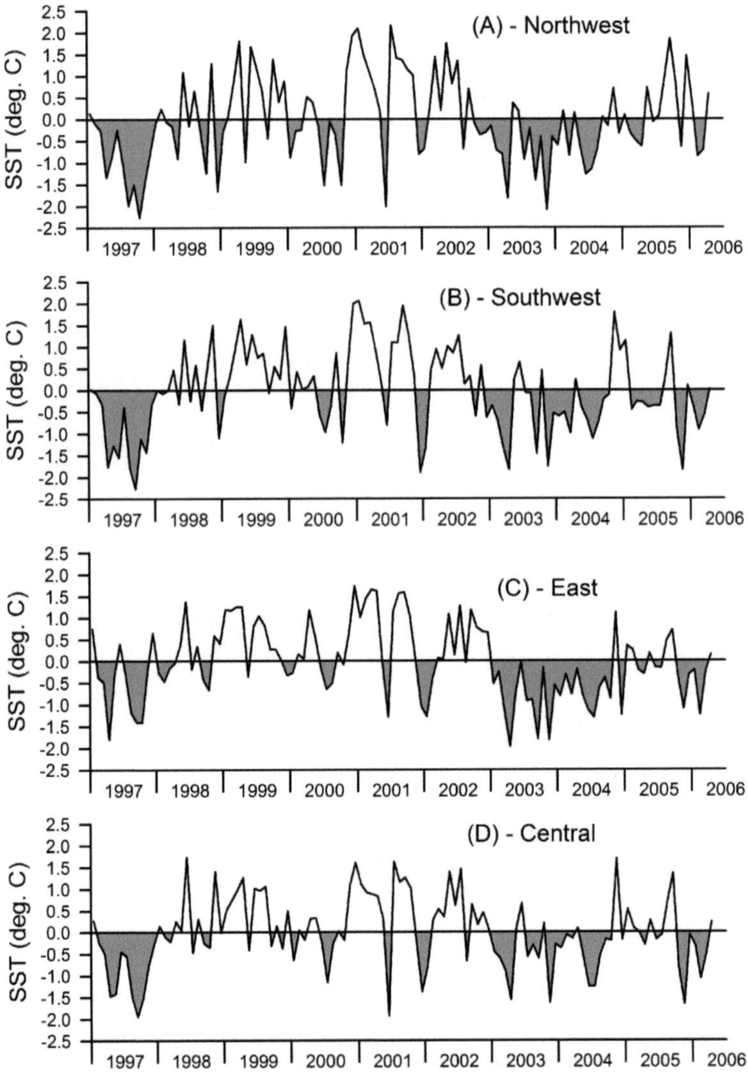

Fig. 5 Interannual variability of SST seasonal anomalies in different Black Sea regions: **a** Northwest; **b** Southwest; **c** East; **d** Central

high in the open Black Sea region (+ 0.7). The time lag of maximum correlation between NINO3 and CHL anomalies was 2.5 years in the shallow NW region and 3.5 years in the other regions (Table 1). The correlations of CHL anomalies with other indices (SOI and NAO) were weaker.

No significant correlation was revealed between CHL and PAR anomalies.

Significant positive correlation between SST and CHL (Table 1) looks surprising. Indeed, in highly stratified Black Sea, phytoplankton growth is

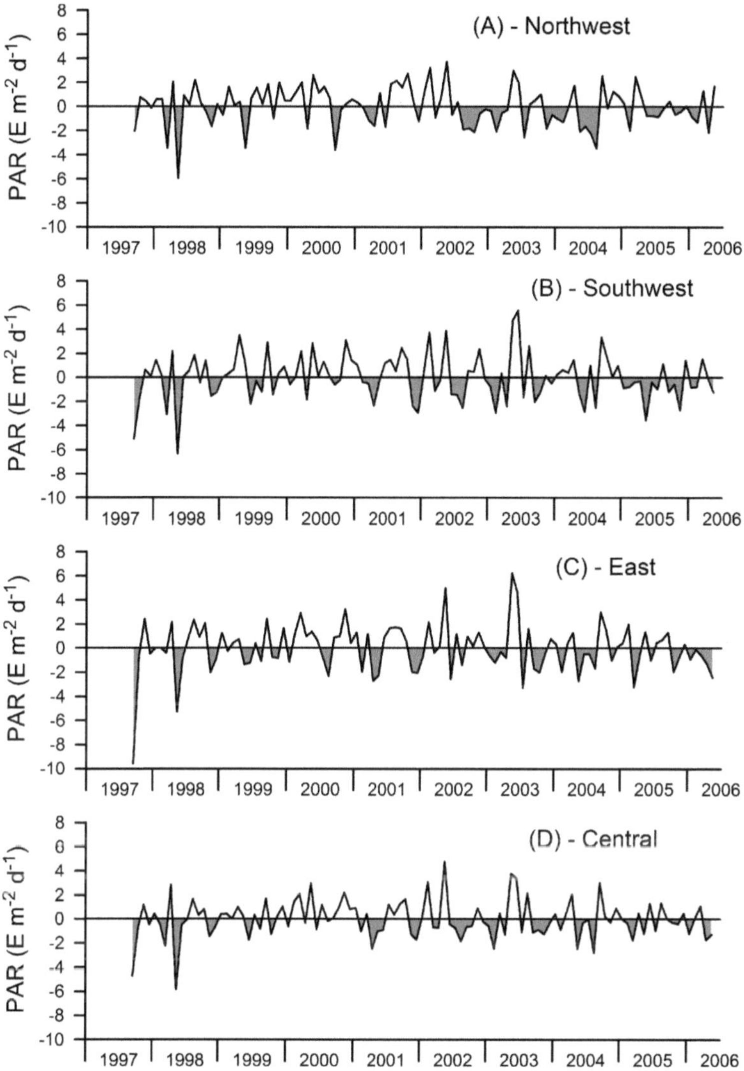

Fig. 6 Interannual variability of PAR seasonal anomalies in different Black Sea regions: **a** Northwest; **b** Southwest; **c** East; **d** Central

limited by lack of nutrients and vertical mixing or upwelling is expected to result in both an increase of CHL and a decrease of SST. This kind of negative correlation between CHL and SST is evident from their seasonal variations (Fig. 3). However, at interannual time scale the correlation between CHL and SST anomalies is absolutely different and requires special explanation.

We consider Danube discharge to be a primary factor modulating SST and CHL in the Black Sea. Increased freshwater discharge stimulates phyto-

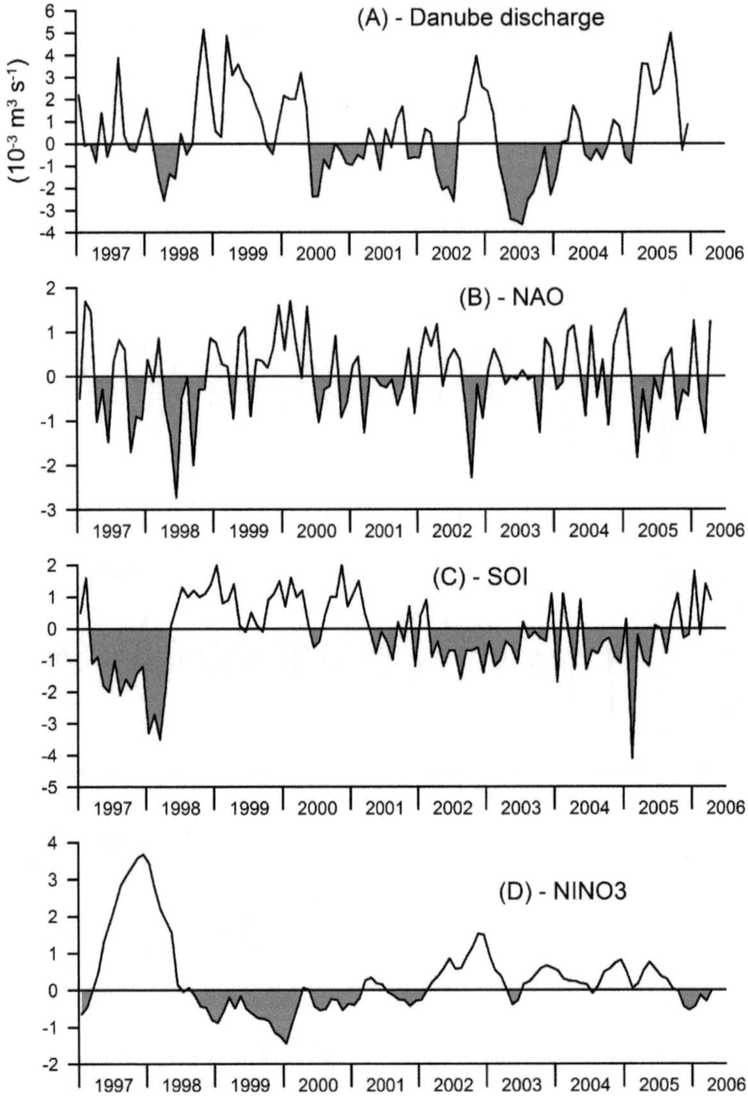

Fig. 7 Interannual variability of Danube discharge seasonal anomalies (**a**) and global climatic indices NAO (**b**); SOI (**c**); and NONO3 (**d**)

plankton growth with nutrients and at the same time enhances water column stratification. In more stratified waters the temperature of the upper mixed layer increases as a result of solar heating of water surface, because less heat is dispersed from the upper mixed layer to cold subsurface layers. As such, positive Danube discharge anomalies result in positive anomalies of both CHL and SST.

Table 1 Time-lagged correlations between monthly Danube discharge seasonal anomalies, climatic indices (NAO, SOI and NINO3), log-transformed CHL and SST seasonal anomalies in different Black Sea regions in 1997–2006. Positive signs of time lags (in months) indicate that the Parameter-1 is leading the Parameter-2 time series

Parameter-1	Parameter-2	Maximum correlation Coefficient	Time lag (months)
Danube discharge seasonal anomalies	CHL (Northwest)	+ 0.338	0
	CHL (Southwest)	+ 0.271	+ 24
	CHL (East)	+ 0.325	+ 29
	CHL (Central)	+ 0.411	+ 29
NAO	CHL (Northwest)	- 0.249	+ 44
	CHL (Southwest)	- 0.246	+ 36
	CHL (East)	- 0.225	+ 39
	CHL (Central)	- 0.274	+ 36
SOI	CHL (Northwest)	- 0.294	+ 48
	CHL (Southwest)	- 0.354	+ 42
	CHL (East)	- 0.340	+ 44
	CHL (Central)	- 0.463	+ 43
NINO3	CHL (Northwest)	+ 0.404	+ 31
	CHL (Southwest)	+ 0.490	+ 43
	CHL (East)	+ 0.433	+ 42
	CHL (Central)	+ 0.700	+ 43
SST (Northwest)	CHL (Northwest)	+ 0.446	- 8
SST (Southwest)	CHL (Southwest)	+ 0.351	- 2
SST (East)	CHL (East)	+ 0.409	+ 4
SST (Central)	CHL (Central)	+ 0.353	+ 6
NINO3	Danube discharge seasonal anomalies	+ 0.394	+ 18

The correlations between Danube discharge anomalies and SST are low; however, they become substantially better when we change the time scale of the analysis from monthly to seasonal. Table 2 illustrates the correlations between SST, CHL and Danube discharge anomalies averaged for cold (October–March) and warm (April–September) seasons. The correlation between SST and CHL anomalies in cold season was low (from 0.084 to 0.220). However, the correlations between SST and CHL anomalies in warm season as well as the correlations between SST anomalies in cold season and CHL anomalies during the next warm season were high in three deep regions (SW, E, and C) and low over the shallow NW region. We speculate that the response of SST to freshwater discharge is immediate during all seasons, in contrast to CHL, which manifests itself only during the summer period of nutrient limitation of phytoplankton growth. Also, this correlation is low in the shallow NW region, where water column is less stratified and the aforesaid scheme does not work.

Table 2 Coefficients of determination (R^2) of linear correlation between seasonally averaged CHL and SST anomalies and annually averaged Danube discharge (DD). CHL and SST are analyzed separately for cold (October–March) and warm (April–September) seasons

Parameters	Regions 1 (Northwest)	 2 (Southwest)	 3 (East)	 4 (Central)
SST (cold season) – CHL (cold season)	0.143	0.112	0.220	0.084
SST (cold season) – CHL (next warm season)	0.078	0.349	0.425	0.548
SST (cold season) – CHL (next cold season)	0.043	0.022	0.241	0.100
SST (warm season) – CHL (warm season)	0.125	0.693	0.373	0.391
SST (warm season) – CHL (next cold season)	0.044	0.029	0.665	0.236
DD – CHL (cold season)	0.605	0.005	0.055	0.043
DD – CHL (warm season)	0.350	0.080	0.005	0.008
DD – CHL (next cold season)	0.000	0.068	0.272	0.129
DD – CHL (next warm season)	0.258	0.021	0.016	0.040
DD – SST (cold season)	0.000	0.161	0.009	0.135
DD – SST (warm season)	0.406	0.200	0.336	0.282
DD – SST (next cold season)	0.091	0.003	0.011	0.030
DD – SST (next warm season)	0.029	0.048	0.112	0.072

Over NW shelf, Danube discharge anomalies are highly correlated with CHL anomalies during cold season ($R^2 = 0.605$) and to less extent during warm season ($R^2 = 0.350$). The correlation between Danube discharge and CHL anomalies during next cold season is low, but CHL during next warm season retains a significant signal of freshwater discharge during the previous cold season ($R^2 = 0.258$). SST over NW shelf is not correlated with Danube discharge in cold season, when freshwater is colder than seawater, but significantly correlated in warm season ($R^2 = 0.406$), when freshwater is warm and its contribution increases SST according to aforesaid schema.

6 Discussion

Comparing seasonal and interannual CHL variability in the Black Sea, we see that the range of seasonal cycle significantly exceeds the range of interannual variations. Standard deviations of log-transformed CHL anomalies in different regions range from 0.21–0.29, that after exp-transformation is equal to 23–34%. It is comparable with the estimation of primary production in the Black Sea sustained by Danube discharge (15–25%) made earlier on the basis of average values of nitrogen discharged with Danube runoff and mean primary production in the Black Sea [25].

The correlation between SST and CHL anomalies in the Black Sea was high during summer and low during winter, but its sign was always positive. In our previous study based on CZCS time series obtained in 1978–1986, a positive correlation was revealed between SST and CHL during winter period [26–28], explained by the difference in meteorological conditions during cold and mild winters. Some authors [29] expect negative correlation between winter SST and CHL, explaining it with shallow pycnocline in the centers of two main gyres during extremely cold winters ("winter upwelling"). The results of this study do not support this hypothesis.

Discussing CHL variations in the Black Sea, we should keep in mind that Level 3 chlorophyll data we analyzed were obtained using a standard CHL algorithm developed for open ocean (Case 1) waters. The Black Sea is attributed to Case 2 waters, where surface color results not only from chlorophyll concentration, but also from total suspended sediments (TSS) and color dissolved organic matter (CDOM), which concentrations are uncorrelated with chlorophyll. TSS and CDOM can result in significant overestimation of CHL [11], especially in the shallow NW Black Sea region. However, the contribution of TSS and CDOM cannot explain the correlations between Danube discharge, SST and CHL anomalies we observe, taking into account the time lags of 2–2.5 years. The persistence of TSS in river plumes is measured in days [30–32]. CDOM is a much more conservative tracer, but its optical properties also fade with time due to a photo-bleaching effect [33]. As such, nutrients sustaining phytoplankton production look like a primary source of correlation between Danube discharge and remotely sensed CHL.

The correlation between Danube discharge and SST anomalies is weaker and less persistent as compared with CHL, because other factors influence water column stratification and the temperature of the ocean surface, i.e., SST. Wind stress is most important among these factors. As such, a correlation can be expected between SST and climatic indices like NAO, similar to correlations we observed in the Caspian Sea [34]. NAO index indicates dominating paths of storms moving eastward from the North Atlantic. A north/south shift of this path results in changes of precipitation (and, in turn, freshwater discharge) and wind mixing of water column. However, in the Black Sea no correlation was observed between NAO, CHL and SST. We speculate that more deep and stratified Black Sea is less sensitive to wind stress than relatively shallow and less stratified Caspian Sea. Also, Danube discharge was not correlated with NAO at monthly time scale, as well as Volga discharge in the Caspian Sea area [34]. The catchment areas of both rivers are too large and the storm paths do not miss them regardless of NAO regimes. At the same time, at annual time scale NAO is better correlated with freshwater discharge, resulting in pronounced variations of the Black Sea level [35].

The influence of ENSO cycle on weather in Europe and other parts of the world is well known [36]. In our study, a significant correlation (+ 0.394) was observed in 1997–2005 between NINO3 index and Danube discharge anomalies with 1.5-year (18 months) time lag. However, it is questionable is this correlation a constant feature of global teleconnections. The analysis of cross-wavelet diagram between NINO3 and Danube discharge anomalies based on longer time series (1960–2005) shows that these two parameters were correlated in 1965–1977, not correlated in 1978–1995, and correlated after 1995. We conclude that the discharge of major European rivers including Danube cannot be predicted on the basis of global climatic cycles like NAO or ENSO. The analysis of these complex and non-linear correlations is beyond the scope of this study.

Acknowledgements The author would like to thank the SeaWiFS Project (Code 970.2) and the Distributed Active Archive Center (Code 902) at the NASA Goddard Space Flight Center for the production and distribution of remotely sensed images, respectively. These activities are sponsored by NASA's Mission to Planet Earth Program. I also thank the NASA Physical Oceanography Distributed Active Archive Center at the Jet Propulsion Laboratory, California Institute of Technology for SST data. Crosswavelet software was provided by A. Grinsted. I thank A.G. Kostianoy and V.N. Mikhailov for supplying the data on the intensity of the Danube River discharge.

References

1. Nezlin NP, McWilliams JC (2003) Remote Sens Environ 84:234
2. Nezlin NP, Oram JJ, DiGiacomo PM, Gruber N (2004) Continent Shelf Res 24:1053
3. Longhurst A (1995) Prog Oceanogr 36:77
4. Longhurst AR (1998) Ecological Geography of the Sea. San Diego, Academic Press
5. Cociasu A, Diaconu V, Popa L, Buga L, Nae I, Dorogan L, Malciu V (1997) The nutrient stock of the Romanian shelf of the Black Sea during the last three decades. In: Ozsoy E, Mikaelyan AS (eds) Sensitivity to Change: Black Sea, Baltic Sea and North Sea. Kluwer Academic Publishers, Dordrecht, p 49
6. Acker JG, Shen S, Leptoukh G, Serafino G, Feldman G, McClain C (2002) IEEE T Geosci Remote 40:90
7. O'Reilly JE, Maritorena S, Mitchell BG, Siegel DA, Carder KL, Garver SA, Kahru M, McClain C (1998) J Geophys Res 103:24937
8. Smith RC, Baker KS (1978) Limnol Oceanogr 23:247
9. Morel A, Prieur L (1977) Limnol Oceanogr 22:709
10. Muller-Karger FE, Hu C, Andrefouet S, Varela R, Thunell RC (2005) The color of the coastal ocean and applications in the solution of research and management problems. In: Miller RL, Del Castillo CE, McKee BA (eds) Remote Sensing of Coastal Aquatic Environments. Springer, Dordrecht, p 101
11. Siegel DA, Maritorena S, Nelson NB, Behrenfeld MJ, McClain CR (2005) Geophys Res Lett 32:L20605

12. Kopelevich OV, Burenkov VI, Ershova SV, Sheberstov SV, Evdoshenko MA (2004) Deep-Sea Res II 51:1063
13. Banse K, English DC (1994) J Geophys Res 99:7323
14. Walton CC (1988) J Appl Meteorol 27:115
15. Oliver MA, Webster R (1990) Int J Geo Inf Syst 4:313
16. Isaaks EH, Srivastava RM (1989) Applied geostatistics. Oxford University Press, New York
17. Beckers J-M, Rixen M (2003) J Atmos Ocean Tech 20:1839
18. Priesendorfer RW (1988) Principle Component Analysis in Meteorology and Oceanography. Elsevier, New York
19. Moore JC, Grinsted A, Jevrejeva S (2005) EOS 86:226
20. Grinsted A, Moore JC, Jevrejeva S (2004) Nonlinear Proc Geophys 11:561
21. Jevrejeva S, Moore JC, Grinsted A (2003) J Geophys Res 108:4677
22. Oguz T, LaViolette PE, Unluata U (1992) J Geophys Res 97:12569
23. Murray JW, Top Z, Ozsoy E (1991) Deep-Sea Res I 38:S663
24. McPhaden MJ (1999) Science 283:950
25. Nezlin NP (2000) Remote-sensing studies of seasonal variations of surface chlorophyll - a concentration in the Black Sea. In: Halpern D (ed) Satellites, Oceanography and Society. Elsevier, Amsterdam, p 257
26. Nezlin NP, Dyakonov VY (1998) Oceanology (English Translation) 38:636
27. Nezlin NP, Kostianoy AG, Gregoire M (1999) Remote Sens Environ 69:43–55
28. Nezlin NP, Dyakonov VY (1998) Seasonal and interannual variations of surface chlorophyll concentration in the Black Sea on CZCS data. In: Ivanov LI, Oguz T (eds) Ecosystem Modeling as a Management Tool for the Black Sea. Kluwer, Dordrecht, p 137
29. Mikaelyan AS (1995) Mar Ecol Prog Ser 129:241
30. Nezlin NP, DiGiacomo PM (2005) Cont Shelf Res 25:1692
31. Nezlin NP, DiGiacomo PM, Stein ED, Ackerman D (2005) Remote Sens Environ 98:494
32. Warrick JA, Mertes LAK, Washburn L, Siegel DA (2004) Cont Shelf Res 24:2029
33. Del Castillo CE (2005) Remote sensing of organic matter in coastal waters. In: Miller RL, Del Castillo CE, McKee BA (eds) Remote Sensing of Coastal Aquatic Environments. Technologies, Techniques and Applications. Springer, Dordrecht, p 157
34. Nezlin NP (2005) Patterns of seasonal and interannual variability of remotely sensed chlorophyll. In: Kostianoy AG, Kosarev AN (eds) The Caspian Sea Environment. Springer, Berlin, p 143
35. Stanev EV, Peneva EL (2001) Global Planet Change 32:33–47
36. Diaz HF, Hoerling MP, Eischeid JK (2001) Int J Climatol 21:1845

Hdb Env Chem Vol. 5, Part Q (2008): 351–374
DOI 10.1007/698_5_089

Published online: 21 July 2007

Biodiversity and Bioproductivity

Zosim Z. Finenko

Kovalevskii Institute for Biology of the Southern Seas,
National Academy of Sciences of Ukraine, 2 Nakhimov av., Sevastopol, 99011, Ukraine
zfinenko@ibss.iuf.net

Abstract Based on abundant published data and results of original studies, we consider the present-day condition of the biodiversity and productivity of the Black Sea ecosystem. Qualitative and quantitative analyses of the flora and fauna compositions are presented. The response of the biota to changes in environmental factors is shown. The particular features of the functioning of the Black Sea ecosystem and the effects of human activity are assessed.

Keywords Benthos · Biodiversity · Fauna · Necton · Phytoplankton · Plankton · Productivity

1 Introduction

Anthropogenic activity, which increases with time, is the main reason for the decrease in the species diversity on our planet. While earlier, the disappearance of large predators was observed in land ecosystems, now it is also encountered in the oceans [1–3]. Water contamination, especially in near-shore regions, results in a decrease in the area of livable space and thus leads to a reduction in species diversity; directly or indirectly, this affects the economy related to fishery, aquaculture, and restoration of environmental quality [4–6]. Species diversity is closely related to the features of the functioning ecosystem and to the processes of adaptation of organisms to environmental conditions. With the growth in species number, trophic diversity increases and trophic interrelations become more complicated. In mature communities, the number of links between the species is extremely high and may change in an unexpected manner, which makes forecasting interspecies relations difficult. In order to understand the evolution of communities, one has to analyze in detail the interrelations between the species under the conditions under which they dwell when entering this or that type of ecosystem.

In the course of community development, the biomass and the production of autotrophic and heterotrophic organisms decrease [7]. Meanwhile, the correlation between the biodiversity and productivity of the communities is not yet clear and is still being discussed. The objective of this chapter is to consider the main results of the studies on species diversity in the Black Sea from the point of view of their interrelation to the production processes and to the ecological and physiological characteristics of the phytoplankton community.

2 Formation of Marine Biota

The flora and fauna of the Black Sea are represented by organisms of different origins. A part of the population is retained from the times of the existence of the Pontian Lake, a sea in which brackish-water fauna and flora were formed. These organisms that used to dwell in desalinated regions of the sea and selected lagoons are referred to as Ponto–Caspian relicts. In the Quaternary, owing to the uplifting of the earth's crust and the formation of the Caucasian Mountains, the Pontian Lake divided to form the basins that subsequently became the Sea of Azov, the Black Sea, and the Caspian Sea. It is assumed that the cold-water fauna of the Black Sea was formed during glacial thawing, when the organisms dwelling in the cold waters of the northern rivers were delivered to the sea in the course of the basin filling. These species are also referred to as boreal relics; they inhabit the deep layers of the sea. After the formation of the Bosporus and Dardanelles Straits, the species of the Mediterranean Sea began to penetrate into the Black Sea; their influence is responsible for the major part of the Black Sea flora and fauna. In addition, a part of the species is permanently supplied with the riverine runoff; they are mostly encountered in near-mouth areas of the sea. More than 100 species were introduced to the Black Sea with the ballast waters of ships; most of them originate from the coastal regions of the North Atlantic (see chapter "Introduced species" in this volume).

3 Species Diversity of Plant Organisms

The flora of the Black Sea is represented by phytoplankton dwelling in the water column and phytobenthos inhabiting the bottom. In deep-water regions of the sea, the biotope for phytoplankton covers the water layer from the sea surface to the lower boundary of the pycnocline; in near-shore regions, it is confined to the layer from the surface to the bottom, where illumination is sufficient for photosynthesis. Phytoplankton is composed of unicellular algae; with respect to their sizes, they may be joined into three groups: picoplankton (0.2–2 μm), nanoplankton (2–20 μm), and microplankton (20–200 μm). Bottom plants include unicellular and multicellular algae. Unicellular algae (microphytobenthos) dwell over the surface of the floor, rocks, and solid underwater objects including the surface of large algae (sea grass).

3.1 Phytoplankton

The first data concerning the phytoplankton of the Black Sea were obtained by L.V. Reingard at the beginning of the last century [8], who identified 46 alga species in the region of the Kerch Strait. After a few decades of research (1922–1947), this list was increased to 144 [9], and then to 272 species [10, 11]. In his generalization of the studies performed before 1965, A.I. Ivanov [12] presented as many as 676 species and intraspecies taxa; G.K. Pitsyk, using published and new data, listed 746 species [13]. Selected scientists suggest that the Black Sea phytoplankton contains about 1000 alga species and varieties [14]. A similar growth in the taxonomic alga composition in the course of its studies is also characteristic of individual parts of the sea, in particular, of its northwestern part. On the whole, the taxonomic composition of the Black Sea phytoplankton is studied in sufficient detail.

During the past 50 years, the extension of the species list is related to the progress in floristic knowledge rather than to the appearance of new species; meanwhile, recently, some previously unknown species have been recognized. These are typical of the Mediterranean Sea and penetrated via the Bosporus Strait either in a natural way [15] or with the ballast waters of ships. It is suggested that the transport of new species with the ballast waters may exceed their natural supply via the Bosporus [16]. Certain changes are also noted in the composition of the freshwater species assemblage delivered to the northwestern part of the sea with the runoff of the Danube, Dnieper, and Dniester rivers; there, Cyanophyta and Euglenophyceae species new for the Black Sea were found [17]. The Black Sea phytoplankton is known to mainly consist of the alga of the ancient Pontian–Caspian basin and the Mediterranean Sea. It features similarity with the flora of other inland seas such as the Caspian and the North seas and the Norwegian fjords. The majority of the species refer to seven divisions: Bacillariophyceae, Dinophyceae, Prymnesiophyceae, Cryptophycea, Chlorophyta, Cyanophyta, Chrysophyceae, Dictyochophyceae, and Euglenophyceae [15, 17]. The results of the studies performed in different years show that Bacillariophyceae and Dinophyceae dominate in the Black Sea; they provide up to 80% of the total species number. According to long-term data, the relations between these two groups of alga change: Bacillariophyceae dominate in some years, while Dinophyceae prevail during the others. In the northwestern part of the sea, in 1954–1960, Bacillariophyceae provided 48% of the total species number, while the share of Dinophyceae was 20%. In 1973–1986, these values were 30 and 40%, respectively [17, 18]. In other regions of the sea, Dinophyceae were less abundant than diatoms [16].

Diatoms are represented by 342 species and varieties; they are widely spread over the entire sea area. They reach high abundance values in nearshore regions, especially in the spring and autumn, though some of them are capable of vegetating all the year round. In the 1950s, in Sevastopol Bay, 19 species of diatoms dominated, while 50 years later, the number of dominating diatom species estimated using the same criteria decreased down to three species [10, 11, 16].

Another group of alga distinguished for its species richness is represented by Dinophyceae and includes 205 species and subspecies taxa. They are noted in plankton throughout the year, but their highest species diversity is confined to the summer period. Among the Dinophyceae representatives, one can encounter both heterotrophic and mixotrophic species. The greatest contribution to the biomass of heterotrophic phytoplankton is made by *Protoperidinium pellucidum* [19].

The algae of the Prymnesiophyceae division are also represented by rather high abundances. This group is not distinguished for high species richness (51 species), but they are widely spread and some of them feature mass amounts. During recent years, the role of Prymnesiophyceae in the total phytoplankton abundance significantly increased and a replacement of the peridinean alga assemblage by a primnesian assemblage was noted. In the summer period, intensive coccolithophore "bloom" is often observed in the open regions of the sea.

As one can see, the Black Sea phytoplankton is rather diverse and includes more than 700 species, among which two taxonomic groups (Bacillariophyceae and Dinophyceae) dominate.

3.2 Microphytobenthos

Microphytobenthos integrates the microalgae dwelling over underwater substrates. The biotope of microphytobenthos is confined to sea depths down to 50 m from the surface. Microphytobenthos makes a significant contribution to the production of organic matter that is created by photosynthesis on the sea floor. The dwelling conditions of microphytobenthos are rather diverse; these organisms inhabit loose and solid sediments, they dwell over the bottom vegetation and the objects that occur under subsurface conditions. In the Black Sea and its lagoons, the specialists in systematics cite about 750–970 species and intraspecies taxa are dominated by three classes of diatomaceous algae: Bacillariophyceae, Fragilariophyceae, and Coscinodiscophyceae [20, 21]. The highest species diversity is characteristic of three diatom groups: Amphora, Nitzschia, and Navicula. Among these groups, the following species are widely spread: *Amphora angusta*, *A. coffeiformis*, *A. hyalina*, *Nizschia hybrida f. hyalina*, *N. lanceolata*, *N. rupestris*, *Navicula ammophila* var. *intermedia*, and *N. directa*. The typical planktonic algae *Sceletonema costatum* spend a part of their lives in the form of spores in silty–sandy sediments or over the surface of milticellular algae. The species diversity of diatoms is relatively constant throughout the year and slightly increases with depth. The abundance and biomass of diatoms range within one to two orders of magnitude depending on the season, region, and dwelling site. The maximal concentrations are mainly observed in the spring period. For example, in Sevastopol Bay and off Karadag, the diatom abundance over rocky substrates changes from 18 to 49×10^6 cells cm^{-2}; in the summertime, these values are by an order of magnitude lower [20]. Over the surface of Gracilarias, it varies during the year from 5 to 366×10^3 cells cm^{-2} with a maximum in the spring period and a subsequent summer decrease [21]. Throughout the year, the biomass of the diatom fouling varies from 0.1 to 0.82 mg cm^{-2}; its mean annual value is 0.15 mg cm^{-2} or 1.5 g m^{-2}. At the surface of mussel shells, the biomass is multifold higher and is comparable to the mean phytoplankton biomass in the euphotic zone.

Altogether, the benthic and planktonic unicellular algae are represented by 1500 species, among which neritic and littoral diatoms dominate.

3.3 Macrophytobenthos

In the Black Sea, there are 304 species of macrophytes referring to four alga divisions and one division of higher flowering plants [22, 23]. Among the alga, there are no endemic species; this implies a relatively young age of the Black Sea flora. The Mediterranean forms that invaded the Black Sea met favorable conditions and formed dense populations. The flora of the northwestern part of the sea differs from the Mediterranean flora and is closer to the flora of the North Sea [24]. The greatest species diversity is characteristic of Rhodophyceae, Fucophycae, and Chlorophyceae, which cover about 70% of the total number of species. In these taxa, the seasonal changes in the species composition are poorly expressed. Chlorophyceae are the first to grow (February), followed by Rhodophyceae (March) and Fucophycae (April). The maximal number of species is observed in the spring months. At depths greater than 25 m, the qualitative composition of macrophytobenthos almost does not change throughout the year [22].

In the years of the maximal anthropogenic load (1960–1975), the species structures of the phytocoenoses suffered significant changes that were manifested in the disappearance of selected alga species or in their replacement by other species. The brown alga *Cystoseira barbata f. Hoppii (Ag.)* has degraded or even disappeared from near-shore phytocoenoses due to its extreme sensibility to contamination and excess amounts of nutrients. The Cystoseira biocoenoses, which involved many alga, invertebrate, and fish species, has disappeared together with its governing species. It was replaced by the Ñladophora, Enteromorpha, and Porphyra phytocoenoses [24]. The area occupied by Phyllofora has sharply decreased. The largest aggregation of free Phyllofora was observed in the northwestern part of the sea at depths of 25–60 m. In the middle of the twentieth century, the area of the Phyllofora field reached 11 000 km^2, while the alga biomass reached about 7–10 Mt wet weight. In the middle of the 1980s, the area of the field and the alga biomass decreased by factors of 20 and 50, respectively [25]. This resulted in a sharp reduction in the number of fish and invertebrate species that dwelled in the Phyllofora thickets.

According to the estimates of many scientists, this phenomenon was caused by the anthropogenic eutrophication that resulted not only in serious economic losses (because of the loss of the raw material for industrial production of agaroid) but also in catastrophic aftereffects for the entire ecosystem of the northwestern shelf.

4 Species Diversity of Fauna

The fauna of the Black Sea is extremely diverse and includes about 2000 species referring to 21 types [25]. Some of these types are represented by several hundred species, while others consist of a few species. The pelagic zone is inhabited by planktonic organisms that dwell in a suspended condition; being free-swimming species not connected to any solid substrate, they are not capable of resisting currents. This group includes zooplanktonic organisms that may be joined in a few size groups: microplankton (< 0.2 mm), mesoplankton (0.2–20 mm), macroplankton (2–20 cm), and megaplankton

(0.2–2.0 m). The animals that can resist currents are referred to as nekton. Meanwhile, it is impossible to trace a distinct boundary between plankton and nekton. Many plankton species transfer to nekton or benthos at different stages of their individual development; this especially refers to the larvae of fish and large invertebrates. In the regions of the continental shelf, plankton includes veligers and larvae of echinoderms, polychaets, and barnacles. The benthal zone (the bottom) is inhabited by zoobenthos that is also distributed over size categories.

4.1 Plankton

Among the unicellular animals, ciliates and other protozoans compose a significant part of zooplankton. Planktonic ciliates involve many nontestaceous forms referring to the Oligotrichida order. Testaceous ciliates of the Tintinnidea order are represented by a greater number of species, though they never develop in mass amounts. With respect to abundance and biomass, species of the *Strombidium* genus dominate. The mass ciliate species *Mesodinium rubrum*, which serves as a marker of eutrophic properties of the water, reaches high abundance and tinctures the water with red and bordeaux. Inside this ciliate, symbiotic algae dwell; they contain pigments that provide the corresponding water coloration. In all, 180 ciliate species were described in the Black Sea; most of them refer to the species widely spread in pelagic zones and estuaries [26]. In the planktonic community, a significant role belongs to *Noctiluca scintillans*; this species is a saprophyte and predator and, with respect to its functional characteristics, it may be regarded as an animal. *Noctiluca* comprises a significant part of the total zooplankton biomass all the year round. Owing to its high abundance, it forms glowing fields in the surface layer of the sea.

Many organisms of mesoplankton cannot dwell in the Black Sea because of the reduced water salinity [27]. Of the 525 copepod species known in the Mediterranean Sea (http:/www.copepod.obs-banyuls.fr), only ten species (or 57 species together with interstitial and brackish-water forms) are encountered in the deep-water regions of the Black Sea; all of them are eurihaline organisms typical of the plankton of marginal seas [28,29]. According to the results of the studies, the list of meso-, macro-, and meroplankton was replenished. At present, it includes 247 freshwater and marine species or 189 species if freshwater species are excluded [29,30]. Coelenterata includes about 40 species; among them are the large *Aurelia aurita* medusas and *Rhizostoma pulmo* that feature a complicated developmental cycle. They spend a part of time on the floor being attached to plant branches and then transfer to plankton. Among the ctenophores, widely known are *Mnemiopsis leidyi* and the sea cucumber *Beroe ovata*, which were delivered to the Black Sea in 1982 and 1997, respectively. During a short period, *M. lei-*

dyi spread first in the near-shore waters and then in the open regions of the sea. By the end of the 1980s, its total stock in the Black Sea reached 780 Mt [18]. The enormous abundance of this predator, feeding mainly on zooplankton, altered both the structure of the planktonic community and the food chains. The abundance of the copepods *Paracalanus parvus* and cladoceras *Penilia avirostris* and *Pleopis polyphemoides* decreased several times and that of Sagitta dropped more than tenfold, while *Acartia clausi* became the dominating species in the near-shore regions of the sea. Instead of the trophic chain "zooplankton–planktivorous fish", the energy flux was directed over the chain "zooplankton–Mnemiopsis". According to the data of FAO (http://www.fao.org/docrep/003/w4248e/w4248e10.htm), the hauls of live aquatic resources and anchovy by the Black Sea countries has been growing from 1970 to 1988 (Fig. 1). Meanwhile, from 1989 to 1991, their sharp reduction has occurred; it coincided in time with the maximal Mnemiopsis development. The annual economic losses of the Black Sea countries because of the ctenophore invasion reached a few million dollars. After the mass outburst, its abundance and biomass gradually decreased. This process had proceeded rather slowly until the appearance of the new invader – the ctenophore *Beroe ovata* in 1997. Under the conditions of the Black Sea, it consumed exclusively Mnemiopsis and, to a lesser extent, Pleurobrachia [31–33]. With its appearance, the abundance of Mnemiopsis began to rapidly drop and the anchovy hauls gradually restored.

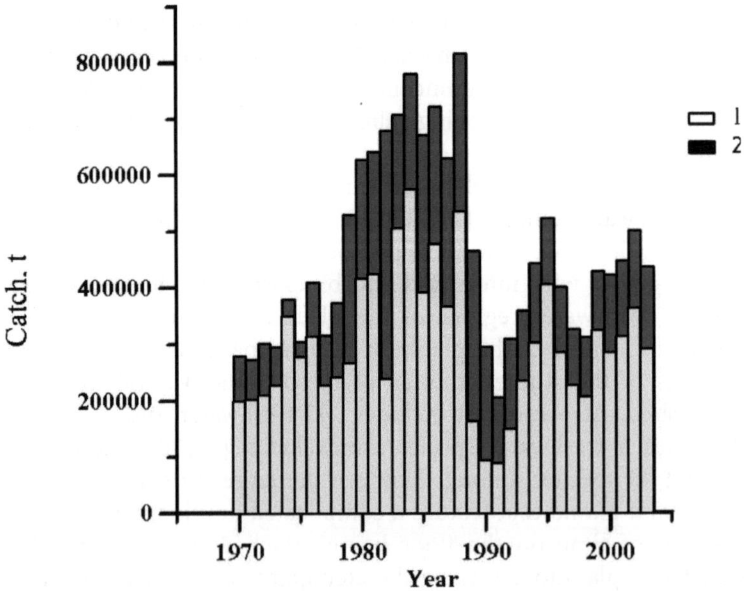

Fig. 1 Dynamics of the catches of (*1*) anchovy and (*2*) biological resources in the Black Sea

4.2
Benthos

The number of species of benthic animals is several times as great as that of zooplankton (Table 1). High species diversity is characteristic of sea worms such as Turbellaria, Nematodes, and Polychaets; they are represented by no less than 450 species. Mass species include Bivalve mollusks (about 100 species), among which widely spread are the mussel *Mytilus galloprovincialis*, the cockle *Cerastoderma lamarcki*, the donacilla *Donacilla cornea*, and the phaseolin *Modiolus phaseolinus*. Gasteropods are represented by 115 species. Among higher crustaceans, the most spread are species of the Isopoda (30 species), Decapoda (40 species), and Amphipoda (110 species) orders [34]. The most spread isopod species is *Idotea baltica*; it dwells in the macrophyte accumulations and in the near-shore thickets. Decapod crustaceans prefer dwelling in shallow-water areas in sea grass thickets, on sandy sediments, and on underwater rocks and stones. Their largest representative – the stone crab *Eriphia verrucosa* – dwells off rocky coasts. Selected shrimp species have an economic value, among them the grass (*Palaemon adspersus*) and stone (*P. elegans*) shrimps. The major part of the benthic fauna (1200 species) has a Mediterranean–Atlantic origin; about 300 species refer to the freshwater and Caspian assemblages [35]. Selected mollusks were occasionally driven to the Black Sea; among them, the widest spread is the gasteropod mollusk *Rapana venosa*, transferred from the Sea of Japan. To date, the list of benthic species is being permanently replenished by species that are new for the science or for the sea [35, 36]. Meanwhile, one can note the absence of selected previously known species in the samples; probably, this results owing to their low recurrence rate. A comparison between the numbers of species registered off the Crimean coasts before 1975 and from 1980 to 1990 shows that, during the past 25 years, the species richness has suffered no significant changes [35].

4.3
Nekton

The ichthyofauna of the Black Sea is characterized by significant species diversity; they refer to two classes – Cartilaginoid and Neopterygii. It numbers about 160 species and subspecies taxa, which is approximately three times less than in the Mediterranean Sea [27, 37]. The major part of the ichthyofauna (60%) consists of species of a Mediterranean–Atlantic origin that permanently dwell in the Black Sea. The ichthyofauna also includes Pontian–Caspian relics (bullhead and sardelle), Boreal–Atlantic relics (sprat, Black Sea salmon, and whiting), and freshwater fish inhabiting near-mouth regions of the sea (carp, bream, pikeperch, bass, and others) (Fig. 2). In the wintertime, selected fish species (mackerel, bonito, tuna, Atlantic mackerel, and common bass) migrate to the Sea of Marmara and to the Mediterranean Sea. The Black

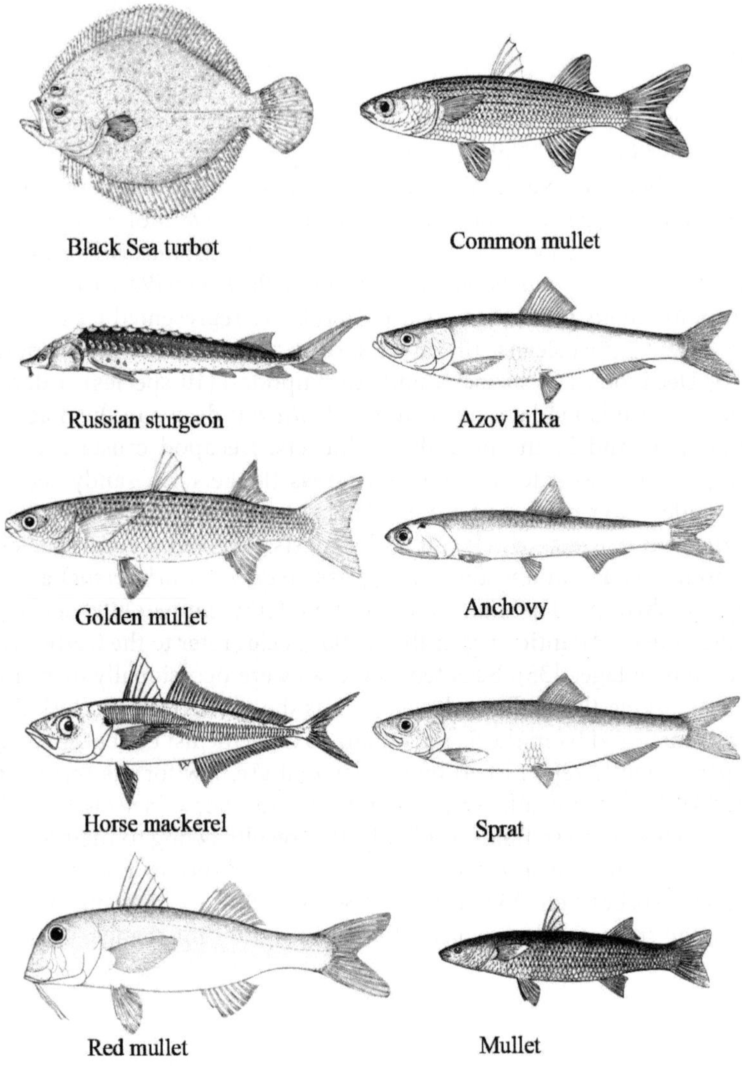

Fig. 2 Principal species of commercial fish of the Black Sea

Sea population of mackerel migrate to the Sea of Marmara for wintering and spawning; then, in the summer, it returns to the Black Sea for feeding and fattening. One part of the population moves northward along the western coasts to the regions of Odessa Bay and Evpatoriya; the other part moves to the Caucasian coast. The hauls of mackerel have reached a few ten thousand tons, but at the beginning of the 1970s, mackerel suddenly disappeared. At present, the mackerel population permanently dwells in the Aegean Sea and never enters the Black Sea [25]. Probably, this is related to the pollution of the northern part of the Sea of Marmara, which fish are not able to overcome.

In recent years, a decrease has been observed in the migration of the bonito *Sarda sarda* and the tuna *Thunnus thynnus thynnus*, which are large pelagic predators. In May, they enter the Black Sea for fattening and spawning and then, at the end of the summer, they pass to the Sea of Marmara and the Aegean Sea for wintering. The main bonito and tuna fishery is performed on the routes of their migration; in successful years, the hauls reach 30–100 kt. However, the hauls of these fish are not stable and have strongly decreased in recent years. Pelagic predators important from the economic point of view also include the Black Sea mackerel, which feeds on small fish and zooplankton. In the Black Sea, it is represented by two subspecies: the small (*Trachurus mediterraneus ponticus*) and large (*T. mediterraneus mediterraneus*) forms. The large mackerel seems to be a hybrid that resulted from crossing of the Black Sea and Mediterranean subspecies [40]. Mackerel forms three local populations: the northern, eastern, and southwestern, different in their localizations, times of reproduction, growth rates, and migration routes. The northern mackerel population dwells off the Crimean coasts including the Kerch region, while the western population is localized near the western coast and in the Bosporus region; in the winter, it migrates to the Sea of Marmara or to the northern coasts of Turkey. The habitat of the eastern population extends along the Caucasian coast from the Kerch region to the coasts of East Turkey. Mackerel winters near coasts at depths up to 100 m. Its lifetime is 8–13 years; the largest mackerel individuals (about 30 cm long) are encountered in the southwestern population.

Planktivorous fish include species important from the economic point of view such as anchovy, sprat, sardine, herring, and sardelle. The most abundant pelagic fish of the open sea, anchovy, is represented by two subspecies: the dark Black Sea subspecies (*Engraulis encrasicolus ponticus*) and the lighter colored Azov subspecies (*Engraulis encrasicolus maeoticus*). The Black Sea anchovy permanently dwell in the Black Sea and forms two populations: the eastern and western ones. The Azov subspecies reproduces in the Sea of Azov and migrates to the Black Sea in the winter. With respect to the haul volumes, anchovy is the principal commercial fish of the Black Sea; in selected years, all the Black Sea countries hauled more than 0.5 Mt of anchovy [40] Meanwhile, owing to the intensive extraction and to the sharp outburst in the abundance of the ctenophore mnemiopsis at the end of the 1980s, the hauls of anchovy strongly decreased (Fig. 1). Due to its high abundance and large size, mnemiopsis won the competition with anchovy, consuming zooplankton, eggs, and fish larvae. After the occasional invasion of the ctenophore *Beroe ovata* that feeds mainly on *Mnemiopsis*, the anchovy catches began to restore and at present they are about 0.3 Mt year^{-1}.

The Black Sea sprat (*Sprattus sprattus*) also refers to mass fish species of the Black Sea basin and occupies an important place in fishery, especially in recent years. This is related to the creation of more perfect hauling instruments and to the decrease in the abundance of predatory fish and dolphins that feed on sprat. The hauls of sprat are performed in the autumn and winter when fish shoals approach the coasts of the Crimea and Caucasus. The major part of the sprat hauls falls on the regions off the Crimean coasts, where they reach about 50 kt $year^{-1}$. Some scientists suggest that, in the Black Sea, there are two sprat populations: the western and eastern ones [41]. Off the coasts of Bulgaria and on the southwestern shelf of the Crimea, the western Black Sea sprat population consists of two subpopulations; one of them dwells in the open waters and the other gravitates toward the coast [40]. The local sprat subpopulations differ in their growth rates, sizes, and reproduction times.

Among the bottom fish, mullet, plaice, gallinule, and sturgeon have commercial value. In the Black Sea, five mullet species are observed; among them, the most widely spread are the mullet (*Mugil cephalus*) and the haarder*Liza haematochila*, introduced in the 1970s from the Japan Sea. These large fish are tasty and have an important commercial value. The reproduction of mullet takes place in the middle of the summer. It refers to those rare species that migrate for spawning to the open sea regions; there, a few tens of kilometers away from the shore, it produces eggs and then leaves the spawning site. In the autumn, when the temperature decrease, the grown-up generations enter the sea and move southward to overwinter off the coasts of Bulgaria, Turkey, and the Caucasus. In the spring, young individuals come back to the northern coasts [25]. Mullet consume epiphyte organisms that they gather from the surfaces of rocks and floor. By the following autumn, mullets reach commercial size with a mass of about 0.5 kg.

In the Black Sea, three mixed mullet populations are recognized. The Caucasian (eastern) population winters off the Caucasian coast and performs migrations for feeding and spawning to the Kerch Strait and partly to the Sea of Azov and its lagoons. The Crimean population winters off the southern coasts of the Crimea; in the spring, one part of the population moves for feeding to Karkinit Bay and the other penetrates to the Sea of Azov via the Kerch Strait. The Balkan population winters in the northwestern part of the sea off the coasts of Bulgaria and in the near-Bosporus region; in the summer, it propagates along the coast toward Odessa Bay [40]. Thus, a complete spatial isolation of different mullet populations is observed only in the winter period. The main mullet fishery is performed in the summer over the routes of its fattening–spawning migrations.

Among the bottom commercial fish, an important role belongs to the plaice–turbot (*Psetta maxima maeotica*), which inhabits the shelf zone along all of the coasts; its maximal accumulations are found in the northwestern part of the sea and in the Kerch region. Within the shelf, turbot prefer sites with smooth sandy or silty–sandy floor at depths of 50–100 m. In shallow-water zones (5–40 m), they appear only in the spring in the spawning period. Turbot feed on small near-bottom fish and benthos. Their mean size is 30–50 cm at a weight of about 4 kg. In the 1950s, the total catch of plaice reached 40 kt; at present, they are an order of magnitude lower.

Among the small near-bottom species, the gallinule *Mullus barbatus ponticus* has the greatest significance. It lives in the near-shore waters forming large accumulations in the areas with soft sediments. With respect to the features of its lifecycle, four geographical subpopulations are distinguished: the Caucasian, the Crimean, the Balkanian, and the Turkish subpopulations. The main hauls of gallinule are performed in the spring and autumn off the Crimean and Caucasian coasts [40].

At the beginning of the twentieth century, more than 50% of the total catches consisted of mackerel, mullet, herring, beluga, and sturgeon, while the proportion of plaice, gallinule, horse mackerel, stellate sturgeon, and sardelle comprised about 20%. By the end of the twentieth century, the eutrophication of the near-shore regions of the sea resulted in a significant reconstruction of all the elements of the sea ecosystem and to changes in the species composition in the fish hauls. The catches of sprat and anchovy increased to 98.5%, while all the other species including sturgeon, plaice, mullet, gallinule, and horse mackerel made only 1% of the total hauls [40].

As one can see, most of the valuable fish species that were caught before have lost their commercial significance. Large pelagic predators were removed from the ichthyofauna, while short-cycling species such as sprat and anchovy reached their mass development. From the 1970s to 1995, a decrease in the species number, including the commercial species, was observed. For example, by 1990, the number of species in the waters off the Crimean coasts has decreased almost twofold and the list of commercial species had lost the small form of the Black Sea horse mackerel, and the catches of anchovy decreased [42]. The economic crisis in the 1990s in the Black Sea countries led to a decrease in the agricultural use of mineral fertilizers, to a reduction of the fishery fleet, and to industrial and municipal contamination. The decrease in the anthropogenic load resulted in signs of restoration of ichthyofauna. For example, off Sevastopol, there were 18 fish species at the beginning of the 1990s and 30 species in 2002 [43]. In the 1990s, 48 species were registered in Odessa Bay; now, the ichthyofauna there numbers 58 sea fish species [44]. These facts confirm the improvement of the general ecological situation during recent years.

4.4 Total Number of Species

The Black Sea is one of the best-studied basins of the world; however, due to various reasons, it is difficult to determine the exact number of species dwelling in it. General information about the species diversity of the Black Sea is presented in Table 1, where the numbers of species of different types and classes of animals and plants known at present are listed. A total of 4073 species attributed to 104 types and classes are included in the list. Flora and fauna are represented by 1800 and 2273 species, respectively.

Table 1 Number of species and taxons dwelling in the Black Sea

Groups	Number of taxa [b]	Number of species	Refs.
Phytoplankton	12	746	[13]
Microphytobenthos	12	750	[20]
Macrophytobenthos	5	304	[22]
Ciliates	9	180	[26]
Zooplankton [a]	14	247	[29]
Macrozoobenthos	26	747	[21, 35]
Meiobenthos	8	522	[36]
Fishes	2	192	[37]
Parasitic organisms	11	305	[38]
Fungi	3	76	[39]
Mammals	2	4	[27]

[a] Meso- and macrozooplankton and meroplankton
[b] System of divisions and classes is applied

Taking into account the limitations of methods for determining the species composition of viruses and bacteria, we may suggest that the total number of species should be somewhat greater. In terms of the species composition diversity, the pelagic zone is poorer than the bottom fauna and flora. The multicellular fauna includes about 2000 species, which agrees with previous estimates [45].

5 Sea Productivity

To estimate the productivity of marine ecosystems, two parameters are usually considered: the rate of the production of organic matter by plant organisms per unit time (primary production) and the specific rate of the increase in the amount of the matter synthesized by organisms or groups of organisms per unit time.

5.1 Primary Production

In Black Sea research, great attention has been paid to studies of the primary production. The first studies performed at the beginning of the 1960s allowed scientists to estimate the levels of the production in characteristic sea regions [46–48]. They showed that the daily production values in different regions of the sea differ by an order of magnitude. The maximal values ($1-3\ \mathrm{g\,C\,m^{-2}\,day^{-1}}$) are typical of the regions with high phytoplankton biomasses and high nutrient concentrations located it shallow-water northwestern part of the sea and in bays and bights, while the minimal values (0.1–0.5 $g\,C\,m^{-2}\,day^{-1}$) are confined to the deep-water part of the sea, where concentrations of nutrients in the upper mixed layer are close to zero and the biomass of phytoplankton is by an order of magnitude lower than that in the shallow-water areas. The studies performed in 1980–1990 confirmed the general features of the spatial variability in the phytoplankton production and the ranges of its changes in deep-water sea regions during the summer and spring periods [49, 50]. Meanwhile, in the northwestern part of the sea, the level of the primary production became somewhat higher than in the 1960s. In the Odessa region, its average value was 1.8 $g\,C\,m^{-2}\,day^{-1}$, while in near-mouth areas it was 3–10 $g\,C\,m^{-2}\,day^{-1}$ [51]. The Danube waters supply great amounts of nutrients together with dissolved and particulate organic matter, thus causing contamination and intensive phytoplankton development. During frequent alga "blooms", especially in the summer period, the production of selected species such as, for example, *Gonyaulax polyedra* in the surface layer alone may reach 1.5 $g\,C\,m^{-2}\,day^{-1}$ at chlorophyll *a* concentrations of 107 $mg\,m^{-3}$ [18].

The first estimates of the annual phytoplankton production were based on the assumption that the measurements performed during the summer minimum and the autumn maximum of the phytoplankton development may be averaged and used for the calculations of mean annual values [47, 50, 52]. However, the studies carried out in the winter–spring period in the western and eastern parts of the sea [53, 54] showed that, in these seasons, an intensive development of diatoms is observed. The winter–spring development of phytoplankton starts in January–February over the domes of vast cyclonic gyres, where the main pycnocline rises toward the sea surface preventing cells from removal from the zone of photosynthesis [54]. Meanwhile, the waters with high nutrient concentrations can easily penetrate into this zone and the phytoplankton community that intensively develops over the pycnocline has no limitation in phosphorus and nitrogen. Under these conditions, in selected years, *Nitzschia delicatissima* dominates, while in the other, large diatoms such as *Pseudosolenia calcar-avis*, which are retained mostly due to their intensive reproduction and high pycnocline position, prevail. The phytoplankton biomass is distributed evenly from the sea surface to the depth of the main pycnocline. M.E. Vinogradov and coauthors note that, according to the observations from a manned submersible, it is clearly seen how the zones of blooming turbid waters, which form layers and locks of greenish fog, are sharply changed by the transparent waters of the pycnocline [18]. Over the domes of the gyres, the primary production level exceeds 1 $g\,C\,m^{-2}\,day^{-1}$ and decreases from the center to the periphery. Thus, it was established that, in February–March, the conditions are favorable for the phytoplankton growth owing to the relatively high position of the pycnocline, sufficiently high illumination, high nutrient contents, and low zooplankton abundance. In April, the seasonal temperature gradient deepens and the vertical transfer of nutrients is suppressed. Therefore, a sharp drop in the phytoplankton biomass and the primary production (0.25–0.60 g C m^{-2} day^{-1}) is observed, the growth rate of diatoms decreases, and diatoms are replaced by the representatives of different alga groups such as Coccolithophores and Pyrrophyta.

A few species of the *Nitzschia* genus and *Rhizosolenia alata* are retained in the 40–60 m layer, forming dense accumulations, which are well visible in the vertical profiles of fluorescence during the summer and autumn [55]. In May–September, a summer species assemblage develops based on the representatives of the Dinophyceae and Prymnesiophyceae divisions. Dinophyceae dominate with respect to the biomass, while coccolithophores dominate with respect to the abundance. During this period, the phytoplankton production is minimal, falling to 0.2–0.4 g C m^{-2} day^{-1}. Meanwhile, in June and July, when an intensive coccolithophore development occurs, the primary production values may reach 1.0 g C m^{-2} day^{-1} [56]. In October and November, the autumn peak of phytoplankton is observed with a domination of the diatoms *N. seriata*, *Cerataulina bergonii*, and *Pseudosolenia calcar-avis* [49, 57] and with a primary production level of 0.4–0.6 g C m^{-2} day^{-1} [58]. On the whole, the high primary production of phytoplankton is provided by a relatively small number of dominating species and, therefore, the species diversity of algae and their production change in opposite directions. According to numerous data obtained in the deep-water areas of the sea from 1970 to 1992, the annual production of phytoplankton was 115–150 g C m^{-2} [56, 59]. Evaluating the correctness of the calculations of the annual primary production, one should have in mind that these values were obtained from averaging individual disconnected measurements performed in different years and seasons; thus, they do not account for the interannual variations in the phytoplankton development, which were registered by direct and remote determinations of the chlorophyll concentrations in the surface layer [60, 61]. The estimates of the production values in deep-sea regions with the use of satellite data for the period 1997–2004 yield higher values of up to 140–270 g C m^{-2} $year^{-1}$ (our unpublished data).

In the phytoplankton development, one can recognize a seasonal succession, whose variability depends on the meteorological conditions over the sea area. For example, the cold winter of 1998 resulted in a decrease in the water temperature in the surface layer down to 7–8 °C and *Proboscia alata* dominated during the spring maximum [49]. The next year, according to satellite data, the water temperature never fell below 9 °C and the species that dominated in the cold winter comprised only a small proportion of the total cell number. In that time, Dinophyceae and Prymnesiophyceae dominated the phytoplankton community, while the primary production value was twice as small as in the cold winter (Fig. 3). One can suppose that, during the mild winter, certain mixing of the water layers proceeded and, as a result, the nutrient content was low and limited the development of diatomaceous algae.

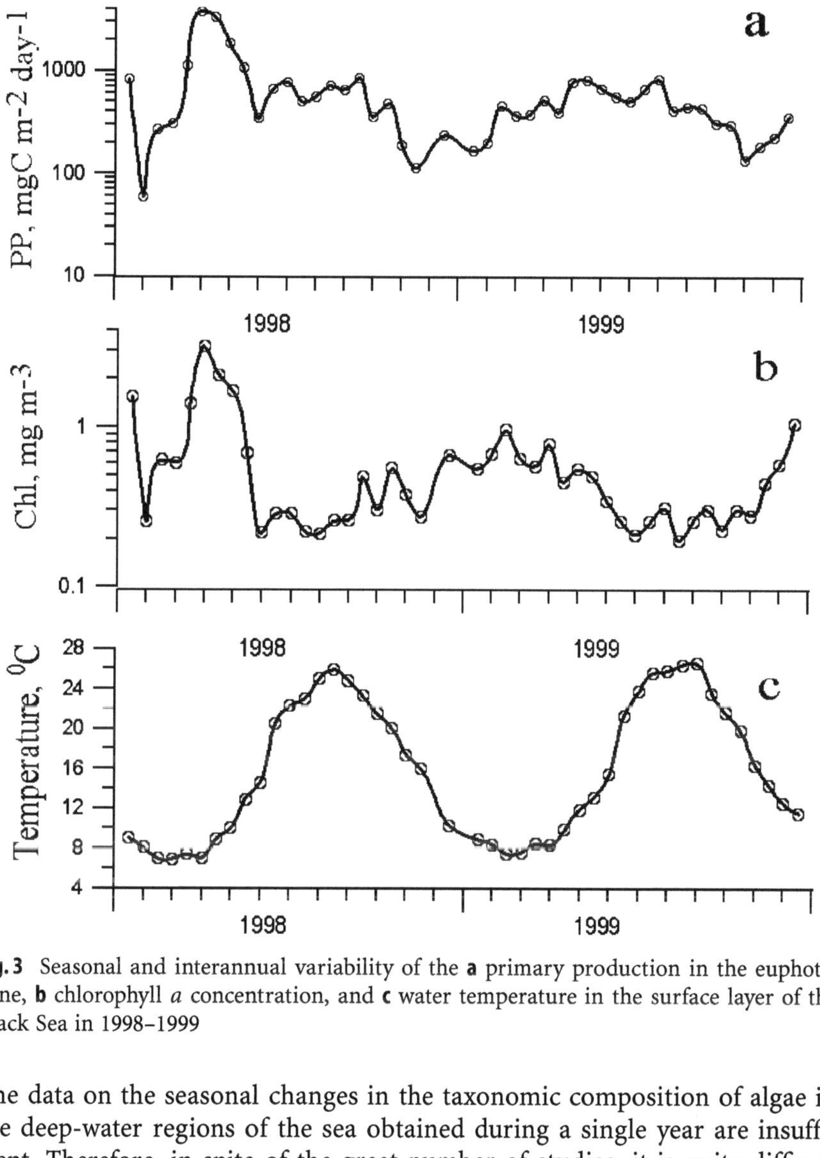

Fig. 3 Seasonal and interannual variability of the **a** primary production in the euphotic zone, **b** chlorophyll *a* concentration, and **c** water temperature in the surface layer of the Black Sea in 1998–1999

The data on the seasonal changes in the taxonomic composition of algae in the deep-water regions of the sea obtained during a single year are insufficient. Therefore, in spite of the great number of studies, it is quite difficult to answer the question: what are the particular features of the phytoplankton succession and the production cycle in areas with different hydrological regimes with respect to climatic fluctuations?

In recent years, in order to study the features of phytoplankton development, satellite observations have been actively applied (see chapter "Seasonal and interannual variability of remotely sensed chlorophyll" in this volume). The use of satellite data allows one to actually observe the spatial changes in phytoplankton on scales from tens to hundreds of kilometers that occur virtually simultaneously, and its temporal changes on scales from a day to a few years. The studies of phytoplankton from satellites are widely spread, although numerous problems related to the conversion of the optical characteristics measured, to the pigment concentrations, and then to the phytoplankton production are not yet resolved. Their solution requires additional information that may be acquired only through direct measurements in the sea.

To study the seasonal and spatial changes in the phytoplankton biomass, we used the determinations performed first with a CZCS radiometer and then with a SeaWiFS color scanner; the results were reduced by inserting a correction for the differences between the satellite data and the direction measurements of the concentrations of chlorophyll *a* and phaeophytine in characteristic sea regions [62, 63]. The results of the measurements showed significant differences in spatial distributions of chlorophyll in the warm and cold periods of the year. In the summer, chlorophyll concentrations decreased from the coasts toward the deep-water part of the sea and their minimal values were registered in the central regions of the eastern and western gyres. In the winter and early spring, the pattern of distribution was opposite: the maximal values of chlorophyll concentration were observed in the central parts of the gyres and the minimal values were confined to the near-shore regions.

According to the satellite (SeaWiFS) data, the seasonal changes in the chlorophyll concentration in the surface layer of deep-water regions in the western and eastern parts of the sea have similar features, with a maximum in February and March and a minimum in July and August. This kind of dynamics is characteristic of subtropical rather than of temperate latitudes. The reason for the intensive phytoplankton development in the winter and early spring lies in the fact that, at that time, the pycnocline in the central parts of major gyres is located at depths corresponding to the zone of photosynthesis. The winter convection involves the layer down to the pycnocline and the phytoplankton community permanently dwells in the euphotic zone, which, at that time, is restricted by the surface layer down to depths of 30–35 m. In addition, the annual cycling of zooplankton is characterized by a maximum in the summer and a minimum in the winter. Therefore, in the cold season, the rate of the alga growth is higher than the rate of their consumption by zooplankton and the metabolism of the planktonic community is shifted toward the autotrophic exchange. Under the summer temperature stratification, the nutrient flux forming deeper layers is suppressed; therefore, equilibrium is reached between the alga consumption by zooplankton and the rate of their growth. This results in a decrease in the phytoplankton biomass down to a certain stationary level. Subsequently, with the destruction of the seasonal temperature gradient and the more intensive nutrient supply to the upper water layers, the biomass of phytoplankton gradually increases and tends toward its winter–spring maximum.

In the near-shore region off Sevastopol and Gelendzhik, three to four maximums of the phytoplankton biomass and primary production are observed in the spring, summer, and autumn [64, 65]. Over 10 years, off the Rumanian coasts in the region of Constantsa, 46 events of water "bloom" related to 15 alga species have been observed; off the Bulgarian coasts, blooms are caused by 24 species [66, 67]. In the region of Odessa Bay, from 1973 to 1997, 135 events of water bloom have been registered; they were caused by 41 alga species [68]. The greatest number of species that cause the bloom are diatoms, although selected Pyrrophyta species such as *Prorocentrum cordatum* and *Gymnodinium sanguineum* may also provide high abundances (224 and 140 million cells l^{-1}). Under high nutrient concentrations, these values of cell abundance may be retained for one or two months. During this time, the species diversity is at a minimum, which results in the extreme instability of the phytoplankton community and the entire ecosystem.

The reason for the mass phytoplankton development lies in the intensification of agriculture in the region of the sea with a sharp increase in the use of fertilizers. This resulted in the excess delivery of nutrients, mostly of phosphorus and nitrogen, with riverine runoff and atmospheric precipitation. The anthropogenic eutrophication caused serious negative aftereffects in the sea ecosystem. This was especially noticeable in the northwestern part of the sea, where, over a few decades, the integrate phytoplankton biomass increased by a factor of a few tens and, for selected species, even greater. Planktonic animals have no time to consume an alga amount that great; therefore, after vegetation, they die and precipitate over the floor. The benthic animals feeding on detritus can neither consume this amount of organic matter and it is destroyed by bacteria that, in the process of their metabolism, utilize all the oxygen dissolved in the water. After the oxygen is gone, the organic matter starts rotting away and is decomposed by anoxic bacteria with release of hydrogen sulfide, which is toxic for most of the bottom organisms. This way, conditions are created for hypoxia formation in anoxic conditions and a hydrogen sulfide contamination of seawater, which results in a mass mortality of organisms.

5.2 Secondary Production

In contrast to the primary production, which is estimated by a wide use of radiocarbon techniques, the methods for calculation of the secondary production are based either on the data for the rate of weight growth at different stages of their lifecycle and their abundance, or on the use of physiological characteristics of the organism such as the daily ration, the proportion of the assimilated food in the ration, and the respiration. The data obtained with these methods allow estimation of the values of production of the main species of copepods and other animals [69]. The mean annual values of the production of crustaceans in the 100-m layer are 120 mg C m^{-2} day^{-1} or about 40 g C m^{-2} $year^{-1}$, which is approximately four times as low as the primary production values. The main proportion of the production (60%) is generated during the summer period, while the shares of other seasons range from 10 to 25%. The interannual changes in the secondary production of crustaceans feature twofold changes. In the near-shore zone, on average, the values of the secondary production of crustaceans are higher than in deep-sea regions by a factor of 1.5. The productions of protozoans and crustaceans are approximately equal, while the sum production of *Sagitta*, *Noctiluca*, and medusas is half that of crustaceans. On the whole, in the Black Sea ecosystem, the production of phytoplankton is rather completely utilized, which is confirmed by the relatively high ratio between the primary production and the production of crustaceans–euriphages, whose average value is about 0.1. Meanwhile, the production of planktivorous fish comprises only 6% of the production of fodder zooplankton [69]. This seems to be related to the fact that *Sagitta*, medusas, and ctenophores are competitors of fish for zooplankton consumption.

6 Conclusions

Studies on the biological diversity and productivity of the organisms dwelling in the Black Sea have a long history and are relatively completely represented in publications. A comparative analysis of the taxonomic composition, abundance, and biomass of different groups of organisms shows that the diversity of taxa and the species richness in the Black Sea are high, though somewhat lower than in the Mediterranean Sea. Meanwhile, despite the advances in systematics, the exact number of species that inhabit the Black Sea is still unknown. The compilation of a complete list is impossible because of the continuous changes in species diversity caused by the instability of climatic conditions, invasion and disappearance of species, and changes in the proportions between the dominating and rare species.

In the development of the populations of planktonic and bottom algae one can trace the seasonal dynamics of the phytocoenosis represented by the alternation of the dominating species and by the changes in their abundance and biomass. In the years different with respect to the climatic conditions, the species composition as well as the dynamics of the abundance and production of algae strongly varies. With respect to the level of the primary production, the Black Sea ecosystem may be referred to as a mesotrophic-type of marine basin. In mesotrophic waters, the phytoplankton consumption by zooplankton, which represents an important food object for fish with short lifecycles, proceeds more intensively than in oligotrophic waters. The coefficient of matter transfer from the primary production to the higher trophic levels is about

10%. The highest intensity of production processes is observed on the shelf in the northwestern part of the sea and in near-shore regions, where the matter transfer between the trophic levels is less efficient. However, these regions are more important for commercial fishery than the deep-sea areas, because here the phytoplankton production is multifold higher and, even at low values of the transfer coefficient, the amount of food for phytophages is greater. The total stock of commercial fish and the volumes of their extraction are subjected to strong interannual changes. They are related to the instability of the climatic conditions, to the interannual variations in the ecosystem productivity, and to the intensity of exploitation of fish resources.

References

1. Carlton JT (1993) Am Zool 33:499
2. Myers RA, Worm B (2003) Nature 423:280
3. Myers RA, Worm B (2005) Philos Trans R Soc Lond B 360:13
4. Costanza R, d'Arge, de Groot R et al. (1997) Nature 387:253
5. Costello MJ (2001) Oceanis 24:25
6. Costello MJ, Bouchet P, Emblow C, Legakis A (2006) Mar Ecol Prog Ser 316:257
7. Piontkovski SA, Landry MR, Finenko ZZ (2003) Oceanol Acta 26:255
8. Reingard LV (1910) Trudy Novoros. Obschestva ispytatel prirody 43:3 (in Russian)
9. Usashev PI (1947) Successes of modern biology 23:265 (in Russian)
10. Morozova-Vodynitskaya NV (1948) Trudy Sevastopol. Biol Sta 6:39 (in Russian)
11. Morozova-Vodynitskaya NV (1954) Trudy Sevastopol. Biol Sta 14:11 (in Russian)
12. Ivanov IA (1965) Research of plankton of the Black and of Azov seas. Naukova Dumka, Kiev, p 17 (in Russian)
13. Pitzyk GK (1979) Systematic composition of phytoplankton. In: Greze VN (ed) Bases of biological productivity of the Black sea. Naukova Dumka, Kiev, p 63 (in Russian)
14. Altuchov DA (2000) Sea Ecology 52:79 (in Russian)
15. Georgieva LV (1993) Phytoplankton. Specific composition and dynamic of phytocenosis. In: Kovalev AV, Finenko ZZ (eds) Plankton of the Black Sea. Naukova Dumka, Kiev, p 31 (in Russian)
16. Polikarpov IG, Saburova MA, Manzhos TV et al. (2003) Microplankton biological diversity in the Black Sea coastal zone near Sevastopol (2001–2003) In: Eremeev VN, Gaevskaya AV (eds) Modern condition of biological diversity in near-shore zone of Crimea (The Black sea sector). Ekosi-Gidrophizika, Sevastopol, p 16 (in Russian)
17. Nesterova DA (2006) Opened regions. In: Zaitsev YP, Aleksandrov BG, Minicheva GG (eds) North-western part of the Black sea: biology and ecology. Naukova Dumka, Kiev, p 175 (in Russian)
18. Vinogradov ME, Sapozhnikov VV, Shushkina EA (1992) The Black sea ecosystem. Nauka, Moscow, p 111 (in Russian)
19. Sukhanova IN, Cheban E (1990) Oceanology 30:979 (in Russian)
20. Nevrova EL (2003) Species abundance of the bottom diatom algae at the Crimeam coast. In: Eremeev VN, Gaevskaya AV (eds) Modern condition of biological diversity in near-shore zone of Crimea (The Black sea sector). Ekosi-Gidrophizika, Sevastopol, p 271 (in Russian)

21. Ryabushko LI (2006) Microalgae of the Black Sea. Ekosi-Gidrophyzica, Sevastopol, p 143 (in Russian)
22. Kalugina-Gutnik AA (1979) Macrophytobentos. In: Greze VN (ed) Bases of biological productivity of the Black sea. Naukova Dumka, Kiev, p 123 (in Russian)
23. Milchakova NA (2003) In: Macrophytobentos. Eremeev VN, Gaevskaya AV (eds) Modern condition of biological diversity in near-shore zone of Crimea (The Black sea sector). Ekosi-Gidrophizika, Sevastopol, p 152 (in Russian)
24. Minicheva GG (1993) Algology 3:3 (in Russian)
25. Zaitsev YP (2006) An introduction to the Black Sea ecology. Aven, Odessa, p 224
26. Kurilov AV (2006) Microzooplankton (Infusoria). In: Zaitsev YP, Aleksandrov BG, Minicheva GG (eds) North-western part of the Black sea: biology and ecology. Naukova Dumka, Kiev, p 224 (in Russian)
27. Zenkevich LA (1963) Biology of seas of the USSR. Izd AN SSSR, Moscow (in Russian)
28. Zagorodnyaaya YA, Pavlovskaya TV, Moryakova VK (2003) Modern zooplankton condition near the Crimean coast. In: Eremeev VN, Gaevskaya AV (eds) Modern condition of biological diversity in near-shore zone of Crimea (The Black sea sector). Ekosi-Gidrophizika, Sevastopol, p 49 (in Russian)
29. Polischuk LN, Nastenko EV (2006) Meso- and macrozooplankton. In: Zaitsev YP, Aleksandrov BG, Minicheva GG (eds) North-western part of the Black sea: biology and ecology. Naukova Dumka, Kiev, p 229 (in Russian)
30. Aleksandrov BG (2006) Meroplankton. In: Zaitsev YP, Aleksandrov BG, Minicheva GG (eds) North-western part of the Black sea: biology and ecology. Naukova Dumka, Kiev, p 236 (in Russian)
31. Finenko GA, Romanova ZA (2000) Oceanology 40:720 (in Russian)
32. Shiganova TA, Bulgakova YV, Sorokin PY, Lukashev YF (2000) Izv Akad Nauk Russia 2:581 (in Russian)
33. Finenko GA, Anninsky BE, Romanova ZA, Abolmasova GI, Kideys A (2001) Hydrobiologia 451:177
34. Kiseleva MI (1979) Zoobentos. In: Greze VN (ed) Bases of biological productivity of the Black sea. Naukova Dumka, Kiev, p 208 (in Russian)
35. Revkov NK (2003) Macrozoobentos. In: Eremeev VN, Gaevskaya AV (eds) Modern condition of biological diversity in near-shore zone of Crimea (The Black sea sector). Ekosi-Gidrophizika, Sevastopol, p 209 (in Russian)
36. Sergeeva NG, Kolesnikova EA (2003) Meiobentos taxonomical composition in the Crimean region. In: Eremeev VN, Gaevskaya AV (eds) Modern condition of biological diversity in near-shore zone of Crimea (The Black sea sector). Ekosi-Gidrophizika, Sevastopol, p 246 (in Russian)
37. Rass TS (1993) Ichthyofauna of the Black Sea and some stages of its history. In: Oven LS (ed) Ichthyofauna of Black Sea bays in conditions of anthropogenic impact. Naukova Dumka, Kiev, p 6
38. Gaevskaya AV, Korniychuk YM (2003) Parasitic organisms as an ecosystems constituent at the Black sea coast of the Crimea. Eremeev VN, Gaevskaya AV (eds) Modern condition of biological diversity in near-shore zone of Crimea (The Black sea sector). Ekosi-Gidrophizika, Sevastopol, p 425 (in Russian)
39. Kopytina NI (2006) List of water fungi species. In: Zaitsev YP, Aleksandrov BG, Minicheva GG (eds) North-western part of the Black sea: biology and ecology. Naukova Dumka, Kiev, p 555 (in Russian)
40. Zuev GV, Melnikova EV (2003) Ecological (intraspecific) diversity of ichthyofauna. In: Eremeev VN, Gaevskaya AV (eds) Modern condition of biological diversity in near-

shore zone of Crimea (The Black sea sector). Ekosi-Gidrophizika, Sevastopol, p 380 (in Russian)
41. Stoynov SA (1953) Bolgarskaya Akad. Nauk Trudy Inst Zoologii 3:90 (in Russian)
42. Oven LS, Shevchenko NF, Giragosov VE (1997) J Ichtyology 37:806 (in Russian)
43. Zaika VE, Boltachev AR, Zuev GV et al. (2004) Marine Ecolog J 3:37
44. Vinogradov AK (2006) General description. In: Zaitsev YP, Aleksandrov BG, Minicheva GG (eds) North-western part of the Black sea: biology and ecology. Naukova Dumka, Kiev, p 305 (in Russian)
45. Zaika VE (2000) Ecology of Sea 51:59 (in Russian)
46. Winberg GG, Muravleva EP, Finenko ZZ (1962) Trudy Sevastopol Biol Station 17:212 (in Russian)
47. Sorokin YI (1964) Izv Akad Nauk SSSR 5:749 (in Russian)
48. Sorokin YI (1982) Black sea. Nauka, Moscow, p 216 (in Russian)
49. Berseneva GP, Churilova TY, Georgieva LV (2004) Oceanology 4:389
50. Finenko ZZ (1988) Primary production in a summer period. In: Zats VI, Finenko ZZ (eds) Dynamics of waters and productivity of the Black sea. Koordinatsionnyi Tsentr Stran-Chlenov SEV, Moscow p 315 (in Russian)
51. Skripnik IA (2006) Primary production. In: Zaitsev YP, Aleksandrov BG, Minicheva GG (eds) North-western part of the Black sea: biology and ecology. Naukova Dumka, Kiev, p 191 (in Russian)
52. Sorokin YI (1962) Dokl Akad Nauk SSSR 144:914 (in Russian)
53. Sorokin YI, Sukhomilin AV, Sorokina OV (1992) Primary production of phytoplankton in the Black sea at the end of winter–beginning spring 1991. In: Vinogradov ME (ed) Winter state of ecosystem of the opened part of the Black sea. Shirshov Institute of Oceanology, Moscow p 72 (in Russian)
54. Finenko ZZ, Krupatkina DK (1993) Oceanology 33:97 (in Russian)
55. Finenko ZZ, Churilova TY, Li RI (2005) Marine Ecolog J 4:4 (in Russian)
56. Vedernikov VI, Demidov AB (2002) Long-term and seasonal variability of chlorophyll and primary production in the eastern regions of the Black sea. In: Zatsepin AG, Flint MV (eds) Multidisciplinary investigations of the Northeast part of the Black Sea. Nauka, Moscow, p 212 (in Russian)
57. Ratkova TN, Kopylov AI, Sazhin AF et al. (1989) Accumulation of diatoms *Nitzschia spp* in the cold intermediate layer of the Black Sea. In: Flint MV (ed) Structure and some functional characters of plankton communities in the Black Sea. Nauka, Moscow, p 105 (in Russian)
58. Stelmakh LV (2006) Phytoplankton growth and zooplankton grazing in the western part of the Black Sea during September and October 2005. Black Sea Ecosystem 2005 and Beyond. Abstracts Istanbul Turkey, p 94
59. Finenko ZZ (2001) Ecology of Sea 57:60 (in Russian)
60. Nezlin NP (2001) Oceanology 41:394
61. Yunev OA, Vedernikov VI, Basturk O et al. (2001) Mar Ecol Prog Ser 230:11
62. Finenko ZZ (1998) Seasonal phytoplankton cycle in the contrastic ecosystem of the Black Sea. Coastal and marginal seas. Unesco, Paris, p 46
63. Kopelevich OV, Shebestov SV, Yunev O et al. (2002) J Mar Syst 36:145
64. Finenko ZZ (1979) Prodution of phytoplankton. In: Greze VN (ed) Bases of biological productivity of the Black sea. Naukova Dumka, Kiev, p 88 (in Russian)
65. Stelmakh LV, Yunev OA, Finenko ZZ et al. (1998) Peculiarities of seasonal variability of primary production in the Black Sea. In: Ivanov LI, Oguz T (eds) Ecosystem modeling as a management tool for the Black Sea. Kluwer, Dordrecht, p 171

66. Bodeanu N (1993) Microalgal blooms in the Romanian area of the Black Sea and contemporary eutrophication conditions. In: Smayda TJ, Shimizu Y (eds) Toxic phytoplankton sea blooms. Proceedings of the fifth international conference on toxic marine phytoplankton. Elsevier, Amsterdam, p 203
67. Moncheva S, Krastev A (1997) Some aspects of phytoplankton long-term alterations of Bulgarian Black Sea shelf. In: Ozsoy E, Mikaaelyan A (eds) Sensitivity to change: Black Sea, Baltic Sea and North Sea. NATO ASI Series B: Environmental security. Kluwer, Dordrecht, p 79
68. Nesterova DA (2001) Algology 11:502 (in Russian)
69. Sorokin YI (2002) The Black Sea. Ecology and oceanography. Backhuys, Leiden, p 875

Hdb Env Chem Vol. 5, Part Q (2008): 375–406
DOI 10.1007/698_5_066

Published online: 13 February 2007

Introduced Species

Tamara Shiganova

P.P. Shirshov Institute of Oceanology, Russian Academy of Sciences, 36 Nakhimovsky Pr., 117997 Moscow, Russia
shiganov@ocean.ru

Abstract Due to increasing human activities such as shipping, deliberate stocking, and accidental introduction, a high number of alien species have become established in the Black Sea over the last century. In addition, global warming facilitates the population increase of thermophilic species and their northward expansion from the Mediterranean. As a result, the Black Sea became a pivotal recipient area for marine and brackish water aliens. It infects all other seas of the Mediterranean basin and the Caspian Sea as a donor. Species that have become abundant in all these seas are euryhaline and rather eurythermic, and are widely distributed in the coastal areas of the world's oceans. As a rule, they are abundant or dominant in their native habitats, where they sometimes cause outbreaks. Such species, with wide environmental tolerance and high phenotypic variability, have developed in high numbers and first became dominant in the Black Sea, and from here they spread to the Sea of Azov and became established in the Caspian and even the Aral Sea. The most euryhaline species also spread south to the Marmara and eastern Mediterranean (mainly the Aegean and Adriatic) Seas. They often greatly affected the recipient ecosystems, first of all the communities in the tropic level they occupy themselves, and thereafter some of them other trophic levels of the ecosystem; and finally, could cause changes in ecosystem functioning and a fundamental rearrangement of the original energy fluxes. The Black Sea became a natural laboratory for invasive biology as recipient and donor area. Some invasions were useful, like the intentional introduction of the gray mullet *Liza haematochila* and the accidental invasion of ctenophore *Beroe ovata*, some harmful, the most dramatic example of alien species effects documented was the invasion of a gelatinous predator, the polymorphic ctenophore *Mnemiopsis leidyi*.

Keywords Alien species · Black Sea · Ecosystem · Shipping

1 Introduction

In the 20th century, and especially in its second half, under the influence of climatic and anthropogenic factors, significant changes have occurred in the diversity of the flora and fauna of the Black Sea. Among the factors mentioned, the occasional and sometimes intentional introduction of alien species of animals and plants is a global phenomenon that has not avoided the Black Sea as well. As a result, the Black Sea became a basin for many alien species of different origins.

First, due to global warming, the process of penetration of Mediterranean species with the Lower Bosporus Current was intensified, representing an example of spreading of thermophilic species in the northern direction. Many of these species are recorded from time to time or are permanently present only in the near-Bosporus region. Their further expansion is most often hampered by the lower salinity of the main part of the Black Sea waters as well as the low winter temperatures. Therefore, most of the Mediterranean species that have penetrated the Black Sea but are found only in the near-Bosporus region are usually not regarded as assimilated established alien species. The near-Bosporus region represents a sort of an intermediate acclimation basin in the route of invasion of Mediterranean species into the Black Sea. Among such species there are representatives of various systematic groups such as phytoplankton, macrophytes, edible holozooplankton, and benthos [1–4]. Meanwhile, selected Mediterranean species of phyto- and zooplankton have been found off the Crimean and northwestern coasts, i.e., beyond the near-Bosporus region, as early as the 1960s and the beginning of the 1970s [5–7]. At present, this process is still ongoing [8, 9] and the number of penetrated and even established species increases each year.

Among the seasonal migrants from the Mediterranean Sea, there are several species of fishes that perform seasonal feeding or spawning migrations to the Black Sea. These species refer to migrants; meanwhile, they play or at least played a significant role in the trophic dynamics of the Black Sea ecosystem.

The penetration of Mediterranean migrants did not damage the Black Sea ecosystem; on the contrary, due to the insertion of these planktonic and benthic species, new edible organisms were added.

The present-day Black Sea, similar to many other seas, is an area of practical human activity. Changes in the composition of flora and fauna are caused by the unintentional delivery of new species with ships, by intentional introduction of commercial species and unintentional release of other species with them, by insertion by aquarium holders, and by the propagation and spreading of species over newly constructed canal systems that connect previously isolated basins. At present, among the above-listed ways (vectors) of penetration of accidental invaders, the principal ways are the transfer with the ballast ship waters and traveling with the fouling communities of ship hulls.

The successful establishment of such species in the Black Sea is favored by natural factors such as the diversity of habitats both in the sea proper and in its bays, lagoons, and river mouths; the favorable food conditions for benthofagous, planktivorous, and predator species; and the existence of free ecological niches because of the low species diversity of the Black Sea flora and fauna both in the near-shore benthic zone and in the pelagic area of the basin.

2 Microplankton

One of the most dangerous alien species of microfauna is the El-Tor strain, which causes epidemic outbursts of cholera; these strains were registered in Kerch and Odessa coastal waters in 1970. In contrast to classical cholera, the El-Tor strain is capable of long-term (over a few months) existence and, most likely, is able to reproduce in brackish-water basins [10]. The way of its penetration to the region is still not clear; meanwhile, starting from the moment of its appearance, has been often recorded in the near-shore waters off selected cities on the coast of the Black Sea and the Sea of Azov [11]. The social and economic losses caused by the El-Tor occurrence on the beaches are obvious, though no estimates have yet been made.

Selected species of the Mediterranean ciliates–tintinnids, which are representatives of microplankton, were found in the northwestern Black Sea in 2002 [12].

3 Marine Fungi

Marine fungi were found in bottom samples in Odessa Bay from northeastern Asia (two species) [13] and an additional five species were recorded in Odessa port in 2006 (Data of B. Alexandrov).

4
Phytoplankton

Phytoplankton species that are new for the Black Sea are annually reported from its various regions. A significant number of phytoplankton species that were previously unknown in the Black Sea (but usually occur in the Mediterranean Sea) were recorded in the near-Bosporus region of the Black Sea [4]. For example, the diatom algae *Fragillaria striatula* Lyngb. and *Thalassiothrix frauenfeldii*, the coccolithophorid *Calyptrosphaera incise* Schill., and the peridinea *Ceratium macroceros*, which are new species for the Black Sea, were registered in the near-Bosporus region as early as the beginning of the 1960s [14]. These species were found at a salinity of 34‰ and a temperature of 14 °C, that is, under the conditions significantly differing from those characteristic of the Black Sea waters.

L.V. Kuz'menko [5] listed a series of species previously unknown in the Black Sea but typical of the Mediterranean Sea, for example, *Dynophysis schuttii* Murr. et Whitt. and *Podolampas spinifer* Okatumura; they were sampled off the southern coast of the Crimea at a salinity of 18–18.5‰. At the beginning of the 1990s, selected species that were new for the Black Sea, such as *Katodinium rotundatum* (Lohm) Fott, *Achradina sulcata* Lohm., and *Pronoctiluca* sp., were found by L.G. Senichkina in the shallow-water area off Yalta. She also recorded *Distephanus octonarius* var. *Polyactis* (Jorg) Gleser and *D. speculum* var. *Septenarius* Jorg, which were previously unknown for the Black Sea [6]. Many of these species were recorded not only in the waters of a Mediterranean origin characterized by a high salinity but also in typical Black Sea waters. Their presence in the upper layers of the Black Sea may be related to the penetration of the waters from the Sea of Marmara followed by their subsequent mixing. As a result, 37 representatives of the Mediterranean phytoplankton were registered in the subsurface layers of the near-Bosporus region of the Black Sea, among them *Syracosphaera cornifera, Ceratium furca var. eugrammum, Pyrocystis hamulus, Pronoctiluca acuta*, and others [4].

In a coastal area of the northwestern Crimea during long-term observations (1968–2002) of phytoplankton development, new species for the Black Sea were recorded, among them diatoms—*Asterionellapsis glacilis* (Castracane) F.E. Round, *Chaetoceros tortissimus* Gran, *Thallassiosira nordenskioeldii* Cleve, *Lioloma pacificus* (Cupp) Hasle, *Pseudonitzschia inflatula* (Hasle), two subspecies of genus *Chaetoceros*, dynophytes—*Dinophysis odiosa* (Pavillard) Tai & Scogsberg [15].

A considerable number of phytoplankton species were recorded in port areas of the northwestern Black Sea [13, 16–18]. Only in the port of Odessa were 15 alien species of phytoplankton recorded, most of the species found were Dinophytes (8), and all of them most likely have a Mediterranean origin [13]. These species occur infrequently and in only a few locations.

In addition to the penetration of euryhaline salinophilic species of the flora, the penetration of freshwater phytoplankton is also occurring, especially in the northwestern part of the Black Sea subjected to the influence of major rivers such as the Danube, Dnieper, and Dniester [18]. However, not all of these species may be regarded as established alien species.

Among the alien phytoplankton species that has established itself and made a significant negative contribution to the phytoplankton community, we must note the diatom alga *Pseudosolenia calcar-avis* (=*Rhizosolenia calcar-avis*), first recorded in the northwestern part of the Black Sea in 1924 [19]. At present, *P. calcar-avis* is a abundant species widely spread in the northwestern part of the Black Sea, which sometimes demonstrates outbursts in its development. The occurrence of *Pseudosolenia calcar-avis* in the deposits of the Sarmatian age suggests its reintroduction to the Azov–Black Sea basin [20]. The diatom algae *Cerataulina pelagica* and *Chaetoceros socialis*, *C. tortissimus*, and *C. diversus*, which have settled in the Black Sea at the beginning of the 20th century, most probably from the North Atlantic, have also become abundant species of the Black Sea. Of them, *Cerataulina pelagica* and *Chaetoceros socialis* feature outbursts in their development in the spring and autumn over the entire sea basin [4]. The recent warm-water established alien *Leptocylindrus danicus* develops only in the summer in the northwestern part of the sea [21].

In all, among the alien phytoplankton species recorded, there are 19 species of diatoms, 19 species of Dinophyceae (of which *Alexandrium monilatum* and *Mantoniella squamata* are potentially toxic species), two species of the green algae, two species of Chrysophyceae, and one species of the Prymnesiophyceae *Phaeocystis pouchetii.* The brackish-water and freshwater phytoplankton species established in the brackish and freshwater bays and lagoons of the western part of the sea [16]. Some toxic alga species capable of forming cysts (resting stages) are among the most dangerous species supplied with ballast waters. The existence of cysts allows them to survive the unfavorable conditions both during transportation and in the new environment. Silty sediments are known to represent the most appropriate substrate for accumulation and maturation of the cysts settled. For the first time for the Black Sea, the quantitative parameters of the cysts of Dinophyta algae were determined when studying the silty grounds of the port of Odessa. Their abundance in the upper 5-mm layer of the sediments varied from from 1.6 to 105.6 million cells. m^{-2}. Most frequently recorded were the cysts of such alien dynoflagellates species as *Gonyaulax, Scrippsiella, Diplopsalis, Oblea, Protoperidinium*, and *Alexandrium*, which successfully germinated under laboratory conditions. Despite no *Alexandrium tamarense, A. affine*, and *A. acatenella* were recognized in plankton, they were assumed as potential established aliens, because cysts of these dinoflagellates were noted in the grounds of the port area [18]. In 2002, water "bloom" caused by the green alga *Prochlorococcus marinus* was registered; for the first time, this alga was observed in the lagoons of the northwestern sea area [22]. Therefore, at present, the process of penetration of alien phytoplankton species, mainly into the northwestern and western parts of the sea, is occurring due to their transfer both with the currents via the Bosporus Strait and with ballast waters. The process of establishment of the species that have already invaded into the sea is also proceeding. Selected species of this kind may temporarily become subdominant species, but, as a rule, they remain rare or are abundant only in selected years, which suggests a high stability of the phytoplankton community of the Black Sea with respect to establishment of alien species [22].

5 Macrophytes

Among macrophytes, the greatest number of Mediterranean species that have probably penetrated with the currents, succeeded in establishment in the near-shore waters of the Anatolian coast of Turkey (27 species). Their proportion reaches 26% of the total number of macrophyte species registered here. Among them, green algae Chlorophyceae, brown algae Fucophyceae, and red algae Rhodophyceae are represented by ten, five, and 12 species, respectively. Off the coasts of Romania and Bulgaria, six new species of Cladophora were recorded together with other green alga species; they refer to euryhaline and eurybiont species and, in addition, serve as indicators of eutrophic waters. These species were probably brought with ballast waters [23].

In 1990, in Odessa Bay, the near-shore euryhaline species of the brown algae *Desmarestia viridis* was found for the first time in the Black Sea [24]. By the winter months of 1994/1995, *D. viridis* had already become an abundant species of the near-shore zone of the bay. During recent years, especially in the cold ones, *D. viridis* was widely spread in the northwestern part of the Black Sea [25].

In the Danube River delta, in addition to marine macrophyte species, two alien species of freshwater ferns were recognized.

The brown alga *Ectocarpus caspicus* is also often referred as an alien species. About 40 years ago, *E. caspicus* was found and described on the Romanian coast in a brackish-water lake connected to the sea [26]; meanwhile, almost simultaneously, another author [27] regarded *E. caspicus* as an endemic species of the Caspian Sea. However, later, this species was referred to the Ponto–Caspian relics of the Black Sea basin [28]. Recently, *E. caspicus* was found off the southern coast of the Crimea as well [23]. In a series of papers [11, 29, 30], this species was mentioned as an invader from the Caspian Sea; this suggestion seems to be wrong, and the origin of this species requires special research, the more so as its habitat is presently expanding. Another example of this kind of error is the reference of the alga *Laurencia caspica* to

Caspian endemics [27]; later, this species was also recorded in the Black Sea on the shelf of Romania [26]. At present, A. Bogani also refers this species to the Ponto–Caspian relics of the Ponto–Azov basin [28].

From a comparison of the list of macrophytes of the Black Sea published in 1975 and the species that were found after 1975, N.A. Mil'chakova reported 38 new macrophyte species, of which 13, 10, and 15 species refer to green, brown, and red ceramian algae, respectively. The most significant change in the macrophyte flora of the Black Sea is related to the almost twofold increase in the number of species of the *Cladophora*, *Ulva*, *Ceramium*, *Polysiphonia*, *Cystoseira* and *Sargassum* genera; many of them play a key role in the bottom communities of the near-shore ecosystems of the Mediterranean Sea. Most of the new species are thermophilic organisms and serve as indicators of the transition zone between the boreal and tropical domains [23].

6 Zooplankton

In 1925, the hydromedusa *Blackfordia virginica*, which was brought from the Atlantic estuaries of North America, was for the first time recorded in the coastal waters of Bulgaria [31]. At present, it represents an abundant brackish-water species in the western part of the Black Sea and in the Sea of Azov. This medusa consumes small zooplankton and thus competes with fish larvae and fries; meanwhile, it never features significant outbursts in abundance and, therefore, is not harmful for the ecosystem. Later, in 1933, another hydromedusa species—the bougainvillea *Bougainvillea megas*—was found in Lake Varna and in the mouth of the Ropotamo River [32]. *B. megas* was also carried from the coastal Atlantic estuaries of North America. At present, it is widely spread in the form of bottom colonies that continuously cover rocks, port constructions, ship hulls, and pipelines in brackish areas of the Black Sea and the Sea of Azov. The negative effect of bougainvillea as a fouling agent is insignificant because of the high water content in the body of the polyp and its low strength [33].

In 1990, in the near-shore waters off the Crimea, five new hydromedusa (Hydrozoa) species (*Coryne pusilla*, *Eudendrium annulatum*, *E. capillare*, *Tiaropsis multicirrata* M.Sars, and *Stauridia producta*) were recognized. Meanwhile, the authors of [34], who found the former four species, were not absolutely sure in the correctness of the identification of at least two species—*Eudendrium annulatum* and *E. capillare* (the colonies encountered were without gonothecas). Later, N. Grishicheva reidentified *Tiaropsis multicirrata* as *Opercularella nana*. A colony of the fifth species, the polyp *Stauridia producta*, which refers to rare species, was found by T.V. Nikolaenko in September 2000 in a sample taken in the region of the exit from Sevastopol Bay. All five species seem to have a Mediterranean origin [29].

The ctenophore *Mnemiopsis leidyi* A. Agassiz was brought to the Black Sea with ballast waters from the near-shore regions of North America at the beginning of the 1980s (Fig. 1). By 1988, *M. leidyi* had spread over the entire Black Sea area and showed an enormous abundance outburst in the fall of 1989 [35]. During the subsequent years, sharp fluctuations in its abundance and biomass were observed caused by temperature and food conditions [36, 37] (Fig. 2).

The development of the *M. leidyi* population in the Black Sea ecosystem led to a decrease in the biomass, abundance, and species diversity of edible zooplankton, fish larvae, and eggs, which are the principal food objects of *M. leidyi* (Figs. 3, 4) [38–41].

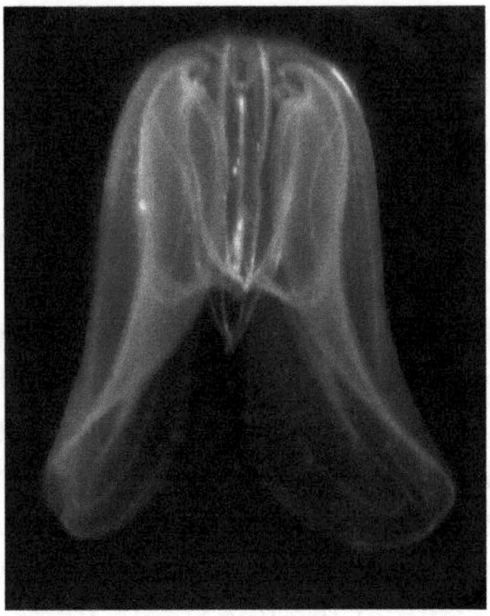

Fig. 1 View of *Mnemiopsis leidyi*

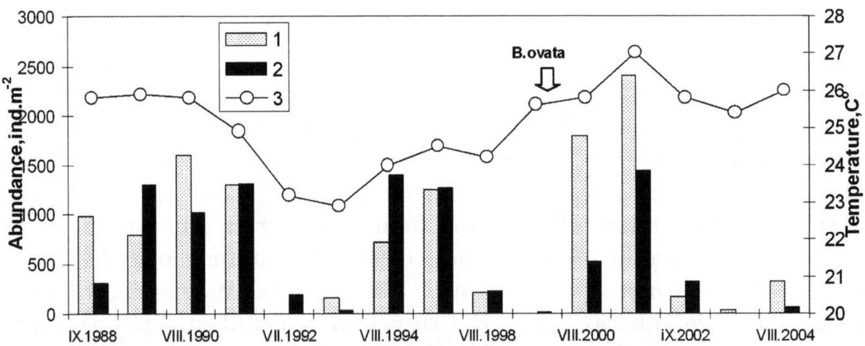

Fig. 2 Interannual variations of the abundance of *Mnemiopsis leidyi* (g m^{-2}) in August–September (abundance was estimated with coefficient for insignificant catchability (in mean 2) after Vinogradov [35]) and average surface water temperatures (direct measurements)

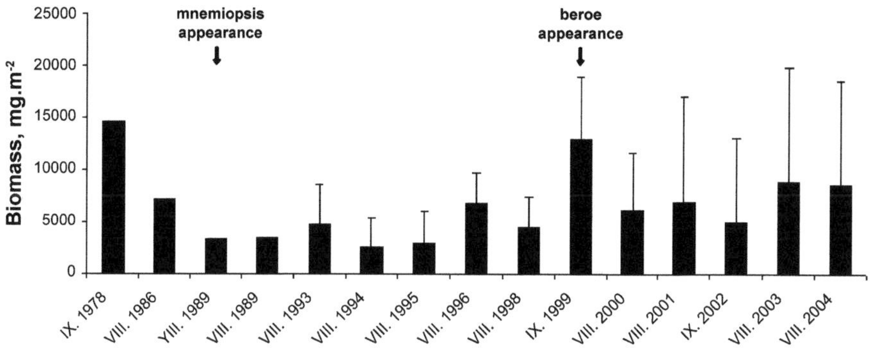

Fig. 3 Interannual variations of zooplankton biomass in August (g m^{-2})

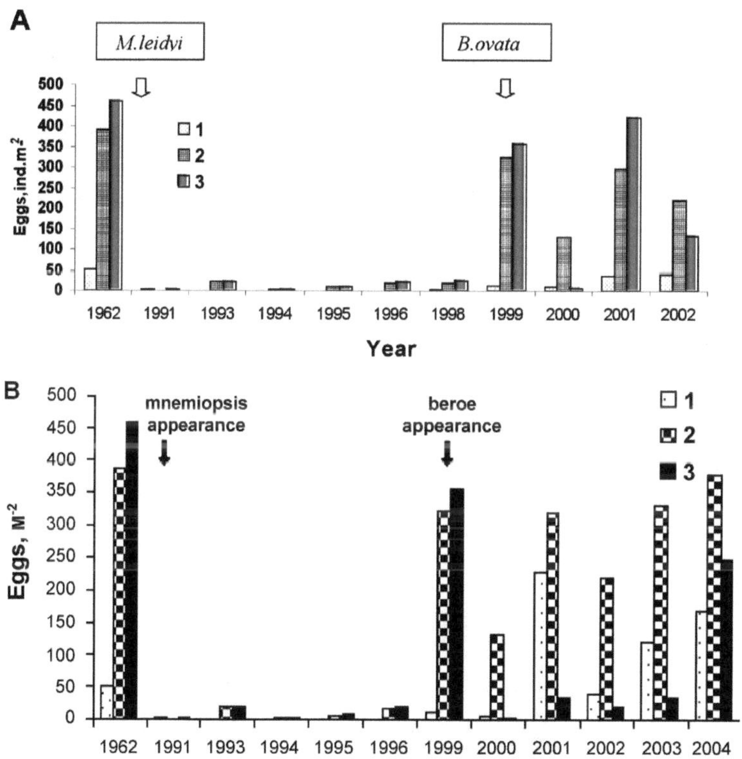

Fig. 4 Interannual variations in the abundance of **A** fish eggs (ind. m^{-2}), and **B** larvae in the northeastern Black Sea in July–August: 1 – scad; 2 – anchovy; 3 – other species

As a result, the commercial fish catches decreased; this especially refers to the planktivorous fishes that are food competitors of *M. leidyi* such as the anchovy *Engraulis encrasicolus ponticus*, the Mediterranean horse mackerel *Trachurus mediterraneus ponticus*, and, to a lesser extent, the sprat *Sprattus*

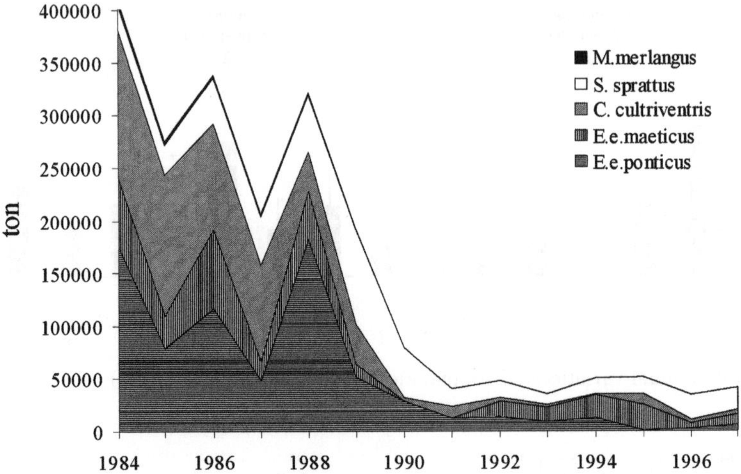

Fig. 5 Fish catch after *Mnemiopsis leidyi* outbreak but before *B. ovata* development

sprattus phalericus (Fig. 5). This also affected the representatives of higher trophic levels—predator fishes and dolphins, who feed mainly on anchovies and sprats [38, 42]. From the Black Sea, *Mnemiopsis leidyi* spread to the Azov and Marmara Seas and was periodically supplied with the Black Sea waters to the Aegean Sea [37, 43, 44]. In 1999, *M. leidyi* was introduced into the Caspian Sea as well, probably from ballast waters of oil tankers [45].

One of the reasons for the high abundance and biomass values of *M. leidyi* in the Black Sea that are never observed in its initial habitats—in the coastal waters of North America, is the absence of an appropriate predator to consume *M. leidyi* and control its abundance [46].

In 1997, a new invader—ctenophore *Beroe ovata* Mayer 1912 (Fig. 6)—was first found in the northwestern part of the Black Sea. This predator feeds on planktivorous ctenophores, first of all, on *M. leidyi* [47]. *B. ovata*, similar to its

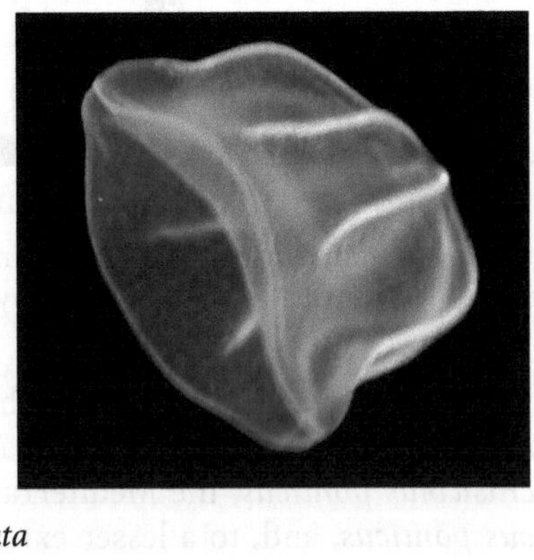

Fig. 6 View of *Beroe ovata*

predecessor *M. leidyi*, was carried with ballast waters from the Atlantic coast of the northern America [48]. In August 1999, the first outburst in the *B. ovata* development over the entire Black Sea was observed [49–51]. The studies of 1999–2006 showed that the development of the *B. ovata* population features a clear seasonal character. The first individuals appear in the middle or late August; in September–October, intensive reproduction proceeds, which leads to a sharp growth of the population; later, it gradually attenuates [52, 53]. The attenuation rate of the *B. ovata* development depends on the food (*M. leidyi*) availability. At significantly reduced *M. leidyi* abundances, the reproduction rate of *B. ovata* decreases and its population correspondingly reduces. Large adult individuals eliminate after spawning. In November, a minor part of the population consisting of individuals of new generation seems to sink down to the near-bottom layers; there, it rests over the winter until the subsequent outburst in the *M. leidyi* development. This kind of behavior is characteristic of the representatives of the *Beroe* genus in other regions as well.

With the appearance of the ctenophore *Beroe ovata* in the Black Sea ecosystem, a new previously absent level in the trophic web was formed. It was represented by a predator consuming *M. leidyi* and capable of significant decreasing the population of the latter. The results of the studies showed that *B. ovata* established in the Black Sea and occupied its niche in the ecosystem. Its seasonal development coincides with the cycling of this species in the near-shore waters of the North Atlantic: the appearance and starting reproduction at the end of the summer–beginning of the autumn and the decay in the late autumn [46].

Beroe ovata radically reduced the *M. leidyi* population and this resulted in the beginning of the process of restoration of the Black Sea ecosystem. The abundance and biomass of edible zooplankton significantly increased (Fig. 3). The species that have virtually disappeared after the *M. leidyi* invasion appeared again. Owing to the low abundance of *M. leidyi* in the first half of the summer, its pressure on the eggs and larvae of the summer-spawning fishes decreased (Fig. 4), while the winter-spawning species had enough time to spawn in the autumn and winter, when only isolated *M. leidyi* individuals existed. Therefore, other trophic levels started to be restored; first, this refers to small planktivorous fishes—anchovies, horse mackerel, and sprat—and their food rations [53].

Meanwhile, although with the appearance of *B. ovata* the pelagic ecosystem began to restore, in selected years, when *B. ovata* was seasonally absent, *M. leidyi* managed to reach high biomass values and significantly impaired the ecosystem under favorable feeding and temperature conditions. However, even in this case, the duration of its impact on the ecosystem was essentially lower: it lasted over one or two summer months instead of 8–9 months before the *B. ovata* invasion.

The most important alien species of mesozooplankton was probably the Copepoda representative *Acartia tonsa*, which was first recorded in the Black

Sea in 1994. Initially, it was regarded to be supplied from the Mediterranean Sea; meanwhile, the studies performed by A. Gubanova showed that *Acartia tonsa* appeared in the Black Sea before its appearance in the Mediterranean Sea (at the beginning of the 1970s) and was seemingly brought from the coastal waters of the North Atlantic. At present, the habitat of *Acartia tonsa* is extending and it was found off Kerch and off Novorossiisk, in the seas of Azov [54] and Marmara [7]. During the cold season of the year, from January to April, no *A. tonsa* is observed in the plankton of Sevastopol Bay, in contrast to the eurythermal Black Sea *A. clausi*. It appears at the end of May when the water temperature reaches 16 °C. From the end of June to August, *A. tonsa* quickly increases its abundance and starts to exceed *A. clausi* in quantitative parameters. According to the data of A. Gubanova, two peaks of *A. tonsa* are observed; they are more distinctly expressed than those of *A. clausi*. Thus, *A. tonsa* replenished the ecological niche of thermophilic zooplankton species in the Black Sea. It probably replaced two local Acarcia species—*A. latisetosa* and small *A. clausi*—but as itself it represents a valuable edible object.

In the coastal waters off the Crimea, the number of the alien planktonic species recorded goes on growing; all of them seem to have a Mediterranean origin. To date, it is not clear whether they will be capable of establishment. Among them, one finds the harpacticoids *Amphiascus tenuiremis*, *A. parvus*, *Leptomesochra tenuicornis*, *Idyella palliduta*, *Ameiropsis reducta*, and *Proameira simplex*, the planktonic copepods *Oithona brevicornis*, *O. plumifera*, *O. setigera*, *Clausocalanus arcuicornis*, and *Scolecetrix* sp., and the species of the Clausidiidae family, *Rhincalanus* sp. and *Oncaea minuta*. Some species were represented by a few specimens; of others, only single individuals were found [9]. Copepod *Oithona brevicornis* is establishing now and becoming abundant off Crimea (A. Gubanova, personal communication). Altogether only two species of ctenophores, two species of Copepoda, two (7) hydromedusae, which have also benthic stage, became established and 59 species of Copepoda were found in the near Bosphorus area, which we did not estimate as established aliens.

7
Benthos

The shipworm *Teredo navalis* (Linne 1758) seems to have performed the earliest penetration of invaders into the Black Sea, which was probably brought by the Greek vanquishers at the Attic times (750–500 B.C.) [55]. Shipworms are among the most widely spread and harmful invaders. Their intensive reproduction rate (up to 2 million larvae per cycle) and high resistance against unfavorable conditions represent the principal factors determining their wide propagation. They travel using wooden hulls of ships making holes with the help of endosymbiotic bacteria. Teredo may survive feeding on wood only, but it also filters and consumes plankton. Over 3 weeks, it may exist under anoxic conditions; it may survive in almost fresh water and in air and overwinter cold winters when wooden constructions are covered with ice. The mean length of the worms is 20–30 cm at a maximum value up to 60 cm. Teredo is a protandric hermaphrodite capable of self-fertilization.

This species was probably followed by spontaneous benthic invaders—the fouling acorn barnacles *Balanus improvisus* and *B. eburneus*, which penetrated into the Black Sea in the 19th century [56]. Balanuses were brought from the coastal Atlantic waters of North America. At present, both of these species, especially the former one, are mass organisms and are widely spread in the near-shore waters of the Black Sea. The negative effect of these species is related to the fact of their fouling over ship hulls, pipelines, piers, and dams. On the other hand, their pelagic larvae form a significant part of the near-shore edible meroplankton. Balanuses are hermaphrodites and, in conditions of the impossibility of cross-fertilization, *Balanus improvisus* are capable of self-fertilizing. In addition, eggs of all the balanuses are fertilized in the mantle cavity and develop there up to the stage of nauplius larva I, which is subsequently released to the water; this provides an additional protection of the progeny. The precipitation of larvae proceeds in a shoal mode, which provides the possibility of future cross-fertilization. Balanuses feature a high growth rate and 1-month-old individuals are already capable of reproducing [57]. These adaptation mechanisms provided the wide spreading and high abundance of *Balanus improvisus* and, to a lesser degree, of *B. eburneus*. One more species of acorn barnacles *Balanus amphitrite* was recorded in Odessa bay in 2001 [13].

The Polychaeta *Mercierella* (*Ficopomatus*) *enigmatica* was first recorded in 1929 in the brackish-water Lake Paliastomi near Poti (Georgia); later, it was also found in Gelendzhik Bay [58]. It dwells in curved calcareous tubes up to 4 cm long, from which it spreads a corolla of its branchial branches. Interlacing tubes form a quaint continuous cover over the surfaces of rocks and other underwater objects such as ship hulls. *M. enigmatica* precisely originates from the brackish-water coastal lakes of India. In 1923, it was found in the Seine River mouth and in the Adriatic Sea; from the Atlantic coasts of Europe or from the Adriatic Sea, it was brought to the Black Sea with ship-fouling organisms. Here, it became widely spread and penetrated into the Sea of Azov and the Caspian Sea. As a fouling organism, it damages ships and hydraulic structures, while its planktonic larvae serve as edible objects.

In 1937, in Dniester–Bug lagoon, A.K. Makarov found a crab species new for the Black Sea; it was identified as the Dutch crab *Rhitropanopeus harrisi tridentata*, which originates from Seider See Bay off the North Sea coast of Holland. Earlier, this species had been brought to Europe form the Atlantic coast of North America. At present, the Dutch crab is widely spread in desalinated areas of the Black Sea; in 1948, it was recorded in the Sea of Azov and in 1957 it was found as far as in the Caspian Sea [59]. It dwells over sandy and silty–sandy grounds of shoals and lagoons. It is intensely consumed by near-bottom fishes such as bullheads, flounders, Black Sea turbots, and sturgeons. It may be regarded as a useful species being an additional food object, the more so because it never competes with local species.

In 1947, off Novorossiysk, the gastropod mollusk *Rapana venosa* (=*R. thomasiana*) was recorded in the biocoenosis of mussels (Fig. 7). It is a predator feeding on oysters, mussels, and other bivalve mollusks. Rapa whelk inhabits the Japan and Yellow Seas, that is, in seas with a rather high salinity (25–32‰) and in relatively brackish waters. It was most likely brought to the Black Sea from the Japan Sea with a ship in the form of an egg clutch attached to its hull [60]. Rapa whelk has successfully reproduced, especially off the Caucasian coast, and in the 1950s it almost completely extinguished the community of the oyster *Ostraea edulis* on Gudauta Bank and then those of the mussel *Mytilus galloprovincialis* and scallops *Chlamys glabra*, that dwelled together with the oyster. Later on, rapa whelk started to destroy mussel colonies off the southern coast of the Crimea, then those off the coasts of Bulgaria, and reached a high abundance. By the beginning of 1970, rapa whelk had spread over almost the entire Black Sea area; at present, it is missing only from the most desalinated areas in its northwestern and western parts. In the 1980s, the intensive commercial fishing of rapa whelk as a valuable food object was initiated, first off the coasts of Turkey and then off Bulgaria. The unlimited commercial catch of the mollusk resulted in a decrease in its abundance; the drop was so crucial that the further fishery became unprofitable [61]. Later, fishing of rapa whelk was started on the Caucasian shelf and in the region of the Kerch Strait, where its total biomass was estimated at 2800 and 6000 t, respectively. The fishing of rapa whelk and the decrease in its stock due to the decrease of its food resources—small bivalves significantly reduced its abundance. The decrease in the stock of the food objects of rapa whelk such as small bivalves also continued with the development of the ctenophore *M. leidyi*, who consumed their pelagic larvae. At the end of the 1990s, the commercial catch of rapa whelk on the Russian shelf of the Black Sea was

Fig. 7 View of *Rapana venosa*

performed only by a single enterprise [62]. After 2001, virtually no Russian organization has performed commercial fishery of rapa whelk. Bulgaria and Turkey continue their fishery activity and export rapa whelk meat to Japan and Korea. On the Turkish coast, there are several factories that specialize in the export production of rapa whelk meat. In the 1990s, the export of rapa whelk by Turkey alone exceeded 1000 t. The decrease in the commercial stress on the rapa whelk population in the mid-1990s seems to result in a growth in its abundance in the Russian part the Black Sea.

In 1995–1997, the studies of benthos in the northeastern part of the Black Sea showed a significant excess of the rapa whelk biomass over the biomass of other species of mollusks. In the Arkhipo-Osipovka–Kerch Strait region, the biomass of edible benthos was only 12 g m^2. The deterioration in the condition of food resources also affected the condition of benthofagous fishes. Only the Black Sea red mullet gut content mollusks were found. In the food of other fishes, benthos made up from 1 to 20% [62].

In 1999, during the survey of zoobenthos off the North Caucasian coast between Gelendzhik and Tuapse, a mass rapa whelk development was noted after the first outburst of the invader ctenophore *Beroe ovata* and the significant decrease in the *Mnemiopsis leidyi* population [63]. The total rapa whelk abundance reached 40 ind. m^2. This intensive rapa whelk development is comparable with its outburst in the 1950s, when the increase in its abundance proceeded under the availability of abundant food resources such as the oyster *Ostraea edilis* [1]. The outburst in the rapa whelk abundance in the 1990s was related to the growth in the provision of food resource such as brushes of the rock-dwelling variety of the mussel *Mythilus galloprovincialis*. In 2002, the abundance of rapa whelk continued to grow and reached its maximum in the near-shore waters of the northeastern part of the sea; later, in this region, because of the absence of food, rapa whelk remained the only representative of mollusks in the biocoenosis of zoobenthos and its abundance started to decrease. In 2005–2006, one could find numerous empty shells of rapa whelk that died from starvation on the coast of the sea. The main factor that restricts the rapa whelk development is the absence of food for its juvenile individuals [64].

No examples of consumption of adult representatives of rapa whelk by fishes or other hydrobionts are known; only its planktonic larvae may be consumed by planktivorous fishes. Rapa whelk is an active predator that consumes valuable representatives of benthos. It inserts significant changes into the structure of bottom biocoenoses and often is the dominating species of the bottom communities being itself an ecological dead-end. Therefore, its commercial extraction is extremely important for reducing the pressure on bivalve mollusks.

Another alien species—the gastropod mollusk *Potamopyrgus jenkinsi*—originates from New Zealand. Meanwhile, *P. jenkinsi* first appeared and inhabited the Atlantic near-shore waters off Europe and then penetrated into the Mediterranean and Black seas. In the Black Sea, it was recorded off the coast of Romania. Now, *P. jenkinsi* is observed in the lagoons of the northwestern part of the Black Sea [1].

In 1966, in the near-shore waters off Odessa, the bivalve mollusk *Mya arenaria* was found [65]. It is supposed that its larvae were brought to the Black Sea with ballast waters of ships either from the North Sea or from the coasts of North America. Mya widely spread over brackish areas of the Black Sea and the Sea of Azov and replaced the local species of the small bivalve *Lentidium mediterraneum*. During the first years after the mya invasion, its biomass had reached a value of 16–17 kg m^{-2} of the floor. The near-bottom oxygen deficiency at depths greater than 8–10 m represented a limiting factor for its further propagation. At lesser depths, where no hypoxy could restrict the mya development, its abundance was high and it often dominated over zoobenthos. Mya juveniles are consumed by near-bottom fishes—gobies, turbo, and sturgeons [29]. While estimating the significance of *Mya arenaria* as a new component of the Black Sea ecosystems, one can separate several aspects. On the one hand, its negative effect is represented by the significant replacement of the local species *Lentidium mediterraneum*, which served as a food for the fries of all the species of near-bottom fishes. On the other hand, mya juveniles proper became a food for adult fishes. In addition, this large alien species significantly enhanced the process of seawater filtration in the coastal zone, which is especially important under the conditions of eutrophication. One more factor has a positive effect. Every storm, when the wind blows from the sea, terminates in a high numbers throwing out of these mollusks from sea depths of 4–5 m onto the beaches, where they are consumed by sea birds—gulls, terns, and other species [25].

The blue crab *Callinectis sapidus*, one of the largest representatives of its order, was first found off Bulgaria in 1967; later, isolated specimens were encountered in the Kerch Strait in 1975, in Varna Bay, and in the Bosporus Strait. They were no more than single findings [25]. The blue crab originates from the Atlantic coastal waters of North America and was brought to the Black Sea and the Sea of Marmara with ballast waters of ships. At present, the blue crab seems to undergo establishment in the Black Sea. In recent years, it was recorded in the near-shore waters off Sevastopol and its abundance is increasing [66].

The nudibranchiate mollusk *Doridella obscura* was first found in the northeastern part of the Black Sea in 1980; later, it was also observed in Varna and Burgas bays in Bulgaria, in the Kerch Strait, and on the southern shelf of the Crimea. Doridella dwells off the Atlantic coasts of Canada and the United States. The studies of its food spectrum in the Black Sea showed that its main ration consists of the Bryozoa *Electra crustulenta* and *Conopeum seurati*. Having a food spectrum that narrow, this mollusk can hardly become an abundant species in the Black Sea and its influence on local fauna should not be significant [67].

One more alien species—the bivalve mollusk *Anadara inaequivalvis*,—was first found by V.E. Zaika on the Caucasian shelf in 1968. This species widely spread over the northwestern shelf of the Black Sea and the southern part of the Sea of Azov [68]. Due to the geometrically closing valves and the presence of hemoglobin in its tissues, anadara is capable of long-term existence under an oxygen deficiency in the near-bottom water layer and to survive hypoxy when other mollusks die. Adult individuals *Anadara inaequivalvis* can hardly serve as a food for fishes because of its thick massive shell; meanwhile, it helps sea purification as a filtrator. Anadara was brought to the Black Sea from the Adriatic Sea; there, it was inserted from the coastal waters of the Philippine Islands in the Pacific Ocean. In 1989, the anadara community almost completely replaced the community of *Chamelia gallina*; meanwhile, anadara was also noted in the bottom communities of the North Caucasian shelf as a subdominant species with a mean biomass of 70 g m^{-2} and an abundance of 10 ind. m^{-2}. In 2000, during a survey in the region between Gelendzhik and Tuapse, accumulations of anadara juveniles with abundances up to 3000 ind. m^{-2} were recorded at depths of 20–27 m. Judging from the size structure of the anadara population, the precipitation of juveniles observed was the first one over at least recent 8 years. In 2001, a high concentration of anadara (250–625 g m^{-2}) were observed in the northeastern part of the sea and intensive development of juveniles was also noted [55]. *Anadara inaequivalvis* is gradually becoming a natural component of the coastal biocoenoses off the Crimea [58].

In 2001, two new alien Bivalvia species were found in Odessa Bay: edible *Mytilus edulis* and *Mytilus trossulus* [13].

Altogether, among benthic alien species were recorded: Oligopaeta-1 species, Polychaeta-11, Mollusca-13, Cirripedia-3, Amphipoda-1, Decapoda-4.

8 Fishes

As mentioned before, selected Mediterranean species of fishes perform regular feeding and/or spawning migrations to the Black Sea. This refers, first of all, to valuable large predator species—the Mediterranean Atlantic horse mackerel *Trachurus trachurus trachurus* (Linnaeus), the Atlantic bonito *Sarda sarda* (Bloch), bluefish *Pomatomus saltatrix* (Linnaeus), the Atlantic mackerel *Scomber scombrus* (Linnaeus), and the Mediterranean mackerel *S. japonicus colias* Gmelin.

The swordfish *Xiphias gladius* Linnaeus (and even its spawning), the bluefinned tuna *Thunnus thynnus thynnus* (Linnaeus), the Mediterranean picarel *Spicara moena* (L), and the European pilchard *Sardina pilchardus* (Walbaum) were sometimes noted in the western and northwestern parts of the sea [69]. In the 1970s–1980s, the abundance of migrating species significantly decreased, and most of the species virtually stopped entering the Black Sea. Meanwhile, during recent years, the conditions for fattening have enhanced owing to increase in the stock of small pelagic fishes after the *Beroe ovata* invasion. As a result, some Mediterranean species again appeared both in the western part of the sea (the mackerels, the bonito, and the bluefish) [70] and in its northwestern part (the horse mackerel, the bonito, the bluefish, the Mediterranean picarel *Spicara moena* (L), the European pilchard *Sardina pilchardus* (Walbaum), the green wrasse *Labrus viridis* (Linnaeus), and triplefin *Tripterygion tripteronotus* (Risso) [71]. In addition, starting from 1999, their feeding area is expanded and new Mediterranean fish species appear; for example, in the near-shore waters off the Crimea, the dorado *Sparus aurata* Linnaeus, the salema *Sarpa salpa* (Linnaeus), and the thick-lipped gray mullet *Chelon* (=*Mugil*) *labrosus* (Risso) appeared and intensely reproduced [71]. Previously, in contrast to the gilthead bream, the thick-lipped gray mullet has never been recorded in the northwestern part of the Black Sea [69]. For the first time, a juvenile of *Chelon labrosus* was caught in October 1981 in Donzulav Bay. In October 1983, shoals of the thick-lipped gray mullet consisting of 10–15 fishes were observed in the waters off Sevastopol. Starting from 1999, the thick-lipped gray mullet has been repeatedly found in the areas off Sevastopol. A specimen of the salema off the Crimea was first noted in 1999 [71]. At present, its abundance in this region is rapidly increasing.

The dorado can be often recorded as single specimens or minor shoals in Balaklava Bay and adjacent near-shore waters. The dorado and the salema probably overwinter in the coastal waters off the Crimea [71].

The Mediterranean umbrine *Umbrina cirrosa* was once found in the Black Sea biospheric reserve in 1962 [72]. In the summer of 1999, one female with eggs with a length (*L*) of 43.5 cm was caught in Pshada Bay [73].

The common eel *Anguilla anguilla* may also be regarded as a migrant. E.I. Drapkin [74] reported 16 catches of eels in the region of Novorossiisk from 1946 to 1964; in 1958, an eel was caught off Anapa. Eels were also noted in the system of Kisiltash lagoons located on the Taman' Peninsula [75].

All the above-listed species are no more than seasonal Mediterranean migrants rather than invaders into the Black Sea. Among the alien species, three species of fishes previously not encountered in the Black Sea were found in the coastal waters of the Crimea. They include two specimens of the barracuda *Sphyraena obtusata* Cuvier 11.5 cm long caught with a bottom trawl in Balaklava Bay in August 1999. This is an Indian–Pacific species, which penetrated as a Lessepsian migrant via the Suez Canal to the Aegean Sea and then, probably, to the Black Sea [71].

A specimen of the northern blue whiting *Micromecisthis poutassou* 15.7 cm long was caught in January 1999 over a sea depth of 60 m off Balaklava (Crimea). It is a typical Atlantic–boreal species widely spread in the Mediterranean basin including the Aegean Sea and the Sea of Marmara; most likely, it penetrated from the Mediterranean Sea. Blue whiting performs long-lasting migrations; it is known as a stenohaline eurythermal species dwelling at salinities no less than 33‰, but was first encountered at a salinity of 18‰. There are two ways of explanation of the appearance of the above two species in the Black Sea: fishes might migrate from the Sea of Marmara or the Mediterranean Sea or, which seems more probable, might be brought with ballast waters. The third species is the coral-dwelling butterfly fish *Heniochus acuminatus*. A specimen 76 mm long was caught by a net in Balaklava Bay in October 2003. It is a typical tropical Indian–Pacific species and the conditions of Balaklava Bay are hardly favorable for it. This fish was most likely delivered with ballast waters [71].

The golden goby *Gobius auratus* Risso, which was first recorded in the communities of near-shore macrophytes off the Crimea in early 1970-s and now it regularly occurs in the northeastern part of the sea and may also be referred to Mediterranean invaders [76].

During the recent years, in the waters off Romania, centracant *Centracanthus cirrus*, which probably also penetrated from the Mediterranean Sea, was observed. To date, it has significantly increased its abundance and now represents a commercial fish in the littoral zone of Romania. In the central part of the sea, its developing eggs were first found in June 1982 [70].

In order to enhance the fishery potential of the basin, attempts to introduce 22 valuable commercial fishes were made; however, only a few of them managed to establish and became fishery objects [76]. The undoubtedly most significant event is the introduction of the large Far Eastern haarder *Liza haematochilos* (Temminck & Schlegel, 1945) (=*Mugil soiuy*) into the Black Sea and the Sea of Azov; it became a valuable commercial species for both of the seas.

The fry and juveniles of the haarder *Liza haematochilos* were brought from the estuaries of the Japan Sea in 1972–1980 and introduced to the lagoons of the Black Sea and the Sea of Azov and directly to the northwestern part of the Black Sea and to the open part of the Sea of Azov. In 1980, it became a widely spread commercial species in the regions off the coasts of Russia and the Ukraine; isolated specimens were also caught off Turkey, in the Sea of Marmara, and even in the Mediterranean Sea [77]. It was supposed that this fish would feed on detritus; meanwhile, its food spectrum widens and, in addition to detritus, it consumes small benthic organisms. In the Black Sea, this introduced species enters the food competition with local mullet species and reduces the abundance of the latter. In the Black Sea basin, the growth rates of the haarder changed; it reached significantly greater size and weight. Both a 1.3–3-fold increase in its growth rate (the corresponding values of mass and size are 3 kg and 65 cm in the Japan Sea and 4.5 kg and 71 cm in the Sea of Azov) and acceleration of its sexual maturation by a year were noted [78]. In 1992, the haarder was added to the list of commercial fishes of the Azov–Black Sea basin. In 1993, the allowable quota was established and the fishery began. The haarder fishery in the Sea of Azov and the Black Sea is performed by Russia, the Ukraine, and, starting from 1999, by Georgia.

Among the intentionally introduced sturgeon species, only the rainbow trout may occur in natural conditions. The environments in rivers and lagoons are favorable for the rainbow trout dwelling; they were episodically caught but their reproduction has never been observed. Most likely, no natural spawning of the rainbow trout proceeds and its population is replenished only owing to the individuals of a artificial origin [73].

Two sea bass species—the Japanese sea bass *Lateolabrax japonicus* (Cuvier) and the European bass *Dicentrarchus labrax* (Linnaeus) were brought for intentional introduction from desalinated waters of the Japan Sea. Both species are from time to time recorded in the northwestern part of the Black Sea. The European bass *Dicentrarchus labrax* is regularly caught in the Black Sea [79], although in small amounts.

In 1963, plecoglos *Plecoglossus altivellis* (Temminck & Schlegel) was brought from the Japan Sea with the purpose of intentional introduction to the western part of the Black Sea.

For commercial aquaculture, large freshwater fishes were brought to the ponds and lagoons of the Azov–Black Sea basin, among them the silver carp *Hypophthalmichthys molytrix* (Valenciennes, 1844), the speckled carp *Aristichthys nobilis* (Richardson, 1846), and two amur species—the black amur *Mylopharyngodon piceus* (Richardson) and the grass carp *Ctenopharyngodon idella* (Valenciennes, 1844). At present, carps became important commercial fishes. However, these fishes do not reproduce under the natural conditions.

In order to reduce the abundance of malarial mosquitoes in the swampy regions of Colchis, intentional introduction of the mosquito fish *Gambusia holbrooki* (Girard, 1859) was made. The mosquito fish was brought from freshwater regions of Italy in 1925 [57] and was successfully introduced into the wetlands of the Black Sea. The mosquito fish dwells in brackish estuaries; meanwhile, it behaved as an euryhaline species and expanded over the Azov–Black Sea basin. At present, it occurs in a wide salinity range from 0 to 15–17‰.

Together with the carp and amurs, the Amur River stone moroco *Pseudorasbora parva* (Schleg) was brought from the rivers of Far East to the Romanian freshwater basins, from which it spread via channel systems to other rivers of the Black Sea basin, then to the rivers of the basin of the Sea of Azov, and then farther to Europe. At present, it represents a widely spread species of weed fishes. This species is extremely eurybiontic and is capable of dwelling in various freshwater basins such as rivers, ponds, reservoirs, wastewater channels, and strongly eutrophicated basins. Its food spectrum is similar to that of juveniles of commercial fishes. Its larvae and fry feed mostly on planktonic crustaceans; adult individuals add benthic organisms to their ration. Thus, it competes with fish juveniles and adult planktivorous and benthofagous fishes. In addition, it extinguishes eggs of valuable commercial fishes [80].

Selected freshwater species penetrated to brackish-water estuaries and the Danube and Dnieper river deltas as a result of the activity of aquarium holders. Among this species is the aquarium sunfish *Lepomis gibbosus* (Linne, 1758), which was brought to Europe from North America as early as in the 18th century [69]. From the ponds into which it was released, it penetrated to major rivers such as the Rhine, the Oder, and the Danube and to related inner basins. In the Danube River delta, it was first noticed in 1949; then it gradually spread along the adjacent coasts of the Black Sea both to the south to the near-shore lakes of Romania and Bulgaria and to the northeast to selected basins between the Danube and the Dnieper Rivers [65]. At present, this fish occurs in the basins of the Tiss and Bug Rivers, in the lower reaches of the Danube, the Dniester, and the Dnieper, in Odessa Bay, and in Berezan lagoon; isolated individuals were recorded in significantly desalinated areas of the Black Sea. It is quite common in numerous channels of the Dnieper River, in the Kakhovka Reservoir, and in the Dnieper–Bug lagoon, from where it penetrated to the lower reaches of the Southern Bug River. At the end of September 2002, in an aquaculture pond located in the North Crimea, more than 200 individuals of sunfish were caught; they seemed to penetrate there from the Kakhovka Reservoir via the North Crimean Canal [79]. This species is euryhaline and eurythermal spreading in the near-shore waters with a salinity of 14–15‰; it survives equally well both the high summer temperatures and the under-ice wintering even in small basins. Its juveniles feed on crustaceans, while adult individuals consume insects and small fishes. It is harmful for the fishery both in natural and artificial basins because it consumes eggs, larvae, and fry of valuable species and is a food competitor for some of them [69]. In its turn, this species is characterized by a low growth rate; it refers to weed fishes and has no commercial value [79]. The process of introduction of the sunfish to the inland basins of the Ukraine and Russia is proceeding both because of the spontaneous expansion of its habitat and due to the occasional transport from fish nurseries with juveniles of valuable species. An analysis of the tendencies of its propagation allows one to expect its further penetration in the eastward direction to the rivers of the basin of the Sea of Azov (Don and Kuban') and related natural and artificial basins and its expansion up to the Volga River. Taking into account the high ecological plasticity of the sunfish and the availability of favorable conditions for it in the south of the Ukraine and Russia, one can suggest that in the forthcoming years it will be capable of making a negative impact on the ecosystems of the inland basins of this region and to cause economic losses in the fishing industry. This is another aggressive invader that requires an immediate monitoring of its propagation and elaboration of adequate measures in order to reduce its abundance [79].

Another aquarium fish—the Japanese oryzia *Oryzias latipes* (Temminck & Schlegel) was also occasionally released by its holders. It was brought from the estuaries of the Japan Sea, introduced, and is now is rather widely spread in the brackish and freshwater areas of the northwestern part of the Black Sea. This fish is also a widely distributed weed fish.

9
Pathways of Penetration of Alien Species

We have distinguished pathways of species penetration into the Black Sea on the base of analysis of the composition of the already established alien species and the regions donors from which they were brought (Figs. 8, 9).

The majority of the established accidental invaders were brought to the Black Sea from the near-shore Atlantic waters of North America; this mainly happened in the 1950s–1990s, although selected species had penetrated earlier. All the alien species of this group are neritic, rather eurythermal, and, most importantly, a euryhaline species widely distributed in the coastal waters of the world's oceans. This pathway is also characteristic of a group of brackish-water species that were accidentally introduced into the Black Sea; this group is represented by the inhabitants of near-shore brackish-water bays and estuaries of the same region. They established in the brackish western and northwestern areas of the Black Sea.

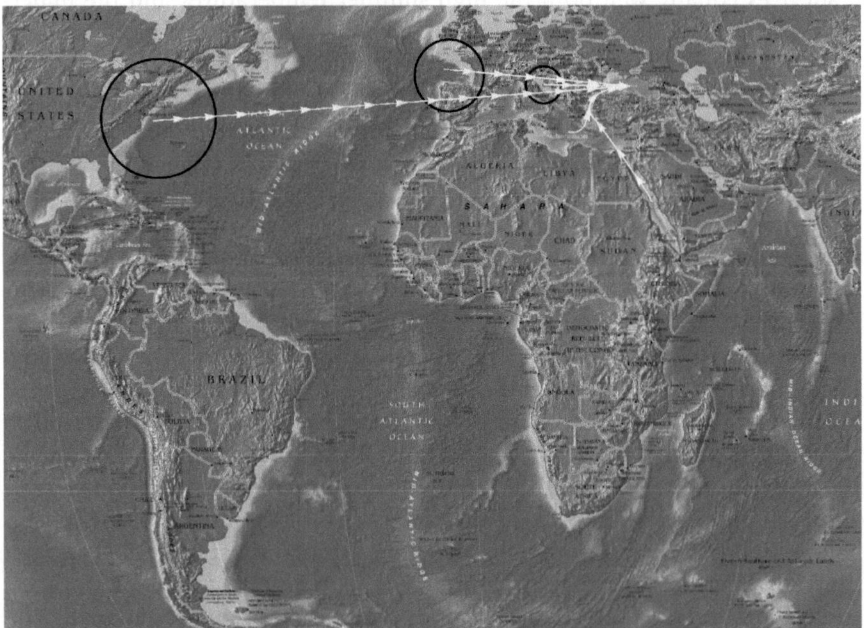

Fig. 8 Pathways of penetration of alien species to the Black Sea

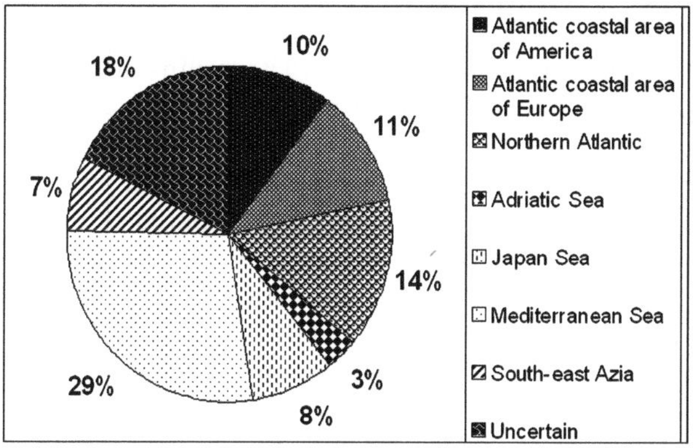

Fig. 9 Share of alien species in the Black Sea with respect to their donor area

One more group of alien species that were accidentally introduced into the Black Sea also features a northern Atlantic origin, although they arrived from the eastern part of the Atlantic Ocean—from the near-shore waters of Europe (Fig. 9).

The third group consists of alien species of a Mediterranean origin; their share is rather high and is gradually increasing over the recent years; meanwhile, not all of these species have already established. These organisms driven with currents and ballast waters represent phyto- and zooplankton, macrophytes, benthic or demersal organisms, and fishes. None of them became an abundant species; their greater number occur only in the near-Bosporus and southern parts of the Black Sea. Selected species penetrated to the near-shore regions off Bulgaria, Romania, and the Crimea also with currents, in the course of their migration, or with ship-ballast waters. Among them, one should note phytoplankton and zooplankton species, ciliates, macrophytes (whose greater part settled in the southern part of the sea), the crab *Sirpus zariquieyi* found off the Turkish coasts, the amphipod *Microprotopus maculatus* recorded off the Crimea, and three fish species, two of which may be regarded as already established species—the golden goby *Gobius auratus* and the centracant *Centracanthus cirrus*.

On the contrary, most of the species brought from the Adriatic Sea established in the Black Sea, created reproductive populations, and became abundant. Among them there are the bivalve mollusk *Anadara inaequivalvis* (*Cunearca cornea*), which was brought to the Adriatic Sea from the Pacific Ocean, and the oyster *Crassostrea gigas*, which was initially brought from the Japan Sea to the Adriatic Sea and then to the Black Sea. Later, the measures on the aquaculture of *Crassostrea gigas* in the Black Sea started; they are still going on [81]. The brown alga *Desmarestia viridis* could also penetrate not only from the near-shore waters of Europe but also from the Adriatic Sea [30]. Mosquito fish *Gambusia holbrooki* was brought for its intentional introduction from the Italian coasts of the Adriatic Sea, to where it had been previously brought from Central America. The fouling polychaete *Mercierella enigmatica* seems to have penetrated not only from the near-shore waters of Europe but also from the Adriatic Sea. The successful establishment of the species brought from the Adriatic Sea is explained, on the one hand, by the lower salinity of some of its regions as compared to other parts of the Mediterranean Sea and, on the other hand, by the winter and summer water temperatures close to those characteristic of the Black Sea; this allows the Adriatic invaders to overcome the conditions of the winter cooling in the Black Sea. In addition, the intensive shipping between the ports of the Adriatic and Black Seas favors the penetration of the organisms from this region. For example, according to the data of B. Aleksandrov, most of the 29 Mediterranean species recognized in the area of the port of Odessa was brought from the near-shore waters off the Italian coast of the Adriatic Sea [13].

The invasion of alien species via the three pathways mentioned is mainly implemented with ballast waters of ships or with the ship hull fouling communities.

The Japan Sea, with its estuaries and rivers, represents one more important source for alien species invasion to the Black Sea; several species were accidentally or intentionally introduced from this region. They are the spontaneously brought rapa whelk *Rapana venosa* and the intentionally introduced haarder *Liza haematochilos* (*Mugil soiuy*). Together with the haarder, three species of fish parasites were also introduced [11]. Aquarium holders occasionally released the aquarium fish Japanese oryzia *Oryzias latipes*, which originates from Japanese freshwater basins. The accidental introduction of alien species from the Japan Sea has become possible after the opening of the Suez Canal, when new navigation routes appeared. Meanwhile, except for rapa whelk, none of the species of the Japan Sea that spontaneously introduced into the Black Sea succeeded to establish in it. Of interest are the findings of Indian–Pacific fish species that might penetrate over this route with ballast waters and, probably, even in the course of their migration: initially to the Aegean Sea as Lessepsian migrants and then to the Black Sea; however, their further destiny is still unknown. During last years, other groups of species of Indo-Pacific origin began to appear in the Black Sea and they were most likely brought from the Mediterranean where they penetrated as Lessepsian migrants, but we cannot exclude their penetration with ballast waters from the Indo-Pacific area.

In addition to the species mentioned, there are others that also introduced from the near-shore waters of the Pacific or Indian oceans after the opening of the Suez Canal; initially, they established in the coastal waters of Europe and the Adriatic Sea and then were driven to the Black Sea. These are the polychaets *Capitellethus dispar* and *Glycera carpita*, the gastropod *Potamopyrgus jenkinsi*, and the decapod—Chinese mitten crab *Eriocheir sinensis*. However, these invaders have not become abundant species in the Black Sea.

The attempt of intentional introduction of five fish species from the estuaries of the Japan Sea, the Amur River, and other rivers of Southeast Asia was not successful: two sea bass species occur as isolated individuals and only the pond carps and amurs dwell in freshwater lagoons, rivers, and deltas of the Black Sea, but their populations are supported by artificial reproduction. Two fern species and strains of *Vibrio cholerae* were brought from the same regions.

One more route was mentioned—the penetration of the organisms from the Caspian Sea [11, 29], though it is hardly probable. For example, the algae mentioned in these publications (*Ectocarpus caspicus* and *Laurencia caspica*), were regarded as Caspian endemic species [27]; meanwhile, later they were recognized in the Black Sea waters off the coast of Romania [26] and referred to as the Ponto–Caspian relics of the Ponto–Azov province [28]. The same authors present the crustacean harpacticoid *Schizoptra neglecta* as an invader from the Caspian Sea; meanwhile, in addition to the Caspian Sea, this species is also mentioned as dwelling in the eastern part of the Black Sea and the Dnieper–Bug lagoon [82]. Thus, most likely, at present there is no invasion route from the Caspian to the Black Sea. Moreover, this route seems to be hardly possible because the species that were capable of establishing in the Ponto–Azov environment at its latest connection with the Caspian had already established in it and are now known as Ponto–Caspian relics. In the Ponto–Azov area, Ponto–Caspian species are common in origin with the autochthonous species of the Caspian and the species that could establish there, did it during the latest connection between the basins [83]. It is hardly probable that some other species from this group (particularly a Caspian-endemic species) is capable of penetrating into the Black Sea. These species were formed under the particular conditions of the Caspian Sea and can hardly establish under the conditions of the higher salinity of the Black Sea. Experimental studies showed that this fauna couldn't resist enhanced water salinity [84].

10 Vectors (Ways) of Alien Species Penetration

Various ways (vectors) of alien species penetration to the Black Sea may be distinguished: intentional introduction of commercial species, accompanied unintentional release of weed species during the latter operations, occasional release by aquarium holders, and penetration via straits, canals, and rivers. Meanwhile, the most important and largest vector whose significance increases each year is shipping. In so doing, while previously mostly the species of ship hull fouling communities invaded, in the recent years, a greater number of organisms are brought with ballast waters. The intensified shipping in the Black Sea increases the risk of appearance and establishment of new species accidentally brought if there is no control of ballast waters and ship fouling.

There are 92 (+32?) marine and brackish-water alien species brought with ships, which may be regarded as actually established. Among the 15 (17) fish species, ten are intentionally introduced, one species was a result of an accompanied unintentional release, and one more was released by aquarium holders; one fish species penetrated via canals and two more entered the sea via the Bosporus Strait. Thirty-seven more species have not established yet (regardless of parasites).

The currents via the Bosporus Strait delivered 59 species of zooplankton, 37 species of phytoplankton, 51 benthic species, and 23 species of macrophytes. Meanwhile, it seems probable that some of the species recognized during the recent years have not been noted previously because of the absence of appropriate studies in the near-Bosporus region aimed at the groups of species mentioned.

Summarizing, we may conclude that at present there are 152 established alien species (including freshwater species) in the Black Sea or 161 species including species doubtful with respect to their establishment. In addition there are 150 species that are recorded as isolated individuals or observed only in the near-Bosporus region; more than 37 species have not established yet. Probably, our generalization does not account for selected species, especially among those recorded as single specimens or observed only in the near-Bosporus region, if the specialists who studied these species did not report them as recent invaders from the Mediterranean Sea. Locally, in the northwestern and western parts of the sea, alien species were recorded whose status was not definitely determined. Among them, the same planktonic and meroplanktonic species brought from the Mediterranean Sea are noted in ballast tanks and port areas. It seems that some of the most euryhaline species of this group may soon establish in the Black Sea.

The invasion rate has been increasing over the last years. More and more Mediterranean species penetrate via the Bosphorus, are released from ballast waters, and establish due to global warming (Fig. 10).

Among the alien species one finds representatives of various ecological, systematic, and functional groups in the Black Sea (Table 1, Fig. 10).

The most negative effect on the Black Sea ecosystem was provided by the invaders–predators that formed abundant populations and featured abundance outbursts under favorable food and environmental conditions. Among them, one finds benthic species (such as rapa whelk, which consumes benthic organisms: oysters, mussels, and other bivalve mollusks) and pelagic species (ctenophore *Mnemiopsis leidyi*, which consumes edible holozoo-meroplankton, fish eggs, and larvae and indirectly affects all the trophic levels of the ecosystem). This group also includes the weed sunfish *Lepomis gibbosus*, which is widely spread and causes increasing damage to the ecosystems of the freshwater and brackish-water area of the Black Sea basin by consuming zooplankton, fish eggs, larvae, fries, as well as small adult fishes.

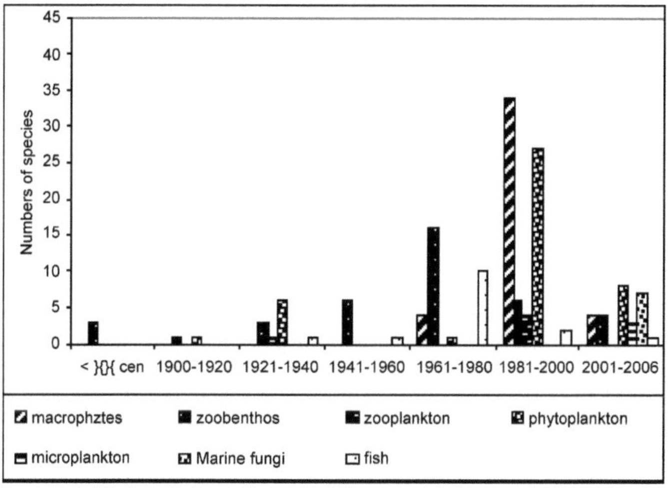

Fig. 10 Invasion rate of alien species in the Black Sea

Table 1 Taxonomical and ecological groups of alien species; *brackets* indicate species that were described as identified but whose status was still uncertain

	Established species	Species occurring only in the near-Bosporus area	Species not established yet
Fungi	7		
Bacteria	1		
Parasites of fish	3		
Infusorians	3 (>)		
Phytoplankton	43	37	
Macrophytes	42		
Kamptozoa	1		
Hydrozoa	2 (7)		
Ctenophora	2		
Polychaeta	11	29	
Copepoda	2	59	> 30
Decapoda	4	1	2
Amphipoda	1	4	
Cirripedia	3		
Bivalvia	8		
Gastropoda	5	18	
Pisces	15 (2)	2	5
Total	152 (161)	150	> 37

The proportions of the ecological groups of the species–invaders are similar to those of the aboriginal species of the Black Sea, especially if all the organisms both established and found as isolated individuals are taken into account. This may be explained, first of all, by the fact that in the course of the long-term geological evolution, particular conditions were formed in the Black Sea and the more the deviation from the normal environment setting the poorer its biological diversity (which depends on the number of species capable of adjusting to the existing conditions) and the higher the abundance and biomass of these species (productivity of the basin). Therefore, one may suggest that, under the existence of the limiting factors that strong (the low salinity, the continent climate, and the presence of the hydrogen sulfide layer), species that are ecologically similar to the aboriginal ones but stronger with respect to their competitive force may introduce into the community.

11 Conclusions

Let us try to analyze all the aliens that have established, both spontaneously and intentionally, in order to clarify which species (and why) became a large and widely spread species and which ones could not be established and are found as singular specimens.

The major part of the species that have established directly in the Black Sea are the widely distributed marine neritic, first of all euryhaline and, to a significant extent, eurythermal organisms. Their ability of inhabiting new areas is genetically provided. The progeny of widely distributed species features a phenotypic polymorphism. The habitats of species, especially of the widely spread ones, as well as the conditions existing in them, are irregular and "patchy" and change in space and with time; therefore, the species that are capable of both expanding their habitats and of existing under the given conditions can follow the best strategy. At the existence of the phenotypic polymorphism, two properties are coupled: the ability of migration and the genotypic variability. The expansion strategy is best manifested for the species existing in overpopulated communities, that is, for the species dominating in the communities or occur in high abundance and capable of providing outbursts. As a rule, these species feature a wide range of tolerance with respect to all the factors, which favors their wider expansion.

The species with the most complete set of the above-listed properties not only succeeded to establish but also became abundant species in the Black Sea and continued their further expansion. They expanded, first of all, to the brackish-water Sea of Azov with currents and ships via the Kerch Strait; selected species invaded to the Sea of Marmara via the Bosporus Strait and to the Aegean Sea via the Dardanelles or to the Caspian Sea with ballast waters or with ship fouling communities. Thus, the Black Sea became a donor basin for the further expansion of the alien species that have established in it to other southern seas.

The ctenophore *Mnemiopsis leidyi* is the most impressive representative of invaders; it is the most aggressive invader, which expanded over all the seas of the Mediterranean basin and over the Caspian Sea and affected their ecosystems. The subsequent introduction of *Beroe ovata* to the Black Sea provided the beginning of restoration of its ecosystem.

For 15 years after the spontaneous introduction of two lowly organized but adjustable gelatinous animals—the ctenophores *Mnemiopsis leidyi* and *Beroe ovata* the Black Sea ecosystem has changed significantly. As a result of the development of *Mnemiopsis leidyi*, the ecosystem, from the lowest trophic levels to the higher ones (fishes and dolphins), was significantly degraded. Meanwhile, after the introduction of the *Beroe ovata*, it started to recover. These events present a dramatic example of the impact on the ecosystem that is provided by the invasion of a single species and, undoubtedly, this process should be thoroughly controlled by man.

References

1. Chuhchin VD (1984) Ecology of Gastropoda in the Black Sea. Naukova Dumka, Kiev
2. Kiseleva MI (1979) Polychaeta of the Black Sea, p 208 (in Russian)
3. Greze II (1966) Distribution of benthos and biology of bottom animals in the Southern seas (in Russian). Naukova Dumka, Kiev, p 33
4. Georgieva LV (1993) In: Kovalev AV, Finenko ZZ (eds) Plankton of the Black Sea (in Russian). Naukova Dumka, Kiev, p 31
5. Kuzmenko LV (1966) In: Kovalev AV, Finenko ZZ (eds) News of systematic of algae AN USSR (in Russian). Botan Ins M.-L., p 51
6. Senichkina LG (1993) Plankton of the Black Sea (in Russian). Naukova Dumka, Kiev, p 32
7. Kovalev AV, Besiktepe S, Zagorodnyaya J, Kideys AE, et al. (1998) In: Ivanov L, Oguz T (eds) Ecosystem Modeling as a Management Tool for the Black Sea. 47:199
8. Senicheva MI (2001) Ecol Sea 62:25 (in Russian)
9. Zagorodnya YA, Kolesnikova EA (2003) In: Conference on the evolution of marine ecosystem under impact of invaders and artificial mortality of fauna (in Russian). Rostov-on-Don, Russia, p 80
10. Burgasov PN, Pokrovsky VI (1986) In: Large medical encyclopaedia (in Russian). T. 27 M Soviet Encyclopaedia (in Russian). p 36
11. Shadrin NV (2000) In: Matishev GG (ed) In: Exotic species in the European seas of Russia (in Russian). Apaty. Kolsky nauchny center, p 76
12. Polikarpov IG, Sabutrova MA, Mànkos LA, Pavlovskaya TV, Gavrilova NA (2003) In: Eremeev VN, Gaevskaya AV (eds) The present state of biodiversity of the coastal zone of Crimea (in Russian). NAN Ukraine, IBSS. ECOCI-Gydrophysics, Sewastopol, p 16
13. Alexandrov B (2004) Marine ecological journal Ukraine T 3. 1:5 (in Russian)
14. Skolka BX, Bodynu N (1963) Rev Biol Acad RPR T 7 1:89
15. Senicheva MI (2002) Ecogogia morya 62:25 (in Russian)

16. Moncheva S, Petrova-Karadjova V, Palasov A (1995) In: Lassus P, Arzul G, Denn E, Gentien P (eds) Harmful marine algal blooms. Lavoisier Publ Inc., p 193
17. Terenko LM, Terenko GV (2000) Ecologia moray. National Academy of Sciences of Ukraine. Institute Biology of the Southern Seas (in Russian). 52 p 56
18. Terenko LM (2003) Conference on the evolution of marine ecosystem under impact of invaders and artificial mortality of fauna. Rostov-on-Don, Russia p 135 (in Russian)
19. Usachev PI (1928) In: Desription of all-Soviet conference of botanical scientists (in Russian). Leningrad, p 64
20. Proshkina-Lavrenko AI (1971) On the flora of the Black Sea diathoms. In: Vodyanitzky VA (ed) Problems of marine biology. Naukova Dumka, Kiev, p 41 (in Russian)
21. Pitzyk GK (1979) The biological productivity of the Black Sea. Naukova Dumka, Kiev, p 63 (in Russian)
22. Zaitsev YUP, Nesterova DA (2003) Conference on the evolution of marine ecosystem under impact of invaders and artificial mortality of fauna (in Russian). Rostov-on-Don, Russia, p 85
23. Milchakova NA (2002) Ecol Sea 62:19
24. Minicheva GG, Eremenko TI (1993) Algology 4:83
25. Minicheva GG (2006) 1st Biannual Scientific Conference. The Black Sea ecosystem 2005 and beyond. Istanbul, Turkey (8–10 May), p 78
26. Celan M, Bavari A (1967) St Cerc Biol Seria Botany 19.N3:215–219
27. Zinova AD (1967) Guide of the green, brown and red algae of the southern seas of USSR. Nauka, Leningrad
28. Bogani A (1980) Journees Etud System et Biogeogr Medit CIESM. Cagliari, p 93
29. Gomoiu MT, Alexandrov B, Shadrin N, Zaitsev YU (2002) Invasive aquatic species of Europe. In: Leppakoski E, Gollash S, Olenin S (eds) Distribution, impacts and management. Kluwer, Dordrecht, p 341
30. Stepanyan OV (2003) Conference on the evolution of marine ecosystem under impact of invaders and artificial mortality of fauna (in Russian). Rostov-on-Don, Russia, p 32
31. Valkanov A (1936) Belezhki varkhu naskhu naskhite brakichny vody (Notes on our brackish waters). God Sofia Univ (in Bulgarian) 32:133
32. Naumov DV (1968) Guide of fauna of the Black and Azov Seas. 1 (in Russian). Naukova Dumka, Kiev, p 56
33. Zaitzev YuP (1998) The most blue sea. T 6. Pub. NO.CEP GEF. New York (in Russian)
34. Grishicheva NP, Shadrin NV (1999) In: Ecosystem processes and service for society. Sewastopol, p 229
35. Vinogradov ME, Shushkina EA, Musaeva EI, Sorokin PY (1989) Okeanologiya 29(2): 293 (in Russian)
36. Shiganova TA (1998) Fisheries oceanography. In: Coombs S (ed) Globec Spec Issue, p 305
37. Shiganova TA, Mirzoyan ZA, Studenikina EA, Volovik SP, Siokou-Frangou I, Zervoudaki S, Christou ED, Skirta AY, Dumont H (2001) Mar Biol 139:431
38. Vinogradov ME, Sapozhnikov VV, Shushkina EA (1992) Ekosistema Chernogo morya (Ecosystem of the Black Sea) (in Russian). Nauka, Moscow
39. Shiganova TA, Niermann U, Gugu AC, Kideys A, Khoroshilov VS (1998) In: Ivanov L, Oguz T (eds) Ecosystem Modeling as a Management Tool for the Black Sea. NATO Scientific Affairs Division. Kluwer, Dordrecht, p 171
40. Tsikhon-Lukanina EA, Reznichenko OG, Lukasheva TA (1991) Okeanologiya 31 N 2:272 (in Russian)

41. Tsikhon-Lukanina EA, Reznichenko OG, and Lukasheva TA (1993) Okeanologiya 33 N 6:895 (in Russian)
42. Shiganova TA, Bulgakova YV (2000) ICES J Mar Sci 57:641
43. Studenikina EI, Volovik SP, Mirzoyan ZA, Luts GI (1991) Oceanologia 31 N 3:722 (in Russian)
44. Shiganova TA, Ozturk B, Dede A (1994) FAO Fisheries Rep 495:141
45. Shiganova TA, Kamakin AM, Zhukova OP, Ushivtsev VB, Dulimov AB, Musaeva EI (2001) Oceanologia 41 N, 4:542 (in Russian)
46. Purcell JE, Shiganova TA, Decker MB, Houde ED (2001) In: Purcel JE, Graham WM, Dumont HJ (eds) The ctenophore Mnemiopsis leidyi in native and exotic habitats: U.S. estuaries versus the Black Sea basin. Hydrobiologia. 451. Kluwer, Dordrecht, p 145
47. Konsulov A, Kamburska L (1998) Trudy Institute of Oceanology, BAN. Varna, Bulgaria, p 195
48. Seravin LN, Shiganova TA, Luppova NE (2002) Zool J 81 N, 10:1193 (in Russian)
49. Shiganova TA, Bulgakova YUV, Sorokin PY, Lukashev YF (2000) Biol Bull 27:247
50. Vinogradov ME, Vostokov SV, Arashkevich EG, Drits AV, Musaeva EI, Anohina LL, Shushkina EA (2000) In: Matishev GG (ed) Exotic species in the European seas of Russia (in Russian), Apatity. Kosky scienific centre (in Russian), p 91
51. Finenko GA, Anninsky BE, Romanova ZA, Abolmasova GI, Kideys AE (2001) Hydrobiologia 451:177
52. Arashkevich EG, Anochina LL, Vostokov SV, Dritz AV, et al. (2001) Oceanology T 41 1:116
53. Shiganova TA, Musaeva EI, Bulgakova YV, et al. (2003) Biol Bull 2:225
54. Gubanova AD (2003) In: Eremeev VN, Gaevskaya AV (eds) NAN Ukraine IBSS Sewastopol, ECOSI-Hydrophysics, p 99
55. Gomoiu MT, Skolska M (1996) GEO-ECO-MARINA, RCGGM.v.1 Danube delta–Black Sea system under global changes impact. Bucuresti – Constanta, p 49
56. Marinov T (1990) Zoobenthos ot bulgarskya sector na Cherno more (Zoobenthos of the Bulgarian sector of the Black Sea) (in Bulgarian). Bulgar Acad Sci Publ, Sofia
57. Zevina GB (1957) Dokl USSR T 113 2:450
58. Marinov TM (1966) Izv Zool Ins BAN 21:139
59. Nebolsina TK (1959) Priroda 6:116 (in Russian)
60. Drabkin EI (1953) Priroda 9:92 (in Russian)
61. Zaitsev YU, Mamaev V (1997) Marine biological diversity in the Black Sea: a study of change and decline. United Nations Publications, New York
62. Studenikina EI, Volovik SP, Tolokonnikova I, et al. (1998) In: Makarov EV, Volovik S, Tutina YE (eds) The main problems of fishery and conservation of fishery of Azov-Black Sea basins. Rostov-on-Don. Azov Institute for fishery. (in Russian), p 67
63. Kucheruk NV, Basin AK, Kotov AV, Chikina MV (2002) In: Zatzepin AG, Flint MV (eds) Comprehensive investigations of the northeastern Black Sea. Nauka, Moscow, p 289 (in Russian)
64. Chikina MV, Kucheruk NV (2005) Oceanology 45 (Suppl 1):176
65. Beshevli LI, Kolyagin VA (1967) On finding mollusk *Mya arenaria* L (Bivalvia) in the northwestern Black Sea (in Russian). Naukova Dumka, Kiev
66. Revkov NK (2003) The present state of biodiversity of the coastal zone of Crimea (The Black Sea). In: Eremeev VN, Gaevskaya AV (eds) NAN Ukraine, IBSS. ECOSI-Hydrophysics, Sewastopol, p 209
67. Roginskaya IS, Grintzov VA (1990) Oceanologia 30(5):855 (in Russian)
68. Zolotarev VN, Zolotarev PN (1987) Dokl RAS USSR 297(2):501 (in Russian)

69. Svetovidov AN (1964) Fishes of the Black Sea. Nauka, Moscow (in Russian)
70. Abaza VL, Boicenco M, Moldoveanu F, Timofte AS, Bologa A, Sburlea C, Dumitrache I, Staicu G, Radu G (2006) In: 1st Biannual Scientific Conference. Black Sea ecosystem 2005 and beyond. 8–10 May. Istanbul, Turkey. p 50
71. Boltachev AP (2006) 1st Biannual Scientific Conference. The Black Sea ecosystem 2005 and beyond, 8–10 May. Istanbul, Turkey. p 114
72. Tkachenko PV (1994) All-Russian conference on the ecosystems of the Russian seas in anthropogenic conditions (including fishing) (in Russian). Astrahan, Russia, p 334
73. Pashkov AN (2005) Ecosystem study of the Sea of Azov and the Kerch strait. 7 (in Russian). Apatity. Kolsky Scientific centre. Matishev GG, et al. (eds) (in Russian), p 263
74. Drabkin EI (1964) Bull MOIP T 69 5:140 (in Russian)
75. Plotnikov GK, Emtyl MH, Abaev YI (1990) Conference on the actual problem of ecology and conservation of nature of basin. Krasnodar, Russia, p 117 (in Russian)
76. Nadolinsky VP (2004) Structure and stock assessment of aquatic resources in the northeastern Black Sea (in Russian). PhD Thesis
77. Pryahkin YUV, Volovik SP (1997) The problems of the fishery and conservation of Azov-Black Sea basin fish. AzNIIRKH. Rostov-on-Don, Russia, p 204 (in Russian)
78. Tzarin SA, Zuev GV, Boltachev AP (1999) Growth of *Pilengas Mugil soiuy Basilewsky*, 1855 (Mugilidae, Pisces). Ecol Sea 48:68 (Review) (in Russian)
79. Boltachev AR (2003) The present state of biodiversity of the coastal zone of Crimea (The Black Sea). In: Eremeev VN, Gaevskaya AV (eds) NAN Ukraine IBSS, Sewastopol. ECOSI-Hydrophysics, p 363
80. Frolova LA (2002) Biology of inner waters: problems of ecology and biodiversity (in Russian). Borok, p 227
81. Zlotarev VN (1996) Mar Ecol 17:227 (in Russian)
82. Birshtein AYA (1968) Atlas of invertebrates. Pishevaya prom, Moscow, p 118 (in Russian)
83. Mordukhai-Boltovskoi FD (1960) Caspian fauna in the Azov-Black Sea basin (in Russian). Moscow, Leningrad. AN USSR publisher, p 288
84. Karpevich AF (1975) Theory and practice of introduction of aquatic animals (in Russian). Piszevaya prom, Moscow

Hdb Env Chem Vol. 5, Part Q (2008): 407–421
DOI 10.1007/698_5_076

Published online: 31 March 2007

Environmental Issues of the Black Sea

Igor S. Zonn[1] (✉) · Dmitry Y. Fashchuk[2] · Anatoly I. Ryabinin[3]

[1]Engineering Research Production Center on Water Management, Land Reclamation and Ecology, 43/1, Baumanskaya ul., 105005 Moscow, Russia
igorzonn@mtu-net.ru

[2]Institute of Geography, 29 Staromonetny per., 109017 Moscow, Russia

[3]Marine Branch of Ukrainian Hydrometeorological Institute, 61, Soviet Street, Sebastopol, 99011, Ukraine

Abstract Intensive economic development of the vast watershed basin of the Black Sea by the former Soviet Union and also Rumania and Bulgaria and inadequate nature conservation efforts led to a situation where considerable progress in industry, power engineering, agriculture, transport, residential utilities and recreational spheres created very unfavorable and in some regions even menacing environmental situations. The Black Sea is characterized by higher environmental vulnerability resultant from it being nearly a completely land-locked basin which is why the sea area is dependent, to a great extent, on the quality of water running into it from nearby land. Coastal regions are characterized by interaction and contradictions of production and social interests being responsible for environmental conflicts arising in sea nature management. Among the key environmental issues of the Black Sea we can name the following: presence of practically all forms and kinds of pollution of the sea and coastal environment, qualitative and quantitative depletion of bioresources, invasion of alien species.

Keywords Anthropogenic loads · Black Sea · Chemical pollution · Coast · Dumping · Invaders

1 Introduction

The environmental issues of the Black Sea are determined by specific effects of the economic development in the littoral states and the resultant economic activities, first of all, in watershed basins of rivers flowing into it. One cannot neglect here the contribution of coastal territories being of great social and economic value due to their natural and climatic conditions. In the historically shaped system of nature management the coastal areas were always of key importance for the economics of the regions although the foci of their development and resource utilization (harbor facilities, shipbuilding and ship repair, cargo handling, fishing and mariculture, tourism and reaction, etc.) have changed with time.

The phenomenon of the sea-shore contact is a potent stimuli for "attraction" of various kinds of economic activities which leads to aggravation of environmental stress [1].

The present-day environmental situation in the Black Sea is very difficult to manage. Industrial development and growing population in the littoral states, increased oil transit, intensive development of navigation, construction, recreational development and dumping are all responsible for extensive water pollution, including in regions of active fishing. The principal sources of sea pollution are river flow; disposal of industrial, domestic and agricultural wastes in the coastal zone, in particular in the areas around large cities and ports; economic activities in the water area (navigation, oil leaks during transportation, dumping, etc.). The results of environmental monitoring revealed considerable pollution of the sea water and bottom sediments with oil products, pesticides, toxic and heavy metals, polyaromatic hydrocarbons and radionuclides causing degradation of the sea ecosystem.

2 Sources of the Black Sea Pollution

Environmental issues of the Black Sea have been shaped in the course of economic development of coastal territories and watersheds of rivers flowing into the sea. Each part of the Black Sea basin has its own environmental problems mostly similar in their consequences and finally affecting the open sea area. The basic issues of the Black Sea ecology are collected in Table 1.

At the end of the last century, as a result of an integrated impact of natural and anthropogenic factors, the environmental system of the sea was damaged seriously and by the estimates of many specialists the Black Sea is now one of the world's most polluted water bodies.

In the post-Soviet period notwithstanding significant economic recession in Russia the load on the environment in the Black Sea region did not de-

Table 1 Black Sea water pollution

Sources	Object of activities	Results of activities	Environmental consequences
Watersheds of rivers flowing into the Black Sea	Industrial enterprizes Agriculture Cities and settlements Navigation	Reduced fresh water flow Disposal of industrial, domestic wastes, wash-off of fertilizers, pesticides, herbicides	Pollution Degradation of river systems Reduction of solid flow
Sea coast	Port facilities protection efforts Cities and settlements Transport infrastructure Agriculture and health resorts Recreational construction and resort areas management Creation of artificial beaches Naval bases	Slide-control and bank- Wastewater disposal Domestic waste disposal Extraction of great quantities of beach material for construction, agricultural and other needs Trampling of beaches Coastline erosion	Pollution Degradation of coastal ecosystems Damage of biodiversity
Sea	Cargo and passenger shipping Tanker traffic Bottom dredging Dumping Naval exercises Hydrocarbon surveys in a shelf zone Separation of some parts of water area Trawling in the course of seafood fishing	Disposal of oil products, debris Overfishing of bioresources Oil spills during accidents Wash-off of heavy metals from bottom sediments Disposal of bilge waters	Microbial pollution Pollution of microbiota Less biodiversity Accidental invasion of foreign exotic species of animals and plants Noise pollution Hypoxia Migration of loose bottom material

crease, but some of the most important parameters even increased. For example, the Krasnodar Territory disposes the largest amount of wastewater—in Russia—into local water bodies and the second largest amount of polluted wastewater. In the last 5 years deterioration of the water quality in natural water bodies has been witnessed over the whole Krasnodar Territory. Similar phenomena were observed in the 1990s on the Russian territory of the Circum-Black Sea area, in the Ukraine, Bulgaria, and Romania as a result of curtailing of the environmental policy and sharp deterioration of the socioeconomic conditions of the population, and growing wearout of fixed assets of operational enterprizes. The state lost control over some industries, in particular, the fishing industry, and as a result the most precious national resources turned into a "golden goose" for criminal groups. Thus, it could be said that the main hazards to environmental stability in these states should be sought within these states.

The scale of the Black Sea pollution problem may be appraised by the following estimates [2]:

- every year approximately 653 thou tons of suspended matter, about 8000 tons of organic matter, about 1.9 thou tons of nitrogen and 1.12 thou tons of phosphorus enter the sea via river flow;
- every year about 33.8 thou tons of suspended matter, 8.8 thou tons of nitrogen, 2.6 thou tons of phosphorus and 24.1 thou tons of petroleum products are discharged into the sea by public utilities;
- about 11.6% of undigested nitrogen and 13% of phosphorus fertilizers, 6% of pesticides get into the Black and Azov Seas with the flow of small rivers from agricultural areas of the Circum-Azov, Circum-Black Sea areas and Crimea.

The most affected in this respect is the northwestern shallowest part of the Black Sea where up to 65% of all living organisms are produced and where the main spawning grounds are found.

The general environmental situation in the coastal regions of the Black Sea is very complicated and is close to critical [3]. The recent decades have witnessed growing pollution of waters with total phosphorus and nitrogen (Danube seaside), petroleum products (nearby Sebastopol and the Georgian coast), detergents and phenols (the southern coast of Crimea), phenols and pesticides (Odessa coast). Here the quality of coastal waters is determined not so much by the source of the pollutants and the width of the continental shelf, but by the nature and intensity of currents in the particular regions.

With the growth in economic potential seen recently in the Circum-Black Sea countries the environmental stress may be aggravated. The main causes include construction of new and rehabilitation of existing sea ports, revival of the merchant and tanker fleet, consolidation of the naval component, construction of oil and gas pipelines and hydrocarbon extraction, development of health resorts and recreation activities.

Among the factors influencing the environmental condition of the Black Sea is the availability of a vast watershed basin that is nearly 5-fold greater than the area of the sea itself; and the greater part of this watershed is occupied by densely populated industrial regions. The Black Sea becomes a terminal collection point for wastes and discharges generated by 170 million people. Permanent pollution sources for the sea are industrial and domestic-municipal wastewater that are only partially subject to mechanical and biological treatment, while they are mostly discharged untreated. Most industrial-domestic wastewaters contain toxic substances of technogenic origin. The total volume of wastewater is approximately 4 km^3 a year or about 8000 m^3 per 1 km^3. Approximately 80% of these wastewaters or 2000 m^3 are transported via rivers into the shallow northwestern part of the sea which accommodates every year 10 mln m^3 of wastewaters per 1 km^3 [4].

The second factor is a lack of normally developed shelf along 70% of the sea coast and also the small size of the self-purification zone because in the Black Sea it is limited by the upper oxygen layer which is 120 to 150 m thick.

3 Pollution of the Water Area

3.1 Pollution with Petroleum Products

Variations in petroleum hydrocarbon concentrations in space and time in the coastal water of the Black Sea are considerable. The general tendency for all regions is their increase from winter to summer and a greater level (by 20–40%) of petroleum hydrocarbons in the surface layer compared to the bottom. Oil pollution in the Black Sea has not reached—so far—the scale of an environmental disaster [5]. Every year the sea receives from 80 to 100 000 tons of oil wastes, or by other estimates [4] from 130 to 170 000 tons (65% with the flow from the Danube and Dnieper; when including the Danube waters three times more than with the Dnieper waters alone). The main sources of these wastes are industrial enterprizes. Their share in the total annual discharge of hydrocarbons varies from 60 to 90% depending on the flow size. In addition, other sources of oil products include carriage by sea involving, on the average, 180 000 vessels every year whose contribution is 12 000–15 000 tons, and the dumping of ground involving 10 000 tons of oil products a year. Every year as a result of accidents at industrial enterprizes and involving vessels, inefficient operation of treatment facilities and loading jobs in ports, the rushed discharge of untreated wastewaters, ballast and bilge waters containing very high quantities of pollutants occur. This is why the most polluted areas include the large ports and harbors and also the Bosporus Strait. The Sebastopol, Gelendjik and Novorossyisk harbors where the abnormally high levels of oil products are recorded not only in the water, but in the bottom sediments as well, deserve a special mention. Some places in the Sebastopol harbor concentrate up to 120 000 tons of oil-containing sediments [6].

After the commissioning in 1964 of the oil base "Sheskharis" and the stationing of a tanker fleet the Novorossiysk harbor was subject to heavy oil pollution.

Hydrocarbon development on the Black Sea shelf is just beginning. Since 2002 French oil concern "Total" has been engaged in hydrocarbon prospecting near Shatksy Val in the deep-water part of the Black Sea. Later it signed a contract with the "Rosneft" Company for joint development and prospecting of oilfields near Tuapse trough. The availability in the sea of environments containing hydrogen sulfide places special requirements on environmental safety which have not been attained so far.

In the open sea greater accumulations of oil spots are found in the intensive shipping lanes (Bosporus-Odessa-Yalta-Batumi) in the zones of quasistationary anti-cyclonic circulation (near Sebastopol and Feodosia) and also nearby large ports. The oil film on the sea surface causes decreased absorption of solar energy interfering with the heat accumulation processes. Despite the decreased quantity of incoming heat the surface temperature in the presence of an oil film increases with increasing thickness of the film. It should be noted that a film 30–40 μm thick is capable of completely absorbing the infrared irradiation. The area covered by the oil film may vary from 0.5% to 5% of the total sea surface. Oil and its compounds produce a toxic effect on fish eggs and fish fry, in other words, on the future biodiversity of the Black Sea. Fish eggs and fish fry reveal much higher sensitivity to oil compared to adult species. In some oil polluted areas of the Black Sea ichthyologists found up to 80% dead eggs of commercial fish. The long-term effects of oil and its compounds even lead to the disappearance of such hydrobionts as crabs, mollusks and algae.

Sea pollution with oil products resulting from shipping increases greatly in the case of accidents. In the last decade alone more than 200 accidents with ships were recorded in the Black Sea.

To illustrate the above assertion the following examples of major accidents are provided.

- In 1979 the tanker "Independenta" under a Romanian flag collided with the Greek vessel "Evriyali", as a result 64 000 tons of oil spilled out into the sea and there was a fire; fragments of the tanker lay for several years on the shore.
- In January 1986 in the Ilyichevsk port there was an accident with the tanker "Uzhgorod"; that time 40 tons of fuel oil spilled into the sea spoiling 4 km of beaches.

Quite often there are accidents in the Bosporus Strait.

- In 1988 the tanker "Blurstar" under a Panamian flag with ammonia onboard collided with the Turkish tanker "Gaziantep", as a result the strait's waters were polluted with a great quantity of ammonia.
- In 1994 there was a collision between two vessels, the "Nassia" and "Shipbroker", as a result of which 20 000 tons of oil spilled into the sea which burnt for 4.5 days; the strait was closed for shipping for several days.
- On March 2, 1997 in the Odessa oil harbor 50 tons of oil poured out from the vessel "Athenian Faith" under a Maltese flag; this accident disturbed the environmental equilibrium in this region for a long time.

- In 1999 the Russian tanker "Volgoneft-248" carrying 43 000 tons of fuel oil had an accident in the Bosporus, as a result nearly 1 ton of fuel oil spilled into the sea. On September 4, 1999 as a result of the breakup of the vessel "Christina Valetta" the Laspi Bay became the scene of an environmental disaster. 1.7 tons of oil products were spilled into the sea. The damage incurred to the marine environment was estimated at more than half-a-million US dollars.

Notwithstanding the high rate of accidents in the area, sea pollution with oil spills during accidents represents only 1% of the total [4].

Recently, the traffic of Caspian oil via the Black Sea has increased. These very important sea carriages enhance the risk of sea environment pollution with oil products.

A complicated situation is observed in the Kerch Strait. Here during oil transportation and reloading at the entrance into the strait from the Black Sea, the spills of oil products as a result of unauthorized discharges of washing bilge waters from tankers and also due to accidents with merchant ships and oil tankers awaiting pilotage and going via a canal are inavoidable. Russia transports Caspian oil bypassing the Kerch-Yenikalsky canal along a parallel shipway. Depths here are no more than 5 m, while the seabed is abound in shells that have survived after World War II. Taking into consideration the planned increase of the draft (loading) of "river-sea" tankers to 4.4 m—for economical reasons—the scale of likely oil spills may grow.

To the west of the renowned beach in Anapa on the southern coast of the Taman Peninsula a terminal for export of a propane-butane mix is under construction, and the creation of a navy base in nearby Novorossiysk is being contemplated, as a result a risk of pollution is growing which is incompatible with the recreational activities. In addition, on the Georgian coast of the Black Sea near Kulevi village a new oil terminal is under construction that will be used for the reloading of tankers with oil and petroleum products delivered to Georgia via railway from Azerbaijan, Turkmenistan and Kazakhstan.

In recent decades the load on the public resort area of the southern coast of Crimea (SCC) has been constantly growing. For long it was thought that the capacity of the SCC unique nature might outbalance the likely technogenic effect on the sea environment. As a result at present up to 37 mln m^3 (about 1 m^3/s) of untreated wastewaters are discharged into sea waters on the coastline from Yalta to Alushta. But approximately the same quantity of untreated storm runoff from nearby mountains and agricultural lands is also added to this water body.

The Odessa region with the urban agglomeration being the greatest on the Black Sea coast holds a specific position in the northwestern part of the Black Sea. It includes three major ports of the Ukraine: Odessa, Ilyichevsk and Yuzhny. In the last decade alone 104 cases of emergency spills of oil products from ships and industrial enterprizes were recorded here. This is where the main routes of cargo and passenger traffic meet, where a high-capacity oil terminal designed for handling up to 40 million tons of oil products a year is under construction, although the conclusions of all state and public environmental researches conducted from 1992 stated the impossibility of construction of oil complexes in the Odessa bay. This undoubtedly aggravates the environmental conditions in the region—the Odessa bay is characterized by a permanently high level of pollution. In 1991–1995 its waters were classified from "dirty" to "exceptionally dirty". According to the available information, over 400 tons of oil products get into the marine environment here every year.

In the last five decades an intensive anthropogenic load on the marine environment in the Sebastopol region has led to a sharp decline in the environmental situation. This process was spurred by construction here of the main Navy base of the Black Sea Fleet; and although recently the quantity of pollutants getting into the Sebastopol bay has somewhat diminished (in 1996–2000 heavy metals from 203.7 tons to 68.3 tons, hydrocarbons from 292.6 to 206.3 tons), the general condition of the Sebastopol bays (and there are 19 of them) [7] is evaluated as critical.

At present the pollution of the eastern part of the Black Sea is evaluated as weak, although there are some regions where the pollution level exceeds the norm 10–15 times. One of these regions is the Batumi bay where household wastes are discharged into the sea practically without any treatment.

Estimates on the basis of the water pollution criterion for the Black Sea have shown that the Danube seaside, the coastal zone of Crimea and Northern Caucasus waters are classified as "polluted" and "dirty". Coastal waters in the Dnieper seaside, in Karkinitsky and Kalamitsky bays and also in open areas of the northwestern part of the sea are classified as "very dirty".

Nonuniformity of the quality of coastal waters in space may be attributed to a combination of various factors (volume and concentration of incoming pollutants, specific features of a shelf area, peculiarities of water circulation). The narrowness of the shelf in the Krasnodar Territory and the availability here of a stream of the Main Black Sea Current going practically along the shore create rather good quality coastal waters despite large volumes and high concentrations of pollutants in wastewaters discharged here from the coast. In the Georgian shelf the stream of the Main Black Sea Current is distanced from the shore and in its coastal margins the quasistationary anticyclonic turnover is formed, thus, facilitating accumulation here of pollutants and, consequently, deterioration of the quality of coastal sea waters.

Still more unfavorable is the situation in the northwestern part of the sea, because this is a shallow area and the effect of anthropogenic factors is the strongest here. Environmental conditions in this region depend, to a great extent, on the chemical composition of the abundant river flow. The main kinds of pollution are hydrocarbons, phenols and detergents. Since the 1980s the quantity of biogenous substances has increased 2–5-fold. For the northwestern shelf area the problem of hypoxia is rather acute; zones with oxygen deficiency are formed here every year. Here three main zones of hypoxia can be identified: Danube, central and Odessa. The growing pollution of the northwestern water area and high fish mortality has caused in recent decades a significant shrinkage of the area populated by the valuable alga phyllophora and mussel shoals, reduction of the total population and biomass of many sponges, crabs, valuable mollusks, shrimps and bottom fish [10].

An unfavorable situation is also formed on the shelf of the Bulgarian coast. Most polluted waters are found in the Burgass bay where more than 100 000 m^3 of untreated wastewaters a day are discharged there containing up to 13 tons of oil products, 3 tons of ammonia, 22 tons of organic matter, which constitutes a great load in this water area. Wastewaters getting into the bay are characterized by redox-toxicity that in the summertime may lead to a sharp growth of pathogenic microflora, outbursts of endemic diseases and fish mortality. The average concentration of oil products in the bay may exceed MAC a dozen times. In the autumn-winter period the concentration of biogenous substances increases 1.5–2 times which causes hypertrophy of the ecosystem of the Burgass bay. It has been determined that pollution of bay waters with heavy metals exceeds the norm considerably [8].

The high intensity of anthropogenic load on ecosystems of the Black Sea affects, first of all, the viability of organisms confined to the phase interface: coastal zone, near-surface and near-bottom layers of the water. Adverse environmental consequences of the Black Sea pollution are revealed in the reduced populations of hydrobionts that were previously widespread.

3.2 Organic and Mineral Pollution

The most widespread type of pollutant in the Black Sea are phosphorous and nitrogen compounds. In fact, the waters of the Danube and Dnieper are the main suppliers of these substances (about 80 and 600 000 tons, respectively). The contribution of the Danube for both substances is 12-fold higher than that of the Dnieper alone [9]. Every year they bring into the Black Sea up to 30% of the total wastewater flow. The contribution of the Krasnodar Territory is 20%, Southwestern Crimea 10%, Odessa and Black-Sea coast of Georgia represent 5%. The runoff of biogenous substances from the coast represents no less than 2%. The highest concentrations of biogenous substances are registered in the coastal waters, while offshore their concentrations decrease. Here the correlation between various forms of biogenous substances is broken for the sea, in general, which leads to pollution of shelf areas with unstable organic substances and to changes of physical and chemical properties of sea waters. The increased accumulations of biogenous substances involve enhanced photosynthesis and eutrophication of the sea.

The Black Sea receives 0.5 to 0.8 km^3 of untreated wastewaters from industrial centers and settlements on the coast. The population of these settlements is approximately 17 million people plus 4–5 million visitors. The population is distributed as follows: Turkey 6.7 mln, Ukraine—6.8 mln, Russia—1.2, Bulgaria—0.7, Romania—0.6 and Georgia— 0.7 mln. Every year an additional 2–4 km^3 of wastewaters get into the sea via river flow. The Danube and Dnieper rivers bring in about 80% of this material. The share of wastewaters may be as high as 3–4% of the total river flow. In addition the inflow from the Sea of Marmara to the Black Sea brings every year organic pollutants comprising 12 000 tons of phosphorous, 190 000tons of total nitrogen and 1 500 000tons of organic carbon [11].

The accelerated turnover of the organic matter and biogenous elements in the production-destruction cycle in coastal waters caused an emergency in the late 1970s. This was the phenomenon called "red tides" connected with development of Peridinean algae *Exuviaella cordata* and others, the life products of which are toxic and may cause mass death of fish and other animals. "Red tides" are observed in the northwestern shelf, including the Danube mouth on the Bulgarian coast near Varna and the Romanian coast where the total mass of Peridinean algae may be as high as 1 kg/m^2. The consequence of this "red tide" is accumulation in bottom waters of excessive quantities of organic matter formed as a result of the death of algae that were involved in blooming.

Oxygen dissolved in the water is used for oxidation of dead phytoplankton remnants. An oxygen deficit causes the death of bottom animals, thus, creating additional sources of organic substances (secondary pollution), the increased oxygen consumption and release of hydrogen sulfide. Under conditions of oxygen deficit the death of zoobenthos and bottom fish is observed over vast areas. Thus, according to [6], in the period from 1973 to 1980 in the Black Sea approximately 60 mln tons of bottom animals, including about 5 mln tons of fish, died due to insufficiency of oxygen.

The main source of pesticides in the Black Sea is agricultural lands. The level of pesticides in sea water has a clear-cut maximum in spring–autumn which coincides with the time of their application in agriculture. According to the adopted norms the sea water should contain no chlorine organic pesticides at all. But still every year the Black Sea receives more than 20 tons and 1 ton of chlorine organic pesticides with the waters of the Danube and Dnieper, respectively. Another source is irrigation and drainage waters widely used on the northwestern and western shore of the sea. Thus, every year up to 1 km^3 of municipal-drainage waters get into the Karkinitsky and Djarylgachsky bays in the northwestern part of the sea. Because of pesticide transfer as aerosols and in the dissolved form higher concentrations are often observed beyond the shelf area as well.

The sea areas with the maximum levels of pesticides do not mostly coincide with the sea areas featuring the maximum levels of oil products, although in some cases the oil pollution may exceed that of pesticides. Several tons of chlorine organic products are transported annually via the Kerch Strait to the Black Sea from the Azov Sea where their average content is an order superior to the Black Sea. In the 1990s the increased content of some kinds of pesticides was registered nearby such ports as Tuapse, Sochi, Novorossiysk, Anapa and Gelendjik where it exceeds MAC several times. In the Pre-Caucasian coastal zone the concentrations of pesticides were 1.0–2.0 ng/l and more (1993–1996). They have an uneven distribution forming lenses with the increased concentrations at some distance from the shore. Nearby Odessa, Sebastopol and near Georgian ports the concentrations of chlorine organic products may reach several dozens of ng/l. Areas with higher concentrations may be found both in the coastal zone and offshore [10].

In the 1990s the Black Sea received annually up to 1000 tons of phenols with the river flow (80% and 20% from the Danube and Dnieper, respectively). Phenols also arrive with industrial wastewaters. Their great quantity is supplied by the oil refinery in Tuapse (about 100 tons of toxic substances every year). Up to several dozens of tons of phenols get into the sea from dumping grounds. The level of this toxic substance in the waters of rivers flowing into the northwestern part of the sea is 4–5-fold more than MAC. The average phenol content near the shores of Crimea and Georgia is 3–6 MAC with the maximums registered in some years reaching 20 MAC. In the vicinity of Ochakov and in the Karkinitsky bay the phenol level may be as high as 17–18 MAC, near Odessa—14–16 MAC, with the maximum observed in some years being much higher.

The annual input into the Black Sea of detergents is an average of 20 000 tons. In this amount the share represented by the Danube is over 30%, by the Dnieper up to 20%, and industrial wastewater discharged from the shore about 40%. The main supplier of detergents arriving via industrial wastewater is the Odessa region, the second place is taken by the Krasnodar Territory. Despite a lower level of detergents in the Danube and Dnieper flows, the northwestern part of the Black Sea remains the most polluted area. Along the Southern Coast of Crimea the detergent level exceeds MAC 2–3 times, while near the Georgian coast it reached in some years 7–9 MAC.

In the northwestern part of the Black Sea the natural environment of brackish lagoons has also changed. According to estimates, there were 14 basic lagoons and estuaries covering approximately 1950 km^2 [2]. Thus, in the largest Dnieper-Bug lagoon the intensive land reclamation works in plavni (flooded areas) and fresh water intake for Dnieper flow regulation led to restructuring of the ecosystem which seriously affected the biological productivity of the lagoon, largely the sturgeons and partially semianadromous fish, and in some cases was responsible for overall death of fauna in the lagoons. In the 1980s the share of agricultural wastewater in the total pollution of the Dnieper-Bug and other lagoons increased greatly. This was connected with the increased application of fertilizers and chemicals in agriculture. It may be one of the probable causes of changes in the population and species composition of flora and fauna in the lagoons.

3.3 Heavy Metal Pollution

In terms of hazard to the life of sea organisms heavy metals are inferior only to pesticides. Every year the Black Sea receives 300 kg of mercury, 290 tons of cadmium, 400 tons of copper, 2200 tons of lead and 14 200 tons of zinc from natural sources [4].

But the main sources of heavy metal pollution are thermal engineering, sea and motor transport, sea ports, ship repair and oil refining works, municipal treatment plants, agriculture and dumping. Heavy metals are brought in with river flows; household and industrial wastewaters affect significantly the coastal areas. River waters polluted with copper, zinc and lead flow into the Black Sea from the Caucasian watershed. Heavy metals also get into the surface film of sea water through aerosol sedimentation [10]. Heavy metals are second only to chlorine organic pesticides and PCB in terms of their negative impact on the water quality and biological communities.

In the surface water layer in the Black Sea shelf the dissoluble forms of heavy metals distribute unevenly. In most cases the strips with their maximum concentration stretch along the shore forming separate lenses. In the Gelendjik and Tsemess bays the zinc level is more than 15 μg/l or three times higher than MAC. The concentrations of copper, cobalt, nickel and lead in the Gelendjik bay are the same as in the shelf. The Tsemess bay is subject to a more intensive anthropogenic load although the higher concentrations of heavy metals are only found close to the pollution sources (the copper and nickel levels exceed MAC two times, while zinc exceeds MAC 9–10 times). Higher mercury content (up to 30–50 ng/l) is registered in the layer 50–100 m. Offshore the average concentration of dissolved mercury is 5–14 ng/l.

Bottom sediments in the coastal zone of the sea may be polluted with copper, zinc, nickel and cadmium. The highest levels of toxic heavy metals are found in the mouths of rivers. The bottom sediments in the Black Sea have a high mercury level—from 0.28 to 0.40 μg/l. In the coastal waters of the Krasnodar Territory the mercury level is 0.15–1.55 μg/l, while its maximum concentrations are registered in the Danube and Dnieper mouth areas. The Danube alone brings annually up to 50–60 tons of mercury, while the Dnieper brings up to 5 tons. The distribution of heavy metals in bottom sediments in the Russian shelf of the Black Sea is not uniform. Their greatest quantities are accumulated in sediments in the deepest part of the shelf where their concentration is 3–5 times higher than in sediments in the shallower part. Toxic metals contained in sea water in the dissolved and suspended forms are accumulated by biota, which may cause poisoning of a food chain and, thus, endanger man's health.

3.4 Dumping, Bottom Dredging

The Black Sea environment is strongly affected by dumping—burial in the sea of various materials and substances, in particular, ground excavated during bottom-dredging works, drill cuttings, industrial waste, construction debris, solid waste, explosives and chemicals. Dumping in the sea is based on the opinion that the sea environment is able "to digest" organic and inorganic substances without perceptible damage to it. However, this capability of the sea water is not infinite, thus, the dumping process should comply with strict regulations.

Dumping of ground excavated during bottom-dredging works in ports and approach channels causes serious pollution of the coastal water areas. The main dredging regions are water areas in the Kerch port with its approach channel and the Kerch-Yenikalsky canal. The dumped ground represents a mix of ground with toxic pollutants that are very hazardous for the ecosystem. They affect species composition and population of various groups of plankton communities, greatly damage the ichthyofauna by extermination of the forage base, cause destruction of spawning grounds and reduction of fattening areas; this has been observed in near-mouth areas of the Danube and Dnieper. Thus, after intensification of ground dumping in the Kerch Strait the quantity of herring arriving during its seasonal migration to the traditional fishing zones dropped which led to a sharp reduction of the catch, while in the 1990s the herring population completely lost its commercial significance. Dumping also affects the benthic organisms. The dumping of materials and the resulting higher turbidity of water maintained for long periods in the near-bottom layer causes asphyxia and death of sedentary forms of benthos.

In the Black Sea basin the annual discharge of dredging materials in the 1970s–1980s was 7–13 mln m^3 reaching 30–35% of the total discharges into the sea in the former Soviet Union, while at present it has dropped to 10%.

In the coastal zone of the Black Sea bottom dredging, drilling, dumping and sand aggradation prevent normal reproduction of bioresources in their habitats, and the normal functioning of valuable recreational zones.

3.5 Invaders

Despite the fact that a special section is devoted to invaders, we, considering the pollution problem, cannot avoid mentioning the biological pollution caused by invasion into the sea of new biological species. The cause of this process is thriving shipping which made it possible for many biogeographical barriers in the sea environment to be overcome by way of active unintentional introduction of various species into new areas. The pace of anthropogenic invasion is growing every year. Some invaders die rather quickly, others fit into the local communities not causing significant damage to them, still others may induce serious restructuring in the community and impair the reserves of commercial species.

In this context the most well-known example in the Black Sea is the sea snail Rapana (*Rapana thomasiana*) that feeds on other mollusks and which exterminated the main oyster shoals near the Caucasian coast and then near the shores of Turkey and Bulgaria.

In the last decade the ecosystems of the Black Sea sustained disastrous changes due to the appearance and mass propagation of jellyfish (*Mnemiopsis leidy*). This species was first found in the Black Sea not far from the southeastern coast of Crimea in 1982; and the basis for its appearance was prepared by man. Overfishing and eutrophication led to the disappearance of higher predatory fish and animals, such as turbot, bluefin tuna and monk seal. A sharply reduced population of plankton-eating fish cleared a niche for jellyfish [2]. By mid-1980 the jellyfish resource had reached 1 billion tons.

The absence of natural hunters led to a sharply increased population of jellyfish eating plankton, fish eggs and larvae in the Black Sea and, as a result, to reduction of the number of commercial fish species from 25 to 5. Fish catches had dropped both near the northern and southern shores of the Black Sea. In Turkey alone during one year the catches dropped from 295 000 tons (1989) to 66 000 tons (1990) [11]. This entailed grave economic and social consequences (by available estimates in the period from the 1980s to the early 1990s alone approximately 300 mln USD were lost due to reduced earnings from the fishery). The share of jellyfish (sea jellies and *Mnemiopsis*) by their raw mass makes now 99% of the total zooplankton.

4 International Cooperation

In the face of likely serious environmental changes occurring in the Black Sea region the littoral countries are attempting to find solutions by seeking international assistance from well-known international organizations and some leading states interested in settlement of environmental issues at the regional or global level. In 1992 the Black Sea countries signed the Bucharest Convention with specific protocols in relation to Black Sea pollution control after two decades of endless discussions about its need. And a year later in 1993 the Black Sea Environmental Program was established. As a result of these efforts two important and concrete documents were elaborated: "Transborder diagnostic analysis of the Black Sea" and "Strategic plan of actions on the Black Sea rehabilitation and protection". By the signing in 1996 of this plan the Black Sea countries undertook to elaborate the general strategy of the Black Sea protection and rehabilitation, management of its shore and marine resources for two decades. However, the last document is an instrument for coordination of only the intent of the Black Sea states to take action on pollution control of the Black Sea. Among the real achievements of the mentioned cooperation mechanism we can name the building of good communication links among international, national and local nature conservation organizations, establishment of national activity centers in the region and others, in other words, provision of only certain conditions for future effective cooperation.

Regarding the not so simple geopolitical situation established in the Black Sea region, mutual contradictions among the Black Sea countries, and growing external influence one cannot expect that the community of the Black Sea countries will cope with the problems of the Black Sea quickly and concertedly. There is a greater likelihood of attainment of separate compromises between neighboring countries and separate partnership with transnational corporations, international financial institutions and non-Black-Sea countries.

References

1. Shlikher SB (1987) Problems of increasing the economic capacity of the sea coast territories. Latvian State University, Riga (in Russian)
2. Pustovoitenko VV (2004) Environmental monitoring of sea areas: Problems of the Black Sea ecology. In: Environmental Safety of Coastal and Shelf Zones and Integrated Resource Management, Issue 11. Sea Hydrophisical Institute, Sebastopol (in Russian)
3. Aibulatov NA (1993) Geoecology of the Black Sea coastal zone. Coastlines of the Black Sea (part of a series of volumes on coastlines of the World). No. 4. Am Soc Civ Eng, Moscow
4. Sorokin Yu (2002) The Black Sea Ecology and Oceanography. Backhuys Publishers, Leiden
5. Zaitsev YuP (1998) The Bluest in the World. United Nations, New York
6. Zgurovskaya LN (2004) Curiocities of the Black Sea. Business-Inform, Sebastopol (in Russian)
7. Grinevetsky SR, Zonn IS, Zhiltsov SS (2006) Black Sea Encyclopedia. M. Mezhdunarodnye otnoshenia, Moscow (in Russian)
8. Keondjyan VP, Kudrin AM, Terekhin YuV (eds) (1990) Practical Ecology of the Black Sea Regions. Naukova dumka, Kiev (in Russian)
9. Fashchuk DYa, Shaporenko SN (1995) Coastal pollution of the Black Sea: sources and present-day level and annual variability. Water Res 22:3 (in Russian)
10. Noosfera M (ed) (2002) Geoecology of the Shelf and Coastlines of Russia (in Russian)
11. Kosarev AN, Tuzhilkin VS, Daniyalova ZKh, Arkhipkin VS (2004) Hydrology and Ecology of the Black and Caspian Seas. In: Atmosphere and Hydrosphere Dynamics and Interaction, M. Gorodetz (in Russian)

Hdb Env Chem Vol. 5, Part Q (2008): 423–437
DOI 10.1007/698_5_064

Published online: 13 February 2007

Socioeconomic, Legal and Political Problems of the Black Sea

Igor S. Zonn[1] (✉) · Sergey S. Zhiltsov[2]

[1]Engineering Research Production Center on Water Management,
Land Reclamation and Ecology, 43/1, Baumanskaya ul., 105005 Moscow, Russia
igorzonn@mtu-net.ru

[2]53/30, Bozhenko ul., Kiev, 03150, Ukraine

Abstract Historically, the Black Sea covered a zone where the Russian, Persian, and Ottoman Empires met. For more than 1000 years, Russia struggled for the Black Sea straits. In the times of the cold war, this area was divided between the West and the East. A number of revolutions of 1989 and 1991 that led to overthrow of the communist regimes in Eastern Europe and disintegration of the USSR proper opened a new chapter in the history of the Black Sea region. It acquires its own shape keeping in mind that the NATO members—Romania, Bulgaria, and Turkey dominate on the western and southern shores of the sea, while new independent states, such as Ukraine, Georgia, and the Russian Federation—on the northern and eastern shores. The process of statehood shaping Ukraine and Georgia is ongoing, and thus, the issues of socioeconomic development of the regional, international-legal, and political issues acquire new facets.

Keywords Black Sea · Legal status · Economics · Navy · Recreation · Ports · Navigation · Transport

1 Introduction

The Black Sea region is one of the most ancient in mankind's civilization and history. The sea itself and its surrounding territories were always in the center of attention due to their rich natural resources, economic significance, and strategic location.

After the breakup of the USSR, the number of states with access to the Black Sea remained the same. But now Russia, Ukraine, and Georgia became independent states and, in fact, started to formulate their interests and rights to the seawater area in different ways. The independent policy pursued by these countries gave a new impulse to development of the Black Sea region. The role of Bulgaria, Romania, and Turkey has also changed. The first two countries are no longer under the "umbrella" of the USSR. Turkey, which already two decades ago was the front line of the Western countries in the Black Sea region, also had to revise its policy and role in the region. The military and political mechanisms of maintaining security here were replaced with economic ones. Turkey ceased to be the front line of rivalry when the foreign-economic trend of the country's development changed and was given a new meaning. This urged leaders to seek mutually beneficial mechanisms of cooperation.

2
Socio-Economic problems

The Black Sea is surrounded by six independent states. One of them is Russia, which has the world's largest territory.

In the Black Sea region, the territories with direct access to the sea comprise 30 administrative units: two provinces (Bulgaria), one territory (Russia), three autonomous republics (Ukraine, Georgia), three regions (not taking into account two regions going out to the Azov Sea) (Ukraine), two *uyezds* (Romania), 19 *vilajats* (provinces) (Turkey).

The Black Sea region plays a very important role in the world economic system; the condition and development of the economics and active involvement of the region in world economic relationships determine to a great extent the affairs in world politics as a whole. General information about the region's countries is given below in Table 1.

From the information about the population and economic potential we can distinguish the three most economically developed countries in the region: Russia, Ukraine, and Turkey. This is confirmed by the per-capita production

Table 1 General information on the states of the Black Sea region [1]

Country	Population (in millions)	Territory, (1000 km^2)	Population density (people/km^2)	GNP billions $	Per capita GNP $
Russia	143	17075.2	8.37	1.12	7832
Georgia	4.9	69.7	70.30	22.8	4653
Bulgaria	7.7	110.9	69.43	48.0	6233
Ukraine	47	603.7	77.85	189.4	4029
Turkey	66.5	780.6	85.19	444.0	6676

of these countries. However, after the breakup of the USSR, Russia lost a considerable part of its Black Sea shoreline, making 4340 km. After Ukraine and Georgia became independent states, Russia's landmass access to the Black Sea narrowed significantly (out of 2413 km of the USSR coastline, Russia had only 475 km). Romania and Bulgaria have only a modest economic capacity, but in having a beneficial economic–geographical position, they play an essential role in servicing the transit links between European countries and the countries of the Near and Middle East. Georgia is now somewhat isolated due to its unsolved internal political problems, and thus its economic capacity is not used to the extent of international requirements.

The total population of the six Black Sea countries is 291.5 million people. Of the largest cities on the Black Sea coast, the most prominent is Istanbul, with a population of 12 million people. It is followed by Odessa, with 1.03 million people. Other large cities are Sevastopol 400 000, Samsun 400 000, Sochi 396 000, Constanta 351 000, Varna 301 000, Novorossiysk 231 000, Trabzon 230 000 and Batumi 140 000.

The Black Sea region is also a very important communication center due to international transport corridor of energy flows from the Caspian basin to the West.

The Black and Azov Seas are intensively used for international shipping, including carriages within a basin and beyond its borders. Here, many large ports can be found, the most significant being located on the northern and northwestern coasts: Burgas and Varna in Bulgaria, Constanta in Romania, Odessa, Nikolayev, Kherson, Ilyichevsk, Mariupol, Sevastopol in Ukraine, Novorossiysk, Tuapse in Russia, and Batumi, Poti, Sukhumi in Georgia. On the northern coast of Turkey, ports such as Samsun, Zonguldak, Sinop, Trabzon, Erdemir are found. Regular shuttle passages along the coal-ore line are made between ports Poti and Mariupol and from Poti, the manganese ores of Chaitura are carried to the "Azovstal" Works, while in the opposite direction, coal from Donetsk (to meet the needs of the Transcaucasian countries) flows. Apart from this, coal and coke from Donetsk are delivered to the ports on the Danube—Ilyichevsk, Kherson, Nikolayev, and reverse—iron manganese ores from Kherson and Nikolayev to Mariupol. The regular routes for transit of agglomerates are maintained between Kerch (Kamysh-Burun) and Mariupol. This is a pioneer route in the world practice of carriage of hot agglomerate with a temperature of 600 °C on ships. A considerable portion in carriages is also given to oil and petroleum product traffic. The main oil traffic flows from Batumi, Tuapse, Novorossiysk and Feodosia to Mariupol, Odessa and Reni and also for export. Light oil products (produced at oil refineries in Odessa, Tuapse and Batumi) are also supplied to these ports from other regions of Russia, Ukraine, and Azerbaijan. Fuel oil is transported in tankers from Batumi and Novorossiysk to various ports of the basin.

Rather weighty in carriages among the ports of this basin is the share of agricultural products: grain, cotton, tobacco, tea, sugar and others. Many freight traffic lines connect the ports of the basin with ports of European countries such as Bulgaria, Romania, Turkey on the Danube.

Before the disintegration of USSR, the intensive passenger traffic (up to 25 000 passengers each year along the Crimean-Caucasian and local lines and also international passenger lines) was observed in the Black Sea–Azov basin. At present, the passenger traffic in the basin has shrank significantly.

Regular freight traffic lines connect the ports of the basin with the ports of Vietnam, Egypt, India, Indonesia, Italy, Yemen, Cuba, Malaysia, Turkey, Sri-Lanka, Japan, and other countries. Recently the number of containers has been escalating. The ports of the basin are connected via feeder lines with the main container ports on the Mediterranean (in Italy, Malta, Greece, and Turkey) and also in the Persian Gulf. Container traffic lines linking the Ukrainian ports on the Danube with the Mediterranean ports also exist.

Large ship-repair and ship-building yards with floating and dry docks as well as workshops provided with modern equipment are found on the Black Sea. Such yards locate in Odessa, Izmail, Mariupol, Kerch, Reni, Kherson, and Kilie. Seaports often become the centers around which industrial regions develop. Port-industrial complexes, for example in Novorossiysk (Russia) and Varna (Bulgaria), are shaped on their basis. In the face of growing export-import needs, Russia has to make considerable investments into refurbishment of its ports (Novorossiysk and Tuapse) and construction of new ones.

The Circum-Black Sea region acquires great importance due to its transport-pipeline capacities, biological resources, and recreation potential.

From 1981, commercial-scale gas production had been conducted in the Azov Sea, and from 1983 in the northwest of the Black Sea from where gas is delivered to Crimea. Gas reserves under the Azov Sea bed are estimated to be 30 billion m^3, while in the northwest shelf of the Black Sea they are 49 billion m^3 [2]. The major Russian oil companies Rosneft, LUKOIL, Gazprom, and the French concern Total Fina Elf undertake geological prospecting drilling of hydrocarbons in the shelf of the Black and Azov Seas. In 2003 Rosneft and Total concluded an agreement on joint exploration and development of oilfields in the area of the Tuapse trough. In the Russian section, the most promising are considered to be this trough and the Shatsky ridge. However, even prospecting (and more so commercial-scale production of hydrocarbons) presents great difficulties here as perspective formations occur at depths of 3 or 4 km from the sea surface (and the sea depths here are approximately 2 km). In addition, it is a known fact that the waters of the Black Sea at depths more than 200 m are aggressive and oversaturated with hydrogen sulfide.

Still in the Soviet times several wells were drilled in the Azov Sea that did not indicate the availability of commercial oil reserves. In 1978, the perspective Gelendjik structure was discovered in the south of the Temryuk Bay, while in 2001, in the North-Temryuk delta was discovered. The main obstacle for development of hydrocarbon production in the Azov Sea is the unsettled legal status of the sea border between Ukraine and Russia in the Kerch Bay and in the sea. In addition, as drilling areas are located in health-resort zones, the issue of environmental safety becomes pivotal.

Prospecting works were conducted in the Bulgarian and Romanian sections of the self. In the Romanian shelf, main expectations are connected with the structures located eastward of Constanta, in the Bulgarian shelf, with an area from the Yemine Cape to the Bulgarian-Romanian state border [3].

Clearly, we can speak only about forecasted hydrocarbon reserves, while at present the Black Sea is used mostly for transit of Caspian hydrocarbons.

Turkey started has taken one of the key positions in energy export to the foreign markets in the Black Sea region. Its role in export of energy from the Caspian region escalated as a result of geopolitical changes that involved changes in transit routes of hydrocarbons to foreign markets.

The territory of Turkey became a link in a large-scale energy corridor that connected countries of this region. Gas-pipeline projects of Blue Flow going over the Black Sea bed and oil pipeline Baku–Tbilissi–Ceyhan may be an example of creation of a new framework for economic cooperation among Black Sea countries. Russia and Turkey are planning to extend the pipeline Blue Flow.

The construction of the oil pipeline Samsun–Ceyhan is underway that will enable oil transit in bypass of Turkish Straits. It is designed to construct 550 km of the oil pipeline going from Black Sea port Samsun to Kyrykkale City. This new pipeline is designed for the transit of 50–70 million tons of crude oil per year. As a result, additional amounts of oil will flow from the countries of the Black Sea and Caspian regions. It is anticipated that the first oil from Samsun will reach Ceyhan as early as 2008.

For more than a decade, the construction of a Trans-Balkan oil pipeline from Burgas (Black Sea coast of Bulgaria) to Alexandrupolis (Black Sea coast of Greece) for delivery of Russian oil from Black Sea terminals to the Balkans is under discussion. For relieving pressure on the Bosporus, this pipeline should be 320–400 km long with a capacity of 22–29.2 bill barrels of oil a year.

But not all countries succeeded in becoming active participants of new energy projects. The Ukrainian oil pipeline Odessa–Brody has been idling from 2001. The main reason is an unsettled problem on financing of the "tube" extension project and also insufficiency of additional volumes of the Caspian oil [4]. In August of 2006, the Ukrainian government declared the pipeline will be operated in a reverse regime until 2008.

The Black Sea region is crossed by major international transport corridors that support considerable freight and passenger traffic among the countries. Here, the four most significant transport arteries of the European continent pass—the so-called Crete corridors—third, fifth, seventh and ninth, as well as corridors Europe–Caucasus–Asia (TRASECA) and north–south (Baltic Sea–Black Sea) ensuring transport links of the Black Sea ports with the countries of Central and Eastern Europe. Of particular importance for the economics of the region is the utilization of these corridors for transportation of transit freight. It is anticipated to increase considerably the international cargo exchange in the direction Europe–Asia (TRASECA) and in the direction north–south. The issue on construction and development of a ring automobile road around the Black Sea approximately 7100 km long is under study now. Its main route will go via Istanbul, Batumi, Novorossiysk, Taganrog, Mariupol, Odessa, Kishinev, Bucharest, Kharkov, and Alexandrupolis.

Evidently, growing freight traffic may be expected along the seventh international transport corridor – the "Danube Waterway". A certain contribution will be made by a new navigation route in the Danube mouth "Danube–Black Sea", built via the neck ("*girlo*") Bystroye. This is facilitated by construction of large ferry lines: "Crimea–Caucasus", via the Kerch Strait; the ferry line "Caucasus–Poti" commissioned in 2004; the ferry line operating in summer between Turkish Trubzon and Russian Sochi.

At present, Ukraine has approved the program of actions aimed at construction and rehabilitation of several reloading complexes (in the ports of Ilyichevsk, Odessa, Yuzhnyi, Mariupol), further development of the existing ferry traffic between Ilyichevsk, Varna and Poti-Batumi [5]. The port facilities in Varna (Bulgaria) are also being extended.

The Black Sea has always been famous for its fishing resources and its significance does not become less at present. Recent years have witnessed wider application in the food and pharmacological industries of invertebrates and algae. In 1940, the catches of fish and dolphins in the Black Sea littoral countries totalled 86 000 t, while in the late 1970s these countries caught 250 000 t of fish only, which is three times as large [6]. Then the catches started dropping quickly and the species composition of the sea "dwellers" lost its diversity. Out of 23 species of commercial fish that might be met not long ago, hardly five survived. Such fish species as mackerel, bluefish, flatfish, goatfish, gray mullet, beluga, sea perch, herring, croaker, scad, silver side, bullhead, etc., have now lost their former commercial dimensions. Fifteen sea fish species, six crab species, and all three species of dolphins are put listed in the Red Book. The monk-seal is also listed in the International Red Book.

The littoral areas of the Black and Azov Seas with their warm climate are of great socioeconomic significance for intensive development of recreation resources. The natural, ethnic, cultural, historical, and other features of the Black Sea area attract many people for rest and tourism. In the Black Sea there are no predatory and noxious sea animals and there is no tidal activity. One can admire vast expanses of sand beaches and wonderful mountain and forest coastal landscapes, and these regions are not far from European countries.

In the Russian part of the Caucasian coastal zone of the Black Sea, whole resort complexes were built intended to accommodate a great number of the people for rest. Thus, Greater Sochi extending for approximately 150 km comprises six large complexes located on mountain slopes facing the sea or constructed directly in the coastal zone. In early 2000, these complexes (Lazarevskoye, Dagomys, Sochi, Macesta, Khosta, and Adler) accommodated more than 3 million people who came here for rest.

In Ukraine, the resort and health-improvement complex is located on the southern coast of Crimea, being 110 km long and 2 or 3 km wide, and also in the Odessa area about 500 km long (from the state border with Romania to the Perekop Isthmus in Crimea). The latter comprises the Odessa recreational complex located in the cities of Skadovsk, Belgorod-Dnestrovsky, and Ochakov, where people can improve their health by taking mineral waters and curative mud from the lagoon.

The Georgian coast of the Black Sea, which had great recreational importance in Soviet times, has sustained great losses due to the ethnic conflict between Georgia and Abkhasia as a result of which resorts such as Gagra, Gudauty, Pitsunda, Sukhumi were practically destroyed. Access to the resorts of Adjaria-Batumi and Poti for visitors from outside has been significantly limited.

In general, it should be said that the resort-tourism industry of the Black Sea coastal zone of the former Soviet Union is in its transitional period from large complexes to small and medium private hotels, rest houses, camping places, and small bungalows. Today, a new kind of rest industry is developing in the Black Sea coastal zone—from aqua parks to windsurfing, diving, etc.

The Turkish coast of the Black Sea, extending for approximately 1500 km, is so far beyond the interest of mass tourism, and is still a kind of a nature preserve. There are no high-tech hotels here so far, everything is done for calm rest. This is a domestic sluggish resort coastal zone, the greenest region in Turkey.

The Bulgarian Black Sea coast for a long time has been famous for its resorts and potential for development of international and domestic tourism. Widely known are such resort complexes as Albena, Zlaty piascy, Druzhba, Slanchev bryag, Nesebr, Burgas, Sozopol and others. Similar resorts, but not so grand, may be found on the Romanian Black Sea coast. These are Mamaja, Constanta, Ephoria-North, Ephoria-South, and Mangalia. A special place in tourism development is taken by the Danube delta with its romantic landscapes and variegated bird's community.

The priority line in cooperation among the countries of the Black Sea region is development of relationships among the Black Sea subjects having direct access to the sea. Recognizing a great number of problems here, their role in development of economic cooperation in the region remains significant. The socioeconomic situation in the region, in general, depends on the ability of the Black Sea subjects to find ways for mutually beneficial cooperation. One such mechanism for multilateral cooperation of the Black Sea states is the Black Sea Economic Cooperation Organization (BSECO), which was established in 1992 by 11 countries—Azerbaijan, Albania, Armenia, Bulgaria, Greece, Georgia, Moldova, Russia, Romania, Turkey, and Ukraine. It is a unique organization in terms of economic cooperation and maintenance of stability in the Black Sea region.

3 International legal problems

In the Black Sea region one can find two blocks of international-legal issues. One of them includes a well-known problem of the Black Sea Straits having a long history. The other block covers the issues that had arisen after disintegration of the Soviet Union and the block of socialist countries and appearance of new independent states in the Black Sea region.

The regime of the Black Sea Straits, understood in the international law as Bosporus and Dardanelles, connecting the Mediterranean with the Black Sea remains one of the most important and complicated political issues of the region. The complexity of this problem lies in the fact that the neck of the Black Sea belongs to one littoral state – Turkey that is why it acts as the "Straights guardian" and the guarantor of observance of the shipping regime via them for foreign vessels.

In the XVIII–XX centuries the Black Sea Straits were a part of the foreign policy of Russia-Eastern problem in the settlement of which all great European states were interested.

The geographical location of the Black Sea Straits defines the key significance of their international regime for security and economic interests of the Black Sea countries, which is connected with freight traffic via them. Before the conclusion of the Kyuchuk–Kainardji Peaceful Treaty in 1774 the regime of the Black Sea Straits was determined by Turkey alone. After 1774, many treaties and agreements were concluded concerning passage of foreign ships via the Black Sea Straits (Russian–Turkish Treaties of 1799, 1805, and 1933, London Convention of 1841 and others). In 1936 in Montreux (Switzerland) the special Convention on Straits Regime was concluded, which was signed by 72 countries, including Russia and Turkey. This convention stated a free passage for merchant ships from all countries via the Black Sea Straits, a procedure of passage for military ships of the Black Sea and non-Black Sea countries, and renounced resolutions of the 1923 Lausanne Conference prohibiting Turkey to fortify the Black Sea Straits. The need of revision of the 1936 convention, which serious drawbacks were revealed during World War II, was recognized by the Berlin (Potsdam) Convention of 1945 and then was stressed more than once in the notes of the Soviet government.

With the breakup if the USSR, the problem of the Straits was subject to restructuring due to some changes that occurred [7].

In view of the absence in some cases of a coordinated policy towards these Straits in the Black Sea countries and other states, Turkey resting upon two international principles, shipping safety and environment protection, gradually transferred shipping via the Straits from international to national jurisdiction. In 1982, Turkey took the way of imposing a technical supervision over the Straits and in 1994, it unilaterally introduced its own Regulations for Shipping via Straits that violated and changed the provisions of the Montreux Convention. Then from 1994 to 1998 Turkey took persistent actions to set the Special Marine Shipping Regulations in the Straits Zone that stated tough rules for some ships, first of all for large-tonnage tankers. As justification for its actions, Turkey raises the issue of menacing environmental disaster, saying that it will do its best to prevent turning the Straits into an oil pipeline and, consequently, those who carry oil from the ports on the Black Sea coast of Caucasus should look for other transit routes. It is quite clear that just at this time Turkey lobbied the construction of the oil pipeline Baku–Tbilissi–Ceyhan. The Turkish authorities stated that the transport load on the 325-km zone, including Bosporus, Dardanelles and the Sea of Marmara, are at the limit of their capacity. Each day 150 ships on average (one ship per 10 min, of which 20 are tankers) pass via the Bosporus, being one of the narrowest and most difficult for navigation straits in the world. After commissioning of the KTK oil pipeline (Caspian Pipeline Consortium), transporting Kazakh oil to Novorossiysk the volume of tanker traffic has increased sizably. The Straits may fail to survive such pressure, and as a result, this may lead to a devastating environmental disaster. According to Turkish statistics, over a period of five decades more than 500 accidents were recorded in the Straits, and 40 of them may be classified as large ones. In general, about 50 000 ships, including 5000 oil tankers, each year pass via the Bosporus dividing Istanbul with its 12 million people into the European and Asian parts.

By concerted efforts of all interested parties, a part of the Turkish restrictions was removed, but at the start of 2000, only double-shell tankers were allowed to pass via the Straits. In 2002, Turkey enforced the instruction specifying that the vessels carrying hazardous freight should pass the strait only in one direction, which resulted in jams on both sides from the Bosporus. Up to 40–50 tankers may be caught in such jams and during storms there is a great danger of ships leaning on each other, which may have grave consequences. Idling time incurs losses not only to private ship owners but also to oil companies. By restricting the ship passage due to safety considerations, the Turkish side itself sustains financial losses and, at the same time, its political image somehow deteriorates. In general, the problems of transport, business and ecology clash in the Bosporus.

The second block of the international legal issues in the Black Sea that had arisen, as it was already said, after the USSR breakup, includes the unsettled issues of the water area delimitation between Russia and Georgia, Russia and Ukraine, and Ukraine and Romania.

Difficulties with delimitation of the Black Sea between Russia and Georgia are connected with the situation around the unrecognized Republic of Abkhazia. The international conflict between Georgia and Abkhazia that originated in the Soviet times in 1989 and that persisted after disintegration of the USSR does not allow the parties so far to meet at the "negotiation table" for resolving the border problems.

Demarcation of the sea borders between Russia and Ukraine in the Azov and Black Seas is also not completed and there is no agreement on delimitation lines 298 km long in the Azov Sea and 22 km long in the Black Sea [8].

The key disputable issue is the present-day legal regime of the Kerch Strait connecting the Azov and Black Seas. The control over the strait is, in fact, a control over the whole Azov Sea. Ukraine dismisses proposals concerning joint control of the strait without its division with a state border. At present, about 8000–8500 ships pass every year via the Kerch Strait, of which approximately 65% sail under the Russian flag. Joint jurisdiction over the Azov Sea is beneficial for Russia because there is some probability that perspective oil and gas reserves may be found in the Azov Sea and the Kerch Straight. Moreover, the Ukrainian part of the sea has greater fish resources. If the Ukrainian proposal on making sea borders is taken, then Russia will have less than 40% of the Azov Sea area, a non-shipping part of the Kerch Straight, and a small sector of the Black Sea. In case of recognition of the state border in the Kerch Straight, Russia will have to agree that the Kerch–Yenikalsky canal belongs to Ukraine. Apart from economic considerations, there are also political ones—the joint use of the Azov Sea and the Kerch Straight enables Russia to control the movement of ships from third countries. The only achievement of negotiations so far is the Declaration about the Legal Status of the Azov Sea, the Kerch Straight and on Delimitation of the Black Sea in which both parties recognized that the Azov Sea is "the internal waters being in joint use of both countries".

The Russian position (Treaty of 1997) is the main political agreement between two countries. The Azov Sea and the Kerch Straight should be the internal sea of Russia and Ukraine with joint use of the water surface for navigation, bioresources beyond the 12-mile zone. The seabed is divided similar to the resource division in the Caspian [9]. However, all these issues are waiting for further discussions. The Ukrainian position is that the border in the Kerch Straight already exists and it goes over water and not only over the bed. This position has its reasoning: this concerns the Tuzla Island claimed both by Ukraine and Russia being a part of the Tuzla bar connecting with the Taman Peninsula.

The Tuzla Island appeared in 1925 due to breech of the Tuzla bar during a strong storm. Natural sedimentation processes on both sides of the Tuzla bar facilitated survival of its remnants in the form of the Tuzla Island. By 1950, the width of the Tuzla waterway (300 m) has increased to 3 km and in the late 1970s it was 4 km. In 2006 its length, depending on the water level in the straight, varied from 6.5 to 7 m, and a maximum width of 500 m. In the early 1970s, after transfer of the continental part of the Crimean Region under the Ukrainian SSR administration, the border between the administrative unit of

RSFSR (Krasnodard Territory) and the Ukrainian SSR was agreed upon. And that time a sea border along the Kerch Straight was made and a part of the Tuzla Island was referred to the Crimean Region. After the USSR breakup Ukraine unilaterally declared this administrative border to be a state border.

One more disputable international legal issue concerns the Zmeiny Island (formerly Akhila, Levka, Belyi, in the Antique times "Fidonisi"). This is one of a few islands in the Black Sea that belongs to Ukraine. It is located 37 km from the Kili arm of the Danube delta and thus is of strategic significance. Its area is 1.5 km^2 and has an elevation up to 40 m a.s.l. In the late 1940s, Romania passed this island to the USSR, but after its disintegration started to lay claims to this island again. According to international law, Zmeiny has all the features of an island and not cliff (Romania insists that it is a cliff and this changes the island status). The island has a 12-mile zone of territorial waters which is not disputed by Romania. The subject of disputes with Ukraine is the division of the oil–gas continental shelf belonging to Ukraine and containing considerable oil and gas resources that were explored in in the 1980s–1990s near an island located beyond these waters.

4 Geopolitical problems

With the changing geopolitical situation in the early 1990s, serious transformations occurred in the Black Sea region. One of the results of the USSR breakup was the emergence of military conflicts (at present "frozen or dormant conflicts") in the Black Sea littoral states and nearby territories as well as sharpening of many issues that earlier were latent or were not developing. Among them: terrorism, organized crime, trade with people, illegal narcotics trade, arms and explosives, illegal migration and also environmental threats. But the main problem of the "Great Black Sea region" is the attainment of political stability that directly depends on settlement of relationships with the unrecognized states. Three self-declared republics in the post-Soviet space, Abkhazia, Southern Ossetia and Pridnestrovie, are located in the Black Sea region. Without settling this problem, the Black Sea region cannot become politically stable and economically prosperous. The socioeconomic situation in the countries of the Black Sea region depends to some extent on what mechanism will be suggested for settlement of the issues of unrecognized formations and whether it will be efficient enough.

Military conflicts in the Black Sea region are to some extent a historical heritage. Century-old relationships of the peoples that lived in the past and live now on the territory of the Circum-Black Sea countries were not simple and were often dramatic and conflicting in nature. This "experience" is felt in the present-day relations among the countries of the region. The typical example is Turkey and Russia. These are two key countries in the Black Sea region and for centuries they had struggled for domination in this strategic area.

At present, the most problem country in the Black Sea region is Georgia. On its territory there are two self-declared formations—Abkhasia and Southern Ossetia, and the leaderships of Georgia and Abkhasia have different outlooks on the future. Abkhasia advocates "associated relationships" with Russia. After presidential elections in Abkhasia that were held in October–December 2004, the Abkhasian elite maintained the earlier attained consensus on pursuance of the pro-Russian course. In any case, both unrecognized formations are not ready for reintegration with Georgia on the conditions that are suggested by official Tbilissi.

To some extent, these conflicts affect affect Bulgaria, Romania, and Turkey. Thus, Bulgaria with great interest watches the situation with the ethnic Bulgarians living in the Republic of Moldova and in Ukraine. Romania is actively involved in settlement of the Pridnestrovie-Moldavian conflict providing information support to Kishinev. Turkish businessmen provide sizable investments into economic development of the self-declared Republic of Abkhasia.

The key role in settling the conflicts in the Black Sea region belongs to Russia. Its political weight and efforts curb the escalation of conflicts by creating conditions for peaceful negotiations.

The leading role in the Black Sea region, including in the settlement of conflicts, is claimed by Ukraine. Its attempts to recapture the initiative from Russia became more persistent after presidential elections in Ukraine in November–December 2004. Various plans for reanimation of GUAM (Georgia, Ukraine, Azerbaijan, and Moldova) appeared, then the commonwealth of the countries of the Caspian-Black Sea-Baltic region was established. Ukraine expressed its readiness to replace Russian peace-keeping forces in the zones of inter-ethnic conflicts in the Southern Caucasus and in the Circum-Dnestr region, having suggested its own (alternative to the Russian) plan for settlement of the Pridnestrovie conflict ("Roadmap of Yuschenko"). For its realization, Ukraine went as far as imposing a customs blockade for cargo from Pridnestrovie.

As a result, the axis of states is now shaping in the Black Sea region ready for closer cooperation with the USA, European Union, and NATO. Ukrainian initiatives integrate into this axis Georgian and Moldova being weaker in military and political respects.

Therefore, a great many of the BSECO member-states are involved to a greater or lesser degree in addressing the problem of unrecognized states formations. However, no real achievements in extinguishing the conflicts in the Black Sea region are visible.

In the recent decade, notwithstanding the lack of international recognition, these unrecognized republics conducted parliamentary and presidential elections, which became a symbolic act, a kind of appeal to the international community. Through such elections they demonstrated the legitimacy of the power bodies supported by the population and also effective work of state power institutes in unrecognized republics.

Special attention should be drawn to the complicated problems existing in the Russian–Ukrainian relations in the Black Sea. The most acute for the Ukrainian side is the problem of the presence of the Russian Black Sea Navy on its territory. Regardless of the treaty concluded between these countries, under which the Russian Navy is entitled to rent navy bases on the territory of Ukraine in Crimea until 2017, the Ukrainian side advances newer and newer demands concerning actual location of forces and means of the Russian Navy on its territory. Such a policy pursued by the Ukrainian side and spearheaded to sharpening of bilateral relations in the region is an attempt to consolidate its cooperation with EU and the USA at the expense of the relations with Russia. In its relations with the EU, Ukraine sees the perspectives of political cooperation overlooking here the issues of political stability in a wider geopolitical context.

Withdrawal of the Black Sea Navy (keeping in mind that Romania and Bulgaria stand for the widened military presence of the NATO countries in the region) will result in destabilization of the political situation in this region. The balance of forces will be broken. As a result, the countries of the Black Sea region will not be closer to regional stability.

The other destabilizing factor in the Black Sea region is the establishment in 2005 of the Commonwealth of Democratic States of the Black Sea–Caspian–Baltic Region. Ukraine initiated and supported this commonwealth considering it as an effective way for the establishment of bilateral and multilateral links and implementation of multiple projects in the region as a means for harmonization of the regional and subregional organizations and forums in the Baltic–Black Sea region, such as the Council of the Baltic States, Central-European Initiative, groups of GUAM countries, Black-Sea Economic Cooperation Organization as well as others.

Establishment of the Commonwealth of Democratic States of the Black Sea–Caspian–Baltic Region spurred disintegration tendencies in the Black Sea region into geoeconomic and geopolitical parts. These are attempts to turn the Black Sea into the front line that will divide the Circum-Black Sea countries by their political and military preferences. On the one side, there are membership of Georgia, Moldavia and Ukraine in GUAM, commitment of Bulgaria and Romania to the idea of NATO extension, active policy in the region of Turkey. On the other side, there is Russia, an active member of Collective Security Treaty Organization (CSTO) that advocates development of relationships within the framework of the single economic space. These differences in approaches to ensuring regional security will affect political stability in the Black Sea region.

One of the key events that radically changed the military–political situation in the Black Sea region was the breakup of the USSR. As a result, the alignment of military forces here has changed dramatically.

Transformation of the geopolitical situation in the region involved the changed roles of some Circum-Black Sea countries in maintaining regional stability. Turkey that in the times of the USSR took one of the key positions in promotion of the interests of the Western countries in the Black Sea region was the main rival of Russia in this region. In the USSR times Turkey had no possibility to influence the Caucasian countries playing the role of the main partner of the USA and NATO in the south. After the USSR breakup Turkey returned to the Caucasian geopolitics. This was facilitated by formation of Turk-speaking independent state Azerbaijan and ethnic self-determination of the Turk peoples in the North Caucasus.

At the same time new independent participants of regional security consolidated their positions as they had no previous commitments. Bulgaria and Romania became consistent supporters of the regional stability via NATO structures taking active efforts for developing of cooperation with this organization. Negotiations of these countries with NATO led to their enhanced role in the region, moreover, that their political orientation met the support on the part of the NATO leadership.

Both the Ukrainian and Romanian Navy will become new partners in the theater of military operations of the USA European Command. Romanian military locations in Constanta on the Black Sea coast will become a base for NATO and USA troops on their movement to the Balkans. Here we mean military bases on the airfield "Mikhail Kogelnichyanu" not far from port Constanta and training grounds Babadag located to the north.

Bulgaria is seeking to become a reliable partner of NATO in the Black Sea region by pursuing consistent foreign policy. Bulgaria also declared that it will give consent to the deployment of US military bases on its territory. It is planned to locate them in regions Novo Selo (testing grounds), Bezmer (both regions are not far from the border with Turkey) and Graph-Ignatievo (airfield near Plovdiv).

Turkey, Romania, and Bulgaria have advanced much in their cooperation with EU being candidates to membership in this organization. In its turn, the EU, through cooperation with the Black Sea countries, has practically reached the "coast" of the Black Sea. For European countries, the Black Sea region acquires an ever-growing importance because the EU is interested in stability over the perimeter of its borders and is also seeking to diminish the risk of disturbance in the hydrocarbon supply. The EU wants to ensure trouble-free delivery of energy from the Caspian region that may become an alternative source for European countries [10].

In the future, the US military bases may appear not only in Romania and Bulgaria but also in Ukraine and Georgia, and it is not accidental that recently US Navy ships have steadily become familiarized with the Black Sea area making regular calling at ports in Romania, Bulgaria, Georgia, and Ukraine, and also taking part in joint military exercises with the fleets of these countries and Turkey.

All littoral states are interested in creating conditions for stable development of the Black Sea region. One of the important steps in this direction is the development of cooperation in the military sphere. For this reason, in 1997 there were organized meetings of commanders-in-chief from all six navies of the Black Sea (Bulgaria, Georgia, Romania, Russia, Ukraine, and Turkey). This decision was taken for consolidation of cooperation on the Black Sea. However, already in 1998 Russia and Turkey put forward an idea about establishment of the quick-action Black Sea Naval Cooperation Task Group "BlackSeaFor". This initiative was supported by Bulgaria, Georgia, Romania, and Ukraine. As a result, the multinational navy group for quick interaction entrusted with flexible functions appeared that is intended for action in emergency situations in the Black Sea solely in peaceful purposes.

The activities of BlackSeaFor are supported by Russia and Turkey that assume it as a barrier for extension of the military presence of the US and its allies by NATO in the Black Sea region and the more so as the USA and NATO countries lobby the extension to the Black Sea of the operation "Active Efforts" that is realized by the North-Atlantic Alliance in the Mediterranean.

Meanwhile Russia and Turkey believe that the Black Sea should be a region of cooperation and not rivalry. The countries of the Black Sea region should seek independently the solutions to the security problems, including those related to the threat of terrorism and proliferation of weapons of mass destruction.

References

1. Grinevetsky SR, Zonn IS, Zhiltsov SS (2006) Black Sea Encyclopedia, M. International Relations, Moscow, p 660 (in Russian)
2. Fashuk DYA, Andreeva TM, Egorov AP, Petrenko OA (2006) Hydrogeochemical Consequences of Offshore Gas Production in North-western Black Sea. RAS Newsletter, Geographic series, # 1. Nauka (in Russian)
3. Vylkanov A, Danov Kh, Marinov Kh, Vladev P (eds) (1983) Black Sea. Collection. Gidrometeoizdat, Leningrad (in Russian)
4. Shevtsov AI (2005) Import–Export Policy of Ukraine in Energy Sphere: Strategic Priorities. Orbita-Servis, Dnepropetrovsk (in Russian)
5. Martynenko VT, Tsymbal NN (2006) Geography of Marine Shipping. Feniks, Odessa, p 243 (in Russian)
6. Zgurovskaya LN (2004) Curiosities of the Black Sea. Business-Inform, Simferopol (in Russian)
7. Nezhinsky LN, Ignatov AV (eds) (1999) Russia and Black Sea Straits (XVIII–XX Centuries). International Relations, Moscow, p 557 (in Russian)
8. Korzun VA (2004) Conflict Use of Sea and Coastal Zones of Russia in XXI Century, M. Ekonomika, Moscow, p 558 (in Russian)
9. Kostianoy AG, Kosarev AN (eds) (2005) The Caspian Sea Environment. Springer, Berlin, Heidelberg, New York, p 271
10. Agency of Defence News (2002) Security Program of the Black Sea 2002, Kiev

Hdb Env Chem Vol. 5, Part Q (2008): 439–452
DOI 10.1007/698_5_095

Published online: 28 September 2007

Conclusions

Aleksey N. Kosarev[1] · Andrey G. Kostianoy[2] (✉)

[1]Geographic Department, Lomonosov Moscow State University,
Vorobievy Gory, 119992 Moscow, Russia
kostianoy@online.ru

[2]P.P. Shirshov Institute of Oceanology, Russian Academy of Sciences,
36 Nakhimovsky Pr., 117997 Moscow, Russia

Abstract This book is aimed at a systematic description of the knowledge accumulated on the physical oceanography, marine chemistry and pollution, marine biology and geology, and hydrometeorological conditions of the Black Sea. It presents the principal particular features of the environmental conditions of the sea and their changes in the second half of the 20th century. At present, the principal problems of the Black Sea are related to the estimation of the intensity of the chemical pollution of the sea and its impact upon the biota. Special attention is paid to the socioeconomic, legal, and political issues in the Black Sea region. The book is based on numerous observational data collected by the authors of the various chapters during sea expeditions, on the archive data of several Russian and Ukrainian oceanographic institutions, as well as on a wide scientific literature, mainly published in Russian editions.

This conclusion completes one more generalization concerning the investigations of the state of the Black Sea environment and the principal trends of the changes observed in the second half of the 20th century. The study performed shows how complicated and diverse natural processes and phenomena are; precisely, their interaction forms the ecological system of the Black Sea. Regardless of the relatively short period that has passed since the appearance of the previous most important publications concerning this basin (they are listed in the Introduction), this new monograph, first, significantly complement many issues assessed, and, second, reflects the opinions of the scientists who prepared this book proper. Modern science as well as its methods and priorities are developing at a fast rate. This requires an adequate response, which, in particular, is reflected in generalizing publications. The new book about the nature of the Black Sea presents the scientific views of the researchers referring to different generations and scientific schools. Thus, they may slightly differ in selected details. Meanwhile, the general concept of the formation of the natural regime of the sea, its hydrological and hydrochemical structure, its biological diversity, and the information on anthropogenic factors keeps uniformity of the basic principles. First, they refer to the interrelation between the natural environment and the society, which is more than characteristic of the urbanized Black Sea region.

The book starts with a brief history of the oceanographic studies in the Black Sea. An analysis of the present-day condition of the Black Sea environment is preceded by a brief description of the Quaternary paleogeography of the basin. This section describes the evolution of the Black Sea basin in the process of its interaction with the adjacent Mediterranean and Caspian seas. During the Quaternary, the character of their connection and water exchange altered depending of the components of the water balance and the altitude of the Bosphorus sill. In this way, the hydrometeorological control over the Quaternary history of the Black Sea is once more emphasized. The main differences between the transgressive basins were manifested in the changes of the salinity values, which determined the type and composition of the fauna of the sea. The subsequent process of the evolution of the Black Sea is traced in the description of its coasts.

Great attention is paid to an analysis of the processes proceeding in the near-mouth areas of the rivers that enter the Black Sea and the Sea of Azov. River mouths are regarded to be the most complicated, variable, and vulnerable geographical objects in the coastal zones of seas. Using new data, riverine runoffs of water and particulate matter to the Black Sea and the Sea of Azov were calculated. To do this, the coast of the Black Sea was subdivided into six sectors: the northeastern (Russian coast), eastern (Georgian coast), southern (Turkish coast), southwestern (Bulgarian coast), northwestern (Romanian and Ukrainian coasts), and the Crimean coast (Ukraine). The natural and anthropogenic changes in the riverine runoff over the past decades were estimated. The present-day overall mean annual water runoff to the Black Sea is 354.5 km^3, while the particulate matter runoff equals 84.0×10^6 t.

The features of evolution, structure, and hydrological regime of the mouths of the major rivers of the region such as the Danube, Dniester, Dnieper, Rioni, Don, and Kuban' rivers are described in detail. Based on an analysis of the present-day condition of the river mouths of the region, principal tendencies in the hydrological, hydrochemical, morphological, and ecological processes in these areas are revealed. In future, one can expect an increase in the anthropogenic stress on the environments of the rivers and their mouths and a stronger influence of sea level rises and sea waves on river deltas. Deterioration of the ecological conditions and strengthening of the eutrophication processes in river mouths and adjacent areas are also probable.

The main distinctive property of the Black Sea is its inland location and high isolation from the World Ocean. Because of this, formation of the sea hydrological regime and water structure is governed by the outer factors: the fluxes of heat, moisture and wind stress via the sea surface, as well as the river runoff. In this connection, the sea is characterized by a high level of its environmental variability. At the same time in different parts of the Black Sea, the influence of the outer factors is very unequal. Therefore, these factors exert a different impact on the formation of hydrological fields and vertical thermohaline structure in the sea. All this confirms the necessity for more detailed and regular observations of hydrometeorological parameters of the Black Sea. Based on the long-term and seasonal data, the basic hydrometeorological features that form the natural environment of the Black Sea were considered in the book. They include climate (regional atmospheric circulation, winds, atmospheric pressure, air temperature, moisture content, precipitation), wind waves, water balance, sea level (multiannual and seasonal changes, storm surges, seishes, tidal oscillations), and sea ice.

When considering water circulation, quantitative and qualitative generalizations and a comparison of the the results of the studies of the Black Sea currents performed using modern observation techniques (autonomous buoys stations, shipborne acoustic measurements, drifter and altimetric satellite observations) with different types of hydrodynamic models were presented. The horizontal and vertical structure of the general water circulation in the Black Sea is shown, together with its seasonal and interannual variabilities. The principal mechanisms for the current formation under external hydro- and thermodynamical forcing are discussed.

The generalization of the results of field and model studies of the Black Sea general circulation (BSGC) allows the following conclusions to be drawn. In the upper 500-m layer, the BSGC consists of the Rim Current along the entire continental slope; several sub-basin cyclonic gyres (SBCGs) in the central area, whose number and position change during the year, and a few nearshore anticyclonic eddies (NSAEs) existing in fixed areas between the Rim Current and the shore over 5–9 months per year. The main source for the generation of the BSGC and its seasonal and interannual variabilities is related to the relative vorticity of the tangential wind stress; the influence of the momentum, and heat fluxes across the sea surface, and via the river mouths and straits which is significantly lower.

In the winter and early spring, the surface BSGC is strongest, in the summer and early fall, it weakens by a factor of 1.5–2 and undergoes disintegration with a domination of mesoscale variability. According to the data of direct observations and diagnostic calculations, in the core of the Rim Current 30–50 km wide, the winter-spring velocities in the upper 100-m layer comprise 0.30–0.80 m s^{-1}; at depths of 200–500 m they decrease 2–4 times; the summer-fall Rim Current is wider and slower by a factor of 1.5–2. Below 500 m, the BSGC is poorly studied; it significantly differs from the surface pattern by low mean velocities (not higher than 0.01–0.03 m s^{-1}).

Satellite imagery and hydrographic surveys suggest a diversity of the hydrodynamical features in the Black Sea. Mesoscale eddies up to 100 km in diameter and jets of different origins propagating over ~200 km away from the coast significantly affect the intra-basin water exchange in the sea, since the width of the deep-water part of the sea is only a few times greater than the size of these structures. For example, water entrainment by the large anticyclonic eddies located over the wide northwestern slope provides the propagation of the desalinated shelf waters rich in nutrients to the deep-water basin of the western part of the sea, and controls the biological productivity in the western part of the Anatolian coastal zone. Mesoscale features (eddies, jets, and filaments) formed over the entire perimeter of the sea equalize the chemical and biological parameters over its area. Mesoscale eddies (anticyclones, cyclones, and vortical pairs) and related jets also affect the structure of the Rim Current and lead to the formation of large meanders of the current, moving its axis away from the coast over great distances, and branching. The lowering and rising of the upper boundary of the hydrogen sulphide zone in anticyclones and cyclones, respectively, may stimulate the ventilation of the anoxic waters of the Black Sea.

Local winds also favor the intensification of the horizontal water exchange in the Black Sea. For example, they cause formation of filaments of coastal upwellings and provide the separation of coastal anticyclones from the eastern coast (under northerly winds) or removal of the shelf waters to the deep-water basin by anticyclones over the northwestern continental slope (under westerly winds).

Consideration of the thermohaline structure of the Black Sea provides new results on the statistical and physical analysis of the historical data of shipborne observations of the vertical profiles of the temperature and salinity of the waters. The general features of the vertical thermohaline structure of the Black Sea waters, the seasonal and interannual variabilities of the horizontal structure of the temperature and salinity in all the main water layers are described. The relations of the large-scale features of the hydrology of the Black Sea waters to external forcing (heat and moisture fluxes across the water surface, river mouths and straits, fluxes of the momentum and relative vorticity of wind) are shown. The generalization of the results of the studies of the T,S-structure of the Black Sea waters and of its seasonal and interannual variability allows the following conclusions to be made.

The T,S structure of the Black Sea waters consists of a few characteristic layers with different thickness; top-down: the upper mixed layer, the seasonal pycnocline (thermocline); the cold intermediate layer (CIL), the main pycnocline (halocline), the isothermal intermediate layer, the thickest deep layer with a slow temperature and salinity increase with depth, and the near-bottom mixed layer. The principal features of this structure are related to the very weak vertical turbulent exchange of the T,S properties between, on the one hand, the freshened surface and the cold intermediate water masses and, on the other hand, the significantly more saline deep water mass.

The seasonal and interannual variabilities of the upper mixed layer, the seasonal pycnocline, and the CIL are caused by the corresponding variations in the heat and freshwater fluxes through the sea surface and in the riverine runoff. The ventilation of the Black Sea waters is restricted to the CIL; it is renewed over the major part of the area in severe winters and in some regions (focuses of ventilation) over the shelf and continental slope of the western part of the sea in other years. The seasonal and interannual variabilities of the main pycnocline are caused by the changes in the flux of the wind relative vorticity. In the layers below the main pycnocline, the interannual variability is caused by the variations in the inflow of the waters of the Sea of Marmara related to the rest of the other components of the external water budget of the Black Sea.

An analysis of interannual variability of the satellite-derived basin-averaged sea surface temperature (SST) values in the period 1982–2002 revealed a mean positive trend of the Black Sea SST of about 0.06 °C year^{-1}. Within this period, the SST trend was slightly negative in 1982–1993 ($\sim$−0.03 °C year^{-1}) and positive in 1993–2002 (0.17 °C year^{-1}). The warming of the Black Sea water in general occurred in all the seasons. The Black Sea warming is consistent with the warming of the World Ocean during the same time period. Similar patterns of changes in the SSTs in 1982–2000 in the closely-spaced Black and Caspian inland seas occurred, which suggests the determining role of climatic factors in the interannual and decadal variability of SSTs in both of the seas.

The interannual variability in the Black Sea SST can significantly influence the climatology and ecology of this semi-enclosed basin. For example, the cold winters of 1985 and 1987 seem to oppose the mass development of the ctenophore *Mnemiopsis leidyi*, which invaded into the Black Sea in 1982–1983. A sharp decrease in its biomass that followed its mass development at the end of the 1980s, occurred in the cold winters of 1992 and 1993. However, in the warm 1995, its biomass grew again. The warming of the Black Sea since 1995 resulted in a weakening or disappearance of the winter (February–March) peak of annual phytoplankton biomass. An unusually long phytoplankton bloom was observed in the warm 1998–1999 and especially in 2001, which was the year with the highest winter and mean annual SST in the period 1982–2002.

Large-scale atmospheric oscillations, apparently, influence the temperature regime of the Black Sea, although there is no unambiguous correlation between them and the character of SST anomalies. The analysis showed that, in each specific case, various combinations of different global (ENSO, NAO, EAWR, warming or cooling, etc.) and regional factors determine the value and sign of the Black Sea SST anomaly.

While analyzing chemical processes in the Black Sea, the general features of the vertical structure and variability of the principal ingredients were considered. The analysis involved the distributions of the main parameters of dissolved oxygen, nutrients, sulphur species, metals, etc. In so doing, the redox layer was in the focus of our attention. The analysis was based on the use of the recent observation data of 1997–2006.

The authors prove that the features of the hydrochemical structure of the sea are closely related to the hydrophysical factors and their variability, first of all, with the intensity of water exchange via the Bosphorus Strait, the Danube River runoff, and the temperature of the cold intermediate layer. The possibility of the relation of the variability in the hydrochemical conditions with large-scale climatic phenomena such as, for example, the North Atlantic Oscillation, is taken into consideration. In the chemical processes that proceed in the redox layer, the importance of the bacterial activity (chemosynthetic production), which is comparable with the photosynthetic processes, in the nutrient balance is acknowledged. In addition to the climatic changes, the redox layer is also influenced by eutrophication. When, owing to eutrophication, the nutrient amounts in the sea grow, the oxygen content in the cold intermediate layer decreases, which affects the position of the redox layer.

During the past decades, the basic features of the biogeochemical structure of the Black Sea have radically changed under the impact of climatic and anthropogenic factors. The amounts of selected nutrients (for example, nitrates owing to eutrophication) have increased, while those of others (such as silicon due to the regulation of the Danube runoff) have decreased.

With respect to the hydrochemical structure, one can distinguish the southwestern part, which finds itself under the influence of the Bosphorus and represents an area of intensive redox processes in a multilayered transition zone. Here, the chemical conditions are extremely instable due to the temporal and spacial variations in the supply of the Bosphorus waters. In other regions of the Black Sea, the hydrochemical structure is mainly formed and maintained by a combination of biogeochemical and hydrophysical processes such as advection, turbulence, sedimentation, etc. and can be explained with 1D-model approach. This leads to the formation of a "chemotropic" structure, where all features of the chemical parameters distribution are closely correlated with the water density.

Observations over the position of the boundary of the hydrogen sulphide zone in the northeastern part of the Black Sea showed that its interannual variations are related to the wintertime temperature conditions and the formation of the balance between the riverine water supply, the delivery of saline waters via the Bosphorus, and the wintertime formation of the cold intermediate layer enriched in oxygen.

The results obtained illustrate the mechanism of the response of the Black Sea natural system to large-scale climatic changes. An analysis of the estimates shows that changes in the sea surface temperature lead to changes in the conditions of the wintertime formation of the cold intermediate layer and of the reneweal of the oxygen content in it. The amount of oxygen in the cold intermediate layer represents a sort of accumulator, which maintains the oxygen consumption for utilization of organic matter and for the downward diffusive flux throughout the year. The interannual changes in the oxygen renewal in the cold intermediate layer result in the changes in the character of the oxygen–hydrogen sulphide structure. Results of direct observations showed that the displacement of the boundary of the hydrogen sulphide zone by a few meters leads to a change in the volume of the oxic waters by 5–10%, which is of vital importance for the ecosystem of the sea.

A special section is devoted to the consideration of one of the most urgent issues about the natural environment of the Black Sea – of its hydrogen sulphide structure. This chapter has a fundamental importance, the more so as various, sometimes "exotic", versions about the reasons for the formation of hydrogen sulphide in the Black Sea, its sources, and the dynamics of the anoxic zone proper appear in scientific literature. The principal source for hydrogen sulphide in the water column is represented by its production by sulfate-reducing bacteria. The residence time of hydrogen sulphide in the water column of the Black Sea of 90–150 years is comparable with the water exchange rate between the oxic and anoxic layers and an order of magnitude lower than the residence times of major salt components of the Black Sea water.

The existence of the Black Sea bottom convective layer (BCL) has important implications for the physical and chemical exchange at the sediment/water interface and at the interface between intermediate and bottom water masses. Two-fold increased vertical gradients of dissolved sulphide at the upper boundary of the BCL suggest the presence of the "anoxic interface" separating entire anoxic water mass dominated by turbulent diffusion from underlying waters of the BCL where double diffusion is the main mixing mechanism.

Temporal variations in the average depth of the chemocline in the Black Sea and the upper sulphide boundary particularly are mainly the result of climatic changes in the density structure of the water column. The upper anoxic boundary location versus density for this basin did not change over the period from 1910 to 1995. However, recent data have shown a prominent increase in sulphide concentrations, as well as nutrient levels, within the anoxic zone supposedly due to anthropogenic impacts or climatic variations.

The review made in the book put forth the importance of mixing and ventilation processes in the Black Sea anoxic zone reflected in sulphide concentrations and its sulphur isotopic composition. The Bosphorus flux cannot be considered as a main factor for deep basin ventilation as suggested by the sulphide budget. Near-shore mesoscale dynamics associated with the propagation of anticyclonic eddies along the Rim Current and their influence on chemocline processes and horizontal exchange between shelf and open waters, as well as pycnocline erosion during exceptionally cold winters, are additional or probably major ventilation mechanisms for the anoxic zone. Precisely these mechanisms that provide vertical exchange in the water column of the Black Sea were considered in the classical studies by V.A. Vodyanitsky (see chapter on the history of the Black Sea exploration and oceanographic investigation) and were supported by other scientists. Global climate change has impact on most of the aforementioned ventilation processes as well as the replenishment of the CIL and its dissolved oxygen content and therefore on the magnitude and direction of the processes within the sulphur cycle in the Black Sea.

Based on abundant published data and results of original studies, the present-day condition of the biodiversity and productivity of the Black Sea ecosystem is considered in the book. The flora and fauna of the Black Sea are represented by 1800 and 2273 species, respectively. A part of the population is retained from the times of the existence of the Pontian Lake, which represented a sea in which brackish-water fauna and flora were formed. In the Quaternary, owing to the uplifting of the earth's crust and the formation of the Caucasian Mountains, the Pontian Lake divided to form the basins that subsequently represented the Sea of Azov, the Black, and the Caspian Seas. It is assumed that the cold-water fauna of the Black Sea was formed during the glacier thawing, when the organisms dwelling in the cold waters of the northern rivers were delivered to the sea in the course of the basin filling. After the formation of the Bosphorus and Dardanelles straits, the species of the Mediterranean Sea began to penetrate into the Black Sea; their influence is responsible for the major part of the Black Sea flora and fauna. In addition, a part of the species is permanently supplied with the riverine runoff; they are mostly encountered in near-mouth areas of the sea. More than one hundred species were introduced to the Black Sea with the ballast waters of ships; most of them originate from the coastal regions of the North Atlantic.

Today scientists suggest that the Black Sea phytoplankton is rather diverse and includes more than 700 species and varieties. In the Black Sea and its lagoons, there are about 750–970 species and intraspecies of microphytobenthos and 300 species of macrophytes. Among the alga, there are no endemic species; this implies a relatively young age of the Black Sea flora. The Mediterranean forms that invaded the Black Sea met favorable conditions and formed dense populations. The flora of the northwestern part of the sea differs from the Mediterranean flora and is closer to the flora of the North Sea. In the years of the maximal anthropogenic load (1960–1975), the species structures of the phytocoenoses have suffered significant changes that were manifested in the disappearance of selected alga species or in their replacement by others.

The fauna of the Black Sea is extremely diverse, and includes about 2300 species, among them fishes (192 species) and mammals (four species). The major part of the ichthyofauna (60%) consists of species of a Mediterranean–Atlantic origin that permanently dwell in the Black Sea. The ichthyofauna also includes Pontian–Caspian relics (bullhead and sardelle), Boreal–Atlantic relics (sprat, Black Sea salmon, and whiting), and freshwater fishes inhabiting near-mouth regions of the sea (carp, bream, pikeperch, bass, and others).

At the beginning of the 20th century, more than 50% of the total catches consisted of mackerels, mullets, herrings, belugas, and sturgeons, while the proportion of plaices, gallinules, horse mackerels, stellate sturgeons, and sardelles comprised about 20%. By the end of the 20th century, the eutrophication of the near-shore regions of the sea resulted in a significant reconstruction of all the elements of the sea ecosystem and to changes in the species composition in the fish hauls. The catches of sprats and anchovies increased up to 98.5%, while all the other species including sturgeons, plaices, mullets, gallinules, and horse mackerels made only 1% of the total hauls. Most of the valuable fish species that were caught before have lost their commercial significance.

According to the data of the FAO, the hauls of live aquatic resources and anchovy by the Black Sea countries have been growing from 1970 to 1988. From 1989 to 1991, a sharp reduction has occurred; it coincided in time with the maximal comb jelly *Mnemiopsis leidyi* development. The economic losses for the fishery of the Black Sea countries because of the ctenophore invasion reached hundreds of millions of US dollars. With the appearance of the new invader, the ctenophore *Beroe ovata* in 1997, the abundance of *Mnemiopsis* began to rapidly drop and the anchovy hauls gradually restored and at present they comprise about 0.3 Mt per year.

From the 1970s to 1995, a decrease in the species number, including the commercial species, was also observed. The economic crisis in the 1990s in the Black Sea countries led to a decrease in the agricultural use of mineral fertilizers, to a reduction of the fishery fleet, and of industrial and municipal contamination. The decrease in the anthropogenic load resulted in signs of restoration of ichthyofauna. For example, off Sevastopol, there were 18 fish species at the beginning of the 1990s and 30 species in 2002. In the 1990s, 48 species were registered in Odessa Bay; now, ichthyofauna there numbers 58 sea fish species. These facts confirm the improvement of the general ecological situation during the recent years.

Due to increasing human activities such as shipping intensity, deliberate stocking, and accidental introduction, high numbers of alien species have become established in the Black Sea since last century. In addition, global warming facilitates the population increase of thermophilic species and their northward expansion from the Mediterranean. As a result, the Black Sea became a pivotal recipient area for marine and brackish water aliens.

On the other hand, some of the Black Sea species expanded, first of all, to the brackish-water Sea of Azov with currents and ships via the Kerch Strait; selected species invaded to the Sea of Marmara via the Bosphorus Strait and to the Aegean Sea via the Dardanelles or to the Caspian Sea with ballast waters or with ship fouling communities. Thus, the Black Sea became a donor basin for the further expansion of the alien species that have established in it to other southern seas.

Alien species often greatly affected the recipient ecosystems, first of all the communities in the trophic level they occupy themselves, and thereafter some of them other trophic levels of the ecosystem; and finally, could cause changes in ecosystem functioning. The Black Sea became a natural laboratory for invasive biology, as a recipient and donor area. Some invasions were useful, like the intentional introduction of the gray mullet *Liza haematochila* and the accidental invasion of ctenophore *Beroe ovata*, some harmful, the most dramatic example of alien species effects documented was the invasion of a gelatinous predator, the polymorphic ctenophore *Mnemiopsis leidyi*. The last one is the most impressive representative of invaders; it is an edificing species, which expanded over all the seas of the Mediterranean basin and over the Caspian Sea and affected their ecosystems.

Over 15 years, after the spontaneous introduction of gelatinous animals, the ctenophores *Mnemiopsis leidyi* and *Beroe ovata*, the Black Sea ecosystem has significantly changed. As a result of the *Mnemiopsis leidyi* development, the ecosystem, from the lowest trophic levels to the higher ones, fishes and dolphins, significantly degraded. Meanwhile, after the introduction of the *Beroe ovata*, it started to recover. These events present a dramatic example of the impact on the ecosystem that is provided by the invasion of a single species and, undoubtedly, this process should be thoroughly controlled by man.

Satellite monitoring of the Black Sea performed during the past decade showed that the seasonal changes in the chlorophyll concentrations in the deep-water regions of the sea featured a distinct minimum in the summer and a maximum in the autumn–winter period. This kind of seasonal change is characteristic of subtropical seas, in which the summertime stratification restricts the nutrient supply and phytoplankton growth. The Black Sea, because of its extreme haline stratification, may be referred to the same type.

The entire surface of the Black Sea was characterized by positive chlorophyll anomalies in 1997–2001 and negative anomalies starting from 2002. In 2002–2003, the negative chlorophyll anomaly was especially sharp on the northwestern shelf. In all the regions of the sea, interannual chlorophyll variations were closely related to the Danube runoff. A more intensive runoff result in a chlorophyll increase on the northwestern shelf and, subsequently, in other regions of the Black Sea.

Selected authors relate the periodical chlorophyll peaks with cold winters, during which the cyclonic circulation is intensified and the pycnocline in the central parts of the sea rises with respect to its usual position thus favoring phytoplankton growth. Along with this, the positive correlation recovered between the SST and chlorophyll indicates the dominating role of the Danube runoff in the interannual dynamics of the Black Sea pelagic ecosystem. First, the Danube waters deliver great amounts of nutrients required for phytoplankton growth. Second, the freshwater runoff enhances the stratification; this provides the retention of phytoplankton cells in the illuminated upper layer and favors the growth of the phytoplankton biomass.

A special analysis of the oceanographic and biological conditions of the Sea of Azov was performed. The interest to the Sea of Azov was always related to its large fish stocks, which are inferior only to that of the Caspian Sea. Previous to the early 1950s, under the natural water regime, the Sea of Azov was distinguished by its extremely high biological productivity. Annual fish hauls (sturgeons, pikeperchs, breams, and sea roaches) in this small sea reached 300 kt. The riverine runoff delivered great amounts of nutrients, 70–80% of which were supplied during the spring flood period. This provided abundant development of phytoplankton, zooplankton, and benthos. The area of the spawning zones related to flooded regions and lagoons in the lower reaches of the Don and Kuban' rivers reached 40 000–50 000 km^2. Along with the good heating, low salinity, sufficient saturation with oxygen, long vegetation period, and rapid cycling of nutrients, these factors provided conditions favorable for ichthyofauna that included up to 80 species

The regulation of the Don (1952) and Kuban' (1973) rivers and the withdrawal of the riverine runoff for reservoir filling caused negative qualitative and quantitative aftereffects in the runoff to the sea, in particular, reduced flooded and spawning areas. In the sea proper, one observes a growth in the vertical temperature and salinity gradients and an increase in the formation of oxygen-deficient zones in the near-bottom layer. In 1987, the presence of hydrogen sulphide was first registered in the lower layers of the sea.

Under the present-day conditions, the amount and composition of the nutrients supplied to the sea radically changed as well as their distribution throughout the year. The major part of the particulate matter precipitates in the Tsimlyansk Reservoir, while its amount delivered to the sea in the spring and at the beginning of the summer significantly decreased; simultaneously, the supply of mineral forms of phosphorus and nitrogen reduced, while the amounts of their organic forms that are hardly assimilated by organisms sharply increased.

Meanwhile, the pollution of riverine and sea waters by different hazardous chemicals such as pesticides, phenols, and, at selected places, oil products also increased. The highest pollution degree is observed in the near-mouth regions of the Don and Kuban' rivers and in the areas adjacent to major ports. These ecological changes resulted in a sharp drop in the biological productivity of the sea. The trophic base for fishes was dramatically reduced and the total fish hauls, especially those of valuable fish species, also decreased.

Total number of aliens in the Sea of Azov comprised 46 species. When analyzing the ecological role of species-invaders in the Sea of Azov, one should first mention the enormous negative effect at all the levels of its ecosystem, fish resources included, caused by the invasion of the predator ctenophore *Mnemiopsis leidyi.*

The introduction of other organisms may be regarded as a positive event. Benthic species such as mya and anadara widely spread over the regions with low oxygen contents unfavorable for other benthos representatives; they provided valuable food resources for benthofagous fishes, while their larvae are consumed by small pelagic fishes. The role of the fouling species *Balanus improvisus* is negative; meanwhile, its larvae are consumed by small pelagic fishes. The crab *Rhithropanopeus harrisi tridentata* also became an additional food object for benthofagous fishes.

The ctenophore *Beroe ovata* is, beyond doubt, a useful invader; unfortunately, according to its seasonal dynamics, it appears in the Sea of Azov too late, when mnemiopsis has already reproduced, widely spread, and undermined the stocks of trophic zooplankton. No positive role of *Beroe ovata* in reducing the mnemiopsis population in the Sea of Azov has been noted to date. Meanwhile, its development in the Black Sea influences the size of the mnemiopsis population; therefore, after the beroe appearance, mnemiopsis enters the Sea of Azov later and its abundance is significantly lower.

The monograph is completed with the assessment of the ecological and socio-economic problems of the Black Sea and the Sea of Azov. The environmental issues of the Black Sea are determined by specific ways of economic development in the coastal states and the resultant economic activities, first of all, in watershed basins of rivers flowing into it. The Black Sea vast watershed basin is nearly 5-fold greater than the area of the sea proper. Its greater part is occupied by densely populated industrial regions. The Black Sea becomes a terminal for wastes and discharges generated by 170 million people.

The present-day environmental situation in the Black Sea is very strenuous. Industrial development and growing population in the coastal states, increased oil transit, intensive development of shipping, harbor engineering, recreational development, dumping are responsible for great water pollution, including in regions of active fishing. The principal sources of sea pollution are river flow; disposal of industrial, domestic and agricultural wastes in the coastal zone, in particular in the areas around large cities and ports; economic activities in the water area (navigation, oil leaks during transportation, dumping, etc.). The results of environmental monitoring revealed considerable pollution of sea waters and bottom sediments with total phosphorus and nitrogen (Danube seaside), detergents and phenols (southern coast of Crimea), phenols and pesticides (Odessa coast), oil products (nearby Sevastopol and Georgian coast), toxic and heavy metals, polyaromatic hydrocarbons and radionuclides causing degradation of the sea ecosystem.

With the growth of the economic potential marked recently in the Circum-Black Sea countries being witnessed in recent times due to construction of new and rehabilitation of the existing sea ports, revival of the merchant and tanker fleet, consolidation of the naval component, construction of oil and gas pipelines and likely perspective of hydrocarbon extraction, development of the health resort and recreation activities the environmental stress may be aggravated.

In 1996 *Strategic Action Plan for the Rehabilitation and Protection of the Black Sea* stated that the state of the Black Sea environment continues to be a matter of concern due to the ongoing degradation of its ecosystem and the unsustainable use of its natural resources.

A decade later, the Black Sea ecosystem continues to be threatened by different chemical pollutants, mainly by nutrients that enter the Black Sea through rivers from land based sources. The Danube River accounts for about a half of the nutrient input to the Black Sea. As a result eutrophication occurs over wide areas of the Black Sea, in particular, in its northwestern part. Inputs of microbiological contaminants with insufficiently treated sewage result in a potential threat to public health as tourism continues to develop in the coastal zone. Satellite SAR imagery and related statistics of oil spills show that oil and oil products continue to threaten the Black Sea environment. It is supposed that almost half of the inputs of oil from land based activities are brought to the Black Sea via the Danube River. Pollution of the sea is rising as a result of illegal, accidental and operational oil discharges from vessels and oil terminals due to increase of oil products transport via the Black Sea. It was recently announced that the Burgas – Alexandroupolis oil pipeline project will transport 35–50 million tons of oil per year, implying a significant rise of tankers traffic between the Caspian Pipeline Consortium oil terminal close to Novorossiisk and Burgas. Experience in satellite oil spill monitoring predicts a notable increase of oil pollution if additional measures on pollution control will not be undertaken.

A decade ago, the state of biodiversity and productivity of the Blak Sea ecosystem led to suggestions that the process of degradation of the Black Sea is catastrophic or even irreversible. However, environmental monitoring, conducted over the past 10 years, reflects continued improvements in the state of the Black Sea ecosystem. These improvements appear to be the direct or indirect results of reduced economic activity in the region, to a certain degree of protective measures taken by governments and international organizations.

It is natural that this book could not reflect all the problems of the Black Sea. Meanwhile, its content makes clear the objectives of the future research. In the temporal range, they extend from the mean condition of the oceanographic fields, which are studied best of all, toward both to mesoscale processes and to the interannual variability of the condition of the marine ecosystem. Among the wide spectrum of urgent problems of the Black Sea we should emphasize the following.

First of all, it is necessary to continue and improve the monitoring of the principal parameters of the Black Sea and the Sea of Azov environment, which is subjected to a strong variability. In so doing, it is very important to use a combination of different methods of research such as traditional, satellite, drifter, numerical and laboratory modelling. Precisely this kind of approach should allow us to obtain reliable results by comparing the data acquired with different techniques.

In physical oceanography, the principal target is the research of various exchange processes in the sea, such as, for example, the water exchange via the Bosphorus (whose role in the Black Sea hydrology can hardly be overestimated) and Kerch straits. Also important are the studies of the exchange between the shelf zone of the sea and its open areas, including the complicated system of shelf mesoscale eddies, which strongly control the water dynamics of the active layer of the sea, and the interaction between regional ecosystems. Finally, the most important problem (of not only regional but also philosophic meaning) is the assessment of the vertical water exchange between the oxic and hydrogen sulphide zones of the sea and its mechanism as well as the estimation of the total residence time of the waters in the sea.

In the field of biology, the most urgent is the estimation of the present-day condition and productivity of the Black Sea biota after the long-term impact of the invader ctenophore *Mnemiopsis leidyi* and of the recent invader, ctenophore *Beroe ovata*.

It is very important to continue investigation and assessment of the impact of the regional climate change, pollution, eutrophication and other human activities on fish stocks and fishing, as well as impact of fishing on the ecosystem state, and to elaborate measures to protect species and habitats. Development of aquaculture could promote restoration of resources and reduction of a fishing pressure on marine living resources and Black Sea ecosystem.

Because of the construction of oil and gas pipelines on the floor of the Black Sea and to the intensification of the geological prospecting and transportation of hydrocarbons in the sea, the issues of the studies of the water dynamics in the deep-sea layers and of the topography and structure of the upper layer of the sediments in the Black Sea basin becomes especially urgent.

Finally, there are two kinds of international legal problems. One of them is the well-known problem of the Black Sea straits, which is one of the key problems of the region. The other set of disputable issues was formed after the decay of the Soviet Union, which involves different unsolved questions of delimitation of the sea area between the Black Sea states. This also refers to the boundaries in the Kerch Strait and in the region of Zmeinyi Island.

It is of vital importance to have international cooperation in scientific research, development of scientific programs, data collection and analysis, reporting and exchange of the relevant scientific and technical information, in sustainable exploitation and conservation of living marine resources, and management decisions aimed at conservation of the Black Sea environment. The integrated and sustainable development of the Black Sea region will require an interdisciplinary approach, which the present book reflects. A set of the described environmental problems has an international dimension. Only by interdisciplinary, international cooperation can the environment of the Black Sea region be conserved for future generations.

Subject Index

MIX
Papier aus verantwortungsvollen Quellen
Paper from responsible sources
FSC® C105338

If you have any concerns about our products,
you can contact us on
ProductSafety@springernature.com

In case Publisher is established outside the EU,
the EU authorized representative is:
Springer Nature Customer Service Center GmbH
Europaplatz 3, 69115 Heidelberg, Germany

Printed by Libri Plureos GmbH
in Hamburg, Germany